Beverly, Paige

TECHNOLOGY AND SOCIAL SHOCK

TECHNOLOGY AND SOCIAL SHOCK

Edward W. Lawless

RUTGERS UNIVERSITY PRESS
New Brunswick   New Jersey

This report was prepared with the support of National Science Foundation Grant No. GI 32102. However, any opinions, findings, conclusions or recommendations expressed herein are those of the author and do not necessarily reflect the views of the Foundation.

Library of Congress Cataloging in Publication Data

Lawless, Edward W
    Technology and social shock.

    Includes bibliographical references.
    1. Technology assessment—Case studies. 2. Technology—Public opinion—Case studies. 3. Technology—Social aspects—Case studies. 4. Social history-1945–   I. Title.
T174.5.L38   301.15'43'6   75-44184

ISBN 0-8135-0781-2

Contents

Tables

Figures

Preface

This book is the product of a project initiated at Midwest Research Institute in January 1972, with the support of the National Science Foundation. It presents the results of a study of the kinds of episodes of public alarm over technology — "social shocks" — that have inspired recurrent major news stories in the media in recent years. I hope that this book will stimulate its readers to think about the implications of social shock caused by technology, and what, if anything, we should do about it. This book is viewed as a pioneering effort, a forerunner of many further studies and more intensive analyses. I hope that it will be interesting and useful to a wide range of readers, including social, psychological, political, physical and biological scientists; industrialists, agriculturalists, lawyers and academicians; consumer advocates; governmental employees in legislative, regulatory and judicial roles; those interested in the technology assessment concept; and those in the news media — in short, informed people in all walks of life.

The author is pleased to acknowledge the contributions to the project of many people, including members of the staff of Midwest Research Institute and other individuals. The backgrounds of major contributors may be of interest to some readers. The author, who served as project leader for the study, is Head of the Technology Assessment Section of MRI's Physical Sciences Division and had training as a chemist. Dr. Harold M. Hubbard, Director of the Physical Sciences Division, Dr. Thomas Bath, who was manager of MRI's Washington Office, and Mr. Robert E. Roberts and Mr. Howard Gadberry, both of MRI's Economics and Management Science Division, aided greatly in conceptualizing the program. Mr. Carl Cunningham, whose background is in sociology and criminalistics, Mr. Lawrence Rosine, whose background is in mass communications media and technology transfer (both of MRI's EMS Division), and Dr. Alan Booth (consultant from the Department of Sociology, University of Nebraska) assisted in formulating the case selection and media approaches. Mr. Charles Mumma, Mr. Gary Kelso, and Miss Anne Aspoas (all of the Physical Sciences Division) helped in many project tasks, including the identification of cases, compilation of information and preparation of drafts. Assistance was also provided on some tasks by Miss Merritt Alden, Miss Kathryn Lawrence, Mr. Daniel Punzak, Mr. Gregg Morgan, and the staff of MRI's Stenographic and Graphic Arts Departments. A special thanks is expressed to Dr. Harold Orel (Department of English, University of Kansas), who made many excellent suggestions during the preparation of the manuscript and reviewed it editorially.

The National Science Foundation made the study possible through Grant No. GI-32102, entitled "Unstructured Technology Assessments—Case Histories of Public Alarm Over Technology," during the period 1 January 1972 to 30 September 1973. The Project Officer, Mr. Joseph Coates, Program Manager, Officer of Exploratory Research and Problem Assessment, RANN Program, has shown continued interest in the study throughout. Mr. Coates's assistance in obtaining many independent reviews of the manuscript and his thoughtful suggestions for presenting the results of the study are greatly appreciated.

The study benefited substantially from the insights of its Advisory Committee, which consisted of the Hon. Emilio Q. Daddario, formerly Congressman from Connecticut, and now director of the Office of Technology Assessment; Dr. Roy W. Menninger, president, the Menninger Foundation; Mr. Dolph C. Simons, Jr., president and publisher, *Lawrence Daily Journal World,* and director, Associated Press Board; Dr. Stephen T. Quigley, director, Office of Chemistry and Public Affairs, American Chemical Society; and Dr. Charles Kimball, chairman of the board of trustees, Midwest Research Institute. These members, together with Mr. Coates who met with the Committee, represent a wide spectrum of experience including (respectively) legislative government; mental health and medicine; public press; trade press and professional societies; research administration; and governmental agencies. The Committee met with the project team twice during the research phase of the study, reviewed project reports and made many valuable suggestions. The author would also like to thank the nearly two dozen persons who reviewed a manuscript of this book for the National Science Foundation and made many extremely valuable contributions and constructive criticisms.

While the study has benefited from these many contributors, all interpretations and subjective statements are those of the author.

EDWARD W. LAWLESS

TECHNOLOGY AND SOCIAL SHOCK

I.

Introduction

An amazing succession of alarms and controversies over technology has filled this nation's mass communications media in recent years—an average of at least one significant new case per month. Do you remember thalidomide, the sleeping pill and tranquilizer that caused birth defects; or the cyclamate sweetener affair, the DDT debate, or the nerve gas disposal controversy? Perhaps you have had a new automobile or a color TV set recalled because of some hazardous defect. In Santa Barbara, they still remember the Great Oil Leak; in New York City, the Great Power Failure; in the Corn Belt, the leaf blight to which corn was susceptible because of genetic engineering; and in Kansas, the AEC's plan to store radioactive wastes in old salt mines. The automobile has not only contributed convenient personal transportation, but has helped pollute our sky, has contributed to urban sprawl, and has proliferated faster than freeways can be constructed. The freeways themselves are accused of causing neighborhood disruptions and decay, as well as being a danger to the bodies and minds of their users. Biomedical research and technology have not only contributed to better human health, but have generated enormous ethical and legal controversies because they have blurred the boundaries of human life for both the aged and the unborn. In December 1972 a jumbo jet crashed in the Florida Everglades with the loss of 101 lives. The pilot, distracted by a minor malfunction, failed to note until too late the warning signal that—correctly—indicated an impending disaster. His sudden, astonished cry of "Hey, what's happening here?" were his last words. These words have already been asked in the one hundred cases of "social shock" over technology that we de-

3

scribe here; they will be asked again and again by millions of Americans unless efforts to utilize technology more wisely are successful.

Our cases have all occurred since World War II, but public concern over technology is not entirely new. The successful science and technology of the eighteenth and nineteenth centuries not only generated much optimism for the future of mankind, but also led to tragic numbers of occupational injuries, sicknesses, and deaths. In large measure, however, those were afflictions of the working classes, and the technological causes were readily understood; the hazards were regarded as acceptable, if not avoidable risks to the worker, and the rest of society was little concerned. In contrast, the nature of the hazards and the people who are concerned are now significantly different, because modern technologies can exert subtle influence far from the point of their use, and modern communications methods can make the discovery of an adverse effect known nationwide almost overnight. People thus learn that technologies, of which they may know little, pose hazards which they may only dimly understand. For the most part, they are dependent on the news media's descriptions of what the problem is and what is being done about it. They may have little confidence in their own ability to judge correctly the actual danger to themselves or how to avoid it. Thus the social shock can be very great.

For the author, growing up in rural Illinois, perhaps the first "technological shock" of this kind was the news stories that the pristine, falling snow was contaminated with radioactivity from a faraway bomb test. For readers with longer memories, perhaps the spectacular crash and fire of the hydrogen-filled *Hindenburg* dirigible in 1937 at Lakehurst, New Jersey, with the loss of thirty-six lives, may have been a similar shock; this event not only brought twenty-five years of commercial airship service to a close, but the dramatic radio and film accounts of it left a legacy that might present, even today, serious psychological barriers to the use of hydrogen as a fuel. Perhaps the first real public shock of this kind was the earlier case of the *Titanic:* the largest, fastest and most magnificent ship ever built, it was the culmination of a century of progress and discovery. Supposedly unsinkable, the *Titanic* hit an iceberg—a relatively small scrap of nature—on its maiden voyage and sank like a ruptured tin can in a mill pond with the loss of 1,517 lives, including those of many famous persons.

The *Titanic* tragedy did lead to many improvements in maritime safety practices—more lifeboats, for example—and we will see that many of the cases described here have also led to remedial action. But as we will see, also, these episodes of public alarm have had many adverse effects, and as a consequence modern Americans have developed a great ambivalence toward science and technology. On the one hand, we are quick to adopt the new products and services that are produced by increasingly sophisticated scientists and engineers, and poured forth with heavy advertising or much publicity by industry and government agencies. On the other hand, we are constantly buffeted by evidence that one or another of these technologies has become a problem, or is being accused of producing unpleasant side effects. Thus, while cases of social shock over technology are not new, we believe this study will show that such cases have become more numerous and vastly more complicated in recent years as our technologies have grown and become entwined with each other and with our culture.

Society's institutions have by no means been standing idly by as these cases of technological shock have occurred. In recent years, for example, we have seen the enactment on the national level of a Consumer Protection Safety Act (1972); an Occupational Safety and Health Act (1970); a National Environmental Policy Act (1969); a Federal Environmental Pesticide Control Act (1972); Amendments to the Food, Drug, and Cosmetic Act (1958); a Poison Prevention Packaging Act (1970); and a National Traffic and Motor Vehicle Act (1966); to name just a few. Other efforts have been made in the state and local governmental area and in the private sector—especially through citizens' participation groups. In general, however, the effort of governmental and regulatory bodies has been largely reactive on a case-by-case (crisis-by-crisis) basis—after a *Titanic* has gone down.

In contrast, and perhaps of great future significance, was the establishment by the Congress in late 1972 (while the present study was under way) of an Office of Technology Assessment. That office is intended to be a direct aid to the Congress itself in the identification and consideration of existing and probable impacts of technological application. A "technology assessment" may be defined as a systematic study of the effects on all sectors of society that may occur when a technology is introduced, extended or modified, with special emphasis on any impacts that are unintended, indirect, or delayed. In other words, a technology assessment is a search for the ways in which a technology might exert influences outside itself; the results can be used in the identification and evaluation of public policy options, in reaching wise decisions in the legislative and regulatory area, and in the expenditure of research funds. Ideally the concept of technology assessment will be used in all the important decision-making levels in the public and private sectors.

The results of the present study present a guide to the kinds and range of problems that we are facing—a ready reference to diverse examples of how alarms have arisen, spread, exerted influences themselves, and been eventually resolved. The brief analysis of the collective cases—in which characteristics, commonalities, and structural elements are examined—will, we hope, stimulate further, more sharply focused and more probing formal studies, and help a wide range of readers in their own efforts to see that our technologies are utilized wisely.

The extent to which our rapidly growing technologies have pervaded our culture—and the growing number of cases of social shock that they have produced—may come as a surprise to many readers of this book; they will not to the governmental authorities who have had to deal with some of these problems, or to the small but growing number of researchers who are trying to find ways to help our societal institutions manage our technology for the common good. But as will be shown here, enormous social and political implications are involved and the assistance of thoughtful persons across the country—and indeed throughout the world—are needed in helping modern society in coming to grips with its technological capabilities. This book will, we hope, be a modest contribution in helping its readers to recognize some of the problems and to participate responsibly and effectively in their solutions.

II.

About the Study

A summary of the objectives and methodology of the study is presented here to help the reader understand its scope and limitations. A brief outline of how the results will be presented may also be helpful.

The primary objective of this study has been to compile a series of short case histories of recent episodes of strong public concern; these cases will illustrate the diverse effects that technology is having on our individual lives, our institutions, and our culture. A secondary objective was to make preliminary analyses of the origins, developments, dispositions, and impacts of these cases in order to determine what characteristics they might have in common and what, if anything, society might or should do differently.

This study is, therefore, historical in its emphasis: it focuses on cases where public concern or alarm has already occurred because existing or proposed technologies and products have been perceived to have serious defects, or to pose unnecessary hazards to man, his environment, or his way of life. We examine what kinds of things have happened to create public concern, and try to show whether or not these events are traceable to inadequate information or foresight, and if they were complicated by policy decisions either before or after the problems arose. We have been particularly interested in the role of the news media in transferring technical information from the scientific community to the public, and in the effects that accounts of technological stories printed in the popular press have had on the public. We have relied less heavily on scientific journals, highly technical analyses, legislative hearings, and reports with limited public circulation, not because what was

7

presented there failed to possess great intrinsic interest, but because the public, in general, was unaware of the technological details as described in those accounts, and took its information from the news media. Our histories, in brief, are descriptions of the cases as we think they were perceived through the eyes and ears of the public. However, documents from authoritative sources were extensively utilized and are frequently referenced for statements of fact.

A detailed description of the research approach is given in Appendix A. Briefly, we established guidelines to help recognize as many different kinds of technologically related alarms as possible, identified well over two hundred cases of public concern (and believe we could have turned up several times that number), used a set of selection criteria to pick cases for study, collected case material from standardized and auxiliary sources, cross-checked the material for accuracy, prepared the case histories, and made a brief analysis of the results.

For purposes of the study, "technology" was broadly defined to include all varieties of applied physical and biological science and engineering, and also basic research that might soon lead to a proposed technological development. The reader has probably already recognized that different cases might "concern" different "publics": a few controversial issues will have a truly national public, whereas the concerned public in other cases may be regional, monolithic, or relatively few in number. In some cases several levels of "concerned publics" may exist, ranging from a tiny number of people whose values are directly threatened to larger numbers of less-affected people who may become interested to varying degrees because of the nature of the threat or news coverage of the case. To simplify the problem, we have assumed that the popular news media are reasonably good judges of what interests the general public, and that if a case involving unpleasant news has been covered by the media, then it may be said to be of public concern.

The study raised the questions of why and when concern or alarm was raised, how was it transferred to the public, and how its degree could be measured. We have used the news media for three purposes—as an aid in identifying cases of social shock over technology, as one of the primary sources of information in reconstructing the case histories, and as a direct means of measuring public concern.

The identification of potential cases for study was made by using a three-way approach: (a) a substantial number of cases were suggested from personal recall by members of the project team, the advisory committee and other interested individuals; (b) a systematic analysis of the basic needs of the individual and society indicated the kinds of threats to human values that might be most likely to produce alarm, and suggested many further cases; and (c) many additional cases were identified by perusal of current and back issues of news magazines and newspapers. A set of selection criteria were then applied to eliminate some cases from further study (for example, they were still ongoing, with the outcome unclear) and to select a group of cases that would be broadly representative of the manner and extent that modern technology is impinging on our lives.

The initial sources of information on the cases were based on the standard indexes for the printed media—although these are not as complete as we would have liked (by comparison, information retrieval procedures for the electronic media are in a barely embryonic state). Searches of the *Reader's Guide to Periodical Literature* and the *New York Times Index* and the *Business Periodicals Index* (since 1958) were supplemented at times by searches of *Facts on File* and *Wall Street Journal Index* and some of the standard scientific abstracts. Information from these searches was supplemented in some cases by other published works, by government reports, and in a few cases by personal contacts.

The transfer of information requires, of course, not only its transmission, but also its reception; the transfer of alarm over technology to the public is no exception. Numerous studies have attempted to determine how the public receives information of different kinds and what reliance the public places on various sources for specific kinds of information. Needless to say, word-of-mouth rumors are at times probably more widely believed than august editorials, and a clever cartoon may create a stronger impression than a front-page headline. Nevertheless, an important role is often attributed to an elite group of "opinion leaders" in our society (sometimes referred to as opinion makers or brokers); the existence of such leaders is based on several studies of the formation of public opinion on national affairs issues, and on a few studies which focused on the adoption of specific technological advances. These latter studies have found that when a new scientific development is introduced to practicing technologists (such as a new drug to physicians, or hybrid seed corn to farmers), it is tried initially by a small number (1–5%) of its potential users, then adopted rapidly by most of the users—who seem to base their decisions primarily on contacts with the innovator group—and then slowly by the remainder of the users who may have little contact with their peers. This adoption pattern, which can be shown graphically as a sigmoid curve, has been observed also when some new technologies have been introduced on the consumer market (for example, synthetic fabrics to replace wool and cotton) where news media advertising presumably plays an important role. In these cases, of course, it is in the user's self-interest to seek out new information on the innovation. In contrast, the roles of the opinion leader and the news media, and the attitudes of the information receiver are much less clearly defined in the transfer of the unpleasant news (that is, the technologically related shocks) under study here. Hence, we have simply assumed that the transfer of information to the public is roughly proportional to the degree of news coverage in the wide range of cases studied. We have utilized a combination of measurements of citations and column inches in selected indexes and publications to calculate an index of concern for each case, as will be described in Chapter V. In short, we have relied heavily on the popular press so that the cases could be described and evaluated as they were presented to the general public, but more technical sources have been freely utilized where needed to supply detail and insure accuracy.

The format of the forty-five diverse case histories has been standardized so that it can serve as a useful data bank of information for interested readers and later researchers: title, abstract, time line (a visual depiction of significant events), a narrative account (background, key events and roles, and disposition), author's comment, and references.

In the case histories, we have tried to place each episode in the context of its time and the public and scientific attitudes then prevailing; to identify the significanct events, people, agencies, or institutions in the development of the episode; to follow the resolution of the controversy; to note what remedial action or other outcome resulted from each alarm; and to identify, if possible, the implications for our society and its institution.

Fifty-five cases are presented in synopsis form. Some of these were still too "open" at the time we selected our cases to be studied in depth, a few duplicated many aspects of selected cases, and others were almost too long and complex to be included. Suggestions for further reading on these cases are included.

Conclusions from the study are presented in Chapter V. The reader may be particularly interested in the summary of news coverage of the cases.

III.

Case Histories

The forty-five histories presented herein can be classified on the basis of more than one organizing principle, but the arrangement employed should be of particular convenience and help to the reader. The case histories appear in the following sequence: the first three cases involve reproduction and genetics; the next twelve, food and medicines; the next three, unique hazards of the x-ray; the next thirteen, various environmental problems; and the next nine, issues created by projects of the federal government. (Some of these last nine may also touch on environmental or health questions.) The remaining case histories illustrate specialized, individual points.

The Prototye Case: Public alarms over technology arise from so many causes and take such varied forms that almost every case has atypical aspects. Certain characteristics appear to be common to many cases and some of these will be described here—using the same format that will be used in the study cases—to help the reader identify key points.

a. *Background:* The case usually begins with a basic scientific discovery or technological advance that may be only remotely related to the ultimate alarm. After a period of time (and this period appears to be shrinking yearly), a technological application of the basic discovery is made (or at least proposed) to satisfy some perceived societal need or activity. Preliminary study concludes that the technology can be safely used, and it then begins a period of growth (or at least intensive and expensive planning in the case of some governmental projects). A few thoughtful people may at this point object to the technology or suggest that it may have undesirable consequences, but they usually lack proof and not much of

11

anything happens: they are frequently burdened by an assumption that they have a conflicting interest (which may be true) or that they are simply "against progress" (some people do oppose most techno-socio changes). More often, their objections are simply drowned out by the reassurances/advertising/propaganda of the industry or government agency involved and by the public's acceptance of the new technology. The technology by this time is experiencing rapid growth, and perhaps even new technologies are spinning off from it.

b. *Key events and roles:* The development of public concern over a technology may, in the simplest case, be a gradual realization by society that previously known and accepted risks are no longer acceptable. More often, however, the concern develops because a previously unknown risk is identified.

The early signs of difficulty frequently arise from changes in analytical methods (statistical, instrumental, chemical, biological, etc.), wherein an existing method is applied as it had not been previously, a technological advance makes a method more sensitive than previously, or a new technology produces an entirely new method. Application of the analytical method to some area that interfaces with the subject technology then yields evidence that it is producing undesirable consequences. These may be: direct, but unanticipated; indirect and the result of accidental events; cause-effect chains-of-event, or interactions with other technologies; or the abuse of the technology.

In many cases these undesirable consequences become observable only after the technology has grown very large,* and in some cases they may not even be objectionable if the technology doesn't pass a critical size. But when one considers that man's technological capabilities now permit him to produce changes ranging from cloudbursts and earthquakes to synthetic diamonds and genetic freaks, then one must assume that the critical size could be passed by almost any technology. The key is the time requirement, which may be a half century or more for a major technology. At any rate, the increasing size of the technology and the curiosity of the scientific community about its effects usually leads to an increasing number of publications about its adverse impacts in the technical literature.†

*This is really a very relative term: the scale may vary from technology-to-technology and from impact-to-impact; for example, only a few incidents involving Chemical Mace caused alarm, whereas millions of cars were needed to make Los Angeles' smog problem critical.
†Assuming that such publications are not prohibited by government security.

A time lag generally occurs, however, before the case gets into the popular press--the length of which depends somewhat on the nature of the undesirable consequence; the lag has probably been getting shorter in the last decade or so as the news media have become increasingly conscious of technological matters. The publications in the popular press about the case may then increase regularly, although still lagging behind those in the technical literature (as in the case of oral contraceptives, Figure 1) and the growth of the technology (as in the case of DDT, Figures 3 and 4). This time lag may be greatly shortened or even eliminated, however, if the undesirable consequence is exceptionally threatening (someone gets hurt) or is keenly articulated by a governmental authority, a public spokesman (a scientist or other highly quotable person), or a particular news medium or journalist. In contrast, the time period is occasionally lengthened, the coverage is minimal, or the threat is minimized, because certain topics are not perceived as very "newsworthy" by the media or are held as "sacred cows." In most cases, however, the concern of a few experts or spokesmen grows into public alarm or shock. Media interest tends to be high if the socioeconomic status of those affected is high or if the topic is related to an already controversial issue.

Relevant governmental regulatory agencies are quickly drawn into the case as the furor mounts (if they have not already been involved), along with any industries, professions, labor unions, commerical interests, military agencies, or whatever, that were involved in causing the alarm.

By this time, the case is usually being widely played in the news media, with the typical story being a composite of incomplete facts* and controversial opinions under a cryptic and overly dramatic headline. Statements of denial, reassurance, or that something is being done about the problem are usually issued by the accused parties and government agencies involved, but these appear to have generally low credibility with the media and possibly with the public at large. An interim or holding action of some sort may be announced, or an extensive public-relations campaign may be launched (particularly if industry is involved).

Some of the accusers, on the other hand, may make claims that

---

*All the facts are not known, of course, and the reporter is further limited by their inaccessibility in some cases, by his publishing or news deadline, by the length of article or air time permitted, and possibly because his minimal scientific training does not allow him to note incongruities or to synthesize those facts that are available into an accurate story.

the danger is even greater than previously reported and that the government or industry is lax in its actions. These charges are usually widely quoted and may raise the issue to a fever pitch. The lawmakers, who have been largely sitting on the fence up to this point, may then demand stern action, and the regulatory agency may initiate a precipitate course—one that may subsequently prove either ill-advised or well-justified. If the pressures become intense, the issue may degenerate into a search for a scapegoat by elements of the government, industry, the media, or citizens groups.

In a fairly typical case, however, the issue does not reach a fever pitch, and the case enters into a period of reasonably serious effort to find out the facts and to effect remedial actions, if need be. This period may involve a presidential panel, a Congressional hearing, a review by the regulatory body (governmental or private sector), or experimental studies by governmental, industrial, university, or independent research organizations.

c. *Disposition:* Depending on the nature of the case, the public alarm may have one of several outcomes. It may serve to bring effective remedial action to a serious problem that had been heretofore neglected by society's formal institutions. It can do this by direct action through the regulatory agencies—the legislature for new laws, the judicial processes, or even the elective processes. Alternatively, it can bring pressure on the offender through the weight of public opinion or through other private-sector options, such as the marketplace, the stock market, labor unions, or the churches and shools. If the public alarm does not bring about remedial action fast enough, frustration and a loss of confidence may well result.

Sometimes the alarm may have been far out of proportion to the true danger. Much of the confusion may have been unnecessary. Perhaps severe economic or psychological "disbenefits" have resulted. One may argue that the scientific reservoir of knowledge about a particular problem has been increased as a result of public concern. But it also seems evident that the avoidance—or at the very least the minimizing—of technological shocks is a valuable goal of modern society.

For the prototype case outlined above, the time line is approximately that shown on page 15. (Note that this is not a *recommended* sequence of events.)

The time scale in the prototype case is not indicated, because it can vary widely from case to case as has already been shown. For a

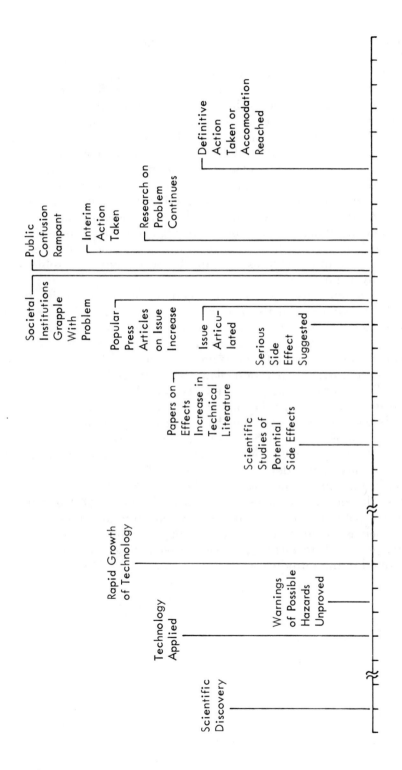

16 TECHNOLOGY AND SOCIAL SHOCK

substantial new technology, one or two decades may pass after it is introduced before a cause for concern is identified. For a really major innovation such as the automobile, some of its more subtle adverse effects become obvious only after several decades. In a few cases, just the proposed use of the technology has stirred immediate protest. In addition, technologies often lead to two or more successive periods of concern as new adverse impacts are identified.

d. *Summary:* In summary then, in many of the cases of this type, new information of an alarming nature is announced and is given rapid and widespread visibility by means of modern mass communications media. Almost overnight the case can become a subject of discussion and concern to much of the populace, and generate strong pressures to evaluate and remedy the problem as rapidly as possible.

The publicity over the case may be beneficial to society; rapid and effective remedial action may follow because of it, perhaps because of the active paticipation of a wide range of citizens and governmental groups. In other cases, however, unnecessary adverse effects on society, or segments of it, may be produced, for example, early, impression-making publicity on the problem might produce serious negative effects in the economy and a "management-of-crisis" situation within government agencies. The public's alarm may even be shown subsequently, on the basis of more complete information, to have been much greater than warranted. In any case the public usually receives a running account of statements and denials, charges and countercharges, and conflicting interpretations of a wide range of technical data as the case develops. The process, as a societal method of determining which path of technological development to follow, might be described as an "unstructured" technology assessment, in contrast to the proposed organized, prospective technology assessments described in the introduction.

Human Artificial Insemination

Abstract

In 1954 and 1958 two court cases created sensations and focused attention on the long-simmering legal, ethical and moral
problems presented by artificial insemination of a woman by a man
(donor) who was not her husband (abbreviated AI or AID). The
courts, legislators, churches and medical scientists have still not
resolved many aspects of these problems, even though an estimated
200,000 Americans have been conceived by this technology.

Background

During the 1930's the techniques of artificial insemination[1] of
women to attain pregnancy were developed into a successful medical science. AI, in which the physician uses a syringe to introduce
sperm-containing semen directly into the cervix of the woman, had
been known for over 160 years.* In the 1890's, AID, the practice of
using semen by a fertile "donor" (produced by masturbation) to
impregnate the wife of an infertile man was pioneered in America.
(Impregnation with the husband's sperm is denoted AIH.) Not until
the 1930's, however, did the so-called "test tube babies" come to
the attention of the general public.[2-7] In part, the interest in human
AI was stimulated by successful development of animal husbandry
techniques of sperm collection and insemination for horses and
cows: an estimated 250,000 mares were inseminated artificially in
1930 in the Soviet Union[8] and the method was becoming widely
used by American breeders and veterinarians by 1940.

* The first recorded successful human AI was performed in England in 1790, five
years after the first successful experiments there with dogs. The first successful human
AI in America was done in 1866. The records do not indicate whether the semen
samples were from the husband or a donor.[1]

17

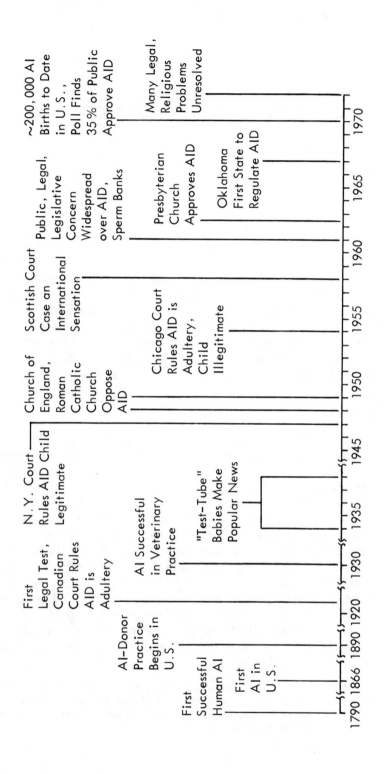

Medical practitioners of AI were made aware of potential legal and moral problems associated with the practice, as a result of a precedent-setting decision by a Canadian court in 1921 which denounced AID as adultery. In addition, the legality of the practice and the legitimacy of the child produced by AI had not been determined by any U.S. court or legislation. The careful practitioner, therefore, in addition to making detailed investigations of the physical and mental well-being of the patient and the carefully selected donor,* would take extensive precautions to match the physical characteristics of the donor as closely as possible to those of the husband, to keep the identities of the donor and couple completely unknown to each other, to obtain the written consent of husband and wife (and sometimes of the donor's wife) and to keep the records of the entire transaction coded and carefully locked up.[5] Interestingly enough, doctors were nearly unanimous in urging that the couple never reveal to the child his origin, although complete disclosure was urged in adoption cases. The AID procedure also posed one further ethical problem for some doctors: the doctor delivering the baby has to sign a birth certificate which gives the name of the child's father. While some doctors falsely listed the husband as the father without qualms, many leading practitioners advocated the ruse of allowing an obstetrician to deliver the baby and then unwittingly list the husband as father. In order to avoid the many potential legal and emotional complications, some AI practitioners mixed semen from the husband with that of the donor before insemination, so that the possibility would exist that the husband was the father.

Technically, the AI procedure was quite simple. Pregnancy was often attained in the first month, with a single insemination during the woman's most fertile days (some physicians preferred two or three inseminations), although many cases required inseminations for several months. By 1945, the practice had become established, with leading practitioners, such as Dr. Alan Guttmacher, claiming that about 55 percent of the AI attempts were successful and that the children produced were well-accepted by the parents. AID could be used not only when the husband was sterile or impotent, but also when he carried some genetic impairment or when he and his wife had blood Rh incompatibility. In addition, the success of AID appears to have later stimulated interest in AIH and in

---

*Frequently a medical student, intern or doctor. He usually received a $25 fee.[7]

adaptations of the semen collection and preparation techniques to help the husband of low fertility attain fatherhood.

In addition to the 1921 Canadian court case, which cast the pall of adultery* over AID, at least three[1] European court cases before 1945 had involved peripheral aspects of AI. One English court had also apparently held[9] AID to be adulterous, while another had, strangely enough, ruled an AIH child to be illegitimate.[1] In general, however, as an American Medical Association spokesman noted[9] in early 1945, "Society has formed no opinion and enacted no law regarding artificial insemination." The first rumblings of what was to be a long legal, legislative and religious controversy were, however, already being heard.

Key Events and Roles

In January 1945 a Circuit Court divorce case in Chicago established the first U.S. legal precedent on AI. A returning veteran's wife claimed she was impregnated by means of AID. Judge Feinberg ruled that AID alone would be legally insufficient for a divorce on grounds of adultery, although he granted the divorce on other evidence.[9] In early 1945 the Roman Catholic Church, however, first suggested that it would hold AI to be adultery[10] and the British Ministry of Health held that AI children were illegitimate.[11] In contrast, a 1947 New York court case (Strnad vs. Strnad), involving a husband's visitation rights with his wife's AID child after the divorce, found no adultery or illegitimacy and ruled that the husband, by his consent, had "potentially adopted or semi-adopted" the child and had visitation rights.[1,12] The moral questions were pursued by the Church of England, which concluded in 1948[13] after a 32-month study, that only AIH was permissible,† and by the Roman Catholic Church, which announced in 1949 that AID was prohibited and that AIH could be practiced only if the semen were obtained after normal intercourse. (Orthodox Jews also dissapprove of masturbation for AI purposes.)

The legal actions stimulated considerable interest by the public and by state legislative bodies; the Federal government was without authority in the matter. Bills were introduced during the 1948–1950 sessions in New York, Virginia, Wisconsin, Indiana,

---

*Adultery was at that time grounds for legal action in most states. Uncertainty existed over who might be charged as an adulterer in an AID case.

†AID was called a "breach of marriage" but not labeled adultery. The report was widely debated in England.

and Minnesota to provide that the child conceived by AID (with the husband's consent) would be considered legitimate* with inheritance rights. None of these bills passed, however. In fact, a regulation by the New York City Department of Health regarding physical examinations and records for semen donors and a resolution by New York State restricting semen collection or sale to physicians, both passed† in 1947, remained as almost the only laws in the world regarding AI for nearly twenty years. The absence of definitive legislation was to cause great concern in 1954 when, after an estimated 10,000 or more children had been born by AID, another court case arose.

The 1954 (Doornbos) case in Chicago's Superior Court involved a divorce and a sole custody claim by the mother of the AID child (with husband's consent), with a further request by the woman that the court approve artificial insemination.[15] Judge Gorman granted the divorce and custody, but denounced AID as adultery on the part of woman, contrary to public policy and morals, and ruled that the offspring was illegitimate. The remarks created a sensation in newspapers throughout the country,[1] and generated much consternation for couples with an AID child and among medical and legal circles. State legislators were in confusion; at least one (in Ohio) suggested a law in line with the Chicago decision, but others wanted liberalization.

In January 1958, a second court case (MacLennon) created an international sensation. The husband in Scotland brought divorce proceedings there against his wife, who had been living in Brooklyn for over a year, but had given birth to a child. The husband charged adultery, but the wife claimed AID. The husband replied that he had not given consent and that the child was born out of wedlock. Judge Wheatly ruled that AID did not constitute "adultery in its legal meaning." The decision was widely discussed[16,17] in Britain: the popular press was generally favorable to AID, but the Archbishop of Canterbury held it was a sin, if not legally a crime, and Anglican and Methodist conferences condemned it. The House of Lords discussed a motion declaring AID adulterous. The case and its aftermath were widely reported in the United States. (See Table I.)

While these cases gained considerable publicity, a number of

*Curiously enough, the Minnesota bill would at the same time call the AID procedure itself unlawful.

†These actions were taken in response to a bit of private enterprise by some New York City students.[1]

other aspects of AI were questioned during the 1950's and early
1960's. With the increasing use of AID, the possibility increases
that half brothers and half-sisters may unknowingly marry. Concerned British doctors had by 1955[17] already placed a limit on
semen donors: they could not father more than one hundred
children.* American doctors seemed less concerned. An important
development was the technique of preserving sperm by freezing
in a "sperm bank," discussed as early as 1951. By early 1954, three
babies conceived from frozen semen had been born.[18] While the
freezing, pooling, and concentration of several semen samples
might help a husband of low fertility, the idea of a sperm bank
posed new problems as well as possibilities: in animal breeding,
choice bulls were soon being used to sire 3,000 calves in a single
year. Would some regular donors for AID practitioners soon be
fathering unheard of numbers of offspring? Or could the sperm
bank be widely used for eugenic purposes and for protection
against mutation by radiation or chemicals as suggested by H. J.
Muller.[19,20] The thought that future parents (and particularly those
of low genetic quality) could use a catalogue to select sperm of
those genetically endowed with great physical, mental, or artistic
abilities created a sensation in the news media in the early 1960's.
For true eugenic purposes (such as increasing the genetic quality of
the human race or at least to protect it from the estimated 20 percent
natural mutation rate) the sperm† must come from truly *superior*
persons and, Muller says,[19] the selection process is too important to
be left to a physician‡ who is interested in helping a barren couple.
Furthermore, public records of the genealogy of the AID child
(abhorred by most AID practitioners) must be maintained for future
generations to use in the selection process.

   In addition to these aspects, the absence of regulatory legislation and moral guidelines on artificial insemination posed a general
problem.

Disposition

   Society has taken no coherent action on the artificial insemination questions: as the practice has increased, many of the problems
mentioned above have recurred time and again on a local level.

   *How this was to be enforced is uncertain.
   †Or later the ova, if techniques for their preservation and implantation are
developed.
   ‡AID practitioners have often stated that the donors were superior.

TABLE I

ARTIFICIAL INSEMINATION PRESS COVERAGE BY
YEAR[a]

	Number of Entries			Number of Entries	
Year	Readers Guide	New York Times Index	Year	Readers Guide	New York Times Index
1934	2		1955	5	1
5	0		6	1	1
6	2		7	1	1
7	1		8	5	10
8	3		9	1	2
9	0		1960	1	2
1940	0		1	10	3
1	3	1	2	3	3
2	0	0	3	2	2
3	1	0	4	1	2
4	0	0	5	2	0
5	3	1	6	3	3
6	1	0	7	3	4
7	0	1	8	1	1
8	3	3	9	4	4
9	2	3	1970	1	3
1950	0	1	1	0	5
1	0	3	2	2	1
2	0	1			
3	3	3			
4	2	3			

[a]Books on AI were published in 1957, 1960, 1964, and 1970.

The United Presbyterian Church of the United States in 1962 was the first major Protestant denomination to approve of AID under a doctor's supervision, and urged Presbyterians to work for uniform state laws to protect the rights of the AID baby.[21] Most other Potestant churches have taken a less definite stand: the Lutheran Church of America took up the issue in 1970 and indicated the final decision was up to the persons involved; the National Council of Churches is no longer concerned with the issue.[12] Some Orthodox Jews apparently consider AID to be adultery, and the Roman Catholic Church continues to oppose it.[12]

The New York State Supreme Court ruled in 1963 that an AID child is illegitimate *even with the husband's consent* (although he then becomes responsible for the child's support!).[22] Three years later, the question of whether AID (this time without the husband's consent) constituted adultery, and thereby grounds for divorce, was raised in New York courts.[23] And in a 1967 California case, the state itself brought suit to make a former husband of an AID mother support the child.[24] In this case, the California Supreme Court[12] held that all children born in wedlock are presumed the legitimate issue of the marital partners, and the husband had to pay. The court also stated that AID did not constitute adultery. The International Congress of Penal Law stated in 1964 that AI should not be considered a crime unless performed without consent of both woman and husband.

Oklahoma became in 1967 the first state to sanction AI and to define the rights of the offspring: full legitimacy with inheritance rights when husband and wife consent in writing. The Oklahoma law requires maintenance of public records (as with adoption papers). Three other states have recently passed legislation[12,25] on AI: Georgia and Kansas grant legitimacy and inheritance rights; Arkansas grants only inheritance. The Kansas law does not restrict the practice of AI to physicians. As of 1970, no state statute prohibited AID.[12]

Despite legal and moral uncertainties the practice of AI (both AID and AIH) has probably increased. By 1966 the U.S. was estimated to lead all other countries* in AI with a total of 150,000 live births,[23] and by 1970 an estimated 200,000 Americans had been so conceived.[12] Nevertheless, a 1969 Harris poll found that only 35 percent of the public approved of AID, even when it was the only means for a couple to have a family, and surprisingly enough, only 49 percent of the men and 62 percent of the women approved of AIH. Recent increases in the practice of AI appear, in part, to be attributable to the extensive practice of vasectomy[25,26] (particularly among sterilized husbands† who have divorced and remarried a young woman) and to the scarcity of babies for adoption[25] (particularly of white babies) because of increasing use of the birth-control pill. The use of sperm banks also appears to be increasing,[26] and their availability for deposit is used as a selling point for vasectomy. The use of AIH has increased significantly to an estimated 20,000 per year.[12] The cost of AI varies among doctors and may depend on the number of inseminations. Medical insurance coverage is uncertain: most of the 1972 Blue Shield plans were said to pay

*Israel, with a population of less than 3 million was second.
†Vasectomized husbands are said to constitute one-third of the AID requests in one doctor's practice.[25]

for AI as a surgical transplant, but some do not on grounds that it is not a necessity.[12] The average payment to the donor appears to have decreased to $20.[12,25] Most major cities appear to have doctors who practice AI.*[12]

A recent summary[12] of the legal and ethical questions over AI included:

• Is the procedure lawful?

• Is a child conceived by AID legitimate?

• Could the doctor, the donor, the husband, or the wife be found guilty of adultery?

• Is it fraudulent or illegal to execute a birth certificate that does not divulge that the mother's husband is not the child's natural father?

• Does the donor have any obligation to the child?

• Is the child an heir of his mother's husband's ancestors?

• What is the legal relationship between an AID child and a naturally conceived child of the same mother and her husband?

• Might a couple, denied the privilege of adopting a child, circumvent adoption laws by resorting to AID.

• If a woman were inseminated without her consent would the physician, the husband, or others be guilty of rape?

Comment

The discovery of the relatively simple biomedical technique of artificial insemination has raised a host of questions that society has yet to resolve clearly, after 160 years of debate. It continues to be a source of contention in our courts and legislative bodies, and among medical practitioners, eugenicists, moralists, church authorities, and the general public. While the present trend appears to be in the direction of increasing acceptance of the practice of AID, it is by no means conclusive. The trend may be reversed by subsequent emphasis on some of the unresolved questions or possibly even by a single highly publicized episode, such as, a court case involving nationally or internationally known individuals.

While the technology of artificial insemination with donors has been the focus of controversy heretofore, widespread use of the technique with fertile husbands may yet pose the greatest challenge to society. For the AI technique has within it, if the reader will excuse the pun, the seeds of a technology for preselecting the

*Use of AI in animal husbandry has also increased remarkably and half of the dairy cows in the U.S. are said to be from AI.[12]

sex of the baby. It is the man's sperm that determines sex: if the ovum is fertilized by a spermatozoan that contains only the X sex chromosome, the baby is a female; if it contains both X and Y chromosomes, the baby is a male. In a day when technology can separate isotopes of uranium and immensely complex biological molecules such as DNA from RNA, the separation of male and female spermatozoa in a semen sample seems not too difficult. And what will be the effect on society when parents-to-be everywhere can choose the sex of each baby? Recent U.S. surveys have indicated that both husbands and wives may choose far more boy babies than girl babies, an event that would lead to an unprecedented ratio of sexes in the population. When the technology develops, will we decide not to choose at all, will our choices change to a more balanced selection, will government regulation of choice be imposed, or will we adjust to a new order of social arrangement? Society may soon be faced with these questions.

REFERENCES

1. Finegold, W. J., Artificial Insemination, Charles C. Thomas, publisher, Springfield, Illinois, 1964.
2. "Babies by Scientific Selection," Sci Am, 150, 124 (March 1934).
3. "Children Provided for the Childless," Newsweek, 3, 16 (May 12, 1934).
4. "Women Without Men," Newsweek, 7, 30 (April 4, 1936).
5. "Test-Tube Babies," Read Digest, 30, 18 (February 1937).
6. "Test-Tube Babies: A Medico-Legal Discussion," Sci Am, 156, 40 (February 1937).
7. "Proxy Fathers," Time, 32, 28 (September 26, 1938).
8. Biol Abs, 6, 10, 2254 (21993).
9. "Artificial Bastards," Time, 45, 58 (February 26, 1945).
10. New York Times, p. 3 (April 9, 1945).
11. New York Times, p. 13 (April 20, 1945).
12. "Babies in Question," Today's Health, 48, 17 (August 1970).
13. "Breach of Marriage," Time, 52, 49 (August 9, 1948).
14. "Doctor's Dilemma," Time, 54, 83 (October 10, 1949).
15. "Test-Tube Test Case," Time, 64, 52 (December 27, 1954).
16. "The Riddle of Birth," Time, 71, 23 (January 27, 1958).
17. "Artificial Insemination—Has It Made Happy Homes?" Read Digest, 66, 77 (June 1955).
18. New York Times, p. 31 (April 6, 1954).
19. Muller, H. J., "Human Evolution by Voluntary Choice of Germ Plasm," Science, 134, 643 (September 8, 1961).
20. "Frozen Fatherhood," Time, 78, 68 (September 8, 1961).

21. "Presbyterians on Marriage," *Time*, **79**, 83 (June 1, 1962).
22. *New York Times*, p. 52 (August 4, 1963).
23. "The Riddle of AI," *Time*, **87**, 48 (February 25, 1966).
24. "The Child of Artificial Insemination," *Time*, **89**, 79 (April 14, 1967).
25. *Kansas City Star*, p. 4E (April 20, 1972).
26. "Sperm Banks Multiply as Vasectomies Gain Popularity," *Science*, **176**, 32 (April 7, 1972).

Oral Contraceptive Safety Hearings

Abstract

In January 1970, Senator Gaylord Nelson's Subcommittee on Monopoly started hearings on the long-simmering dispute over the safety of prescription oral contracptives and the problem of whether users of "the Pill" were being adequately informed about potentially harmful effects. The Pill, introduced in 1960, was being taken by an estimated 8.5 million women in the United States, and had sales of over $100 million annually. At the widely publicized hearings, critics of the Pill caused a sensation by claiming that it had not been adequately tested and was the potential cause of blood clots and perhaps cancer. Numerous subsequent witnesses, including spokesmen from the medical field, the Food and Drug Administration, and family planning and population control organizations, adamantly asserted that the Pill was safe and that doctors were telling their patients all they needed to know. A poll soon indicated, however, that 18 percent of the users of the Pill had quit, and controversy arose over the public concern the hearings were generating. One critic claimed that the hearings had resulted in 100,000 unwanted pregnancies—the so-called "Nelson babies"—although this estimate could not be verified by subsequent birth data. As a result of the hearings, manufacturers were required to furnish cautionary information to all users of the Pill. The U.S. birthrate dropped to an all-time low by 1972, an effect widely attributed to use of the Pill.

Background

Starting in about 1930 a series of sex hormones were isolated in sufficient quantities to identify their chemical structure. The properties of these compounds were of immediate interest to medical

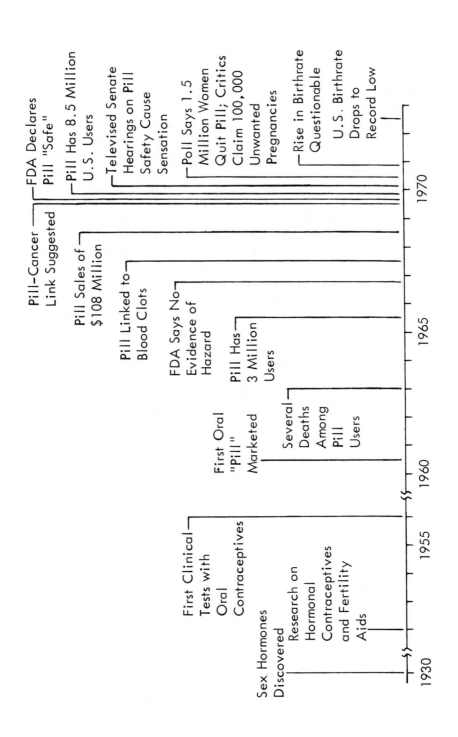

Sex Hormones Discovered

Research on Hormonal Contraceptives and Fertility Aids

First Clinical Tests with Oral Contraceptives

First Oral "Pill" Marketed

Several Deaths Among Pill Users

FDA Says No Evidence of Hazard

Pill Has 3 Million Users

Pill Linked to Blood Clots

Pill Sales of $108 Million

Pill–Cancer Link Suggested

FDA Declares Pill "Safe"

Pill Has 8.5 Million U.S. Users

Televised Senate Hearings on Pill Safety Cause Sensation

Poll Says 1.5 Million Women Quit Pill; Critics Claim 100,000 Unwanted Pregnancies

Rise in Birthrate Questionable

U.S. Birthrate Drops to Record Low

1930   1955   1960   1965   1970

researchers; those hormones that were available and some of their chemical derivatives were subjected to study, primarily in small animals. By about 1950 interest had developed in the use of progesterone, one of the two female hormones, as a treatment for infertility and also as an oral contraceptive in humans. During 1952–54 Syntex Pharmaceutical, Ltd., and G. D. Searle Company, developed methods of making powerful progesterone derivatives* in research quantities. The first clinical investigations of the efficacy of oral contraceptives in humans was begun in Puerto Rico in 1956 by Harvard's Dr. John Rock. During the late 1950's, Searle, Syntex, and a third pharmaceutical company, Parke, Davis and Company, engaged in a feverish race to perfect and test their oral contraceptives and to reach the market first.† A cyclical dosage regimen and "combination" tablets were developed; the latter contained progesterone, to regulate the menstrual cycle, and estrogen, the second female hormone (or an estrogen derivative), to suppress ovulation. The action of this combination was to mimic the hormonal environment of pregnancy itself during the normal ovulation time.

On June 23, 1960, Searle received from the Food and Drug Administration approval to market the first oral contraceptive, Enovid—a product that FDA had approved in 1957 for use by doctors in treating a variety of female disorders. Enovid was rapidly followed by other products from Ortho Pharmaceutical, Syntex, Parke-Davis, the Upjohn Company, Mead Johnson and Company, Eli Lilly and Company, and a second product from Searle. In all, fourteen products were on the market by mid-1967, and by 1970 eight U.S. pharmaceutical companies were selling a total of thirty-six products under 12 brand names.[1] The competition for the market was fierce and promotional efforts were extensive and enthusiastic—particularly among gynecologists, obstetricians, and other physicians who might recommend these prescription drugs to their patients.

The advent of oral contraceptives brought a flurry of articles in the popular press‡ and acclaim from those recommending family planning or population control. Physicians and women tested the oral contraceptive—"the Pill" as it quickly became known—in

*The wild Mexican yam proved invaluable as the starting material in progesterone synthesis.
†Government support of contraceptives research before 1960 had been negligible.
‡In contrast, the mention of other contraceptive devices had been almost tabu previously.

large numbers; by 1965 over 10 percent of the 30 million women of child-bearing age in the United States were users, and by 1967 6.5 million women in the United States and 6.3 million abroad were estimated to be "on the Pill." By 1968 the U.S. pharmaceutical companies' sales of the Pill were $108 million and production was about 10 million lb/year.[2]

The Pill was not without its critics, however. In addition to those who objected on moral and ethical grounds to contraception, or to the mass prescription of *nonmedicinal* drugs by the medical profession, questions soon began to arise about the safety of the Pill. Many women reported unpleasant side effects from taking the Pill, and as early as 1961 reports started to circulate in medical circles that several users of the Pill had suffered serious or even fatal circulatory problems, including pulmonary embolism, thrombophlebitis (blockage of blood vessels in the extremities) and stroke. In September 1962, the Searle Company organized a conference in Chicago to discuss nine deaths among users. By 1963 the FDA assembled an ad hoc committee to look into the thromboembolic disorders among *Enovid* users (about 350 reports), but it confined its attention to fatal cases. It concluded* that a retrospective study on such a small number of deaths (31) was statistically unreliable,† and recommended a controlled, prospective study.[1]

Concern among some medical researchers was expressed during the early and mid-1960's, but little of this found its way into the popular press. According to some analysts[3] the news media extended what has been called "a diplomatic immunity" to the Pill and even carried considerable propaganda in its favor. The Searle Company had, in fact, established a "Bad Press" committee that had the duty of immediately counteracting any adverse publicity on *Enovid*. The FDA and the American Medical Association officially authorized laudatory articles in popular magazines. Thus, when a Senate Committee on Government Operations Subcommittee (headed by Hubert H. Humphrey) heard testimony that the FDA's decision to let *Enovid* be sold as a contraceptive was based on tests with only 132 women for not more than thirty-eight months,‡ the media made almost no mention[3] of it—despite the sharp contrast

*Ref. lc., p. 7235.

†The data had indicated a 50 percent increase in deaths from this disease among users.

‡In the Puerto Rican studies, the best available, only sixty-six women had used the pill for twelve to twenty-one consecutive cycles and sixty-six women for twenty-four to thirty-eight consecutive cycles.

with the manufacturer's contention that *Enovid* had been extensively tested. On the other hand, a July 3, 1964, article in *Life* by the
FDA's top physician, Dr. Joseph F. Sawdusk, Jr. (who subsequently became a Vice President at Parke, Davis and Company),
called the Pill "safe, when given under a doctor's supervision," and
another reassuring article in *This Week* (the Sunday newspaper
supplement) on July 12, 1964, by Dr. Edwin J. DeCosta was
labeled "AMA Authorized." In addition, Dr. Alan F. Guttmacher,
president of Planned Parenthood, Dr. Rock, and other medical men
gave frequently quoted reassurances of the Pill's safety in such
publications as *Good Housekeeping* (February 1966[1c]). But by 1966,
concern over the potential of the Pill to cause blood clotting or
other illnesses* had increased to the point that the FDA directed its
Advisory Committee on Obstetrics and Gynecology to study the
problem. The study concluded in August 1966 that evidence had
not proved these compounds unsafe. This finding was largely translated in the news media as a seal of approval.[3] And while Dr. Louis
Hellman, Chairman of the FDA's committee, termed the report a
"yellow light of caution," Dr. Guttmacher called it a "complete
green light."

A change within FDA had already been initiated, however,
when long-time Commissioner George P. Larrick was replaced in
January 1966 by Dr. James L. Goddard. In early 1967, Goddard
indicated that the original safety data on *Enovid* might not meet
current FDA requirements and in April warned that the side effects
of the Pill were grossly underreported.† (U.S. Senate Hearings, Ref.
lc., p. 7148.) A few days later, the British Minister of Health disclosed in Parliament that a still unreleased report of the Medical
Research Council indicated that use of the Pill would cause only a
slightly increased risk of blood clotting. His remarks created considerable excitement and confusion in the British press until the
report itself was released in May. Called a preliminary report, it
linked the Pill with thrombolic disorders including death but received scant attention in the American press: the *New York Times*
gave the story only twelve printed lines and *Time* magazine (which
in April had run a cover story‡ on the Pill entitled, "Freedom from
Fear") ignored it, while *Newsweek* coupled it in three paragraphs
with a report by Cleveland doctors that nine Pill users had suffered

---

*In addition, Dr. Frank B. Walsh of Johns Hopkins had reported in November
1965, sixty-three cases of eye and nervous system damage to users of the Pill.
†*Ref. lc., p. 7148.*
‡A Time-Life, *Birth Control*, was also nearing publication.

symptoms of stroke and some reassuring remarks by an FDA spokesman.[4] But during 1967 the number of popular articles on the Pill began what was to be a three-year rise, lagging behind a similar rise in technical publications as shown in Figure 1. More important was the fact that the Pill began to lose some of its "diplomatic immunity" from the media.[5] For example, a *Newsweek* article was headlined, "Warning Signs," and in December *Time* carried[6] "The Pill and Strokes." The Pill was headed toward a first-class scare episode.

Key Events and Roles

On April 27, 1968, the prestigious *British Medical Journal* published the final reports to the Medical Research Council and commented favorably on the quality of the work: the results of these retrospective studies clearly indicated a cause-effect relationship between use of the Pill and thrombosis and pulmonary embolisms (clots of the veins and lungs, respectively).* Almost immediately, the FDA summoned the American manufacturers of oral contraceptives to a meeting in Washington: they agreed to a major revision of the uniform package labeling and a warning statement (effective July 1, 1968) about the dangers of clotting. The FDA did not, however, require that the warning statement be given to the consumers of the Pill—only to the pharmacists and (it was hoped) to the prescribing physicians. The British report and the FDA's reaction caused surprisingly little immediate publicity in the media.[7-11]

Three developments were occuring, however, that caught the media's interest and focused attention on hazards of the Pill: (a) some women taking a new drug used to restore fertility after use of the Pill started giving multiple births; (b) law suits led by Raymond Black's $750,000 suit against Searle for the death of his wife from a blood clot were being brought against the manufacturers because of the Pill's side effects; and (c) rumors began to circulate that a report linking the Pill to cancer was being suppressed. The FDA therefore directed its Advisory Committee to initiate yet another study of the Pill.

In May 1969 the case of Black *vs* Searle went to trial in the U.S. District Court at South Bend, Indiana. The suit charged that Searle

*Death rates (per million) among Pill users and nonusers, respectively, in two age groups were: thirty-nine and five (35–44 years); fifteen and two (20–34 years). One out of every 2,000 women on the Pill suffered thromboembolic disease.

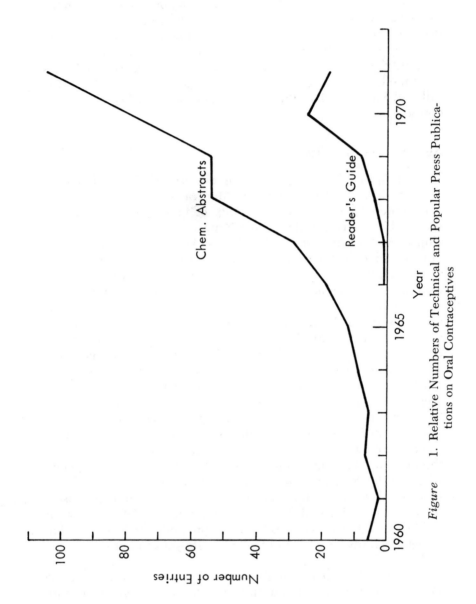

Figure 1. Relative Numbers of Technical and Popular Press Publications on Oral Contraceptives

was negligent, had breached the implied warranty of its product, and was liable for the Pill's known hazards that caused Mrs. Black's death. The jury was presented with a complicated case of medical technology, including conflicting testimony by expert witnesses.[12] In a key ruling, the judge decided that Searle could be held accountable only for what they knew about the Pill's hazards at the time of death in 1965, thus making inadmissible the later British data on blood clots—data which Searle was acknowledging in its bulletins in 1969. Searle won the verdict, but the jury added a plaintive recommendation that Searle be required to advise both doctors and patients in instructive literature of possible hazards of phlebitis and thrombolic and embolic phenomena.*

Proponents of the Pill hastened to point out that women suffer a natural increase in the risk of blood clotting because of pregnancy and birth.[13] Thus the risk of thromboembolism for nonpregnant women under 40 years of age is small, while the risk is slightly higher for pregnant women than for women on the Pill (0.74–0.55 in cases per 1,000 women per year), and much higher during the 3 weeks after delivery (3–10 cases per 1,000). On the other hand, critics countered, other contraceptive methods such as diaphragms, intrauterine devices, and condoms for the male don't increase the risk of clotting at all.

In late July 1969, the *British Medical Journal* published the long-rumored paper on the Pill-cancer study. Two New York researchers, Drs. M. S. Melamed and H. Dubrow, headed a study† of uterine "carcinoma *in situ*." The study had started in the fall of 1965 and was eventually to involve over 34,000 women. Rumors that the study was developing evidence of a link between the Pill and cervical cancer began to circulate in early September 1968 after the American Cancer Society held a closed meeting to review evidence on the Pill; the rumors were reflected on Wall Street in a dip in stock prices. The FDA's Advisory Committee also reviewed the incomplete Melamed-Dubrow results in November, but the FDA Commissioner, Herbert Ley, said they neither proved nor disproved a link. Shortly thereafter, the research group summarized their findings in a manuscript that was submitted to the *Journal of the American Medical Association. JAMA* reviewers requested extensive revision of the paper, then agreed to publish it only if it was further revised or if it was accompanied by a rebuttal

*Ref. lc., pp. 7031–7037.
†In cooperation with Planned Parenthood of New York City under grants from U.S. Public Health Service and National Cancer Institute.

from its critics, who questioned the experimental procedures and doubted the validity of the results. The rumors now included a charge that the AMA was suppressing the report under pressure from the drug companies, and these were not lessened when Melamed and Dubrow submitted the revised manuscript to the *British Medical Journal* and had it accepted.

The report, which concluded that users of the Pill had a slightly higher incidence of cervical cancer than those that used the diaphragm as a contraceptive, caused considerable excitement in medical circles, but not an undue amount of coverage by the news media.[13-15] Considerable public confusion was generated, however, when some critics of the report claimed that diaphragm users should not have been used as control groups—the diaphragm was said to offer a protective barrier against a virus or carcinogenic substance in the male's semen or smegma—while others claimed that the reported tissue changes were to a "precancerous condition" and not to cancer itself.

The public's confusion was further heightened[15-18] when the FDA's Advisory Committee on Obstetrics and Gynecology issued its "Second Report on Oral Contraception" on August 1, 1969. The report said that a relationship between thromboembolic disorders and the Pill had been established, and noted that cancers had been produced in five species of laboratory animals by estrogens, so that research on the possible carcinogenicity of the Pill was top priority. However, the carefully worded conclusion noted that no effective drug could be absolutely safe and that the Committee found the ratio of benefit-to-risk sufficiently high to call the Pill "safe."

Despite the many unanswered questions, the Committee's report did appear to relieve public anxiety for a time. But on December 22, 1969, Senator Gaylord Nelson (D-Wisconsin) released a press statement that his Sub-committee on Monopoly of the Senate Select Committee on Small Business would hold hearings to explore the question of whether users of birth control pills were being adequately informed concerning the Pill's known hazards. Although this particular subcommittee may seem like an unusual forum for hearings on the Pill's safety, it had been conducting a long series of hearings* on the drug industry—to the point that the Pharmaceutical Manufacturers Association had charged that Nelson was conducting a vendetta against them with rigged hearings, and

*These hearings had actually been started by Senator (and 1956 Vice Presidential candidate) Estes Kefauver in the 1950's. Nelson's subcommittee contained nine very powerful Senators.

the drug industry, in turn, was charged with pouring money into Wisconsin to defeat the Senator. The announcement caused considerable comment[19] and concern for the pharmaceutical industry.[20] The hearings themselves were to be a sensation.

The hearings opened on January 14 and 15, 1970, with nationwide news media coverage (including live television), then continued on January 21–23, before breaking for a month before the second session. The first witness to testify was Dr. Hugh J. Davis of Johns Hopkins' Department of Obstetrics and Gynecology, a well-known critic of the Pill, advocate of the competitive IUD, and coinventor of a new ring device not yet on the market. Davis was followed during the first two days of the hearings by seven other doctors, only one of whom was considered "pro-Pill"—Dr. Robert W. Kistner, coworker with Dr. Rock of Harvard. The testimony during these two days related to the technical safety of the Pill rather than to the question of whether women were being sufficiently informed of potential hazards. Although most of the data presented had been previously discussed in the technical literature and at meetings, the recitation of the known and suspected hazards of the Pill by Dr. Davis and the other witnesses generated a great many scare news stories around the country. They stressed such statements as "Estrogens are to breast cancer what fertilizer is to the wheat crop," "Cancers do not develop overnight; shall we have millions of women on the Pill for 20 years and then discover it was all a great mistake," and "If the Pill was a food there would be enough evidence to consider removing it from the market." References to other emotion-laden issues like thalidomide, cyclamates and DDT, were the focus of interest of many news stories. The general tone[16,21–26] was that the Pill might cause blood clotting, stroke, heart disease, cancer, loss of fertility and sexual appetite, or genetic damage. These fears were strengthened when it became known that on January 12, Dr. Charles C. Edwards, Acting Commissioner of FDA had sent a "Dear Doctor" letter to 324,000 of the nation's physicians noting the hazards of the Pill, urging them to advise their patients, and enclosing a proposed revised labeling statement. The story made the front page of many papers on January 20.[27]

During January 21–23, a dozen additional witnesses were heard, including Dr. Hellman of FDA's Advisory Committee, two or three more "anti-Pill" witnesses, and others who offered more "neutral" or merely technical testimony. In addition, the hearings were disrupted on January 23 by members of the Women's Libera-

tion Movement; they objected that the subcommittee had not invited any women to testify, and contended that possibly unsafe
contraceptives were being foisted off on women for the profit of the
medical and drug industries and for the convenience of men. This
testimony, summarized with the preceding two days' evidence by
the news magazines, and did little to lessen the claim by critics that
the hearings had been "stacked" against the Pill.[28]

The massive publicity of "bad news" after years of "good
press" for the Pill caused much confusion among the public and
probably also among hundreds of Pill-prescribing doctors who
were suddenly besieged by their female patients seeking information or reassurance. Almost immediately, the American Medical
Association, the American College of Obstetricians and
Gynecologists, and the Planned Parenthood—World Population issued statements that the Pill was safe under a doctor's supervision
and that the hearings were causing unnecessary panic.

Amost as soon, however, claims began to circulate that many
women were going "off the Pill," that requests for the Pill at clinics
and from private practitioners were down, while requests for other
contraceptives had gone up. Critics of the hearings claimed that
some women who had stopped were already pregnant, and seeking
abortions. A Gallup Poll was commissioned by *Newsweek* magazine
to determine attitudes and reactions of U.S. women to the Pill, and
the results were rushed into print in the February 9 issue.[29] The
results: of the Pill users (approximately 25 percent of the total
questioned), 18 percent said they had quit and another 23 percent
said they were thinking about quitting; one third* of the 18 percent
said they quit because of the hearings; and 66 percent of all Pill
users said they had never been told about possible hazards by their
physicians.

Those who favored the Pill selectively quoted the results of
this and similar[30,31] polls. An article entitled, "Pregnancies Follow
Birth Pill Publicity," in the *New York Times* by long-time Pill
adovcate Jane Brody tended to link all those who had quit, or were
thinking about it, with the hearings.[31] The *Newsweek* story itself
carried the 18 percent figure in the title.[29]

Thus the second session of the hearings (February 24–March 4)
included not only substantial pro-Pill testimony and technical criticism of earlier testimony, but it carried an air of controversy over
the effects of the first session, a decided overtone of criticism by
some witnesses of the nature (and even the existence) of the hear-

*This would be about sixty-eight women out of the total 895 women interviewed.

ings and of the committee chairman, Senator Nelson. The more critical witnesses were Dr. Elizabeth B. Connell of the Family Planning Services of Columbia University, Dr. Guttmacher of Planned Parenthood, and Mrs. Phyllis Piotrow of the Population Crisis Committee. Dr. Connell decried the lack of balance in the hearings. She stated that doctors were already seeing the first of the pregnancies of women who panicked after the January hearings and that an increase in illegal abortions and abortion-caused deaths were sure to follow. Dr. Guttmacher and Mrs. Piotrow both indicated that 18 percent of the Pill users had discontinued abruptly after the January hearings. Senator Nelson objected strongly that according to several studies, 40 to 70 percent of the Pill users ordinarily stopped anyway, for various reasons, even before the hearings. He asserted that the *Newsweek* poll actually showed only 6 percent stopped because of the hearings, that many of these would switch to other methods, and that critics were overlooking the important finding of the poll that two-thirds of the Pill users had never been adequately informed by their doctors in the first place. *Time* quoted[32] Dr. Guttmacher telling Nelson, "I don't think that anyone is saying that your hearings are impregnating women, but the adverse publicity has caused many women to quit the Pill," and also quoted committee member Senator Robert J. Dole (R-Kansas), as hoping that the resulting babies will not all be named for subcommittee members.* Mrs. Piotrow, however, estimated[1b] that the hearings were causing over 100,000 unwanted babies, which she dubbed the "Nelson babies." Senator Nelson discounted the validity of the estimate and noted that the statements of Connell, Guttmacher, Piotrow and others had been prepared barely two to three weeks after the opening day of the hearings; he added that the numbers being thrown about were not based on any statistically significant data. His efforts were to little avail, however, and Washington gossip soon resounded with the quip that there were going to be a lot of little babies named "Nelson" in a few months.

Disposition

Dr. Charles C. Edwards, Commissioner of the Food and Drug Administration, testified on the last day of the hearings, March 4,

*The published record of the hearings[1b] does not show Dr. Guttmacher making this statement and records Senator Dole's statement as, "I do not know if you can blame all these babies that are born on the members of the committee or not." The contrast raises the question: Were the proceedings of the hearings edited by governmental participants before their publication?

1970. At the conclusion of his broad testimony, he unveiled a proposed statement, "What You Should Know About Birth Control Pills," that the FDA was considering as a leaflet to be included in every package of oral contraceptives produced. The degree to which the 600-word statement described potential hazards such as blood clots and cancer astonished advocates of the Pill. The eight manufacturers of the Pill, whose sales were already slipping,[2] and the Pharmaceutical Manufacturers Association, who saw a threat to all their pharmaceutical products, voiced their displeasure,[33] and the American Medical Association (AMA) promptly attacked the FDA for interfering in the doctor-patient relationship and said the statement could possibly lead to malpractice suits.[3,4] Even White House pressure was alleged.[33] Within three weeks the FDA revealed[34] that it was revising the still unpublished warning leaflet to a 96-word statement that eliminated most of the clinical detail and all mention of cancer. It was immediately described as "watered down" by those who wanted a strong statement. The controversy had grown so large by this point, that the Secretary of Health, Education and Welfare, Robert H. Finch, and the Surgeon General, Dr. Jessie Steinfeld, became involved; in a press conference on April 7 they released a mild 120-word proposed statement that was officially published in the *Federal Register* three days later. The AMA immediately pledged an all-out fight against any proposed package insert, but in September 1970 the FDA officially ordered that the approved warning statement be sent to every doctor and inserted in every package of the Pill sold.[35]

Despite the intensity of the media coverage of the Pill's hazards during the early days of the hearings, the story faded substantially from sight by midyear, although a few good summaries of the hearing results appeared in magazines that appealed to special audiences.[36] Reassuring articles on the Pill's safety were published by spokesmen for Planned Parenthood,[37] in *Today's Health*, a lay periodical of the AMA found in most doctors' waiting rooms,[38] and in *Reader's Digest* by Dr. Connel.[39] As a result of the publicity, Searle did lose a suit in April over the death of a woman from a blood clott and was assessed $251,000 damages, according to *Science*.[40] Two hundred such suits were pending in 1970, according to the *Wall Street Journal*.[30]

The safety of the Pill in regard to circulatory problems, was apparently improved[36] substantially during 1969–70 as many companies introduced new formulations with much lower levels of estrogen and progesterones. Searle's new *Enovid* version, for example contained only $1/3$ the progesterone and $1/4$ the estrogen that were in the original. The Pill-cancer link has apparently not been proved to date.

And what of the "Nelson babies"? The hypothesis that the hearings would cause 100,000 unwanted babies cannot be statistically documented, and the question received scant attention by the media after the hearings. Although some clinics and doctors reported a number of such "scare" pregnancies, and the monthly birth rates during November 1970 through February 1971 were slightly higher than in the same period the year before and the year after, the increased rate had actually started in about June (only four months after the hearings started) and the rate in January 1971 actually fell below that of January 1970, as shown in Table II. The birthrate data* (Figure 2) over the long term show that a pronounced 10-year decline had leveled off† about 1967 and then resumed in 1970 (the dollar sales of the Pill started to decline after 1968[2]). At any rate, the U.S. birth rate dropped to an all-time low by 1972—a result widely attributed to use of the Pill,[41] although economic factors probably contributed, as they did during the depression of the 1930's, before the Pill was invented.

TABLE II

MONTHLY BIRTHRATES BEFORE AND AFTER THE HEARINGS

Month	Birthrate (per 100,000 population)			
	1969	1970	1971	1972
January	16.6	17.4	17.1*	15.9
February	17.7	17.6	18.0*	15.9
March	16.7	17.4	17.5	—
April	17.1	17.4	17.1	15.0
May	17.2	17.2	16.7	15.4
June	17.4	18.2	16.8	15.5
July	18.4	18.8	16.8	15.0
August	18.8	19.1	17.5	16.1
September	19.1	19.6	18.7	16.6
October	18.0	18.8	17.4	15.6
November	17.7	18.2*	17.3	—
December	17.7	18.6*	16.3	—

*Months in which impacts of hearings should have been greatest.

*Bureau of Census data.
†The slight rise during 1967–70 may have resulted from other factors (such as age distribution, economics of the Vietnam War) rather than to the Pill controversy.

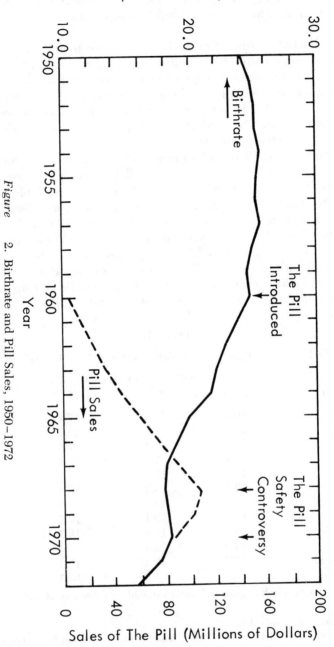

Birthrate per 1,000 Midyear Population

Birthrate

The Pill Introduced

Pill Sales

The Pill Safety Controversy

Year

Sales of The Pill (Millions of Dollars)

Figure 2. Birthrate and Pill Sales, 1950–1972

Comment

The controversy over the safety of the oral contraceptives reflects a series of breakdowns in the operation of some of our most important social institutions. The companies manufacturing the Pill appeared far more concerned with rushing their products to a lucrative market and in beating down any adverse reports, than in investigating their products' safety as well as efficiency. The FDA obviously allowed the products to come on the market with grossly insufficient long-term safety testing and allowed the largest mass experiment in medical history to begin with little supervision. The practicing obstetricians and gynecologists often did not take the time to keep up with the voluminous technical literature, and received much of their information from the glowing promotional material with which they were being deluged. Many U.S.organizations supporting the cause of population control failed to note or advise of the adverse effects of the Pill; this was done consistently by the British, with only 10 percent of the number of Pill users of the U.S. Meanwhile, American women, in increasing numbers, were consuming these potent tablets with little information on their potential hazards, but under the constant stream of reassuring articles by advocates of the Pill and by the news media, which had largely adopted the social engineer's view that the Pill was a godsend in population control.

The technical details of the hazards of the Pill were difficult not only for the layman to understand, but also for the news media, the Congress and the courts. When the hearings opened, they revealed a great polarization between the critics of the Pill, as it was being used, and the wholehearted advocates of the Pill. Many members of the news media immediately focused on the "bad news" portion of the testimony because the advocates' side had already been well reported and was no longer "news." Conversely, when polls showed that many women were concerned, a considerable polarization among news reporters and magazines* was also revealed, and many blamed the committee or its chairman. While the claim that Senator Nelson stacked the hearings has numerous supporters, the fact remains that one could not hear all the testimony in one day, and if the hearings were held in executive session, a cover-up might

*Many news magazines had apparently adopted unannounced positions on how they would handle stories on contraceptives during the 1960's. Even at the height of the controversy, much difference can be noted in coverage. We found that during the height of the controversy (1969–70) *Newsweek* carried ten articles on the Pill compared with four in *Time*.

44 TECHNOLOGY AND SOCIAL SHOCK

be charged on such a sensitive issue. Some of the Senator's critics did little to demonstrate their own objectivity. The testimony as a whole demonstrated the way in which iatrogenic (physician-caused) disease might arise; "Specialists do not see their own mistakes," as someone put it. The "Nelson babies" appear to be few and the net result of the hearings appears to have been beneficial for the medical and pharmaceutical industries, the FDA and the public. Despite the fears of the AMA, the "doctor-patient relationship" seems hardly to have been affected at all by the required new warning insert (which is so short that most manufacturers simply preface it to the "directions for use" statement in the package).

REFERENCES

1. "Competitive Problems in the Drug Industry," Hearings Before the Subcommittee on Monopoly of the Select Committee on Small Business, U.S. Senate on "Present Status of Competition in the Pharmaceutical Industry:" (a) Part 15, January 15, 16, 21–23, 1970, Oral Contraceptives (Vol. 1); (b) Part 16, February 24, 25, and March 3, 4, 1970, Oral Contraceptives (Vol. 2); and (c) Part 17, Oral Contraceptives (Vol.3) Appendices.
2. "Side Effects: 'Pill' Sales Slump," *Chem W*, **106**, 26 (April 1, 1970).
3. "The Pill: Press and Public at the Experts Mercy," Morton Mintz, *Columbia Journ Rev:* Winter 1968–69, pp. 4–10; and Spring 1969, pp. 28–35.
4. "Warning Signs," *Newsweek*, **69**, 82 (June 5, 1967).
5. Reference 1c reproduces 35 articles that appeared during 1964–69. Two widely read "pro-Pill" articles not reproduced were in *Parents Mag* (October 1967) and *Family Circle* (January 1968).
6. "The Pill and Strokes," *Time*, 46 (December 29, 1967).
7. "The Pill—Is There Danger," *Today's Health*, 24 (May 1968).
8. "New Warning About Birth Control Pills," *Good H*, 16 (August 1968).
9. "Caution on the Pill," L. Lasagna, *Sat R*, **51**, 64 (November 2, 1968).
10. "The Terrible Trouble with the Birth Control Pills," and letters, *Ladies Home J* 84, 43, (July 1967) and discussion 84, 92 (November 1967).
11. "Why They Quit the Pill," *McCalls*.
12. "Controversy Over the Pill," Bill Surface, *Good H*, 170, 64 (January 1970).
13. "The Pros and Cons of the Pill," *Time*, 58 (May 2, 1969).
14. "Doubts About the Pill," *Newsweek*, **73**, 118 (May 19, 1969).
15. "Pill and Cancer; Findings of Melamed-Dubrow Study," *Newsweek*, **74**, 59 (August 11, 1969).
16. "Pill: Cloudy Verdict," *Newsweek*, **74**, 90 (September 15, 1969).
17. "Birth Control Pills: Safe But—," *US News*, **67**, 10 (September 15, 1969).

18. "Why There is Growing Concern About the Safety of the Pill," *Good H*, **170**, 129 (January 1970).

19. "Researcher Predicts Ban on the Pill," Barbara Yuncker, *New York Post*, 3 (December 22, 1969).

20. "The Pill Goes to Washington," *Bsns W*, 31 (January 10, 1970).

21. "The Pill on Trial," *Time*, **95**, 60 (January 26, 1970).

22. "Perils of the Pill," *Newsweek*, **75**, 21 (January 26, 1970).

23. "The Pill: Is It Safe—Or a Threat to Health?" *US News*, **68**, 10 (January 26, 1970).

24. "Confusion on the Pill," *New Repub*, **162**, 10 (January 31, 1970).

25. "Amber Light for the Pill," *Newsweek*, **75**, 48 (February 2, 1970).

26. "The New Doubts About the Pill," *Life*, **68**, 27 (February 27, 1970).

27. *New York Times*, p. 1 (January 20, 1970).

28. "Did the Pill Get a Fair Shake from Nelson?" *Chem W*, **106**, 21 (January 28, 1970).

29. "Poll on the Pill—18% of U.S. Users Have Recently Quit," *Newsweek*, **75**, 52 (February 9, 1970).

30. "The Pill on Trial: Adverse Reports Lead Many Women to Switch Birth Control Methods," *Wall St J* (February 2, 1970).

31. "Pregnancies Follow Birth Pill Publicity," Jane E. Brody, *New York Times* (February 15, 1970).

32. "The Pill On Trial (Con't)," *Time*, **95**, 32 (March 9, 1970).

33. "Pill Caution," *Time*, **95**, 46 (April 20, 1970).

34. *New York Times*, p. 8 (March 24, 1970).

35. *New York Times*, p. 25 (September 10, 1970).

36. "The Pill—Do Its Benefits Outweigh Its Hazards?" *Consumer Rep*, **35**, 314 (May 1970).

37. "Pill Is Safe," Edward T. Tyler, M.D., *Look*, **34**, 65 (June 30, 1970).

38. "What You Should Know About the Pill," *Today's Health*, **48**, 9 (September 1970).

39. "The Pill In Perspective," by E. B. Connel, *Read Digest*, **97**, 118 (October 1970).

40. "Woman Wins Pill Suit," *Science*, **168**, 451 (April 24, 1970).

41. "The Revolution in Birth Control Practices of U.S. Roman Catholics," C. F. Westoff and L. Bumpass, *Science*, **179**, 41 (January 15, 1973).

Books on The Pill

The Therapeutic Nightmare by Morton Mintz, revised and reissued as *By Prescription Only*, Houghton-Mifflin Company, Boston, Massachusetts, 1967.

The Doctors' Case Against The Pill by Barbara Scamon, Peter H. Wyden Company, New York, New York, 1969.

Drugs in Our Society, Paul Talalay, Ed., Johns-Hopkins Press, 1964.

The Pill: An Alarming Report by Morton Mintz, Beacon Press, 1970.

Southern Corn Leaf Blight—A Genetic Engineering Problem

Abstract

A virulent outbreak of Southern corn leaf blight, made possible by the use of a newer method of producing hybrid seed, occurred in the U.S. in 1970. It damaged a significant part of the corn crop in northern as well as southern states, resulting in widespread losses for farmers and increased prices for consumers. In 1971, the Department of Agriculture encouraged farmers to plant much larger corn acreages to offset the expected blight losses. But because of the unusual efforts of seed corn producers to produce resistant varieties and because the blight was generally less virulent, a large corn surplus was produced and prices fell sharply. The corn leaf blight raised many questions concerning government involvement in agriculture and the consequences of widespread genetic engineering.

Background

Corn is by far the largest single crop grown in the United States; the supply of corn-derived food products (such as meat, eggs, and milk), the price of most other agricultural commodities, and the financial interests of farm machinery manufacturers, the railroads, Wall Street, and entire rural communities are directly tied to the successful production of corn by the nation's farmers. Since the introduction of hybrid corn in 1933* and the increased use of chemical fertilizers and pesticides, annual corn production had

*By 1950, hybrid seed corn was being sown on more than 90 percent of the corn acreage.

46

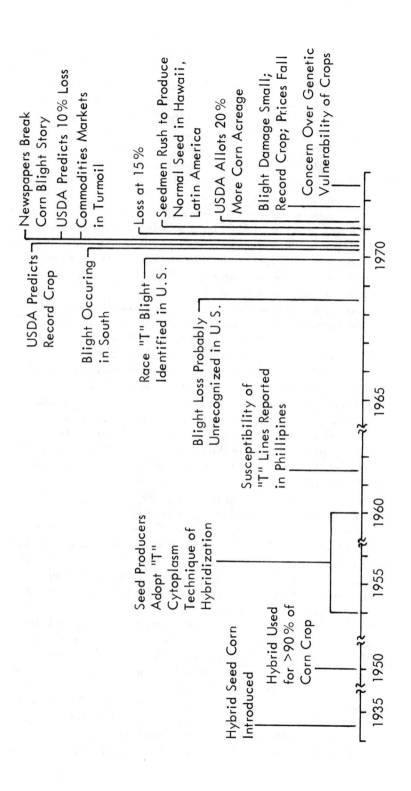

Hybrid Seed Corn Introduced

Hybrid Used for >90% of Corn Crop

Seed Producers Adopt "T" Cytoplasm Technique of Hybridization

Susceptibility of "T" Lines Reported in Phillipines

Blight Loss Probably Unrecognized in U.S.

Race "T" Blight Identified in U.S.

Blight Occuring in South

USDA Predicts Record Crop

Newspapers Break Corn Blight Story

USDA Predicts 10% Loss

Commodities Markets in Turmoil

Loss at 15%

Seedmen Rush to Produce Normal Seed in Hawaii, Latin America

USDA Allots 20% More Corn Acreage

Blight Damage Small; Record Crop; Prices Fall

Concern Over Genetic Vulnerability of Crops

1935 1950 1955 1960 1965 1970

grown to an all-time high of 4.76 billion bushels in 1967. Over half of this corn is produced in the Midwestern "corn-belt" states, although some is produced throughout the South and the East; in all, corn is planted on over 60 million acres each year.

Corn has been a generally hardy plant because of its wide genetic variability. Although it is susceptible to damage by several pests such as rodents, insects (particularly the corn root worm and the European corn borer), and smuts, these can usually be controlled by modern agricultural practice and the available hybrid seed. A minor disease of U.S. corn was the fungus *H. Maydis* or Southern corn leaf blight, which had been known since at least the 1920's, and was troublesome in the Southern states in years when there were prolonged hot, humid conditions.[1,2] Cool weather limited northward spread of the disease.

An understanding of this case requires some knowledge of the pollination procedures for producing hybrid corn, as compared to the older "open-pollinated" corn. It is basically a two-step process. First, the seed breeder produces an inbred strain of corn, a genetically uniform strain which has certain desirable characteristics, although it is not usually a high yielding corn itself. The production of an inbred strain involves careful selection and hand pollination and is costly. Second, the inbred is then genetically "crossed" with a second inbred, that is, the pollen from the tassel of one strain is allowed to fertilize the ear of the second strain. The seed kernels of the cross then have an exceptional "hybrid vigor" and produce high yields. This vigor is, however, not passed on to the next generation (to the F_2 seed) and the farmer normally buys new hybrid seed each year. The "single cross" seed is expensive and most of the seed sold is actually a less costly "double cross," a seed with four inbred parents. Double-cross corn is more stable over a range of conditions. The agricultural schools, the U.S. Department of Agriculture (USDA), and the seed corn companies have developed thousands of inbred strains and hybrid seeds suitable for a large variety of soil and weather conditons or for special purposes.

Until about 1950, the seed corn companies produced the hybrid seed by alternately planting about every fourth row in the field with the second strain: the tassels of the other three rows were removed by hand to assure cross-pollination. This hand labor and also the subsequent mechanical detasslers were expensive, however, and during the 1950's a new technique of developing hybrid seed was introduced—the male-sterile method. In this technique, a genetic mutation—the so-called Texas male-sterile T-cytoplasm (as compared to the normal N-cytoplasm) has been incorporated into one of the inbred strains used in crossing; the tassel does not produce fertile pollen and does not have to be removed in producing hybrid seed. The resulting seed would not normally produce corn yields for farmers, however, because it also would be male-sterile; therefore, the seedmen used an inbred (or single cross) which contained a genetic "restorer" factor as one of the parents. Alternatively, they blended a quarter to a third of the N-cytoplasm seed in with the T-cytoplasm seed to insure field pollination. The seed corn companies produce about ten million bushels of hybrid seed annually, and its use has been estimated to give farmers an average gain of 25 percent in their yields.

Key Events and Roles

In 1961, scientists in the Philippine Islands reported in *Philippine Agriculture* on the susceptibility of corn lines with T-cytoplasmic male sterility and in 1965 published, in the same journal, conclusive evidence of its extreme susceptibility to the fungus which causes Southern corn leaf blight.[1,2] This early warning signal was largely unheeded in the United States,[1] although one attempt to duplicate the early results was unsuccessful in 1963 at the University of Illinois.[2] In 1968, some 200,000 pounds of corn seed was lost in the Midwest because of an ear rot which is now thought to have been the Southern corn leaf blight, but it was not then recognized: the blight had previously damaged only the leaves and was not normally a problem so far north. In 1969, a greater-than-normal susceptibility to Southern corn leaf blight was noted in seed fields and in hybrid test fields by corn company experts and some agricultural college scientists.[1,3] In addition, an association between T-cytoplasm and a less serious yellow leaf blight had been previously noted, and some seedmen planned to switch back to the old detasseling method for the 1970 seed production crop. In the fall of 1969, A. L. Hooker of the University of Illinois demonstrated in greenhouse studies that the T-cytoplasm corn was extremely susceptible to Southern corn leaf blight.

By February 1970, during the winter seed-growing season in Florida, the susceptibility of T-cytoplasm corn was further discovered.[1,4] In the spring and early summer, the blight began to appear in Georgia, Alabama and Mississippi, and by June it was being recognized, in some quarters, as a serious threat—particularly among those who realized that over 80 percent of the corn in the United States might be susceptible. The "corn belt" had not been affected by the blight, however; the USDA was predicting a record

harvest of 4.8 billiong bushels[5] on July 1 and was still predicting
4.7 billion bushels on August 1.[6]

Despite the USDA's optimistic forecast, signs of blight were
appearing in the Midwest and corn futures prices started rising
early in August.[7] On Sunday, August 16, the newspapers broke the
story of the Southern corn leaf blight on page 1; the USDA was then
estimating that at least 10 percent of the crop would be destroyed.[7]

The report sent a shock throughout the nation. The impact on
the Chicago Board of Trade was tremendous; grain futures soared
on frenzied record trading[8,9] as prices shot from $1.30 per bushel up
to $1.58 per bushel. Other grains and livestock also advanced. The
next day an estimated 191.6 million bushels of corn were traded,[9]
but prices broke and declined sharply[10] as many of the "smart
money" traders (who had bought low in early August) unloaded on
the "amateurs."[11] By August 24, the price had dropped to $1.46 per
bushel, but continued to fluctuate thereafter as various spokesmen
raised or lowered the estimates of blight damage.

The initial news accounts of the blight emphasized (apparently
on the basis of information supplied by USDA) that the blight was a
new virulent mutant strain of Southern corn leaf blight—Race T as
compared to the old Race O—which had just popped up in the
South and was being windborn (via the spores of the fungus) into
the Midwest. Subsequent studies have shown that it was already
present in the Midwest in 1968 and 1969, had been known at Pennsylvania State University since 1955, and had been prevalent in
many parts of the world since the 1950's.[3] Almost none of the early
accounts pointed out that it was the seed growers' use of the
T-cytoplasm corn to avoid detasseling costs that permitted the
blight; some accounts even indicated that the new blight was so
virulent it attacked previously resistant hybrids.

At any rate, the Race T spread faster than Race O and could
tolerate lower temperatures: the blight spread into Kentucky, Ohio,
Indiana, and Illinois in July and into Wisconsin, Minnesota and
Canada in August. Whereas Race O had affected only the leaves;
Race T also attacked the stalk and, more important, the ear; infected
corn had sooty black spots. The farmers were completely helpless:
the one known effective chemical fungicide (Zineb) was far too
expensive for anything but special seed-corn use.[12] The USDA
broke a long silence[11] on September 11, and announced its forecast
was down 9 percent to 4.4 billion bushels. A month later it was
down another 5 percent to 4.2 billion. The actual yield was about
4.0 billion bushels—a 15 percent loss costing the farmer $1 billion

despite the higher prices.[1] Fortunately, the corn which was produced was not toxic to animals, even though infected with the blight, and could be used normally.[12]

The second round: By late 1970, considerable worry was developing over whether the blight might be even worse in 1971. As early as September 28, 1970, the*Wall Street Journal* was warning[13] that some 60 percent of the seed corn (the T-cytoplasm corn) to be planted the next spring might be susceptible to the blight. Soon the general public[5] as well as the nation's farmers[14] were learning that the 1971 seed corn supply would consist of 21 percent N-cytoplasm corn (blight resistant), 42 percent T-cytoplasm corn (susceptible), and 37 percent N and T blends (partially susceptible). As the 1971 season approached farmers (who had been generally unaware of the seed companies' common practice of blending until they had observed in 1970 that about one stalk of corn out of four was resistant to the blight) were confused[15] over the blended seed and the seed companies' various claims. Would the seed grow or not? The Midwestern farmer was not happy to hear that most of the resistant seed would go to the South[4] where it was most sure to be needed. Furthermore, the farmer could not even plant his own seed—as he would have thirty years earlier in the "open-pollinated corn" days—because it might be the susceptible variety and result in a total loss, rather than the 25 percent or so loss which he might otherwise incur with normal F_2 seed.

The seed companies, on the other hand, made vigorous efforts to produce more N-cytoplasm seed* for the 1971 crop by rushing to the warm climates of Hawaii, Mexico, and Argentina, as well as Florida. Despite these efforts, the seed supply, as planting time approached, was apparently still nearly 40 percent of the T type,[16] but farmers were reassured that it would not necessarily produce a blighted crop.[15]

The USDA was also greatly concerned about the blight expected in 1971, and announced in March that farmers would be allowed to plant 20 percent more acreage in corn than normal while still remaining eligible for price supports.[17] The USDA was particularly fearful because the reserve corn stocks which had cushioned the 1970 losses had been much reduced. The National Aeronautics and Space Administration, which was planning to place an Earth Resources Technology Satellite in orbit in 1972[18] and had made a

*State and Federal laws apparently prohibited the companies from making $T-F_2$ blends, using a resistant F_2 seed.

trial aerial observation of blighted fields in 1970,[19] joined with the
USDA in planning an extensive program of aerial surveillance for
1972. (Color pictures would be taken of test areas periodically from
an aircraft at about 50,000 ft.[19,20]) The Midwestern farmers, on the
other hand, were concerned not only about a recurrence of the
blight, but about overproduction of corn if the blight did not
develop.[17]

During May and June of 1971 the corn crop was watched
closely, with corn futures prices rising or falling as various reports
came in that blight had been sighted or that a near-record crop was
shaping up. By July the USDA reported that, although the blight
had appeared in 581 counties in 28 states, the level of infection was
light.[21] Except for seed growers, whose return to the N-cytoplasm
corn had created a desperate need for new detasseling equipment
and hand labor,[22] the blight proved not much of a problem in 1971.

On August 28, the USDA forecast an all-time record corn crop
of 5.3 billion bushels.[23] Prices of corn immediately started falling,
reaching $1.11 per bushel in September[24] (compared to $1.38 in the
spring) as farmers prepared to take their crop to market. In
November the full extent of the overproduction became evident: a
5.55 billion bushel yield, up 35 percent from 1970.[25] In addition to
lamenting their 1971 losses, farmers were despondent that corn
prices would not even reach $1.05 per bushel in 1972.

Disposition

Farmers were much displeased with the USDA for having encouraged increased corn acreages and asked for changes.[26] By
mid-October, Secretary of Agriculture Hardin proposed a program
to raise farm subsidies by $600 million or more in 1972 (to a total of
$1.5 billion) in an effort to reduce production of surplus corn and
other feed grains.[27] During the late fall and winter of 1971–72 the
Congress, Secretary of Agriculture Butz, and the Nixon administration debated various plans, such as price supports, for the 1971 crop
(which was already moving rapidly into the hands of dealers). The
USDA ultimately purchased 1.4 million bushels of corn from Midwestern markets to reduce slightly the 600 million bushel surplus[28]
and set the 1972 acreage at 4 percent below the 1971 level[29] (still 16
percent above 1970 levels). In 1972, N-cytoplasm corn seed was
again available and corn leaf blight was no longer of economic
concern to the nation.

The lesson taught by the blight was not forgotten, however, and a tremendous concern developed over the unsuspected side effects of wholesale genetic engineering. The National Academy of Sciences—National Research Council soon started a broad study of the possible genetic vulnerability of other important agricultural crops.[30,31] The study noted that society was demanding a great uniformity in food and fiber crops, for ease and economy in growing, harvesting, processing, handling and marketing. The report concluded that since this physical conformity was being attained through genetic uniformity, the nation's crops were acquiring a great genetic vulnerability to disease which should be avoided by a return to more genetic diversity. Thus the corn leaf blight disaster has sparked much thought and research on genetic vulnerability of plants and animals and renewed interest in the wild ancestors of domesticated crops such as corn.

Comment

The corn blight episode had its origins in the development by agricultural geneticists of the novel male-sterile corn, and its widespread adoption by the seed companies under USDA approval before its vulnerability was determined. Early warning signals went unrecognized by the experts. When the blight struck, the USDA was slow in recognizing it or at least in announcing it, and knowledgeable traders apparently capitalized on the wild board-of-trade fluctuations produced by the announcement and subsequent statements by government figures. The news media, particularly the farm magazines failed to indicate that the blight was other than "an act of God," that seed company policies and agricultural genetics had contributed to the problem. In view of the 15 percent decrease in corn production in 1970, the USDA's decision to raise corn acreage in 1972 was realistic (and even a 20 percent increase could be defended), but it contributed directly to a huge corn surplus and falling prices for the farmer. The extent of the corn blight episode focused much needed attention on the dangers of genetic engineering without adequate foreplanning and study. As this case shows, the male sterile technique had been introduced and tested over nearly a twenty-year period, yet the rapidity and severity of the blight outbreak caught agricultural experts by surprise.

REFERENCES

1. Tatum, L. A., "The Southern Corn Leaf Blight Epidemic," *Science*, **171**. 1113 (March 19, 1971).

2. Hooker, A. L., "Southern Leaf Blight of Corn—Present Status and Future Prospects," *J Environ Quality*, **1**, 244 (1972).

3. "How Bad Will It Be?", *Farm J*, **95**, 42 (February 1971).

4. "A Threat to U.S. Food Supply," *US News*, **70**, 71 (May 17, 1971).

5. "As a Killer Disease Hits a Major Crop," *US News*, **69**, 60 (October 26, 1970).

6. "In the Wake of a Corn Blight," *US News*, **69**, 62 (August 31, 1970).

7. *New York Times*, p. 1 (August 16, 1970).

8. *New York Times*, p. 47 (August 18, 1970).

9. "Grain Futures Hit the Sky," *Bsns W*, 22 (August 22, 1970).

10. *New York Times*, p. 57 (August 19, 1970).

11. "Whiplash," *Forbes*, **106**, 58 (September 15, 1970).

12. "Corn Blight Threatens Crop," *Science*, **169**, 961 (September 4, 1970).

13. *Wall Street J*, (September 28, 1970).

14. "Corn Blight: What About Next Year?", *Farm J*, **94**, 26 (November, 1970).

15. "Corn Blight: Confusion Over Blended Seed," *Farm J*, **95**, 33 (January 19, 1971).

16. "Blight-Infected Seeds Will Sprout and Grow," *Farm J* (April 23, 1971).

17. *New York Times*, p. 24 (March 21, 1971).

18. "ERTS Could Aid Crop Blight Fight," *Aviat W*, **93**, 50 (December 21, 1970).

19. "How They'll Spot Corn Blight From 11 Miles Up," *Farm J*, 33 (February 1971).

20. "NASA to Aid Study of Corn Leaf Blight," *Aviat W*, **94**, 19 (April 19, 1971).

21. *New York Times*, p. 28 (July 19, 1971).

22. *New York Times*, p. 5III (July 11, 1971).

23. *New York Times*, p. 33 (August 29, l971).

24. *New York Times*, p. 57 (October 1, 1971).

25. *New York Times*, p. 71 (November 12, 1971).

26. *New York Times*, p. 1III (October 3, 1971).

27. *New York Times*, p. 1 (October 19, 1971).

28. *New York Times*, p. 33 (December 14, 1971).

29. *New York Times*, p. 35 (January 28, 1972).

30. Wade, N., "A Message From Corn Blight: The Dangers of Uniformity," *Science*, **177**, 678 (August 25, 1972).

31. *Kansas City Star*, August 20, 1972 and p. 10, January 17, 1973.

The Great Cranberry Scare

Abstract

On Novemeber 9, 1959, Secretary of Health, Education and Welfare Arthur S. Flemming, announced at a news conference that some cranberries produced that year had been found to contain trace residues of the herbicide amitrole, which, he said, could produce cancer in rats. Sales of all cranberries were not being banned, but Mr. Flemming indicated that to be on the safe side, housewives should not buy any. This announcement, coming just 17 days before Thanksgiving and prior to the annual holiday peak in cranberry consumption, sent a convulsive shock through the American public and the cranberry industry. The announcement and the subsequent controversy were widely publicized, and cranberries almost disappeared from store shelves and restaurant menus. A crash program to test cranberries from all parts of the country produced nearly enough "approved" cranberries by Thanksgiving to meet the much-reduced demand.

The episode faded quickly from the news, but not from the cranberry industry, which wondered for the next year or so if it could survive, or from certain agencies of the government which sought to relieve some of the damage done by paying an indemnity to growers and by stimulating consumption. Ultimately, the diets of millions of Americans were affected as the school lunch program, the armed services, and the Veterans Administration hospitals added more cranberries to their menus. The key element in the case—the carcinogenicity of amitrole—is still controversial a dozen years later.

Background

Cranberries, the traditional Thanksgiving and Christmas dessert, are grown on wet, low-lying soils (bogs) in the East (Mas-

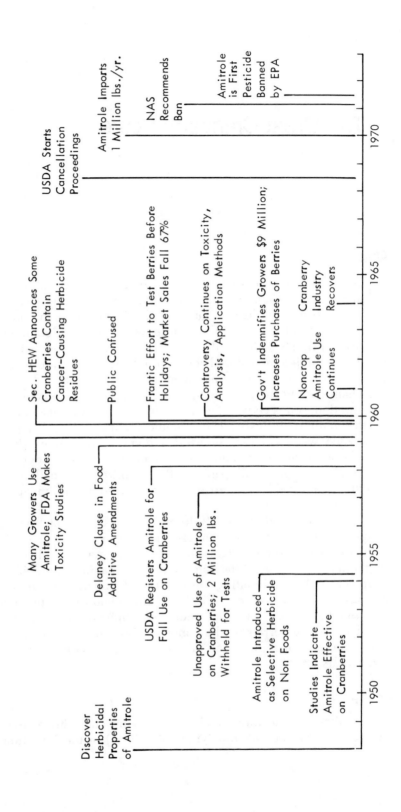

Discover Herbicidal Properties of Amitrole

Many Growers Use Amitrole; FDA Makes Toxicity Studies

Delaney Clause in Food Additive Amendments

USDA Registers Amitrole for Fall Use on Cranberries

Unapproved Use of Amitrole on Cranberries; 2 Million lbs. Withheld for Tests

Amitrole Introduced as Selective Herbicide on Non Foods

Studies Indicate Amitrole Effective on Cranberries

Sec. HEW Announces Some Cranberries Contain Cancer-Causing Herbicide Residues

USDA Starts Cancellation Proceedings

Public Confused

Amitrole Imports 1 Million lbs./yr.

Frantic Effort to Test Berries Before Holidays; Market Sales Fall 67%

Controversy Continues on Toxicity, Analysis, Application Methods

Gov't Indemnifies Growers $9 Million; Increases Purchases of Berries

Noncrop Amitrole Use Continues

Cranberry Industry Recovers

NAS Recommends Ban

Amitrole is First Pesticide Banned by EPA

1950   1955   1960   1965   1970

sachusetts, 50 percent; New Jersey, 10 percent), the Midwest (Wisconsin, 30 percent), and the Northwest (Washington and Oregon, 10 percent). Because weed infestations cannot be controlled mechanically, interest developed early in the use of chemical control methods. By the 1950's, however, even the best of the chemicals, such as salt, compounds of arsenic, copper or iron, kerosene or oils, and paradichlorobenzene were difficult and costly to apply in the large amounts required and not too effective. The wartime discovery of 2,4-D had spurred an intensive search for new herbicides during the late 1940's; in 1953 and 1954 an investigation of several new organic herbicides for use in cranberry bogs was started at the New Jersey Agricultural Experiment Station by the personnel of the United States Department of Agriculture, Argicultural Research Service (USDA, ARS) and Rutgers University.

By far the most effective of the new compounds was 3-amino-1,2,4-triazole (also called amitrole and 3-AT),[1] which was introduced as an experimental selective herbicide in 1954 by Amchem Products, Inc. In 1955 American Cyanamid Company began commerical production of amitrole at its plant in Welland, Ontario, and on March 23, 1956, amitrole was "registered" by the USDA for weed control on noncropland, that is, it was approved by the USDA for a specified use under specified conditions as required by the Federal Insecticide, Fungicide and Rodenticide Act of 1947. Although this registration did not include use on cranberries, excited cranberry researchers in New Jersey,[2] Wisconsin,[3] and Massachusetts[4] were soon telling the industry that amitrole was very effective on cranberries when applied between growing seasons and it could even be applied directly on growing cranberries because of its selectivity.[5]

In early 1957, American Cyanamid applied to the USDA for a registration to sell amitrole for use on cranberries, apples, corn and soybeans on a "no residue" basis, since the Food and Drug Administration (FDA), had not set a maximum amount, or "tolerance" of the pesticide that would be allowed in foods. The FDA had been recently required to set such tolerances (under the 1954 amendment of the Food, Drug and Cosmetic Act of 1938) for all pesticide products designed for use on or in human or animal foods, and was authorized to seize and destroy foods with pesticide levels above the tolerance.* The registration was not granted by USDA, how-

---

The regulation of food quality was originally vested in the USDA under the Pure Food Act of 1906, but soon after passage of the 1938 legislation, the FDA was separated from USDA and given the responsibility of assuring that the public reever, on the advice of the FDA, which felt that insufficient information was available on the chronic toxicity of amitrole.

Nevertheless, about one-third of the cranberry growers in the Northwest and a few in Massachusetts proceeded to use amitrole on the 1957 crop of cranberries. The national cranberry growers cooperative, Ocean Spray Cranberries, Inc., (which represented about 75 percent of total production) took note of this practice and warned its members that amitrole was not an approved product. The FDA also learned of this unauthorized use in October 1957. Nearly 3 million pounds of cranberries (2.4 percent of the crop) were identified by the association as being produced under amitrole treatment, and at the urging of the FDA were quietly set aside—frozen—while awaiting results of residue or toxicity data on the herbicide. A big part of the problem was that an analytical method for testing the berries for amitrole had not yet been developed.

On January 30, 1958, the USDA granted American Cyanamid a registration for the *post*-harvest application of amitrole on cranberry bogs. This action followed a December registration for pre-planting application on corn land, and led the company to open an amitrole-formulating plant at Princeton, New Jersey. The FDA refused in April, however, to grant Amchem's request for a 0.7 ppm (part per million) tolerance on cranberries, apples, and pears, because the toxicity data were still judged incomplete. Amchem then withdrew its petition to avoid having a zero tolerance established officially; in practice the no-residue status remained and postharvest application to cranberry bogs was legal.

Amitrole was apparently used rather extensively in 1958 on cranberries, but it should not have been used until after the 1958 crop was harvested. No contaminated lots were reported, but very little checking seems to have been done. At any rate, the future for the cranberry growers and consumers looked bright and hardly indicative of the impending great cranberry scare.

---

ceived wholesome foods. The inspection powers over foods, however, were largely left behind in UDSA, and the overlap in responsibilities was conducive to conflicts in the enforcement area between USDA and FDA, which became a part of the Department of Health, Education, and Welfare in 1954.

*The acute toxicity was known to be low. The LD_{50} in rats has been variously quoted as 24,600, 14,000 and 1,100–2,500 mg/kg, approximately that of table salt. Under the 1954 legislation, the burden of proof of a chemical product's safety (when used as specified) had been shifted from the government to the manufacturer.

Key Events and Roles

In late 1958 the Congress passed the Food, Drug, and Cosmetic Act Amendment of 1958, which provided new legislation regarding additives in foodstuffs. This amendment contained a feature which was soon to become well known to much of the American public, i.e., the so-called Delaney Clause,* which states:

"That no additive shall be deemed safe if it is found to induce cancer when ingested by man or animal, or if it is found, after tests which are appropriate for the evaluation of the safety of food additives, to induce cancer in man or animal."

While pesticides are not ordinarily thought of as food additives, this legislation was soon to involve amitrole. The chronic toxicity studies being conducted by American Cyanamid† and by the FDA itself,[6] were producing some unusual results: rats fed amitrole in their diets developed enlarged thyroid glands which, with high dosages for long periods (100 parts per million for over sixty-eight weeks), developed into tumors. The diagnoses of these tumors varied, according to American Cyanamid's subsequently published account,[7] with some pathologists naming them as nonmalignant adenomas and others as adenocarcinomas (that is, cancerous). Even at 50 and 10 parts per million some adenomas were observed. The FDA and American Cyanamid scientists were apparently unaware of concurrent studies by Alexander[8] which were showing amitrole to have antithyroid action (that is, to be a goitrogen).

American Cyanamid and Amchem, in early 1959, submitted their data in a new petition for a tolerance of 1 part per million on cranberries, apples, and pears. The FDA subsequently ruled that amitrole was a carcinogen, and denied the petition (it was withdrawn in June[6]). FDA authorities also advised the cranberry growers association of their conclusion; the FDA indicated it would be keeping a close eye on cranberries, but still did not have a procedure for testing the 3 million pounds stored from the 1957 crop. No registration cancellation was enacted by the USDA,‡ however, and amitrole's use was still permitted under the original conditions.

Many growers, on the other hand, apparently concluded that a

---

*After its author, Representative James Delaney (D–New York).

†The tests were being conducted by an independent organization, Hazelton Laboratories.

‡The USDA did deny a registration to Amchem in July 1959 for its new herbicide, Amitrol-T (a mixture containing thiocyanate) on the basis that it was a carcinogen.

registration change for the better was going to be made, and had already proceeded to use amitrole in preharvest (spring) application. The record is not clear at this point as to whether the USDA, the Agricultural College, the growers association, or the manufacturers and distributors of amitrole were influential in their decisions, but the FDA was soon getting word that much amitrole had been used, either correctly or incorrectly, in the Northwest. The growers association developed a grower's affidavit form to certify that cranberries to be shipped had not been treated during the growing season.

By October the FDA had its analytical method working; the Seattle, Washington, district office went into a testing program. Contamination was soon demonstrated in part of the 1957 crop, and the growers association started a program to bury the whole 3 million pounds in Oregon and Washington. Stories on this strange development and of the increasing concern among area growers started to appear in local papers, and inquiries were soon reaching to Washington. Then in early November, the Seattle office reported to headquarters that shipments of the 1959 cranberry crop were also contaminated with amitrole. By November 7, even the Secretary of Health, Education, and Welfare Arthur S. Flemming was consulting with FDA Commissioner George P. Larrick, and FDA officials were conferring with the growers association executives over what to do about the cranberry situation. Although the growers association urged otherwise, the government officials decided to make the situation public.

On Monday, November 9, just seventeen days before Thanksgiving Day, Flemming called a news conference and announced that at least two shipments of cranberries produced in Washington and Oregon in 1959 contained residues of a weed killer that could produce thyroid cancer in rats. Sale of all cranberries was not banned, Mr. Flemming said, but 3 million pounds of the Northwestern crop for 1957 were being buried and the 1958 and 1959 crops from all growing areas would be investigated. In short, the message was that a buyer could not tell bad cranberries from good in the stores, that anybody buying cranberries would be doing so at his own risk, and that housewives should be on the safe side and not buy any, unless somehow they could be sure the berries were not tainted.

When this stunning announcement hit the headlines in newspapers,[9,10] magazines[11-22] and television news programs across the country, it created tremendous confusion in the American

public. Housewives, supermarkets, and restaurants swept cranberries off their shopping lists, shelves, and menus. The cranberry
industry was hit hard as the prospects of losing the Thanksgiving
and holiday market boded an economic disaster, a $40–50 million
crop loss. Canneries shut down and pickers were laid off. Growers,
particularly those who had been careful to use amitrole in accor-
dance with directions and those who had not used it at all, were
incensed. Industry lawyers and government officials spent hours in
conference over what to do. In particular, the industry wanted
blanket approval of *canned* berries already on store shelves, because these were almost entirely from the 1958 crop, not the 1959
crop. The government maintained, however, that these had not
been adequately tested, even though they had been on the market
for a year.

In fact, the FDA's analytical method* for determination of
amitrole residues in cranberries was not even to be published until
Thanksgiving week of 1959; it concluded a statement on special
labeling of cranberry products from 1958 and 1959 crops.[25] The
lengthy procedure was said to be sensitive in the range of
0.005–0.250 parts per million, and berries with *more than 0.15
parts per million* were judged contaminated. Spokesmen for the
cranberry growers claimed that even the contaminated berries had
such extremely low levels of amitrole that a person could not possibly consume enought of them to be harmed.[11] A spokesman for
American Cyanamid stated that a human would have to consume
15,000 lb of contaminated berries daily for several years to get an
amount of amitrole proportional to that fed to the rats.[14] A New
England physician stated,[14] and subsequently published in the
Journal of the American Medical Association,[26] the view that the
amount of amitrole on cranberries was harmless and pointed out[14]
that amitrole was a natural constituent of several vegetables such as
mustard, cabbage, turnips, and broccoli, which were not being
banned. The cranberry industry complained that it was being sin-
gled out for special regulation. Pressure began building on regulatory officials.

The FDA had immediately begun a crash program to test cranberries from all parts of the country in an effort to get the berries
back on the market before Thanksgiving. At his next news conference on November 16, Secretary Flemming announced that 3.5
million pounds of berries (337 lots) had been found free of con-

*An adaptation of earlier colorimetric procedures.[23,24]

tamination, but that 80,000 pounds (4 lots) had been seized—including 25 cases from Wisconsin in addition to those from the Northwest—and 13 other lots were being rechecked. Testing continued at a feverish pace (with one hundred inspectors and sixty chemists according to one story[17]), but it was becoming obvious that not nearly all of the 70-million-pound crop could be tested in time for Thanksgiving.

The cranberry industry was already claiming that the Government's action was leading to a $100 million disaster. Charges were being leveled that the bureaucrats had bungled, that the cranberry was being used in a power struggle between HEW and USDA,[12,18,20,21,27,28] and that HEW was using cranberries as a lever on legislation regarding the exemption of color additives from the Delaney clause, then pending before the Congress.* The case was also becoming a political issue; each department had its partisans, and national elections were less than a year away. Political figures suddenly became quite fond of cranberries—particularly those from cranberry-growing states. Senator William Proxmire (Wisconsin) defended growers and denounced the "blanket condemnation of the entire industry." Representative Frank Thompson (New Jersey) said, "Irresponsible government at its worst," and presidential hopefuls, Senator John F. Kennedy (Massachusetts) and Vice-President Richard M. Nixon, and USDA Secretary Benson were photographed consuming cranberry products to show their backing for the industry and belief in the wholesomeness of the product.

On the other hand, the FDA and HEW spokesmen were explaining that buyers could not tell just how dangerous the contaminated berries were, and that they were protecting the public. The key issues, of course, were whether a pesticide residue should be considered to be a food additive, subject to all legislation pertaining to additives, and whether the Delaney Clause could be applied to residue levels below those at which carcinogenicity had been demonstrated.

As the testing continued, some additional lots of berries were found to be contaminated, possibly including some from bogs where the approved post-harvest method of amitrole treatment had been used. Uncontaminated berries were released and labeled "Examined and Passed," and "Certified Safe," and enough berries

*In addition to the cranberry affair, Secretary Flemming on December 13, 1959, had banned the use of diethylstilbestrol in poultry on the basis that it was a carcinogen and left detectable residues.

became available by Thanksgiving to meet the greatly reduced demand. By January 1960, the FDA had cleared a total of 33.6 million pounds of cranberries and seized[6] only 30 lots (325,800 lb)—less than 1 percent of the crop.* For the public, the great cranberry scare was over, but its after effects were to last much longer.

Disposition

After the holiday season, the Government turned its attention to the effects of the cranberry scare: on January 26, 1960, Secretary Flemming appeared before the House Committee on Interstate and Foreign Commerce.[6] On March 30, President Eisenhower's press secretary announced that the USDA would offer to pay† up to $10 million in indemnities to cranberry growers that had sustained losses through no fault of their own.[29] About $9 million (approximately two-thirds of the loss) was actually paid, according to an industry source.[30] In addition, government agencies, at the urging of legislators from cranberry-growing states, moved to help the faltering cranberry industry during the next few years. The USDA listed cranberries on its "plentiful foods" program. The school lunch program, the armed services, and the Veterans Administration hospitals added more cranberries to their menus. The National Canners Association, the Supermarket Institute, and others, helped push sales. The industry developed new cranberry products (such as cranapple juice) and reoriented its advertising. Although the industry wondered for the next three or four years if it were going to survive, the government aid and its own efforts proved successful. The net result was that the industry was probably better off by the late 1960's because of the crisis.[30]

As a result of the cranberry problem, the concepts of "no residue" and "zero tolerance" were being redefined by scientists and regulatory officials. A National Academy of Sciences—National Rsearch Council committee was appointed at USDA/FDA request to study the matter. In June 1965, the committee reported that both concepts, as employed in the registration and regulation of pesticides, were scientifically and administratively untenable and should be abandoned, especially since continually improving analytical techniques were capable of detecting smaller and smaller residue levels. On April 13, 1966, the USDA and FDA accepted

*About 10% of the total 1959 crop was eventually banned, according to some estimates.
†Payments would be made under a 1935 law.

these recommendations to the extent possible: the concepts could not be abandoned under the law, but instead a finite tolerance would be required henceforth for all pesticide uses where there was a "reasonable expectation" that small residues on food or feed might result. Registration would not be granted if a finite tolerance had not been established.

Several critical technical questions concerning amitrole remained after the crisis, including the validity of the analytical method, the carcinogenicity of the substance, and its future herbicidal use on cranberries or elsewhere. In April 1960, the FDA announced[31] a new "rapid" method of analysis (which was sensitive in the 0–0.10 parts per million range) and in 1961 reported[32] a "modified" procedure for sample preparation. The "modified" method was then found[33] to yield amitrole residue values in cranberries *an order of magnitude higher* than the "rapid" and "original" methods. This astonishing increase, attributed to the contribution of metabolites of amitrole that were lost in the "rapid" method,* spurred further studies. These, in turn, soon indicated[34] that traces of amitrole and/or its metabolites remained in cranberry bushes, roots, and fruits (0.39 parts per million in the fruit) twelve months after application, even by the USDA-approved method of use.

The identities of these presumed metabolites were not known, and of course, no one could say whether they were carcinogenic or not. But if the FDA was going to use this test method on cranberries, then clearly amitrole could not be used any time on them; in January 1963, the registration for post-harvest use of amitrole on cranberries was cancelled by the USDA.

Although amitrole was now banned on cranberries, its use had become something of a cause célèbre among regulatory officials. In 1961 the original American Cyanamid toxicity study results were published, in part,† by Jukes and Shaffer,[7] who noted that the

---

*The original FDA method[25] was a colorimetric procedure utilizing measurements at 500 mμ, although the absorption peak for pure amitrole is 455 mμ and the literature method[24] used 525 mμ. The "rapid" method used a slightly different agent, but like the original method, used an alumina chromatographic column in the workup. The "modified method avoided the alumina column (which was the point where the presumed metabolites were being lost, based on the results of paper and cellulose column chromatography). The metabolites' absorption maximum was at 465–470 mμ.

†Private communications of more complete American Cyanamid data have been published.[36,37] Hodge, et al.[37] found no evidence that amitrole could produce skin cancer when injected or applied to mice, and noted that American Cyanamid reported no tumors in dogs fed amitrole for 1 year.

tumors produced in rats disappeared when amitrole was discontinued and questioned whether it should be characterized as a carcinogen. On the other hand, a little-known Russian publication reported in 1962 that heavy doses of amitrole in rats (0.25 or 0.5 grams a day for 1 year) produced adenomas and carcinomas of the thyroid and tumors of the liver. In 1964, the FDA included amitrole in its Market Basket Survey of American foods. Then in a major study[38] of industrial chemicals and pesticides, the National Cancer Institute (NCI) adopted amitrole as a known carcinogen control—a rather surprising selection in view of the conflicting data and possibly a reflection of amitrole's notoriety in the cranberry scare.

But before these studies were complete, the USDA moved in February 1968, to ban orchard and cropland uses of amitrole (amitrole was judged a carcinogen). The FDA would not set a finite tolerance and reasonable expectation existed that there would be only small residues on food. American Cyanamid and Amchem protested and obtained the appointment of a National Academy of Sciences advisory committee to review the USDA's decision, as provided under the law. The committee met in December 1970, a few days after the birth of the Environmental Protection Agency, in which the registration and regulation of pesticides were consolidated. This consolidation was intended to resolve the long-standing conflict and division of responsibilities between USDA and HEW/FDA.

By this time, some results of the NCI study had been published as a "preliminary note" by Innes et al.[38] These investigators characterized each of the 120 test substances only on the basis of whether they were tumorigenic and did not try to establish carcinogenicity.† In the case of amitrole, however, a footnote added that thyroid cancers* were observed when massive doses had been fed to mice over long periods (0.1 percent of body weight per day from seven to twenty-eight days of age and 0.2 percent of diet from fourth week to about fifty-sixth week). The note by Innes, et al., became controversial almost immediately, not only because of the results on other important industrial and agricultural chemicals (such as DDT), but also because of the techniques employed. The amitrole data appears not to have been completely convincing to USDA, which, in March 1970, listed it as one of a series of compounds needing further carcinogenic study.[39]

†About three paragraphs of the note were devoted to explaining how difficult and uncertain it was to identify malignant tumors from nonmalignant ones by the bioassay method used.

*The method of diagnosis was not given, nor were controls mentioned.

Despite evidence that amitrole is nonpersistent (it degrades rapidly in normal soils, but not in cold, wet cranberry bogs), the FDA's failure to detect a single amitrole residue in four years of the Market Basket Study of American foods, and a complaint that the "reasonable expectation of residues" was a scientifically vague description, the NAS-NRC committee concluded that because amitrole was an accepted carcinogen, its use on food crops of any kind—even for poison ivy on the floor of an apple orchard—should not be permitted.

In 1971 the Environmental Protection Agency cancelled the registration for use of amitrole on all remaining croplands.* The news release stated[41] that the cancellation was given because amitrole *may* cause *tumors* in experimental animals, although no mention was made of *cancer*. The original American Cyanamid data were the only numbers quoted, and EPA said no clinical effects had ever been observed in man or animals when amitrole was used according to directions. The use of amitrole on noncropland is still permitted, and although it is still produced and used in Europe, American Cyanamid terminated production at the end of 1971. The last aftershock of the great cranberry scare was over in the United States.

Comment

The cranberry scare was a culmination of several propitiously timed factors. The growers applied the promising new herbicide in a manner for which it was not approved (or before it was approved), although subsequent improvements in analytical techniques indicate that even the approved application left detectable residues. The USDA and the agricultural chemical manufacturers apparently had not considered the new food additive legislation (with the Delaney Clause) as being applicable to trace residue levels. On the other hand, HEW/FDA was anxious to establish wide precedents under the new law, as shown by Secretary Flemming's news conference declarations that FDA declined to set tolerance for any amount of a chemical in foods if the chemical produced cancer when fed to test animals,† and that HEW would oppose omission of

*In 1970, imports of amitrole into the U.S. (primarily from American Cyanamid's plant in Canada) had been nearly 1 million pounds.

†Two reviewers have commented that the Delaney Clause was not administratively envoked in the cranberries case. This is largely a matter of semantics, however, since Secretary Flemming specifically referred to it in his public announcements.

the Delaney Clause from the color-additives bill pending before the Congress.

The dates on which the evidence of possible carcinogenicity was reported, and on which an adequate analytical method for determining the herbicide residues in cranberries was developed, combined to bring the decision point close to Thanksgiving. The use of the news conference format to announce the cautionary "non-ban" gained a tremendous news coverage that successfully dramatized for the public and the Congress the import of the Delaney Clause. The resulting public confusion and concern appear to have been far beyond the actual danger posed. After the scare had become full-blown and a feverish testing program was being undertaken, the FDA made the informal decision to consider all lots with less than 0.15 parts per million amitrole as being uncontaminated. Thus, unwittingly, the FDA set a finite tolerance in practice (if not in regulation)—the thing it had refused to do previously in rejecting the manufacturer's application, and thus one of the reasons for the episode.

Finally, the general difficulty of determining the carcinogenicity of a chemical compound is well illustrated in this case: a dozen years after the scare the major evidence is still a footnote in a preliminary report of a study in which no effort was made to determine if the tumors would regress when amitrole feeding was stopped. As this case illustrates, in the areas of carcinogenicity and food contamination, society clearly needs much better measures of the potential hazards of our technologies before they are introduced.

REFERENCES

1. Meggitt, W. F., and R. J. Aldrich, "Amitrol for Control of Redroot in Cranberries," *Weeds*, 271 (1958).

2. Aldrich, R. J., "Preliminary Evaluation of Weed Control in Cranberries," *Proc Amer Cranberry Growers Assoc*, 86th Annual Meeting, pp. 7–10 (1956).

3. Dana, M. N., "Weed Control in Wisconsin Cranberry Marshes," *Proc Northeastern Weed Control Conference*, 12, 50 (1956).

4. Demoranville, I. E., and C. E. Cross, "Newest Cranberry Weed-Killer, Aminotriazole," *Cranberries—The National Cranberry Magazine*, 22(12) (April 16, 1957).

5. Dana, M. N., "Effect of Amitrol Sprays on the Growth and Development of the Cranberry," *Weeds*, 277 (1958).

6. "Aminotriazole in Cranberries," report presented by Secretary of

HEW, Arthur S. Flemming, to the Committee on Interstate and Foreign Commerce, U.S. House of Representatives, January 26, 1960.

7. Jukes, T. H., and C. B. Shaffer, "Antithyroid Effects of Aminotriazole," *Science*, 132, 296 (1961).

8. Alexander, N. M., "Antithyroid Action of 3-amino-1,2,4-triazole," *J Biol Chem*, 234, 148 (1959).

9. *New York Times*, p. 1 (November 10, 1959). The *Times* eventually carried 46 stories on the cranberry incident.

10. The *Kansas City Star* carried six stories on page one or two during November plus one additional story.

11. "Cranberry Growers Reel Under Pre-Holiday Blow," *Bsns W*, 31 (November 14, 1959).

12. "Cranberries Not Key Issue?" *C&EN*, 37, 21 (November 16, 1959).

13. "Ordeal of Cranberry Growers," *Bsns W*, 30–31 (November 21, 1959).

14. "The Cranberry Scare—Here Are the Facts," *U.S. News*, 47, 44–45 (November 23, 1959).

15. "The Cranberry Affair," *Newsweek*, 54, 35 (November 23, 1959).

16. "The Cranberry Boggle," *Time*, 74, 25 (November 23, 1959).

17. "Mercy, Ma! No Cranberries," *Life*, 47, 28–33 (November 23, 1959).

18. DuShane, G., "Cranberry Smash," *Science*, 130, 1447 (November 27, 1959).

19. "The Cranberry Boggle (Cont'd)," *Time*, 74, 16 (November 30, 1959).

20. "Those Cranberries," *New Repub*, 141, 3 (November 30, 1959); *Ibid.*, with correspondence, 28–29 (December 28, 1959).

21. "The Cranberry Sensation," *C&EN*, 37, 7 (November 30, 1959).

22. "What Follows the Cranberry Crackdown," *Farm J*, 83, 33 (December 1959).

23. Sund, K. A., *J Agr Food Chem*, 4, 57 (1956).

24. Green, F. O., and Feinstein, R. N., *Anal Chem*, 29, 1958 (1957).

25. *Federal Register*, p. 9543, November 28, 1959.

26. Astwood, E. B., "Cranberries, Turnips and Goiters," *J Am Med Assoc*, 172, 183 (1960).

27. "A Lesson From Cranberries," *Consumer Rep*, 25, 47–48 (January 1960).

28. "There's More to the Cranberry Incident Than Just Cranberries," *Consumer Bul*, 43, 20 (January 1960).

29. "United States to Pay Indemnity for Cranberry Losses," *Science*, 131, 1033–34 (April 8, 1960).

30. Private communications, Cranberry Institute, September 8, 1971, and Ocean Spray Cranberries, Inc., September 9, 1971.

31. *Federal Register*, p. 3072, April 9, 1960.

32. Storherr, R. W., and J. Burke, *JAOAC*, 44, 196 (1961).

33. Storherr, R. W., and J. H. Onley, *JAOAC*, **45**, 382 (1962).
34. Onley, J. H., and R. W. Storherr, *JAOAC*, **46**, 996 (1963).
35. Napalkov, N. P., "Tumor Inducing Action of Amitrole," *Labor Hyg Occupat Dis*, **6**, 48 (1962) (in Russian).
36. Falk, et al., *Arch Environ Health*, **10**, 847 (1965).
37. Hodge, H. C., et al., *Tox Appl Pharm*, **9**, 583 (1966).
38. Innes, J.R.M., et al., "Bioassay of Pesticides and Industrial Chemicals for Tumorigenicity in Mice: A Preliminary Note," *J Nat'l Cancer Inst*, **42**, 1101 (1969).
39. "Data Needs for Certain Compounds," PR Notice 70–78, March 10, 1970, Pesticides Regulation Division, ARS/USDA.
40. Corneliussen, P. E., "Residues in Food and Feed," *Pesticides Monitoring J*, **2**(4), 140 (March 1969).
41. "Amitrole Use Halted," *C&EN*, **49**, 35 (July 5, 1971).

The Diethylstilbestrol Ban

Abstract

Diethylstilbestrol (DES), a synthetic sex hormone, has generated over twenty years of confusion and public concern. Introduced medicinally for humans in the 1940's as a treatment for menopausal disorders, spontaneous abortions, and prostate gland cancers, DES was to find its largest use in agriculture as an animal growth stimulant. In 1950, DES received its first notoriety when the heads of stilbestrolized chickens were fed to mink and produced sterile females. After DES had already been government-approved and widely adopted for fattening cattle and sheep, as well as poultry, it was found in the late 1950's to be a carcinogen: a ban was imposed on its use in poultry because residues could be detected in the meat, but not on its use in cattle or sheep, where residues had not been detected. In 1971, when 75 percent of the beef in America was being treated with DES and after the presence of minute traces of DES in the meat had become known, a frightening discovery was made: daughters born of women treated in pregnancy with DES were developing a rare type of vaginal cancer fifteen to twenty years later. After 2 more years of controversy, the use of DES in meat animals was banned, but even as this furor raged, the U.S. Food and Drug Administration was approving DES for use as a "morning-after" birth control pill.

Background

Diethylstilbestrol (DES), a synthetic chemical,* was discovered in 1938 to have biological activity closely resembling that of estradiol, a natural female sex hormone that had been identified just

*Trans HOC$_6$H$_4$ − C(C$_2$H$_5$) = C(C$_2$H$_5$) − C$_6$H$_4$OH. Also called stilbestrol.

70

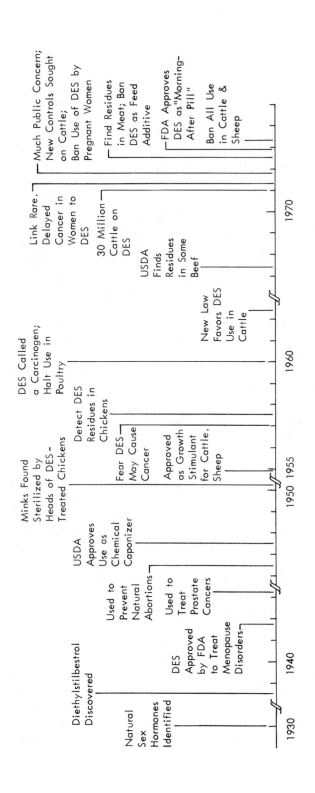

a few years earlier. In other words, DES was a synthetic sex hormone. In 1941, the U.S. Food and Drug Administration (FDA) approved its use for estrongenic hormone therapy, and it was immediately touted[2] as an aid for millions of women troubled by menopausal disorders. The synthetic material could be made cheaply compared to the natural estrogen, and a dozen or more drug companies would make it. It would sell at about $200 a pound, but treatment would require only 0.5 milligrams per day (less than one cent). In 1949 the FDA also approved the use of DES to prevent spontaneous abortions and miscarriages in women. The public was soon hearing, in addition, that this wonderful new substance would be a boon to men, as a control[3] and possibly a cure[4] for cancer of the prostate gland, and possibly an aid with mumps.[5] DES soon developed substantial veterinary and bioresearch uses, such as inducing estrus and lactation in animals.

The big news, however, was the discovery that it could be used to fatten chickens—a use approved by the USDA in 1947. A small pellet of DES (15 milligrams) implanted under the skin at the base of the head of a young rooster dissolved slowly and acted as a chemical castrator: it soon turned him into a plump capon-like bird. The caponettes, as they were called, as well as pullets treated with DES, gained weight faster than regular chickens or surgically castrated birds. Any unused DES would be removed from the bird with the head and neck at slaughter.

The disposal of poultry dressing wastes is, of course, a problem. These wastes had been adopted, however, as a source of meat by the mink-ranching business, which was burgeoning during the late 1940's as prices of mink pelts reached $30 apiece or more. But suddenly, female minks started becoming sterile at nearly thirty ranches. The cause: DES residues in the heads of stilbestrolized chickens.[6] The results: a claim for damages against the U.S. Department of Agriculture (which established the implantation procedures), rumors that DES was producing "caponized" men with large breasts (from eating too many chicken necks, or as a result of medical treatment given to convicts), and a ban by the Canadian government on the sale of stilbestrolized chickens.

DES was not banned for U.S. poultry, even so, and studies were already underway to explore possible uses on other animals. Researchers at the Iowa Agricultural Experiment Station soon discovered that DES could be administered in a feed supplement for cattle better than as a pellet implant.[7] With dosages of 5–10 milligrams of DES per day per steer, weight gains increased by 35 percent, and feed costs were reduced by 20 percent. In December 1954, the USDA approved the use of DES feed supplements for cattle and implants for sheep. The announcement was widely publicized in agricultural channels.[8-15] Its impact on the meat animal business has never been equaled: within three months 2 million head of cattle were on DES.* Although farmers and meat buyers were uneasy about the quality of the beef that the new technology might produce, the field results were soon showing that even if DES-supplemented feed cost $7–10 per ton more, it could lead to savings of up to 10 percent. The only drawback was the USDA requirement that the DES feed be stopped two days before marketing the animal—a nuisance for many producers.

By late 1955, DES was being combined with antibiotics[15] and in January 1956 farmers were told that DES was also approved as an implant for cattle.[16,17] Although some unexpected side effects of this wholesale use of a potent hormone had occurred—such as the inadvertent contamination of other feeds in a mill processing the cattle supplement,[18] and the decrease in the litter size of hogs eating DES-fortified feeds[19]—farm news about DES continued to be exciting through 1958.[20,21] Indications were few that a long controversy was about to begin over the wonderful growth stimulant.

Key Events and Roles

A substantial concern over the presence of potential cancer-causing agents in human foods began to develpop after World War II. By 1954 two widely used chemical additives—coumarin,† a vanilla-like flavoring, and safrole,‡ a root beer flavoring—were being called carcinogens and their use was soon discontinued under pressure of the FDA. Then, in early 1956, four doctors warned[22] that DES might induce cancer (as excess amounts of natural sex hormones were believed to do). In 1957 Representative James Delaney (D-New York) inserted a charge in the *Congressional Record* that DES was a carcinogen.[23] The FDA promptly denied the charge,[23] but attention had now been directed to the

*Eli Lilly and Company made a pre-mix (under license by the Iowa State College Research Foundation), which was then sold to commercial feed manufacturers across the country.

†Coumarin is a natural constituent of the South American tonka bean and other plants. It was the first natural perfume to be synthesized from a coal tar chemical. Synthetic vanillin, itself, was introduced in 1875.

‡Safrole is a natural constituent of the sassafras tree, the roots and bark of which were widely used to make a tea in America since frontier times.

popular feed supplement, and the National Cancer Institute (NCI) began a study of DES.

In 1958, the Congress amended the Food and Drug and Cosmetic Act of 1938 to give the FDA more powerful control over food additives. A feature of this new legislation, which was to become most controversial in following years, was the so-called Delaney Clause, which stated that no additive shall be deemed safe if it is found to induce cancer when ingested by man or animal (see "The Great Cranberry Scare").

By late 1959, evidence had accumulated that DES produced cancers in test animals when fed to them over long periods, and also that improved analytical techniques could detect DES residues in parts other than the neck of treated chicken (20–30 parts per billion were found in the livers and kidneys and 35–100 parts per billion in the skin fat). Although the total amount of DES in a whole chicken might be only 25 μg (millionths of a gram), and from five to twenty-five chickens would be needed to obtain the amount in a daily dose given to women in hormonal therapy, nevertheless the NCI had reported that as little as 0.07 μg had induced breast cancer in mice. These data, therefore, caused great consternation at the FDA and its parent organization, the Department of Health, Education and Welfare (HEW). In fact, FDA/HEW was already embroiled in a controversy over residues of a possibly carcinogenic herbicide in cranberries (see "The Great Cranberry Scare").

Officials of HEW, FDA, USDA, the chicken growers, and the DES-feed manufacturers all conferred to determine what to do about the use of DES on chickens, and about the millions of existing birds that were alrady on DES.* On December 10, the Secretary of HEW, Arthur S. Flemming announced the results of an agreement: sale and use of DES for chickens would be halted, and the USDA would buy and remove from the market an estimated $10 million worth of treated poultry.

The announcement made headlines[24-26] and raised complaints from chicken growers, but the public was not quite as shocked as it had been with the cranberry announcement a month earlier, possibly because the timing was less critical and an arrangement had been made to remove suspected products from the market in an orderly fashion. Twelve million pounds of caponettes were eventually withdrawn from the market,[27] and seizures of chickens sus-

---

*About 17 million DES-treated chickens were estimated to reach market annually—about 1 percent of those sold.

pected of containing DES occurred as late as 1962,[28] but the chicken scare faded rather quickly from the news.

The 1959 ruling did not affect the use of DES in cattle or sheep; the FDA explained that residues had not been found in the meat of these animals. On the other hand, the ban in chickens caused much apprehension among cattle growers. By 1962, they had generated enough force in the Congress to enact special legislation regarding meat animals: in essence it said that DES could be used because label regulations that were "reasonably certain to be followed in practice" would preclude residues in the meat. In practice, the USDA and FDA apparently avoided further checking for DES residues in beef until 1965.[29] At any rate, the cattlemen's news reports on DES were favorable through the 1960's, and it was adopted more and more widely. By 1970, nearly 75 percent of all beef was produced with DES—30 million head per year. In fact, the FDA reportedly[30] permitted the dosage level in cattle feed to be doubled in 1970 to a rate of 20 milligrams per day per animal.* Thus, the outlook for DES had apprently never been better as 1971 began, and Americans were totally unprepared for the outburst of concern that was imminent.

In April 1971, Dr. Arthur L. Herbst and co-workers at Boston's Massachusetts General Hospital published in the prestigious *New England Journal of Medicine* evidence linking rare vaginal cancers in young women with the use of DES by their pregnant mothers to prevent abortions and miscarriages fifteen to twenty-two years earlier. The full implications of this discovery, however, were not brought forcefully or even very rapidly to the public's attention. A story on the original article was buried deep in the *New York Times*,[31] as was one on the New York Health Commissioner's proposed ban on the prescription of DES to pregnant women.[32] Although the "hormonal time bomb" aspect was noted in a Time story in August,[33] the news media more eagerly emphasized[34-37] the discovery that DES might be used as a "morning-after" contraceptive pill, as was reported by Dr. Lucile Kuchera of the University of Michigan Health Service.†

But on October 28, as a result of a suit brought jointly by several citizens' groups,[38] the USDA revealed that, in fact, detectable

---

*Although the 48-hour DES-free diet period remained in effect, this approval is surprising in view of the accepted carcinogenicity of DES and the spectacular cyclamate ban which had embroiled FDA/HEW during 1969 and 1970.

†DES was similarly reported to prevent pregnancy in rape victims by Drs. Takey Crist and Cecil Farrington at the University of North Carolina.[37]

residues of DES had been found in beef samples, at least during 1966 (1.1 percent of 1,023 samples) and 1967 (2.6 percent of 495 samples), but apparently none in 1968 (545 samples), 1969 (505 samples), 1970 (192 samples) or in 1971 samples to date (6,000 planned).[29] The USDA had not made the 1966 or 1967 information public, however, and the suspicion that USDA had suppressed it raised Congressional concern immediately. FDA Commissioner Charles C. Edwards then moved quickly[38] to increase the mandatory DES-free diet period before slaughter to 7 days, on the claim that some cattlemen had not been sufficiently punctual on the previous 48-hr period.

Almost as rapidly, however, came moves in early November by Senator William Proxmire (D–Wisconsin) and Representative Ogden R. Reid (R–New York) to introduce legislation to ban the use of DES in cattle and sheep,[39] and hearings on DES by Representative L. H. Fountain's (D–North Carolina) Subcommittee on Intergovernmental Relations. DES' proponents contended that such a ban would raise the price of beef by 3-½ cents per pound—a total cost increase of $300 to $460 million per year to the American meat buyers. They also claimed that only infinitesimal residues were present in only a few livers, and that this amount posed no hazard.

Critics responded, however, that the USDA's analytical method was sensitive only down to 10 parts per billion, whereas DES caused tumors in mice at only 6.5 parts per billion, that a "no effect" level had not been discovered, and that DES had been banned from animal feeds in twenty-one countries already. In addition, reports were circulating that the 1971 testing program had found nearly a dozen animals that showed DES residues in their livers, with some as high as 37 parts per billion.[29,40,41] On November 9, the FDA proposed new label regulations that would contain a warning to pregnant women,[42] and on January 4, 1972, the USDA officially established the seven-day withdrawal period on meat animals.[43]

This new regulation required that animals brought to slaughter must be accompanied by a "certificate of compliance" if they had been fed DES. Now if the cattle feeder had failed to observe the DES diet-free period under the old forty-eight hour limit, a comparable amount of time might still be consumed in transit and stockyards marketing before slaughter. But under the new seven-day rule, the meat packers were clearly dependent on the good faith of the cattleman observing the limit. Critics contended that if the forty-eight-hour limit was unenforceable, then so was the seven-day limit because the producers tended to watch the market prices and to decide to sell suddenly when they thought it was right. Thus, if the large stockyards and meat-packing companies really wanted to guarantee that animals were DES-free, they would have to maintain the animals themselves for two or three times longer than previously, a step which would mean that facilities would have to be expanded or throughputs would have to drop. The alternatives were to test every carcass, to trust the cattlemen, or make the cattlemen stop using DES entirely. On March 10, the FDA proposed a ban on the use of DES in liquid form in animal feeds on the grounds that DES leftover in feed-mixing equipment was contaminating supposedly DES-free feeds, and that DES residues were now being found in nearly 1 percent of animals sampled—twice the level found in 1971.* Dry DES mixes and pellet implants would still be permitted, however. By June, nearly 2 percent of the animals tested in 1972 had been found to contain DES, and awareness was increasing[44] of the long-term vaginal cancer hazard. The FDA disclosed that it would provoke public hearings on DES to determine whether it should be banned, or whether alternative methods of controlling DES residues in the meat supply were available.[45] These hearings were to involve both liquid and dry mixes of DES.

This action raised a storm of protest from both the advocates and proponents of DES. On the one hand, Representative William J. Scherle (R–Iowa) introduced legislation to permit the use of animal drugs such as DES, unless scientific tests showed that the amounts present as residues in the meat were actually carcinogenic. On the other hand, Senator Proxmire called the FDA action another tardy step, and Representative Fountain said the FDA already had enough evidence to ban DES, was using the hearings only to stall, and that it was hardly surprising that public opinion polls showed widespread distrust of government.[46] While cattlemen and DES manufacturers protested that little danger existed, FDA's antagonists claimed that the public was in great danger from DES.[29,47,48] While FDA Commissioner Charles C. Edwards and USDA Secretary Earl L. Butz implied that DES was not a known hazard and that a ban would be made only because it was required by the fourteen-year-old Delaney Clause, National Cancer

---

*A chromatographic analytical technique sensitive to 2 ppb was introduced in late 1971 or early 1972.

Director Frank Rauscher and other cancer researchers maintained that the Delaney Clause must be upheld, and that no one could tell just how much damage the DES residues were causing.

Disposition

On August 2, 1972, the FDA capitulated and reversed its field: it announced cancellation of the proposed hearings and a ban on not only the use, but according to news accounts[49] even the manufacture of both liquid and dry forms of DES feed pre-mixes.* This ban was effected on the basis that new USDA tests on six steers showed by radiological analysis methods that liver residues of 0.5 parts per billion could be detected seven days after a single 10-milligram dosage.

The new regulations did not impose a recall of shipments or cessation of the use of existing stocks of DES-containing feeds, did not call for a complete halt to feeding DES until January 1, 1973, and did not ban pellet implants at all. The partiality of the ban and the FDA's rationale for it created further distrust[51] and confusion.[52] Its mere existence prompted some Congressmen and industry and government officials to ask whether an outmoded Delaney Clause should be revised.[53] In contrast, the House Government Operations Subcommittee scheduled hearings on FDA's foot dragging,[52] and the Senate Labor and Public Welfare Committee quickly approved legislation which would ban use of DES immediately.

By this time the story of the DES-induced delayed vaginal cancers had reached the front page.[54] In addition to the Boston area cases, 840 pregnant women had been given DES without even knowing it† in a 1950–1952 study of abortions conducted by the University of Chicago's Department of Obstetrics and Gynecology. In September the Senate approved a bill to ban DES in feed immediately, and in implants by year's end.[55] The outlook for the $2-million DES market was dimmed even further as the focus shifted to the implants, which were actually cheaper on the basis of DES needed, but more expensive overall to the producer because of increased labor costs for injecting each animal. The USDA

---

*Although several news accounts described the ban in this way, the law apparently does not allow a ban on the manufacture.

†Medical ethics and practice at that time did not require that the patient give "informed consent" to experimental treatment, as is required now. The Chicago studies actually found that DES was of little therapeutic value in pregnancy, and such use declined thereafter.

quickly instituted a study of DES residues in implant-treated animals—the sampling program had never included these previously. A spokesman for the American National Cattlemen's Association was even then expressing fear that the implants were also doomed.[56]

Even as the ultimate ban on DES as an animal growth stimulant was shaping up, and even as new evidence linked DES not only to the rare vaginal cancers but also to delayed cases of the more common cervical cancers,[57] Ralph Nader's Health Research Group and others[57] were reporting that DES was being dispensed as a "morning-after" birth control pill for the coeds at the student clinics of several of the major universities. The FDA had not yet approved this use! The startling revelation led the FDA to reveal in January 1973, that it was studying the matter.[58] But instead of instituting a ban, as many people expected, the FDA announced on February 21, that the "morning-after" pill would be approved[59] as a prescription drug. The effective dosage would require 25 milligrams of DES twice daily for 5 days (beginning within 72 hours after intercourse) or a total of 250 milligrams DES—about 50,000 times more than one might encounter in a pound of beef liver. FDA Commissioner Edwards stated that an advisory committee had concluded that this use of DES was not a significant risk, and that the pill's unpleasant side effects (nausea or vomiting) would relegate it to an "emergency" rather than routine contraceptive.[57] In addition, the pill's advocates claimed that if the DES pill worked there wouldn't be a baby daughter to get cancer, and if it didn't work the woman might seek an abortion anyway. Thus, the "morning-after" pill was approved by the medical authorities.*

In stark contrast, however, pressure against the use of DES in meat animals continued to build. New, more sensitive tests then found that implant-treated animals contained DES residues of 40–120 parts per trillion, and that DES residues were detectable in the livers even 120 days after implantation. On April 25, the FDA announced an immediate and total ban on DES for beef cattle and sheep.[60]

The industry complained that the nation's meat bill would be boosted $1.8 billion per year and proceeded to take the matter to the courts. Nine months later, the United States Court of Appeals on January 24, 1974 overruled FDA by ruling that the agency must

*Interestingly enough, critics of the university clinic programs charged[57] that the coeds were not being given sufficient information on possible side effects of DES to give their "informed consent."

hold public hearings on the issue.[61] While FDA prepared for the hearings, it also had to prepare guidelines for the cattlefeeders to resume use of DES. The saga of DES continues to rage as additional experiments and tests are performed on it.

Comment

The discovery of the synthetic sex hormone, DES, initiated a series of technological applications followed by almost as many belated discoveries of unpredicted side effects. To a considerable measure, these recurrent episodes followed successive advances in the analytical capabilities to detect ever-smaller levels of residue—from parts per million to parts per billion to parts per trillion. Even so, the slow pace at which the FDA and the USDA moved against this potent compound, with the evidence on hand, is in marked contrast to that used against the less potent herbicide involved in the cranberry scare. The contrast rather clearly reflects the much greater economic and political power of the beef producers compared to that of the cranberry growers, as well as the greater importance of meat in the American diet. The long and torturous efforts to accommodate the technologies that developed from the relatively simple DES molecule has caused great public confusion and a significant loss of confidence in many of society's institutions, including the FDA, the USDA, the Congress, cattle ranchers, and the medical profession. The DES controversy sharpened the debate over the Delaney Clause, and tended to set the FDA/HEW against the National Cancer Institute and its advocates in seeking a change of this legislation. The FDA's 1973 decision to ban miniscule amounts of DES in meat is strangely juxtaposed by its decision to approve the heavily-laced "morning-after" pill—almost as if two unacquainted officials from different departments were speaking through the mouth of the Commissioner. And one cannot help but wonder if society will not soon decide that both decisions should be changed.

REFERENCES

1. Dodds, E. C., L. Goldberg, W. Lawson, and R. Robinson, *Nature*, **141**, 247 (1938).

2. "Help for Women Over Forty," *Read Digest*, **39**, 67 (November 1941).

3. *New York Times*, p. 29 (December 15, 1943).

4. "The Case of Benjamin Twaddle," *Time*, **54**, 34 (August 29, 1949).

5. "Synthetic Female Hormone Helps Men With Mumps," *Sci Digest*, **26**, 48 (September 1949).

6. "Case of the Barren Mink," *Time*, **57**, 65 (February 19, 1951).

7. Burroughs, W., C. C. Culbertoon, J. Kastelic, E. Cheng, and W. H. Hale, "The Effects of Trace Amounts of Diethylstilbestrol in Rations of Fattening Steers," *Science*, **120**, 66 (July 9, 1954).

8. "Stilbestrol OK'd For Fattening Beef," *Farm J*, **78**, 34 (December 1954).

9. "Two Million Head on Stilbestrol," *Farm J*, **79**, 38 (March 1955).

10. "Stilbestrol Continues to Look Good," *Farm J*, **79**, 15 (April 1955).

11. "Latest Stilbestrol News," *Farm J*, **79**, 16 (May 1955).

12. "Latest Feeding Results," *Farm J*, **79**, 14 (June 1955).

13. "Feed Stilbestrol Next to Dairy Cows?" *Farm J*, **79**, 56 (June 1955).

14. "Stilbestrol-Fed Cattle: How They're Selling Now," *Farm J*, **79**, 16 (August 1955).

15. "New Stilbestrol Mixtures Available," *Farm J*, **79**, 18 (October 1955).

16. "Implants Look Good in Steer Trials," *Farm J*, **79**, 22 (October 1955).

17. "OK Stilbestrol Pellets for Beef," *Farm J*, **80**, 22 (January 1956).

18. Hadlow, W. J., E. F. Grimes, and G. E. Jay, Jr., "Stilbestrol Contaminated Feed and Reproduction Disturbances in Mice," *Science*, **222**, 645 (October 1955).

19. "Hormone Balance May Affect Litter Size," *Farm J*, **80**, 45 (December 1956).

20. "Stilbestrol For Milkers?" *Farm J*, **80**, 55 (August 1956).

21. "Now—Better Carcasses with Stilbestrol," *Farm J*, **82**, 64 (March 1958).

22. *New York Times*, p. IV9 (January 29, 1956).

23. *New York Times*, p. 10 (February 23, 1957).

24. *New York Times*, p. 1 (December 11, 1959).

25. "Hormones and Chickens," *Time*, **74**, 32 (December 21, 1959).

26. "Row Over Pure Food—Why All the Cancer Scare?" *US News*, **47**, 34 (December 28, 1959).

27. *New York Times*, p. 23 (May 5, 1960).

28. *New York Times*, p. 47 (February 10); p. 15 (February 22); and p. 14 (February 27, 1962).

29. Wade, N., "DES: A Case Study of Regulatory Abdication," *Science*, **177**, 335 (July 28, 1972); with correspondence, *Science*, **178**, 117 (October 13, 1972).

30. "Carcinogen Doubled in Cattle Feed," *Prevention*, 192 (January 1971).

31. *New York Times*, p. 53 (May 2, 1971).

32. *New York Times*, p. 50 (June 20, 1971).

33. "Hormonal Time Bomb?" *Time*, **98**, 52 (August 2, 1971).
34. *New York Times*, p. 19 (October 26) and p. IV14 (October 31, 1971).
35. "Morning-After Pill," *Sci N*, **100**, 293 (October 30, 1971).
36. "Morning-After Pill," *Newsweek*, **78**, 74 (November 8, 1971).
37. "The Morning-After Pill," *Time*, **98**, 67 (November 8, 1971).
38. *New York Times*, p. 25 (October 29, 1971).
39. *Facts on File*, p. 898, November 11–17, 1971.
40. "What You'd Better Know—About the Meat You are Buying," *Today's Health*, **49**, 37 (December 1971).
41. "Court Suit Asks Ban on Stilbestrol," *Farm J*, **95**, 25 (December 1971).
42. *New York Times*, p. 21 (November 10, 1971).
43. *New York Times*, p. 40 (January 5, 1972).
44. *Kansas City Star*, p. 2 (March 10, 1972).
45. *Kansas City Star*, p. 14A (June 17, 1972).
46. *Kansas City Star*, p. 6 (June 22, 1972).
47. Hunter, B. T., "Diethylstilbestrol—A Known Cancer-Inciting Drug in Meats for Americans," *Consumer Bul*, **55**, 17 (August 1972).
48. Wellford, H., "Cancer-Causing Chemicals in Meat," *Read Digest*, **101**, 134 (December 1972). [This article first appeared in *Atlantic* (October 1972).] Wellford's book, *Sowing the Wind*, Grossman, New York, 1972, also took up the DES matter.
49. *Kansas City Times*, p. 1 (August 3, 1972).
50. *Fed Reg*, 37, 15747 (August 4, 1972).
51. Wade, N., "FDA Invenst More Tales About DES," *Science*, **177**, 503 (August 11, 1972).
52. "Hormone Ban Spreads Confusion," *Bsns W*, **29** (August 12, 1972).
53. "Cancer Clause Change," *Chem W*, **111**, 22 (August 16, 1972).
54. *Kansas City Star*, p. 1 (August 29, 1972).
55. *Kansas City Times*, p. 8A (September 21, 1972).
56. *Wall Street J*, p. 26 (November 8, 1972).
57. *Kansas City Times*, p. 6A (December 21, 1972).
58. *Kansas City Star*, p. 12A (January 31, 1973).
59. *Kansas City Times*, p. 14A (February 22, 1973).
60. *Kansas City Star*, p. 10C (April 25, 1973).
61. *Kansas City Star*, p. 11D (February 3, 1974).

The Cyclamate Affair

Abstract

On October 18, 1969, the Secretary of Health, Education, and Welfare, Robert Finch, announced at a press conference that cyclamates, the widely used, FDA-approved, diet sweeteners, had been found to cause cancer in rats in long-term feeding studies; therefore, all cyclamate-sweetened diet sodas were to be off the market by January 1, 1970, and the multitude of cyclamate-containing foods were to be off grocers' shelves by February 1. The announcement had strong adverse economic effects on the nearly $1 billion dietary beverage and food industry, and emotional effects on the public. The Food and Drug Administration's procedures before the announcement were widely condemned, and HEW's subsequent vacillating decisions in response to the storm of protest which erupted led to considerable public confusion.

Background

In 1937, Michael Sveda at the University of Illinois accidentally discovered the sweet taste of sodium cyclohexylsulfamate, a synthetic chemical. Mindful of the artificial sweetener, saccharin, which had come into use thirty years earlier, he obtained a patent on the compound. Tests during the 1940's showed that "cyclamate," as it is now called, was thirty times sweeter than sucrose (cane sugar) and lacked the unpleasant aftertaste many people found with saccharin (which in turn was 300 times sweeter than sucrose). Like saccharin, the cyclamates* were found to be nonnutritive sweeteners with a potential market as a sugar replacement, particularly in low-calorie dietary foods or diabetic foods. Furthermore, the cyclamates could be produced cheaply from cyclohex-

*Includes cyclamic acid and its sodium and calcium salts.

83

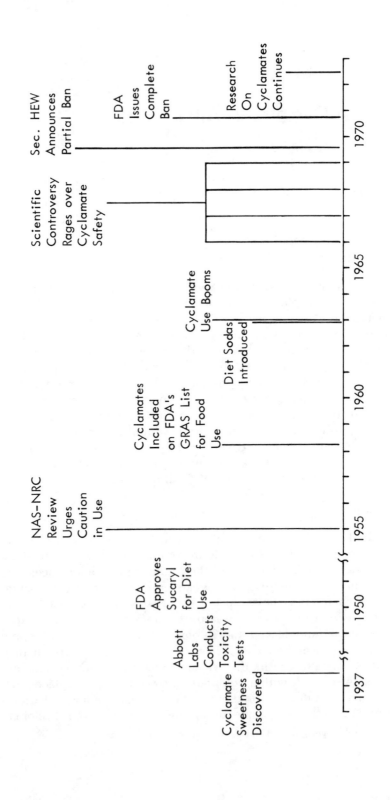

ylamine, an industrial chemical used in rubber manufacture, according to a 1942 patent assigned to the Du Pont Company.

In 1950, the Food and Drug Administration first approved for human consumption the drug use of a cyclamate product—Abbott Laboratories' Sucaryl tablets for diabetics and others who must restrict sugar intake. According to a recent review[1] the safety test data actually supplied with Abbott's application were regarded as inadequate by FDA. Approval was granted, however, on the basis of FDA's own 1948–1949 feeding studies, although much later it became evident that FDA ignored clear implications in the data of tumorigenic activity.[1-2]

Abbott's Sucaryl was a successful product and by the time the original patents expired competing brands of cyclamates appeared around the world. By 1957 the U.S. production of cyclamates reached about 1 million pounds per year.

The FDA had commissioned the Food Nutrition Board of the National Academy of Sciences–National Research Council, to review cyclamates in 1955; it warned of uncontrolled distribution of foodstuffs containing cyclamates, but no FDA restrictions were apparently considered since the use of cyclamate products was essentially confined at that time to diabetics and severely obese persons. In 1958, however, a most important change occurred; the Food, Drug and Cosmetic Act was amended to give FDA increased power to regulate food additives, that is, an applicant must henceforth furnish test data to FDA showing that the additive was safe before using it. The several hundred additives already in use posed a special problem which was resolved by allowing their continued use if the FDA could determine that they were "generally recognized as safe" (GRAS). The FDA's procedures in developing its GRAS list have been the source of much critical discussion (and the entire list has not been published),† but at any rate the first list proposed in 1958 and issued in November 1959[3] contained the cyclamates, heretofore considered drugs. Suddenly the cyclamates were approved for wholesale use in foods.

The use of cyclamates as low-cost sweeteners by the food industry increased markedly, and by 1963 annual production was 4 million pounds. This was only the beginning, however, as production soared to 16.5 million pounds by 1968 under the introduction and vigorous marketing of dietary colas and other beverages and a host of foods, frequently with no mention of the cyclamate content

*Nor apparently have the names of the members of the NAS committee which helped develop the GRAS list been made public.

on the label. Cyclamates were to be eventually used by 75 percent of American families, with or without their knowledge.[1]

The Food Drug and Cosmetic Act Amendment of 1958 also contained another feature of particular pertinence to the cyclamate case, the so-called Delaney Clause (see p. 59), which banned cancer-causing food additives. The practice has been to prohibit an additive or contaminant in food if it causes cancer at any dosage (no matter how large) under any conditions or by any test procedure. The Delaney Clause received wide publicity in 1959 during The Great Cranberry Scare (see p. 55).

Key Events and Roles

In 1962 the thalidomide tragedy (see p. 140) sparked concern over the safety of all drugs and also food additives. During 1964 reports from Japan, Austria and France, as well as from the United States, suggested that cyclamates had a number of physiological actions, such as increased blood pressure in test animals. Again in 1965 the FDA commissioned the NAS to review cyclamates, and again the FDA concluded that they were safe, that adverse data were not substantial, or that they were unreliable in some cases because the studies had been supported by the sugar industry.

A 1966 Japanese report[4] on cyclamates was, however, to start an enduring uproar: for many people, cyclamates were not totally excreted as previously supposed, but were converted back to cyclohexylamine, which was known to be toxic. These results, soon confirmed, prompted further studies of cyclohexylamine and cyclamates, including a recommendation by Britain's Food Additives and Contaminants Committee in 1967 that cyclamate studies were urgently needed. In 1968 the FDA apparently started its own laboratory studies and commissioned a third NAS-NRC review of the available literature. An interim report of this study in late 1968 cited test results[5] showing that massive feedings of cyclamates (5 to 10 percent of total diet) caused retardation of growth in rats and pigs, and evidence of damage to the kidneys, liver, gastrointestinal tract, adrenals and thyroids of other animals. In humans, the review quoted, severe persistent diarrhea occurred in subjects taking more than 10 grams of cyclamates per day, and cyclamates which were sometimes formulated in pharmaceuticals could prevent body absorption of at least one common antibiotic. Further data indicated that cyclamates were not very effective in weight-control programs anyway.

On the basis of this interim report and on preliminary laboratory results, discussed below, the FDA in a December 13, 1968 press release said "totally unrestricted use of the cyclamates (commonly used as artificial sweeteners) is not warranted at this time." The FDA reduced its recommended upper limit for daily cyclamates intake from 5 grams per person (an adult) to a level of 0.05 g/kg of body weight. Thus for a 150-lb adult the new ceiling would be on 3.41 g cyclamate; for a 60-lb child it would be only 1.37 g. The new levels did not pose any particular problem for persons using cyclamate tablets for dietary purposes since nearly 100 tablets per day could still be used. The big impact, however, was on the soft drinks, because most of these contained 0.5–0.7 grams per 12-ounce bottle or can, and a child could exceed the daily limit with only two bottles or cans of low-calorie soda per day. For the first time the cyclamate issue really came to the public's attention.[5–11] Articles containing cyclamate content charts for dieters soon appeared.[5]

The laboratory results which concerned the FDA (but which were not mentioned in the December 1968 announcement) included those of Stone et al.[12] on cyclamates and those of FDA's Legator et al.[13] on cyclohexylamine which demonstrated chromosome breaks—and therefore a potential mutagenetic damage—in human and animal cells. Also not mentioned, but presumably also of concern, were other FDA studies by Dr. Jaqueline Verrett, who as early as March 1968, according to one account,[1] had reported that when cyclamates were injected into eggs, 15 percent of the embryos which were subsequently grown contained physical deformities. None of these results, however, prompted the FDA to remove cyclamates from the GRAS list, even though Dr. Verrett's results—and the clear similarity with thalidomide—were said to have been brought to the personal attention of FDA Commissioner Herbert Ley, Jr., in April 1969.[1] The FDA did begin to consider requiring that the labels of cyclamate products carry an ingredients statement and possibly also a warning, although this was not effected.

The FDA not only accorded low significance to the data of their own Drs. Legator and Verrett, but also may have slowed the publication of the data. Dr. Legator's data were published in September 1969,[13] nearly a year after the implications had first been discussed.[14] On October 1, 1969, Dr. Verrett discussed her results on a televised interview on a Washington, D.C., station, and warned of the potential danger of cyclamates. The interview, which had apparently been approved by a FDA deputy commissioner,[1] created an immediate sensation: the media recognized the analogy between Dr. Verrett and Dr. Kelsey, heroine of the thalidomide tragedy. It also drew an immediate rebuttal from Commissioner Ley, who said[15] "Cyclamates are safe within the present state of knowledge and scientific opinion available to me." But Dr. Ley revealed on a national television news program on October 2 that he had asked the NAS-NRC to review Verrett and Legator's work and to report back to him within thirty days. The Secretary of Health, Education, and Welfare, Robert Finch, became involved, and indicated that more data were needed before making over-zealous warnings to the news media, but that he felt the FDA had vacillated on the decision.[16] Finch indicated further that because of the cyclamate confusion, substantial reorganization of procedures and personnel in the FDA (which was under his direction) was inevitable. Abbott Laboratories, one of the world's largest cyclamate producers, declared[15] "Cyclamate is safe for human consumption as presently used in the diet." Senator Warren Magnuson, from the sugar-beet State of Washington, raised the specter of another "thalidomide scare" from cyclamates in his syndicated column and on network television.[16]

Despite the apparent reassurances by FDA and Abbott, within two weeks—on October 18, 1969—Secretary Finch and Surgeon General Jesse L. Steinfeld announced at a press conference new* evidence that cyclamates had been found in long-term feeding studies to cause cancer in the bladders of rats. The "new" evidence was, in fact, published[17] as a joint communication by researchers at Abbott and FDA together with Dr. Ley and Dr. Steinfeld, and followed by one week the publication of similar results from a possibly more definitive† study, using the controversial bladder implant technique.[18] The indications of cancer clearly placed cyclamates in violation of the Delaney Clause and Secretary Finch announced that they would be removed, therefore, from the GRAS list, and could no longer be used in common foods and beverages. Cyclamate-sweetened beverages were to be off all store shelves and out of vending machines by January 1, 1970, and the multitude of cyclamate-containing foods were to be off grocers' shelves by February 1, according to the Secretary's timetable. The only excep-

*Said to be obtained since October 5 and presented to NAS-NRC on October 17.
†For some strange reason the Abbott-FDA studies had used *mixtures* of cyclamate and saccharin and even cyclohexylamine rather than the pure substances.

tions were to be dietary items which would be available by a doctor's prescription.

The cyclamate issue and Secretary Finch's announcement, although it emphasized that there was "no evidence that they have indeed caused cancer in humans," and that he was acting under the requirements of the law, received much media attention[19-23] and has had strong and long-lasting effects. The initial reaction of consumers was a mixture of alarm, confusion and disbelief, since they had been consuming the diet sodas in large amounts without apparent ill effects, and had cold six-packs right there in their own refrigerators.

The reactions of the financial community and the food and drink industries were far greater: almost every company connected with cyclamates took a temporary shellacking on Wall Street, and every company immediately started to refigure its marketing positions as the $750 million low-calorie market vanished. Abbott Laboratories, which produced over half of the cyclamates in the United States (Chas. Pfizer and Company and Miles Laboratories were other major cyclamate producers), announced that cyclamate earnings contributed only a few pennies to its $2.35 per share stock dividends, but launched an ambitious cyclamate-defense campaign to protect its scientific integrity.[16] The $420 million diet soda industry, which consumed 50 to 70 percent of all cyclamates, was hard hit, but ignored Abbott's request for help in the defense. Coca-Cola, whose Tab and Fresca together claimed 30 percent of the diet soda market but contributed only 5 percent of the company earnings, was not hit as hard as Pepsi-Cola, whose Diet Pepsi had 25 percent of the market, or Royal Crown, whose Diet-Rite was the market leader and constituted one-fourth of the company's sales. Almost overnight, however, the soft drink producers reformulated their products (usually with a saccharin-sugar mixture) and in extensive sales promotions proudly advertised "No Cyclamates."

Other groups affected by the ban were the fruit canners, producers of other cyclamate-sweetened products (ranging from cereals to desserts and soft drink powders), and the supermarkets. A problem for all these groups was how to handle unsold inventories when the ban became effective, but this problem was particularly bad for the fruit canners who had just finished packing a year's supply. To complicate matters, many of the fruit canners and specialty-diet item producers were relatively small companies heavily dependent on the cyclamate products.

Disposition

The storm of protest over the banning of the cyclamate-containing foods caused some back-pedaling in HEW. On November 15,
1969, the deadline for the removal of such foods from store shelves
was extended from February 1 to September 1, 1970, to avoid an
estimated $100 million loss, and on November 20, 1969, Secretary
Finch announced that such foods (but not beverages) could be sold
to diabetics and the obese without a doctor's prescription after
September 1, provided they were labeled as a drug, to which the
Delaney Clause does not apply. This action was not, however, without a counterprotest about FDA procedures, and in June 1970,
Representative L. H. Fountain's House Subcommittee on Intergovernmental Relations investigated.[24] An FDA Medical Advisory
Group, which had previously suggested that cyclamates remain
available to the obese and diabetics, was then reconvened: it concluded that the recommended daily dosage of cyclamates be lowered from the 780 milligrams per day (set in January 1970) to only
168 milligrams per day,[25] a level which would make cyclamates
nearly useless as sugar substitutes. On August 14, 1970, the FDA
reversed its field another time and issued a *total ban* on cyclamate
use in foods, drinks or drugs.[25-26]

On the financial side, the diet soda companies generally
charged off losses of several cents per share, although Royal
Crown's earnings dropped substantially from 1968 to 1969.[27] Diet
foods were helped by the delay,[27] but were estimated[25] to suffer
losses of $31.5 million. Most diet food and drink products were
reformulated and by mid-1970 sales were estimated to be 70 percent of the prescare level.[27] A bill to reimburse the diet food and
drink manufacturers and independent grocers for actual losses suffered by the federal ban passed the House of Representatives in
1972, but has not cleared Senate committee. Estimates of outstand-
ing claims ranged from $120 million to $500 million.[28]

On the administrative and political side, the FDA's procedures
in developing the GRAS list and in evaluating safety data eventually drew President Nixon's attention[2] in October 1969. FDA
Commissioner Herbert L. Ley lost his job in early 1970, apparently
because of Secretary of HEW Finch's displeasure at his handling of
the case, and Finch soon left HEW for a quieter position as a Presidential assistant. An ominous shadow in the cyclamate case was

the spectre that saccharin,* sucrose and a host of other food additives—each a political issue—would be banned before the GRAS list could be defended.

Much controversy was also aroused[29] in the scientific community over HEW's initial ban announcement, and on the FDA's criteria for carcinogenicity testing. In fact, the cyclamate issue reopened an argument (which arose during coal-tar color-additives hearings of 1960) between the National Cancer Institute, staunch supporter of the Delaney Clause, on the one hand, and many members of the NAS-NRC's Food Protection Committee, the American Medical Association's Council on Foods and Nutrition, and the FDA itself on the other.[29-30] The latter groups have urged modification of the Delaney Clause to permit a greater exercise of scientific judgment. The argument is likely to remain unresolved for some time.

The public is now learning[30] that the evidence that cyclamates cause cancer is being contradicted by new, more extensive toxicological tests. Cyclamates may return someday!

Comment

The cyclamate affair was basically a product of four interacting factors. The first factor was the Delaney Clause and either a difficulty in interpreting it or an enthusiastic enlargement of its scope by interested parties in the National Cancer Institute and (in earlier days) in the Department of Health, Education and Welfare. Thus, while the legislation excludes additives that have been found to cause cancer when ingested or "after tests which are appropriate for the evaluation of the safety of food additives," just what constituted an "appropriate test" was left to the regulatory experts; they have generally translated it as "any test." Thus, a susceptible strain of test animals could be force-fed or injected with any dose of a test substance below that which would kill them. If cancer, or something which looked like cancer, was observed, then the substance was banned, even if no evidence existed that it could cause cancer at body levels normally encountered.

The second factor was FDA's curious set of internal procedures, which permitted the generation of a long, if somewhat secre-

---

*In fact, saccharin was reported in 1970 to produce bladder cancer by the implant technique; it has not been banned, but has been recently removed from the GRAS list.

tive, GRAS list that included cyclamates on the basis of minimal supporting data. These procedures provoked accusations that the FDA tried to suppress negative evidence of its own employees about cyclamates. The third factor was the practice of the soft drink companies, in aggressively pushing their heavily cyclamate-laced products on the public before sufficient knowledge about their effects was really known. The final factor was that although an attempt was made to bring about an orderly ban and recall, the Secretary of HEW announced the ban at a press conference, while maintaining that there was no evidence of hazard to humans. This technique had been employed previously with little success, and appears almost guaranteed to bring about much public alarm and confusion when a widely used product is involved. Government agencies have yet to learn how to handle such a conflict well, but the subsequent gradual phase-out of diethylstilbestrol shows improvement. (See "The Diethylstilbestrol Ban.")

REFERENCES

1. Turner, James S., *The Chemical Feast*, Grossman Publishers, New York, New York, Chapter 1, "Cyclamates," 1970.
2. "The Lessons Cyclamates Teach," *Consumer Rep.*, **35**, 59 (January 1970).
3. "Food Additives—Substances That are Generally Recognized as Safe," *Fed Reg*, 9368 (November 20, 1959).
4. Kojima, S., and H. Ichibagese, *Chem Pharm Bull Japan*, **14**, 971 (1966).
5. "A New Look at Cyclamate Sweeteners," *Consumer Rep*, **34**,280 (May 1969).
6. "Warning on Sweeteners," *Newsweek*, **72**, 48 (December 23, 1969).
7. "Low-Calorie Sweeteners," *Time*, **93**, 37 (January 3, 1969).
8. Gardner, W. D., "Safety of Sweetner, Questioning Effects of Cyclamates," *New Repub*, **160**, 13 (January 4, 1969).
9. "Artificial Sweeteners Suspected of Causing Cell Damage," *Chem*, **42**, 6 (January 1969).
10. "Artificial Sweeteners of Questionable Safety," *Consumer Bul*, **52**, 12 (February 1969).
11. Davide, R. C., "How Safe are No-Calorie Sweeteners," *Read Digest*, **95**, 77 (July 1969).
12. Stone, D., E. Lamson, Y. S. Chang, and K. W. Pickering, "Cytogenetic Effects of Cyclamates on Human Cells *in Vitro*," *Science*, **164**, 568 (May 2, 1969).
13. Legator, M. S., K. A. Palmer, S. Green, and K. W. Petersen,

"Cytogenetic Studies in Rats of Cyclohexylamine, A Metabolite of Cyclamate," *Science*, **165**, 1139 (September 12, 1969).

14. "Low Dose Findings are in 'Danger Zone'," *Med Wor N*, 27 (Novemeber 15, 1968).

15. "Bitterness About Sweets," *Time*, **94**, 79 (October 17, 1969).

16. "The Big Brouhaha Over Sweeteners," *Bsns W*, 98 (October 18, 1969).

17. Price, J. M., C. G. Biava, B. L. Oser, E. E. Vogin, J. Steinfeld, and H. L. Ley, "Bladder Tumors in Rats Fed Cyclohexylamine or High Doses of a Mixture of Cyclamate and Saccharin," *Science*, **167**, 1131 (1970).

18. Bryan, G. T., and Erdogan Erturk, "Production of Mouse Urinary Bladder Carcinomas by Sodium Cyclamate," *Science*, **167**, 996 (1970).

19. The *New York Times Index* contains 37 entries on cyclamates in 1969 and 13 in 1970.

20. "HEW Bans the Cyclamates," *Time*, **94**, 84 (October 24, 1969).

21. "Diet Industry Has a Hungry Look," *Bsns W*, 41 (October 25, 1969).

22. "Cyclamate's Sour Aftertaste," *Time*, **94**, 79 (October 31, 1969).

23. "Is Your Food Safe: Dispute Over Testing," *US News*, **67**, 44 (December 15, 1969).

24. "Cyclamates: House Report Charges Administrative Alchemy at HEW," *Science*, **170**, 419 (October 23, 1970).

25. "FDA Extends Ban on Cyclamates," *Science*, **169**,962 (September 4, 1970).

26. "Total Eclipse for Cyclamates," *Time*, **96**, 60 (August 31, 1970).

27. "A Fatter Outlook for Diet Foods," *Bsns W*, 93 (June 6, 1970).

28. "Cyclamate Claim OK'd," *Kansas City Star*, p. 7 (July 25, 1972).

29. Wade, N., "Delaney Anti-Cancer Clause: Scientists Debate an Article of Faith," *Science*, **177**, 591 (August 18, 1972).

30. *Wall Street J*, p. 1 (July 2, 1973).

MSG And The Chinese Restaurant Syndrome

Abstract

Monosodium glutamate (MSG), a food additive widely used to enhance flavor, became a subject of public interest in mid-1968 when it was linked to the "Chinese Restaurant Syndrome," (that is, headache, dizziness, or other symptoms, in some patrons of Chinese food establishments). Much greater public concern was aroused in 1969 when it was reported that MSG had caused brain damage when injected into newborn mice in large doses, because MSG was being widely used in human prepared baby foods. Baby food manufacturers, under public pressure, stopped using MSG in their products. A subsequent investigation by the National Academy of Sciences reported MSG to be a safe food additive for adults but not necessary for babies. However, some scientists of the NAS committee itself have since been charged with "conflict of interest" because they had received previous support from the food additives industry. The debate over the safety of MSG continues.

Background

Glutamic acid, an amino acid present in most proteins, is a normal constituent of a host of natural foods. Certain foods that are rich in glutamic acid, such as seaweed and soy sauce, have been used in the Orient for centuries as seasoning. Although glutamic acid had been studied chemically since its discovery in 1886, Dr. Kikunae Ikeda, a professor of physical chemistry at the Imperial University of Tokyo first recognized its value as a seasoning in 1908. Professor Ikeda found that the meatlike flavor imparted to

94

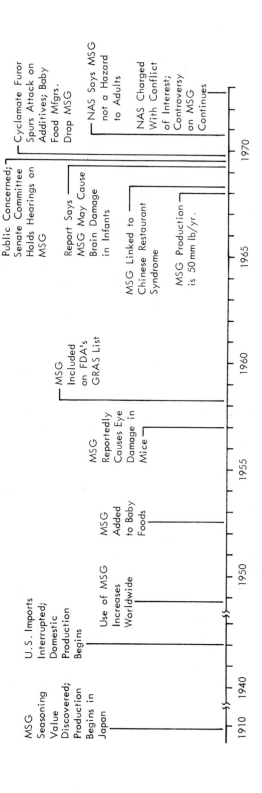

MSG Seasoning Value Discovered; Production Begins in Japan

U.S. Imports Interrupted; Domestic Production Begins

Use of MSG Increases Worldwide

MSG Added to Baby Foods

MSG Reportedly Causes Eye Damage in Mice

MSG Included on FDA's GRAS List

Public Concerned; Senate Committee Holds Hearings on MSG

Report Says MSG May Cause Brain Damage in Infants

MSG Linked to Chinese Restaurant Syndrome

MSG Production is 50 mm lb/yr.

Cyclamate Furor Spurs Attack on Additives; Baby Food Mfgrs. Drop MSG

NAS Says MSG not a Hazard to Adults

NAS Charged With Conflict of Interest; Controversy on MSG Continues

1910   1940   1950   1955   1960   1965   1970

soup stocks made from a seaweed was due to its content of glutamic acid.

The glutamic acid is conveniently handled and used as the sodium salt, monosodium glutamate hydrate, or MSG.* Industrial production of MSG was begun by the Japanese firm of Suzuki and Company, Ltd. immediately following the discovery that it was a flavorful seasoning. Its use as a non-nutritive flavor enhancer has steadily increased for sixty years.[1] After seventeen years of sales experience, Suzuki attempted to establish a joint venture with an American manufacturer for its production in the United States, but failed at that time. Production of MSG in Japan peaked in 1938, because of the Sino-Japanese conflict and came to a halt during World War II.[1] The use of MSG, which had been much greater in Japan and the Far East than elsewhere before World War II, expanded rapidly after the war as the demand increased throughout the world. The Japanese MSG industry was reestablished immediately after the war, and MSG exports contributed to the rehabilitation of the nation's economy.[1]

In the United States some Japanese MSG had been imported prior to World War II, for use in the canning industry, and the outbreak of the war stimulated a domestic MSG industry. The technology of the previous Japanese-American venture was utilized, and the enterprise was merged into the International Minerals and Chemical Corporation.[1] Manufacturing methods for MSG depend on the isolation of glutamic acid from natural sources, either from proteins containing glutamic acid as peptides or from beets.[1] Vegetable proteins (such as those found in soybeans, peanuts and corn) have frequently been adapted as raw materials because of their high content of glutamic acid and low cost. The leading domestic producer, International Minerals and Chemicals Corporation, manufactured MSG from sugar beets. IMC was credited with 85 percent of the bulk and retail sales of MSG—a 40 million pound per year production item in the United States by the mid-1960's. The company aggressively marketed its retail product, "Accent," with a television budget said to be $3.44 million per year.[2]

MSG was used in the United States as a "flavor enhancer" in thousands of common packaged food products, including nearly all canned and dried soups, frozen dinners, meat products, sauces, salad dressings, and seasonings. Because MSG was included in the Food and Drug Administration's list of "Generally Recognized As

*Chemical formula, $NaC_5H_8NO_4 \cdot H_2O$.

Safe" (GRAS) food additives in 1958, its presence in retail foods did not have to be noted on the label.* No limit existed on how much could be added, but processed foods usually had less than 0.5 percent MSG.

Key Events and Roles

In early 1968, a mysterious malady labeled "Chinese Restaurant Syndrome," or "CRS" was described in the prestigious *New England Journal of Medicine*[3] and other papers. The malady, according to Dr. Robert Ho Man Kwok, an American-born Chinese at the Food Nutrition Department of Harvard University, consisted of a combination of symptons such as headaches, burning or tingling sensations, facial pressure, chest pains, dizziness, and fainting; it seemed to strike people who had just dined in Chinese restaurants. The story was of mild public interest,[4] but when MSG was implicated[5] as the cause of "CRS", it received front page coverage in the *Wall Street Journal*.[6] The MSG, said to be used generously by Chinese cooks, was shown[7] to exert a "dose-effect" relationship, but to have a variable threshold dose among individuals—as little as 5 grams is adequate to produce a response in susceptible individuals, but 25 grams would not in others.[5,7] Published in February 1969, results of the study[7] that confirmed the "CRS"-MSG link was also widely reported in the press,[8] and the link has continued to be of popular interest[9,10] and some controversy.† But the real public concern over MSG was just about to begin.

In May 1969, Dr. John W. Olney reported[11] that after MSG had been injected subcutaneously into newborn mice, brain lesions, marked obesity, and female sterility resulted. Olney, who noted that eye damage had been observed in similar studies as early as 1957, used rather massive single injection doses, ranging from 0.5 to 4 milligrams per gram of body weight‡ (comparable to a total of 1.5 to 12 grams in a newborn human infant). His methods, results, and interpretations led to a long series of critical comments and

*Subsequently, mayonnaise, French dressing, and salad dressing were the only exclusions.

†For example, two Italian scientists have theorized that the epidemic of "CRS" cases was a result of autosuggestion on the basis of experimental studies.[9] In addition, only patrons of North American Chinese restaurants were affected, although MSG was widely used in all western countries and Japan.[10]

‡Milligrams MSG per gram body weight of the mouse. By comparison, oral ingestion of aspirin at 4 milligrams per gram of body weight would probably cause death.

further studies with mice, rats, monkeys, gerbils, and man, to try to assess the hazards of MSG.[12-19] But Olney's initial data indicated that it was the still-developing nervous system of the newborn that was affected by MSG; when it was pointed out that prepared baby foods (including those of the Gerber Products Company, the H. J. Heinz Company, and the Beech-Nut Company) contained unspecified amounts of MSG, a public furor erupted.

In July 1969, the Senate Select Committee on Nutrition and Human Needs, under the chairmanship of Senator George S. McGovern (D—South Dakota), opened hearings on MSG. The first witness to testify in the well-reported[20-23] hearings was Ralph Nader, who was followed by Dr. Olney and other scientists, including spokesmen for the American Medical Association (AMA) and the FDA, and finally spokesmen for the baby food manufacturers. Nader questioned not only the safety of MSG in baby foods, but also the high level of salt that was added to please the mother's taste.* Olney testified that MSG is potentially harmful and nutritionally unnecessary. The AMA spokesman indicated that seasoned baby foods were not a significant health hazard, while baby food and MSG manufacturers insisted that no scientific evidence existed that showed that MSG was harmful to humans in the amounts normally consumed. The FDA spokesman, Commissioner Herbert L. Ley, Jr., made the strongest impression on the committee, however. In his July 22 testimony, his "position paper" cited four FDA studies which he said demonstrated that MSG was safe, but almost immediately Ralph Nader charged that only preliminary results were available from two of the tests and that the other two had never been performed. Dr. Ley did admit that one of the tests had not been started until after his testimony and attempted to explain the other discrepancies, while conceding that the FDA had made an "inexcusable error."[25,26] The result was that the FDA's case was badly undermined.[25]

The Senate committee requested that FDA restudy the MSG question and on October 22,[27] just three days after the cyclamate ban, and amid speculation that the FDA itself would be investigated over the GRAS list, the FDA announced that it would restudy MSG. This announcement (which came just a few days after publication of a study indicating brain damage by MSG in an infant monkey[13]) created a new wave of public concern.[28-30] The baby

---

*The MSG was also added to please the mother's taste, since babies' taste buds are too underdeveloped to distinguish a meat flavor.

food companies still maintained that their products were safe, even though it had been revealed in July that the chief of one research laboratory had recommended to his company that they drop MSG. On October 23, *The New York Times*[29] gave front-page coverage to pleas by Dr. Olney and Dr. Jean Mayer (a Harvard researcher and President Nixon's advisor on nutrition) that MSG be removed from baby foods. The following day Gerber, Heinz, and Beech-Nut all announced that they would voluntarily suspend use of MSG in baby foods pending the results of further safety studies, while maintaining that the inconclusive data which had been presented had unnecessarily alarmed and confused the public.[30]

Disposition

While the baby food battle was temporarily resolved, concern continued into 1970[31-32] over the use of MSG in other foods and in those Chinese restaurants, which were using about 5 grams MSG per meal according to one report.[31] In January 1970, the National Academy of Sciences—National Research Council (NAS-NRC) began a study of MSG safety (under a $13,500 FDA contract[32]), and in March the New York State Assembly banned the sale of baby food with MSG in that state. By this time the annual U.S. production of MSG was estimated at 50 to 60 million pounds and imports added another 7 to 8 million pounds.[35,36] At a bulk price of 42 cents per pound, MSG was about a $25 million a year business.[36]

In August 1970, the NAS committee announced the results of its MSG study: theflavor enhancer carried no hazard for adults and the risk for babies was extremely small, but its use in baby foods should be discontinued because it did not confer any benefits.[38] The committee recommended that its use in all foods be indicated on the label. The report received relatively little attention in the popular press, but as the often-conflicting experimental results of studies on MSG continued to appear in the technical literature[15-19] the composition of the NAS-NRC committee itself was questioned. Charges that the committee contained too many scientists with conflicts of interest, because of their food industry relations or research support, were raised by Dr. Olney and became of interest in September 1972 to Senators Charles H. Percy and Gaylord Nelson of the Senate Select Committee on Nutrition and Human Needs.[39] The method by which NAS picks its committee members has not been made clear yet, but the possibility that an industry-dominated committee had conducted a "whitewash" as charged, suggests that

the entire NAS scientific advisory panel system will be investigated. The FDA had already decided to utilize the Federation of
American Societies for Experimental Biology, rather than the NAS
for further advice on about 600 items on the GRAS list.[39] In the
meantime, International Minerals and Chemical Corporation asked
FDA in mid-1971 to reaffirm the GRAS status of MSG on the basis
of new toxicity studies which IMC said showed the flavor enhancer
to pose no risk to human infants. In late 1971, IMC sold its Accent
division.

The controversy over the safety of MSG in baby foods inspired
at least three do-it-yourself books on the home preparation of baby
food.

Comment

The still incompletely resolved controversy over the safety
of MSG as a food additive illustrates the problem which society
has in trying to evaluate chemicals or other technologies that may
pose a very slight risk of serious damage to a large part of the populace. It illustrates also how much public concern can be generated
by one or two articulate individuals, particularly when the subject
has a strong human interest appeal to the media. Unlike the cyclamate case which was topical at the same time, MSG did not involve
a question of carcinogenicity. But the brain damage to infants was
potentially as bad as or worse than cancer in the public mind, and
the FDA's handling of the MSG case did little to inspire confidence
in either its administrators or its technical staff. Furthermore, the
suitability of the members selected for the NAS-NRC MSG evaluation committee was challenged and serious questions were raised
about NAS procedures and conclusions. This fact illustrates one
further problem: it is often difficult to get "instant expert" consultants versed in the details of a technical area who are not open
to charges of preconceived notions or conflicts of interest.

The MSG case is unique in that it was initially presented to the
public as something humorous rather than potentially dangerous
i.e., the Chinese Restaurant Syndrome. The original communication in the *New England Journal of Medicine* was more "cute" than
one usually finds in that prestigious journal and the mass media
eagerly picked up Dr. Ho Man Kwok and his new syndrome.

REFERENCES

1. "Monosodium Glutamate," *Encyclopedia of Chemical Technology,* Second Edition, Vol. 2, p. 198, Interstate Publishers, 1963.

2. "Accent—Drug on the Market," *Sales Manag,* 31 (September 1, 1968).

3. *New England J Med,* correspondence, April 4 and May 16, 1968.

4. *New York Times* (May 19, 1968).

5. (a) Schaumburg, H. H. and R. Byck, and (b) Ambos, M., N. R. Leavitt, L. Marmorek, and S. B. Wolschira, *New England J Med,* **279,** 105 (1968).

6. *Wall St J,* p. 1 (July 11, 1968).

7. "Monosodium L-Glutamate: Its Pharmacology and Role in the Chinese Restaurant Syndrome," *Science,* **163,** 826 (February 21, 1969).

8. *New York Times,* p. 58 (February 22, 1969).

9. Bernarde, Melvin A., *The Chemicals We Eat,* p. 131, American Heritage Press, New York, New York, 1971.

10. "Chinese Restaurant Syndrome," *Chem,* **42,** 4 (September 1969).

11. Olney, J. W., "Brain Lesions, Obesity, and Other Disturbances in Mice Treated with Monosodium Glutamate," *Science,* **164,** 719 (May 9, 1969).

12. "Monosodium Glutamate," letter by F. R. Blood, B. L. Oser and P. L. White, with reply by J. W. Olney, *Science,* **165,** 1028 (September 5, 1969).

13. Olney, J. W., and L. G. Sharpe, "Brain Lesions in an Infact Rhesus Monkey Treated with Monosodium Glutamate," *Science,* **166,** 386 (October 17, 1969).

14. "Monosodium Glutamate: Specific Brain Lesion Questioned," letters by C. U. Lowe and by M. R. Zavon with reply by J. W. Olney and L. G. Sharpe, *Science,* **167,** 1016 (February 13, 1970).

15. Adams, N. J., and A. Ratner, "Monosodium Glutamate: Lack of Effects on Brain and Reproductive Function in Rats," *Science,* **169,** 673 (August 14, 1970).

16. Bazzano, G., J. A. D'Elia and R. E. Olson, "Monosodium Glutamate: Feeding of Large Amounts in Man and Gerbils," *Science,* **170,** 1208 (September 18, 1970).

17. Arees, E. A., and J. Mayer, "Monosodium Glutamate-Induced Brain Lesions: Electron Microscopic Examination," *Science,* **170,** 549 (October 30, 1970).

18. "Monosodium Glutamate Effects," letter by J. W. Olney with reply by N. J. Adams and A. Ratner, *Science,* **172,** 293 (April 16, 1971).

19. Reynolds, W. A., N. Lemky-Johnston, L. J. Filer, Jr., and R. M. Pitkin, "Monosodium Glutamate: Absence of Hypothalmic Lesions After Ingestion by Newborn Primates," *Science,* **172,** 1342 (June 25, 1971).

20. *New York Times,* p. 51 (July 16, 1969).

21. *New York Times,* p. 20 (July 18, 1969).

22. *New York Times,* p. 26 (July 19, 1969).

23. *New York Times,* p. 11 (July 29, 1969).

24. "MSG Use in Baby Food is Defended," *Oil Paint & Drug Rep,* 5 (August 4, 1969).

25. "Salting the Pure Food Mine," *Sci N,* **96,** 295 (October 4, 1969).

26. Turner, J. S., *The Chemical Feast,* Grossman Publishers, New York, pp. 88–95 and 193–197, 1970.

27. *New York Times,* p. 36 (October 23, 1969).

28. "Baby Food Additives May be Next to Go," *Bsns W,* 114 (October 25, 1969).

29. *New York Times,* p. 1 (October 24, 1969).

30. *New York Times,* p. 1 (October 25, 1969).

31. "Chinese are in a Stew over MSG," *Chem W,* **105,** 19 (December 10, 1969).

32. "FDA Orders Investigation Into MSG Safety in Foods," *Oil Paint & Drug Rep,* 3 (January 19, 1970).

33. "The MSG Controversy," *Vend,* **24,** 37 (February 1, 1970).

34. "The Facts About Monosodium Glutamate," *Good H,* **170,** 162 (February 1970).

35. "Monosodium Glutamate," *Consumer Bul,* **53,** 16 (March 1970).

36. "Imports Hurting MSG," *Chem W,* **107,** 28 (July 15, 1970).

37. *New York Times,* p. 37 (March 24, 1970).

38. "MSG in Baby Foods: Risk Extremely Small," *Oil Paint & Drug Rep,* 3 (August 10, 1970).

39. "Academy Food Committee: New Criticism of Industry Ties," *Science,* **177,** 1172 (September 29, 1972).

Botulism and Bon Vivant

Abstract

Between July and November 1971, three widely publicized reports of botulin poisoning incidents in the United States created public alarm about the safety of canned foods and focused attention on the food canning industry and methods for preventing food poisoning. Following one death and a serious illness linked to botulin poisoning in one of its products, Bon Vivant Soups, Inc., was forced into bankruptcy in July 1971. Company operations were resumed under federal guidance in November 1972.

Background

Botulism is a rare type of food poisoning caused by ingestion of foods containing a potent neurotoxin which is formed during the growth of a commonly occurring microorganism—a rod-shaped bacterium called *Clostridium botulinum*.[1] This organism is widely distributed in nature and occurs both in cultivated and forest soils, bottom sediments of lakes, streams and coastal waters, the intestinal tracts of fish and mammals, and the gills and viscera of shellfish.[2]

Botulism should not be confused with the two more common forms of food poisoning, staphylococcus and salmonella; these come from spoiled foods such as custards and mayonnaise, chicken and egg products. These forms of food poisoning occur far more frequently than botulism, but they are seldom fatal.[1]

Although botulism has been known since ancient times, only within the last two hundred years has there been sufficient knowledge to recognize it as a foodborne disease. Botulism was first accurately described in southern Germany in the early nineteenth century after a study of more than 200 cases of poisoning from

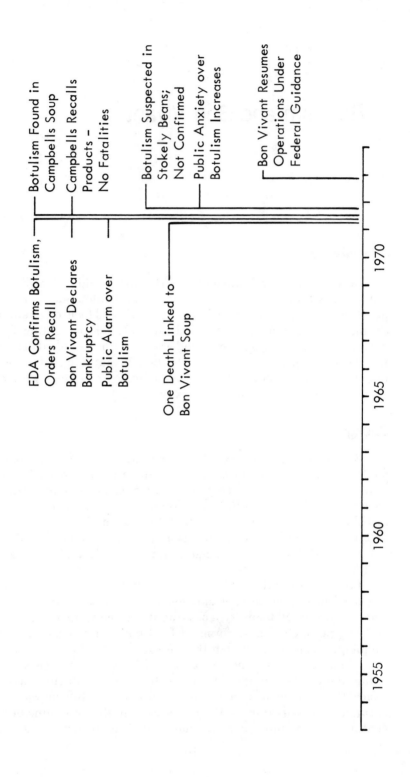

sausages. The term "botulism" (derived from the Latin word "botulus" for sausage) was first applied to this syndrome in 1870; the causative agent was first isolated and described by Emile P. Van Ermengem in 1896. He named the organism *Bacillus botulinus* (now *Clostridium botulinum*).

These bacteria spores, which are unavoidably present in raw fruits, vegetables, and other foods are themselves harmless, and after being ingested along with food they pass innocuously through the human digestive tract.[2] A poison is generated only when these spores start to germinate, and germination can only occur under anaerobic conditions (that is, in the absence of free oxygen), such as in home-canned foodstuffs or commercially produced vacuum-sealed cans or plastic packages. During spore germination botulin toxin, the deadliest poison known to man, is produced; this poison was one of the seven agents revealed in July 1971 to be in the U.S. Army's arsenal of biological weapons at Pine Bluff, Arkansas.[1] Because this poison is produced by the growth of the organism in the food material before it is consumed, botulism is a disease of intoxication and not strictly a foodborne infection.[2]

After being introduced into the body, the poison enters the bloodstream and then attacks the nervous system. Symptoms of botulism poisoning include double vision, difficulty in speaking and swallowing, and muscle weakness.[3] Blurred vision usually occurs eighteen to thirty-six hours after tainted food has been eaten and is followed in turn by the other symptoms. Death, when it occurs, is usually caused by paralysis of the respiratory system or by pneumonia.

From 1899 to 1969, there were 1,696 recorded cases of botulism—which resulted in 959 deaths.[3] Although fatalities have been on the decline since the development of antitoxins in the 1940's, the mortality rate for the 1960's was still 23 percent. During the period 1960–1971 there were about forty-eight fatal poisoning cases in the United States attributed to botulism.[3]

The technical literature[2] indicates that botulin toxin has been found in a variety of foods, such as canned corn, peppers, green beans, beets, asparagus, mushrooms, ripe olives, spinach, tuna fish, chicken and chicken livers, luncheon meats, ham, sausage, lobster and smoked fish. Foods involved in outbreaks of botulism in the United States between 1899 and 1969 include: vegetables—395 outbreaks of which 362 involved home-preserved foods; meat—forty-four outbreaks with thirty-six home-preserved; milk and milk products—seven outbreaks, five home-preserved; fish and seafood—forty-eight outbreaks, thirty-three home-preserved; and fruit and pickles—thirty-five outbreaks with thirty-four home-preserved.[2] Thus, it is apparent that most of the outbreaks can be traced to home-preserved foods, and therefore could *affect only a few people* in each instance.

Food technologists are convinced that botulism poisoning should never occur in commercially canned* foods.[4] The cause of botulism is usually improper food processing methods. When adequate commercial cooking methods and equipment are used, botulism is easily preventable. The food industry employs extensive heating techniques to kill the *botulinum* spores, and consequently commercial products are rarely contaminated. Since 1950, for example, Americans have used some 500 billion cans of food; during that time there have been only three incidents of botulism— resulting in four deaths—from U.S. commercially canned foods.[5]

Key Events and Roles

On June 30, 1971, a New York banker, Samuel Cochran, died from what was diagnosed as botulism poisoning twenty-eight hours after eating Bon Vivant vichyssoise soup for dinner.[3] His wife also became critically ill after eating the soup, but her life was saved by an antitoxin rushed from an out-of-town laboratory.[6]

The incident was widely publicized by the news media and created much public alarm. *The New York Times,* for example, had twenty-seven news entries on the subject during July 1971 and several articles on botulism appeared during the same month in popular magazines.

As the alarm went out, the cannery began an intensive nationwide search for the remaining 6,444 cans of vichyssoise from the suspect production lot.[5] Shortly thereafter the FDA ordered a wider recall for *all* of the company's products—including those distributed under other labels. The recall task was complicated because Bon Vivant had manufactured four million cans of food (mostly soup) that year not only under its own name, but also under thirty-four other labels under contracts from large stores in Chicago, New York, Boston, and other cities.[7] The recalled samples were tested at FDA laboratories around the country—with technicians working extended shifts to test as many cans as possible for the toxin. The test procedure involved preparing a solution of filtered soup from each

*The tin-plated can for preserving food was patented in 1810 in Britain.

can for injection into mice. Samples which killed mice were contaminated with botulin toxin.[3] Of the first 324 cans of vichyssoise recalled and tested by the FDA in July 1971, five were found to be contaminated with deadly botulin.[6,7,9] Four contaminated cans of soup had been recalled from one store in the Bronx, New York.[3] This contamination was attributed to improper processing at the Bon Vivant cannery in Newark, New Jersey.[7]

Spot checks of other products in the plant showed that up to 2.8 percent of the cans were either deformed or abnormal.[7] One FDA official described the situation as "too many defective cans in too many other products."

Realizing that Bon Vivant might still be making and shipping deadly soup, the FDA had promptly dispatched a man to the factory with orders to halt operations.[3] Although the FDA tries to inspect all food plants once each year, officials discovered that this cannery had not been inspected by the FDA since May 1967, or by the State Health Department since October 1966. The USDA, however, regularly inspected meat products made in the plant. An FDA inspection of the Bon Vivant plant immediately following the news of the poisoning revealed flaws in the canning practices that had apparently allowed *Clostridium botulinum* to grow in some products.[3]

The company was a "family" business (40 employees in 1971) which had operated for 108 years.[3,8] Prior to this poisoning case, the company had never encountered any serious problems. What disturbed the owners most was FDA's statement about the company's "totally inadequate and incorrect manufacturing and record-keeping procedures."[8] The owners were convinced, and hoped to prove in court, that the tragic error was confined to five cans in one batch (V-141, prepared May 21, 1971), and that the government showed undue, unjust haste in condemning all their products.

As a result of the massive recall and publicity Bon Vivant, Inc., declared bankruptcy on July 27, 1971.[9] Information about the company's products remained newsworthy months after the official closing of the plant under bankruptcy. On September 16, 1971, a team of federal investigators declared that none of the company's products "were safe for consumption by man or animal."[10]

An interesting incident related to the poisoning case occurred in Kansas City, Missouri, during the summer of 1972. On July 29, 1972, it was reported that more than 450 cans of allegedly contaminated soup had been stolen from a warehouse of a grocery store chain.[11] According to the regional FDA director, the cans of soup, all bearing the chain's label had been held since October 1971 under federal seizure in the building. The soup had not been destroyed because a claim had been filed by the chain denying the allegation made by the FDA. The cans of soup were discovered to be missing when a U.S. marshal was sent to the warehouse, which was already being razed, to remove the soup to another location. FDA immediately warned the public not to eat any of this soup that might come into their possession.[11]

On July 31, 1972, an assistant U.S. attorney stated that eight FDA inspectors, several FBI agents, three U.S. marshals and the U.S. attorney's investigative team in Kansas City would be detailed to find the missing soup cans.[12]

Concern over the disappearance of the soup cans increased in August 1972 following a newspaper report, that some of the missing cans were being passed among Kansas Citians who were unaware that their use might cause serious illness or death.[13] Furthermore, some of the soup had turned up as far away as California. The FDA reported that some of the missing cans had been donated to Crosslines, a church organization concerned with the social welfare of the poor. Kansas City families were asked to look for soup cans with the suspect label.

Following these alarming reports of missing soup cans which might be poisoned, no reports of any kind were noted in the local newspapers concerning any illness or death which could be linked to the soup. The available information indicated that the soup cans had not been contaminated.

On August 22, 1971, The Campbell Soup Company reported that a few cans of chicken vegetable soup prepared by a company plant was contaminated with deadly botulin, and that it was recalling its soup from 16 southern and southwestern states and several Caribbean-area countries.[14] This incident was the first known occurrence of botulin in Campbell's 102 years of operation —but it was the second such discovery in the food industry in as many months. Fortunately in this case, the company beat the consumer to a can which had been packed on July 15, 1971. After passing routine taste and laboratory tests, it had been stored along with 230,000 other cans for a fourteen-day incubation period. During that period, reports of flavor problems in at least two earlier days' production had caused salesmen to send in samples and then to begin recalling those cans. Similarly, a few July 15 cans were found to be tainted. Again, salesmen started picking off samples from dealer's shelves and shipping them to the plant. The badly swollen can containing botulin was found by this procedure.

Following this discovery of contaminated soup, the company

sent many salesmen into sixteen states to collect all chicken vegetable soup—especially from the July 15 production lot.[14] Warnings were spread by all communications channels to every newspaper. Within a few days, more than two-thirds (about 154,000 cans) of the July 15 soup batch had been recovered. Meanwhile, there were no reports of illness, and continued testing by Campbell failed to identify any more poisoned cans.

The soup company, which had cooperated closely with government officials, refused to speculate on how the incident occurred.[14] The president stated that "we thought we were fail-safe—we have to become more fail-safe." The director of corporate services reported that "This is the first occurrence in over 100 billion cans, and we found it because of our own diligence."

The advent of this second soup episode caused a great deal of publicity and during the summer of 1971, the rare poison called botulin suddenly became a household fear.[8] As a result of these incidents, both the factory methods and the federal inspection procedures were being questioned.

Still another botulism scare developed in October 1971, when a case of possible food poisoning of a Marine captain and his son was reported by the news media. A suspicion of botulism in canned green beans led to a recall of 15,000 cans—the FDA issued an urgent warning to consumers who might have the canned beans on their shelves.[15,16] Thus a third company was gravely concerned over the alleged possibility of botulism contamination in their canned products.

On October 31, 1971, a newspaper article[17] stated that an official Navy news release had reported that the father and son were completely well and that "no clinical, bacteriological or laboratory evidence exists to support a diagnosis of botulism." A spokesman for the cannery said that his company wanted the FDA to admit officially there was no botulism in their product, so that his company could immediately stop the recall. The spokesman said, "It has probably cost us millions of dollars on the basis of the insult that has happened to our product; for a hundred years we've sold [our] products completely free of any question of wholesomeness." Subsequently, the FDA quietly rescinded this recall, but apparently never officially acknowledged that it was a false alarm.

Disposition

As a result of these poisoning incidents, food companies tightened production control and self-inspection procedures to provide

additional safeguards for the prevention of botulism poisoning. In
hearings in the U.S. House of Representatives in August and September 1971, the actions of FDA, USDA, Bon Vivant, and Campbell
Soup were extensively reviewed.[19]

The news media reported[18] on November 14, 1972, that the
cannery that originated the scare had resumed operations under
federal guidance. An FDA statement said the company now has "all
information necessary to operate in strict conformance with good
manufacturing practices" that will prevent further botulism poisoning. The company, which has changed its name to Moore and Company, Inc., estimated that its products would be on consumer shelves
by the end of 1972. The owner of the firm said the new operations
will be under the FDA, which "feels our opening again was an
excellent idea."

Apparently the siege of poisoning scares had ended, but the
public would remember the alarm and anxiety which had been
created by these episodes.

Comment

Botulism has all the qualities needed to inspire fear; it is a rare,
deadly, and usually undetected poison when present in canned
foods. When cases of botulism poisoning occurred in the past, they
usually arose from consumption of individually contaminated containers of home-preserved foods, and have thus involved only a few
persons at a time. Commercially canned foods have rarely been
contaminated with botulism poison; if they had been, the modern
technology of mass production, distribution and marketing of foods
might have produced poisoning in thousands of people at a time.
Thus, when a single death was traced to botulism-contaminated
soup the alarm was immediate and of nationwide proportions; the
alarm was made even greater by the FDA's recall of all products
carrying the company's label and the revelation that the company
also made products under thirty-four other labels. The effects of
such an event were sufficient to send a small company into bankruptcy. Reinforced by the lesser scares over botulism in products of
two large firms, the episode created serious suspicions of the food
industry by the public. It may also have created strong and effective
sentiments for more stringent regulation of food technology, and
certainly it inspired the food industry to redouble its efforts to make
a good safety record even better.

References

1. "This is Botulism—and a List of Simple Precautions to Follow," *Life*, **71,** 30 (September 10, 1971).

2. "Botulism" *FDA Papers*, 16 (November 1971).

3. "When Americans are a Swallow Away from Death," *Today's Health*, **49,** 40 (September 1971).

4. "Botulism: An Unnecessary Menace," *Consumer Rep.*, **36,** 540 (September 1971).

5. "How Much Danger in Canned Foods," *Read. Digest*, **100,** 49 (June 1972).

6. "Death in Cans," *Time*, **98,** 36 (July 19, 1971).

7. "The Deadly Soup," *Newsweek*, **78,** 67 (July 19, 1971).

8. "The Canned Menace Called Botulism," *Life*, **71,** 26 (September 10, 1971).

9. *New York Times*, p. 4 (July 27, 1971).

10. *New York Times*, p. 36 (September 16, 1971).

11. *Kansas City Star*, p. 1 (July 29, 1972).

12. *Kansas City Star* (July 31, 1972).

13. *Kansas City Star* (August 1, 1972).

14. "Double Trouble at Campbell Soup," *Bsns. W.*, 26 (August 28, 1971).

15. *Kansas City Star*, p. 2 (October 30, 1971).

16. *New York Times*, p. 34 (October 30, 1971).

17. *Kansas City Star*, p. 16A (October 31, 1971).

18. "Food Oversight—Food Inspection," hearings before the Subcommittee on Public Health and Environment of the Committee on Interstate and Foreign Commerce, U.S. House of Representatives, on "Oversight of Food Inspection Activities of the Federal Government." August 3, 4, and September 10, 13, 14, 1971.

19. *Kansas City Times*, p. 3B (November 14, 1972).

The Fish Protein Concentrate Issue

Abstract

In 1962 the Food and Drug Administration banned the U.S. sale of fish protein concentrate (FPC), a highly nutritious and low-cost food supplement which is prepared from whole fish and which the United States had been exporting for seven years. Following extensive development and promotional efforts by the Department of the Interior, the FDA reversed its earlier ruling in 1967 and approved fish flour as an additive for human consumption in the United States. Efforts to date by private industry to meet FDA specifications on this product have not proved profitable.

Background

Half of the world's people are undernourished, and their most crippling deficiency is in protein, the basic building block of the human body. The absence of protein in the diet can cause mental retardation, stunted growth, and early death. In the long search for cheap and abundant new sources of food, fish have frequently been hailed as one of the more hopeful possibilities.[1] Traditional methods of handling and preparing individual fish were not well suited, however, to mass production methods, and attention turned to methods of producing a refined version of fish meal, which has long been produced in large quantities from whole fish for animal feed.[2]

By 1950, the VioBin Corporation of Monticello, Illinois, had developed a process to manufacture such a product for human consumption. It was a fine grayish power (a fish flour) prepared by grinding up whole fish and then extracting the fats and water from the proteinaceous materials with a solvent (isopropyl alcohol). This product was almost completely tasteless and odorless, and had a high content of animal proteins; it was called fish protein concen-

112

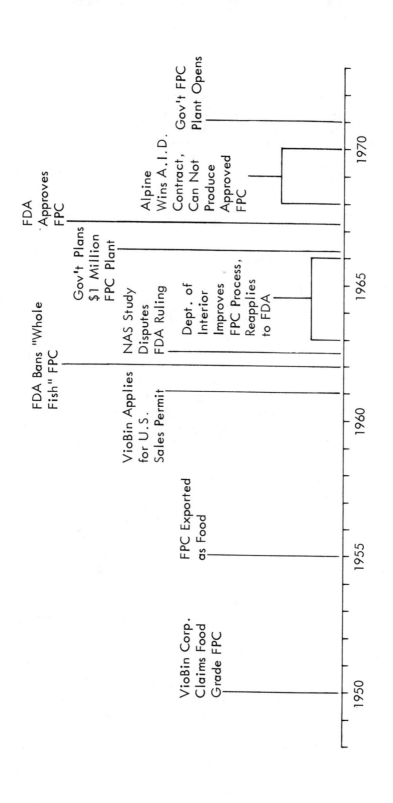

VioBin Corp. Claims Food Grade FPC

FPC Exported as Food

VioBin Applies for U.S. Sales Permit

FDA Bans "Whole Fish" FPC

NAS Study Disputes FDA Ruling

Gov't Plans $1 Million FPC Plant

Dept. of Interior Improves FPC Process, Reapplies to FDA

FDA Approves FPC

Alpine Wins A.I.D. Contract, Can Not Produce Approved FPC

Gov't FPC Plant Opens

1950   1955   1960   1965   1970

trate (FPC). Government scientists have estimated that 10 g of FPC a day could satisfy the animal protein needs of the average human being.[2] The FPC was stable, did not require refrigeration, and was inoffensive as an additive to a variety of foods. In addition, the FPC could be made cheaply from fish such as hake that ordinarily do not find their way to the dinner table. Furthermore, a general belief existed at that time that the world's supply of fish was inexhaustible. For these reasons, FPC was hailed as a possible miracle solution to the world's nutrition problems.

Key Events and Roles

VioBin Corporation started exporting fish flour (FPC) in 1955.[1] In 1961 the company sought FDA approval of its fish flour for domestic distribution. Although VioBin expected only a modest market in the U.S., where protein-deficient diets are not a major problem, such approval promised to help convince skeptical overseas purchasers of their product's efficacy. In contrast, the FDA in January 1962, banned the use of whole fish flour in the United States.[3] The FDA rule that no matter how well the FPC might be sterilized in processing, it was considered unacceptable because it included every part of the fish—head, tail, guts, and all. The FDA did approve flour made from cleaned fish, but that product was not economically attractive.

The FDA's action had many repercussions. In particular, foreign countries buying FPC felt the United States was sending them a product unfit for human consumption. Shortly after the FDA's 1962 ruling, at the request of the Interior Department, A National Academy of Sciences study was begun. The NAS study[4] concluded that fish flour did not deserve FDA's strongly worded disapproval, but noted that more research on fish flour was needed to control the quality and the solvent residues from the production process.

The Bureau of Commercial Fisheries in the Interior Department then undertook an extensive research program on FPC in its laboratories.[4] Preliminary results indicated that a high protein concentrate could be made from whole fish and still be pure.[5] The research was then accelerated at a pilot plant using Atlantic hake as raw material; the Bureau product contained about 80 percent protein and 20 percent of beneficial minerals. A sustaining NAS committee acted to advise the Bureau on its research work. Marine biologists estimated that the unharvested fish in U.S. coastal waters alone, if converted into fish flour, would supply enough animal protein to supplement the deficient diets of about 1 billion people for three hundred days at a cost of 0.5 cents per person per day.[4]

In December 1965, the Marine Protein Resource Development Committee of NAS reported the results of a review of three years of research by the Bureau of Commercial Fisheries on FPC.[4] The report concluded that "fish protein concentrate, from whole hake, as prepared by the Bureau's process is safe, nutritious, wholesome and fit for human consumption." An application was filed, by the Department of the Interior with the FDA in March 1966 to obtain certification of the Bureau FPC product as safe for human consumption.

Before the FDA decision, however, Senators Warren Magnuson of Washington and E. L. Bartlett of Alaska,[5] introduced bills to provide for further research on fish flour, and in November 1966 President Johnson signed into law a marine protein concentrate act which established a five-year research program and provided funds for construction of at least one $1 million plant.* The Secretary of Interior was authorized to lease a second plant and was provided money for operations and research at both plants. The primary objective was to develop the most economical processes for reducing the kinds of fish that are in abundant supply—but which are not widely sought as human food—to a highly nutritious and stable marine protein concentrate. The research was also aimed at finding ways to use such protein concentrate to fortify many foods without changing their taste or texture.

FDA's major objection at this time to the FPC developed by the Bureau of Commercial Fisheries was that it contained too much fluoride—enough to cause mottling of the teeth if a child were to consume large quantities of the product over a long period of time.[6] In February 1967, however, the FDA approved for human consumption FPC produced by both the method developed by the Bureau of Commercial Fisheries and by the VioBin Corporation.[7] The FDA categorized FPC as a *food additive*. According to FDA's regulations governing production, however, fish bone would either be cut out or removed by screening to minimize the fluoride content of the FPC. According to this ruling,[1] the only fish that could be used was hake and related species; the product had to contain 75 percent protein and not more than 100 parts per million of fluoride.

---

*Government activity in marine protein concentrate was the basis for controversy: private commercial plants which produced fish meal, had been operated successfully for several years, and were interested in the FPC market.[5].

The FDA approval was less than whole hearted, however; FDA barred use of FPC by food processors as an enrichment additive or in formulated foods unless the vendor proved that such use would not be deceptive to consumers.

In addition, sale of FPC in the United States was restricted to one-pound packages,[1] a regulation which made the handling and use too expensive for domestic food manufacturers. Export of FPC in bulk was permitted, however, which again raised the old fears abroad that the United States was exporting food which we wouldn't consume at home.

Nevertheless, by the fall of 1968, development and testing of fish protein concentrate indicated sufficient promise that the U.S. Agency for International Development (AID) entered into a pioneering contract with Alpine Marine Protein Industries* under which the company was to produce 970 tons (at a contract price of 42 cents a pound) of concentrate to be distributed abroad for the poor in Chile and Biafra.[8–10]

Disposition

The high hopes for this pioneering project were shattered in January 1970, when AID terminated the contract because only seventy of the first 525 tons processed by Alpine met FDA standards for high enough levels of protein content.[8] FPC had again suffered a setback: the only commercial venture to make FDA-approved FPC in the United States had yet to prove itself. In 1971, the company closed its plant.[2]

In April 1971,[2] the National Marine Fisheries Service of the Commerce Department opened the government's first experimental $2 million plant in Aberdeen, Washington, to produce FPC. The plant is expected to grind up fifty tons of boned hake a day, which, minus fat and water, will produce about seven tons of FPC. The purpose of the plant will be to supply FPC to domestic companies interested in experimental marketing and to agencies for distribution abroad. The primary goal will be to furnish a demonstration of how to operate an FPC plant, in order to encourage private firms to get into the act.

*Alpine used one of Viobin's processing plants which had been in operation since 1959 and could convert 70 tons per day of fish into meal and by-product oil.

Comment

The discovery of a new, nutritious and low cost proteinaceous food in a food-starved world might be expected to receive an intensive development and rapid adoption. The fact that the United States did not approve FPC for sale at home, but permitted its export for consumption abroad, raised international questions of its efficacy. The FDA's conclusions regarding the wholesomeness of the FPC, based primarily on aesthetic considerations (because the whole fish is used), was disputed in part by an NAS study. These conclusions may be reversed in the near future as growing food shortages and rising prices bring political pressures to bear.

REFERENCES

1. "Protein For Everybody," *Time*, **89**, 67 (March 17, 1967).
2. "Fish Flour: Protein Supplement Has Yet to Fulfill Expectations," *Science*, **173**, 410 (July 30, 1971).
3. *New York Times*, p. 28 (January 25, 1962).
4. "Fish Flour: FDA Approval Likely on Improved Product," *Science*, **152**, 738 (May 6, 1966).
5. "Fishing For a World Market," *Chem W*, **99**, 79 (December 3, 1966).
6. "FPC: Nearing Approval?", *Sci N*, **90**, 474 (December 3, 1966).
7. "FPC Gets OK at Last," *Sci N*, **91**, 138 (February 11, 1967).
8. "Setback For a Supplement," *Sci N*, **97**, 90 (January 24, 1970).
9. "Fish Protein Progress," *Sci N*, **98**, 270 (September 1970).
10. *New York Times*, p. 29 (January 15, 1970).

The Fluoridation Controversy

Abstract

In 1944 the Public Health Service began a ten-year test program in two cities to determine whether the fluoridation of public water supplies (adjustment to one part fluoride ion per million parts water) was an effective method of providing improved dental health care. The early results looked so promising that in 1947–1950 many cities, first in Wisconsin and then in other states, started fluoridation. After 1950, an extensive controversy developed over the safety and legality of adding fluoride to public drinking water. The controversy continued long after the procedure was shown to be an effective and economical method of reducing the incidence of dental cavities and led to the defeat of referendums proposing fluoridation in many communities during the 1950's and 1960's. By the later 1960's, however, many of the large cities and a few states had adopted fluoridation. Over 72 million people were estimated to be drinking fluoridated water by the end of 1968 and others received fluoride in topical application to the teeth by their dentists or in fluoride toothpastes.

Background

Scientific research on fluoridated water was largely completed before World War II. Investigation into fluorides began in 1902, not in an effort to reduce dental decay, but to find the cause of stained or mottled teeth of people living in Colorado Springs, Colorado.[1] A young dentist there, Dr. Frederick C. McKay, became intrigued with the problem and devoted his off-hours to it for the next three decades. As he identified other areas of the country in which the tooth stain appeared, he was told by local residents that "something in the water" caused it. Fluorine is an unusual element and ordi-

118

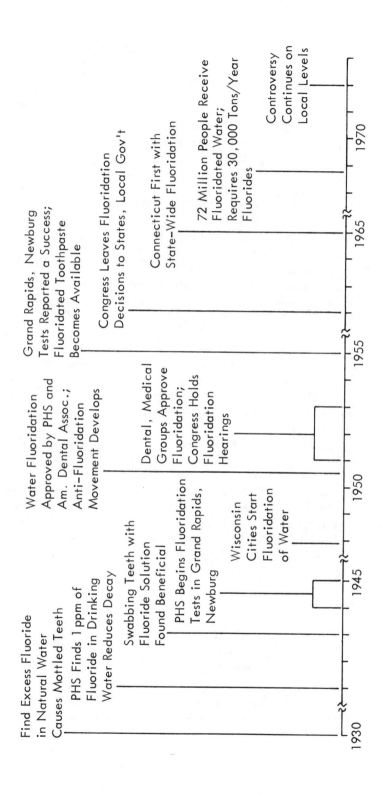

nary chemical analyses of that time did not look for it: dozens of analyses failed to identify fluoride ion in the water supplies in question.

The cause of mottled teeth was determined in 1930. Dr. McKay sent a sample of water from Bauxite, Arkansas, a mining town supplying aluminum ore, to H. V. Churchill, chief chemist of the Aluminum Company of America,* (ALCOA).[1] His laboratory conducted a very complete analysis and found a significant amount of fluoride. Analyses were then made of the water from other towns where many residents had mottled teeth; fluoride was found in every case. In 1930 also, a team of Arizona scientists conducted experiments in which they fed fluoridated water to rats in an effort to duplicate mottling; they succeeded and published their findings immediately after the McKay-Churchill paper.

Dr. McKay had first commented on the lack of decay in mottled teeth in 1908, and again in 1925 and 1926.[1] Other researchers noted the same correlation in the 1920's, but no one undertook serious study of this point. In the 1930's, Henry Dean, a dentist of the U.S. Public Health Service, completed the research on this last link. By 1941 Dean had collected a huge amount of data and was able to make a quantitative judgment: 1 part per million of fluoride ion in the water supply reduced tooth decay sharply without any presence of mottling.

Dr. John W. Knutson, a dental surgeon of the U.S. Public Health Service reported[2] in December 1943, that topically applied fluoride solutions (for example, swabbing teeth with a 2 percent solution of sodium fluoride) could reduce tooth decay of school children by about 40 percent. Actual experiments had proven this level of effectiveness as compared with a control group of children in the same schools. These studies indicated that fluoride treatment of teeth could prevent caries but could not arrest dental decay once it had started.

Key Events and Roles

In 1944, the U.S. Public Health Service (PHS) was ready to begin the experimental addition of fluoride to drinking water.[1] Grand Rapids, Michigan, began fluoridations that year (under a

*ALCOA's willingness to undertake this analysis was probably sparked[1] by charges from a local doctor that aluminum cooking ware was poisonous: unless some other cause could be found, Bauxite's plight might otherwise have served as an indictment of aluminum.

ten-year test program) with nearby Muskegon as the control city. The only issue was whether artificially administered fluoride acted in the same way as fluoride naturally present; except for dental health and mottling, no other differences had been found between persons drinking water with or without fluoride. A similar test was begun at Newburgh, New York, shortly thereafter, under the aegis of the State Department of Health.[3]

The Newburgh test called for adding one part of sodium fluoride to every one million parts of the city water supply. Kingston, a nearby city of similar size, was used as a control. The plan specified that 1,000 school children in both Kingston and Newburgh would have their teeth examined and charted at the start of the ten-year test. Thereafter, examinations were to be made each year of all the 3,500 children in each city. Periodic and final comparisons of the records were expected to determine the efficacy of the fluoride additions. Observers expected[4] that by the end of the test, dental decay rates in Newburgh would rate between those in Galesburg, Illinois (1.5 parts per million of natural fluoride in the water), where school children averaged only two and a half cavities each and those in Michigan City, Indiana (0.5 parts per million of natural fluoride), where the children had an average of ten cavities apiece.

At this point, a group of Wisconsin dentists became anxious to move from the experimental stage and begin mass fluoridation. Madison, Wisconsin, began fluoridation later in 1947 and was soon followed by other cities.[1] Many scientists disapproved of this impetuous action before the Grand Rapids and Newburgh experiments were completed. The PHS at first withheld its support for the Wisconsin fluoridation program, but approved it in 1950; the American Dental Association approval followed a year later. By that time, half of the cities in Wisconsin were already fluoridating their water. The National Research Council approved water fluoridation in 1951;[5] the International Dentist's Association and the American Medical Association followed suit in 1953.[6]

These organizations had intended to wait until the final results were in from the experimental cities, but the early results were so impressive that approval was given. Three years after the project began in Newburgh, New York, the PHS reported a one-third drop in tooth decay. Similar reports were made for other cities over the next few years. The final results in Newburgh showed a 59 percent decrease in dental defects. The data clearly indicated that mass fluoridation of drinking water in the United States could reduce

dental bills by hundreds of millions of dollars annually, and
fluoridation quickly excited public health professionals across the
country. In 1955 the news media reported[7] the successful results of
the Newburgh water fluoridation studies.

The innovation of fluoridating drinking water appeared to offer
a significant health benefit at modest cost. Evidence was available
that the addition of minute quantities of fluoride to community
water supplies(1 to 1.5 parts per million of fluorine) was safe and
decreased dental caries by as much as 65 percent.[8] Authorities estimated by 1955 that the expense of such fluoridation ranged from 9
cents to 12 cents per person per year for the average city. This
innovation was economical compared with the cost of having
fluoride applied directly to children's teeth by a dentist and was of
course much more economical (and less painful) than the cost of
filling dental cavities.

Despite these indications of great benefits, the fluoridation of
public water supplies was at the same time developing into one of
the most controversial issues of the 1950's. A national network of
antifluoridationists had come into being about 1950 and the issue
was soon rephrased in terms of the rhetoric of the era.[1] Some opponents referred to water fluoridation as another example of
socialized medicine; others as a communist conspiracy; and many
as a hazardous practice or medical treatment being forced on them.
"After all, fluoride is used as a rat poison," was a frequently heard
cry.

The Delaney committee of the U.S. House of Representatives
held hearings on fluoridation in 1952.[1] The committee's report (issued in August 1952) expressed concern about the possible toxic
effects of fluoridated water, particularly on the aged and chronically
ill and urged communities contemplating action to take a conserva-
tive attitude toward fluoridation. Dental authorities objected to this
premise and issued articles attacking the reservations of the committee. These authorities noted that, in effect, the Delaney committee report was a deferred minority report on the action of Congress
itself which had earlier approved an appropriation of funds for
fluoridation of the public water supplies in the District of
Columbia.*

The proponents of fluoridation were frequently successful on
the state level; many states passed enabling legislation permitting
local governments to adopt fluoridation. Local public health officers

*Representatives Delaney (N.Y.) and Miller (Md.) were opponents of fluoridation.

and local dental associations at the municipal level became strong proponents of fluoridation, while the PHS continued to carry out a national public relations campaign and to hold conferences on fluoridation.

The published data indicate that 1952 was the peak year for adoption of water fluoridation.[1] During that year, communities in most mid-western states, most border states, and two western states—a total of twelve states in all—adopted fluoridation. The following year, fluoridation spread into the Northeast, and then into the Far West and Deep South.

The opposition was getting organized, however, by the end of 1952, and when fluoridation was brought up for consideration in "unfriendly" territory, its chances of adoption were slim.[1] In 1954, a bill was introduced in the Congress to ban completely all water fluoridation, and new hearings were held. For the next eight years, the adoption of fluoridation slowed. In city after city the local dental society or health department would press for fluoridation only to find the opposition appearing to testify before the city council, demanding a referendum, distributing handbills, and generally conducting an effective political campaign. The number of adoptions dropped quickly while the number of referenda grew until it was equal to the number of adoptions. According to voting tabulations made in 1959, fluoridation had been rejected in 227 communities and approved in only 162.

Disposition

By 1960, nearly a decade after being approved by the nation's health authorities, fluoridation had been adopted by somewhat less than one-third of all cities with populations of 10,000 or more.[1] In 1962, the Congress via a House Rule subcommittee effectively decided that the fluoridation issue was a matter to be handled by lower levels of government; it agreed that water fluoridation prevents decay, but concluded that the issue should be decided by the states and local communities rather than the Federal government.[9]

The number of persons, sizes of cities, and water systems served by the major fluoridation agents as of December 1965, are shown in Table III.[10] The fluoride source was one of several metal fluoride salts (such as sodium fluoride) or a complex fluoride (such as fluorosilicic acid or sodium fluorosilicate). The liquid fluorosilicic acid was the agent of choice for large cities, and convenient sodium fluoride was used almost entirely by cities of less than

TABLE III

CHEMICAL AND FEEDER USED IN FLUORIDATING WATER SUPPLY SYSTEMS, BY RANGE OF POPULATION SERVED[a]

Population Size of Water Supply Systems	Number of Systems Fluoridating	Sodium Fluoride		Sodium Fluorosilicate		Fluorosilicic Acid	Ammonium Fluorosilicate		Calcium Fluoride	Other Adjusted Natural Fluoride, and Not Specified
		Dry	Solution	Dry	Solution	Solution	Dry	Solution	Solution	
1,000,000-over	5			1		4				
500,000-999,999	13	4		11	1	2				
250,000-499,999	17	5		10	2	1				
100,000-249,999	44	3	1	25	2	10				1
50,000-99,999	59		1	45		8		1		
25,000-49,999	137	20	8	74	10	18	1	1		5
10,000-24,999	319	35	26	185	9	48		1		15
5,000-9,999	324	43	42	160	12	51		1	1	14
2,500-4,999	300	24	83	114	11	62		2		4
1,000-2,499	323	17	136	57	12	82		1	1	17
Under 1,000	125		79	3	5	37				1
Not Specified	26			1		1				24
Total	1,692	151	376	686	64	324	1	7	2	81

[a] As of December 31, 1965.[10]

50,000 people. In 1965, Connecticut became the first state to pass legislation requiring fluoridation of all municipal water supplies,[11] and was soon followed by Illinois, Minnesota, and Michigan. Several other states followed suit within a few years. The number of people drinking fluoridated water approximately doubled during the 1960's as many big cities began fluoridating for the first time: an estimated 72 million people in nearly 2,000 communities in the U.S. received fluoridated water in 1968.[12] By 1969 water fluoridation was using an estimated 30,000 tons a year of the fluorides (approximately a $5 million market) and temporary shortages were developing.[13]

The organizations advocating fluoridation of drinking water had also grown over the years. In November 1969, the World Health Organization endorsed water fluoridation,[14] and fluoridation was being practiced in many countries of the world. Despite the vote of confidence from medical, dental, and scientific authorities in the United States and abroad, fluoridation continued during the 1960's to lose frequently in the ballot boxes. In fact nearly 150 communities discontinued fluoridation after once adopting it. In 1969 two bills were introduced in the Congress to bar use of funds of the Department of Health, Education and Welfare for fluoridation of public water supplies. The antifluoridation spokesmen widened to include well-known writers such as John Lear, science editor of the *Saturday Review* and James J. Kilpatrick, nationally syndicated columnist.

A new issue in the antifluoridation movement also evolved in 1970, spurred by environmentalist, Barry Commoner, and consumer advocate Ralph Nader. The thrust of this issue was that the average citizen's total fluoride exposure was unknown, that is, from the air (as industrial pollutants), from food (from the use of fluoride-containing fertilizers) and from drinking water (from farmland runoff and the municipal supplies), and that the possible toxic effects of such exposure were also unknown. In New York a bill was introduced in the state legislature in August 1971 calling for a moratorium on water fluoridation pending detailed and extensive environmental studies.[14]

In 1971 also, the Congress took up the fluoridation issue again via hearings on a bill pertaining to children's dental health, and the Congressional Research Service issued a lengthy review of the fluoridation issue in early 1972.[15] But by 1973 the issue appeared to be again waning as over 95 million Americans consumed water with either natural or added fluoride and nine states required fluoridation of city water supplies.

Comment

Few, if any, issues of our time have sparked such prolonged and heated controversy across the land as that of fluoridation of water as a method of providing dental care. It has been the subject of a dozen books, of probably more public opinion polls than any other technology-related topic (see Refs. 16–18), and of serious sociological[19] and legal[20] studies, to say nothing of the hundreds of referendums and dozens of hearings. The long controversy probably arose in part from the rush with which it was adopted before tests were complete and in part from the suspicions of public officials (particularly in the early 1950's) and scientists. Once aroused, the ideologically-motivated opponents pursued their goals tirelessly, while the public health and medical community was largely united in support of fluoridation, and the public was frequently confused. On the local level especially, the scientific merits of the argument had little significance as emotional themes often predominated.

Two very recent developments are worth noting. Recent experiments indicate that fluoride is an essential element in the diet—at least in mice, which suffer greatly reduced fertility when placed on low fluoride diets.[21] The second development is in the recent use of lithium salts as a mental health treatment in the manic phase of schizophrenia—for which they seem very effective. The possibility has already been raised that public drinking water supplies might be used to insure the proper amounts of this element in the diet as a means of improving mental health. Quite obviously, a "lithiation" issue could spring up to replace fluoridation.

REFERENCES

1. Crain, R. L., et al., *The Politics of Community Conflict; the Fluoridation Decision*, The Bobbs-Merrill Company, Indianapolis, Indiana, 1969.
2. "Putting Fluoride on Teeth About 40% Effective in Reducing Amount of Decay During Year," *Sci N*, **98**, 10 (December 17, 1943).
3. "For Better Teeth," *Bsns W*, 44 (April 22, 1944).
4. "Ten Years for Teeth," *Time*, **43**, 44 (April 24, 1944).
5. *New York Times*, p. 43 (November 29, 1951).
6. *New York Times*, p. 5 (August 3, 1953).
7. *New York Times*, p. 1 (December 13, 1955).
8. "Save Money and Teeth," *Today's Health*, **33**, 13 (January 1955).
9. *New York Times*, p. 22 (March 22, 1962).
10. *J Am Waterworks Assoc*, 440 (April 1967).

11. *New York Times*, p. 1 (May 29, 1965).

12. *C&EN*, **46**, (September 30, 1968).

13. *Chem W*, **104**, 25 (January 25, 1969).

14. *New York Times*, p. 3 (November 29, 1971).

15. Quimby, F. H., and C. C. Bennett, "Fluoridation: A Modern Paradox in Science and Public Policy," Congressional Research Service, Library of Congress publication no. R-150-D, 72-26-SP (February 1, 1972).

16. Davis, M., "Community Attitude Toward Fluoridation," *Public Opinion Quarterly*, p. 474, Winter (1959).

17. Gamson, W., "The Fluoridation Dialogue: Is It an Ideological Conflict?" *Public Opinion Quarterly*, p. 526, Winter (1961).

18. Sapolsky, M., "The Fluoridation Controversy: An Alternative Explanation," *Public Opinion Quarterly*, p. 240, Summer (1969).

19. "Trigger for Community Conflict: The Case of Fluoridation," edited by B. D. Paul, et al., *Journal of Social Issues*, **17**, (4), pgs. 13–25 (1961).

20. Clark, R. E., "Fluoridation: The Courts and Opposition," *Wayne Law Review*, **13**, p. 339 (1968).

21. Messer, H. H., W. D. Armstrong, and L. Singer, "Fertility Impairment in Mice on Low Fluoride Intake," *Science*, **177**, 893 (Sept. 8, 1972).

Salk Polio Vaccine Hazard Episode

Abstract

In 1955, the newly developed Salk vaccine against the dread disease, poliomyelitis, was rushed into the market, and promptly caused numerous cases of paralytic polio and a few deaths. The cause was traced to deficiencies in the mass production process, which failed to kill the virus completely. The process was remedied, but only after a great public scare. As a result of this incident, the Secretary of Health, Education and Welfare, the Surgeon General, and the Director of the National Institutes of Health, resigned and the NIH Laboratory charged with regulating vaccines was reorganized. Nonetheless, an astoundingly similar incident occurred upon adoption of the later oral polio vaccine in 1962. The Salk and Sabin vaccines were ultimately successful, and paralytic polio cases in America had dropped from 14,000 in 1955 to thirty-four by 1967.[14]

Background

Poliomyelitis or polio can cause permanent crippling or death among its victims, particularly among infants and little children—hence the earlier name, infantile paralysis. Although polio is contagious, its occurrence in epidemic proportions is of surprisingly recent origins compared to many diseases: the first recorded epidemic (44 cases) occurred in Sweden in 1887, and the first outbreak in the United States (119 cases) occurred in Vermont in 1894.[1] In 1908, polio was determined to be caused by a virus, but before a prevention could be developed, the United States was hit in 1916 by a massive epidemic: 27,000 cases and 6,000 deaths (of which 2,000 were in New York City). In 1921 the nation was further shocked when Franklin D. Roosevelt, unsuccessful vice-

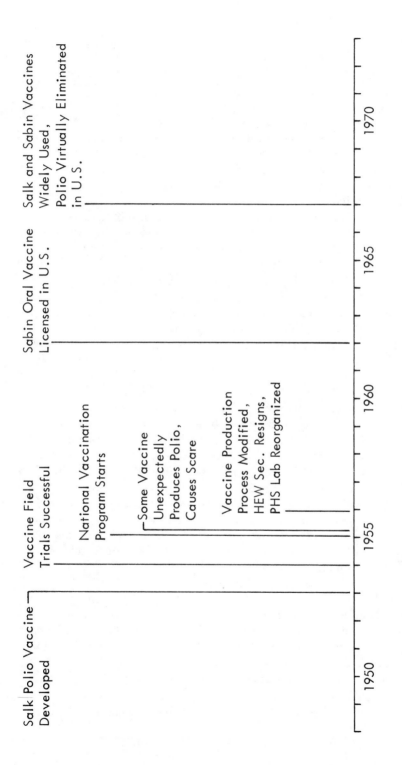

presidential candidate the previous year, was stricken with the "children's disease" at the age of 39.

After Roosevelt's election as governor of New York in 1928 and as President in 1932, the nation became even more aware of the dread disease. In 1938 an independent organization, the National Foundation For Infantile Paralysis (NFIP) was established to help fight polio: with the vigorous endorsement of President Roosevelt, the NFIP's annual March of Dimes campaign began collecting millions of dollars for research on the problem.

Since Edward Jenner's discovery of a smallpox vaccine in 1796 and the introduction of the hypodermic syringe during the Civil War, great strides had been made in public vaccination programs in the United States. Polio, however, was proving to be a difficult disease to fight and all proposed remedies had failed. Trials of a proposed vaccine in 1934 were worse than a failure: the vaccine caused polio cases. Polio was poorly understood and, unlike all other contagious diseases, appeared to have a predilection for striking hardest among the upper and middle classes and in the more advanced countries—just where public health precautions were greatest.

Two major breakthroughs occurred in 1949: Dr. J. F. Enders and coworkers at Harvard University discovered* a method of growing polio virus in tissue culture; another group at Johns Hopkins University demonstrated that three basic types of polio virus existed, to which at least fourteen of the apparently dozens of virus strains could be assigned. In 1952 Dr. David Bodian at Johns Hopkins and Dr. Dorothy Horstmann at Yale discovered that polio virus circulate in the bloodstream for some time before entering the nervous system and causing polio. These three discoveries generated much excitement among polio researchers, the big pharmaceutical companies, and at the NFIP, which prepared for a crash program to make a vaccine.

But this drive would be very costly, and here a major technological decision was required: should the NFIP put its money into development of a "killed" vaccine or a "live" vaccine? Each method had its staunch advocates. The former, which used a carefully killed virus that retained the ability to cause the generation of antibodies, offered the chance of a rapid development of a safe polio vaccine, although it might not be permanent, thus necessitat-

---

*Drs. J. F. Enders, T. H. Weller, and F. Robbins were awarded a Nobel prize for this discovery in 1954.

ing periodic "booster shots." The live vaccine, which used a specially cultured virus strain of low virulence, offered the chance of a permanent polio vaccine, but might require a longer development time. The NFIP chose to put the largest portion of its funds behind the killed vaccine under the direction of a young scientist, Dr. Jonas E. Salk,* Professor of Bacteriology and director of a virus research laboratory at the University of Pittsburgh. This decision was over the strenuous objection of the live vaccine advocates, particularly Dr. Albert B. Sabin of the University of Cincinnati. Dr. Sabin also continued to receive NFIP support† and, in addition, pharmaceutical companies were soon supporting research on each type of vaccine. The race for the vaccine was on and the competition was fierce. And none too soon: the incidence of polio had been rising ominously and the death rate in 1952 was the highest it had been since the 1916 epidemic.

Before the end of the year Salk had selected one strain of each type of virus, had developed a method of killing them with formaldehyde solution, had tested the vaccine on monkeys, and had started his first human experiments. Meanwhile, Dr. Hilary Koprowski had begun experiments feeding children attenuated live vaccine developed by Lederle Laboratories.

Key Events and Roles

Dr. Salk's first trials were carried out cautiously and on a small scale, using a group of children in Pennsylvania who already had polio antibodies.[1] The results were encouraging and were reported in the *Journal of the American Medical Association* of March 1953.

Before the Salk article had appeared, however, much of the information in it had already been made public:[1] apparently, the story was "leaked" to the press after Salk discussed his results at a closed meeting of the National Foundation's Immunization Committee in January 1953. Salk was discomfited at seeing unexpected headlines in the national press about his as yet unpublished work. As far as the National Foundation was concerned, the news was a heaven-sent boost for the March of Dimes program; at a compaign dinner, they announced "tremendous progress." The journalists, naturally, wrote enthusiastic stories. Dr. Salk became alarmed and

---

*Dr. Salk was concluding a $1.5 million, 3-year study for NFIP in which 119 polio strains had been classed in the three types. The study required 30,000 monkeys.

†Said to total over $850,000 by 1955.

obtained permission to hold a press conference during which he could put the matter into better perspective. He also appeared on radio and television to explain his views. Although he spoke with studied moderation, the public expected him, as a scientist, to be modest and cautious; they admired the modesty and disregarded the caution. As a result, Salk's appearance had precisely the opposite effect of that intended. The American public now had more than the promise of a cure for a dread disease: they had a hero.

During 1953, Dr. Salk vaccinated for the first time a group of people who had no polio antibodies.[1] The results were good, with satisfactory antibody response and no complications, which made it seem like no more than a matter of time before commercial vaccine could be produced. It now seemed reasonable to organize a mass trial of the vaccine—a public health effort on a prodigious scale.

The National Foundation for Infantile Paralysis took charge of the mass trial of Salk vaccine.[1] Dr. Thomas Francis, Jr., Salk's former chief at the University of Michigan and a man with years of research on killed viruses, was selected by the Foundation to conduct the field trials.

Five pharmaceutical houses agreed to make the vaccine according to Salk's specifications: Eli Lilly and Company; Parke, Davis and Company; Pitman-Moore Company; Cutter Laboratories; and Wyeth Laboratories. Some, however, apparently found the process difficult; in several early lots the virus had not been completely killed.[1] Although the unsatisfactory lots were rejected, this development was nevertheless alarming, particularly to the live-virus advocates such as Dr. Sabin.

A large-scale experimental vaccine field trial was conducted from April 26 to July of 1954,[1] using relatively small batches of Salk vaccine which was carefully produced and tested.[2] The 1,829,916 children who took part were all entered with the voluntary agreement of their parents.[1] The trial was divided into two parts known as "observed" areas and "placebo" areas: over 200,000 were injected with vaccine and a similar number with an inert substance, although neither group knew which injection they received. The vaccine came from just two sources: Lilly and Parke, Davis.

The collection and analysis of field data was time-consuming, and there was some doubt whether the results would be on hand before the 1955 polio season. The vaccine producers who had supplied the 1954 material approached the Foundation for guidance. Because they could not afford to keep their vaccine production plants idle for too long the vaccine producers wanted a guaranteed order, or they would have to make other plans. If the eventual report on the field trials was favorable (as the Foundation was confident it would be), then the demand for vaccination would be clamorous and overwhelming. If the Foundation failed to provide a supply of vaccine, every polio death of 1955 would be laid at its door.

It was a dilemma for the Foundation. By normal standards of government administration, it should never have happened to the Foundation at all, because the government's National Institutes of Health Biologics Control Laboratory would be required to license any vaccine administered on a public health, rather than on a trial basis, and no license could be issued until the results of the trial were known. Yet, by the time the results appeared it would be too late to make vaccine for 1955. To keep the vaccine companies in operation, the Foundation in 1954 ordered 27 million doses of vaccine, costing $9 million, in advance of the Francis report on the field trials.

A most unconventional procedure was used in the report of the vaccine field trials.[1] Under what has been described as an "undignified circus atmosphere," Dr. Francis read the report at a special meeting on April 12, 1955 at the University of Michigan to which all leaders of virus research, science, medicine, and public health were invited. The meeting was also televised, by closed circuit methods, to movie theaters all over the country, where doctors were admitted to watch it. Newsmen ignored an agreement to hold release of the news until the start of the meeting; even before Dr. Francis began to speak the news was on the street. The American public was told, in a few words, that the experiment was a triumphant success. The vaccine was, to quote Dr. Francis, "safe, effective, and potent."

The actual report as read by Dr. Francis concluded that:

• The vaccine was effective in the majority of the cases. Against spinal paralysis (the most common type), it was 60 percent effective, and against bulbar paralysis 94 percent effective.

• Efficacy varied for different types and from batch to batch of vaccine.

• The vaccine did not prevent nonparalytic polio. (This was an advantage since this type of polio does no permanent harm and gives good immunity against paralytic disease.)

• There was no evidence of any danger as a consequence of giving the vaccine.

The Department of Health, Education and Welfare, which had

done relatively little to develop the polio vaccine, still had to license it before public distribution. The HEW Secretary, Oveta Culp Hobby, and her public health advisor, Surgeon General Leonard A. Scheele, were now only too happy to take an active part in the affair.[1] On the date of the Francis report, a ceremony was also held in Washington, D.C., in which Mrs. Hobby, in front of the press and television cameras, issued licenses to six companies* to produce the vaccine. A wave of rejoicing swept the entire nation—it appeared that polio, the dread killer and crippler of children, had at last been conquered.[3] On April 22, President Eisenhower presented a special citation to Dr. Salk and the NFIP.

On April 12 and 13, 1955, the National Institutes of Health had cleared the first batches of polio vaccine for use in the 1955 season. These batches, produced by Lilly and Cutter, were soon followed by batches made by the other manufacturers. But the demand was tremendous. No one had anticipated that a distribution-control program would be needed, but the demand quickly surpassed the supply. By April 22, Congressional pressure forced President Eisenhower and Secretary Hobby, to arrange a conference with the six manufacturers. The conference agreed upon a voluntary allocation plan for areas and age groups, as outlined in the follow-up Presidential press conference which followed.[4]

Public elation soon gave way to a sense of dissatisfaction when the news began to spread that the quantity of vaccine available would certainly be inadequate.[1] Polio was such a frightening disease that the public began to see itself as a community awaiting the visitation of the plague. They had been assured that the means to eliminate the scourge were available, and yet they were being denied it. Most of the blame was directed toward Mrs. Hobby, whose job the public apparently perceived was to see that vaccine supplies were available.

Then, suddenly, an ominous note developed: starting about April 20, the Public Health Service received reports of twenty-four cases of paralytic polio in children who had already received the Cutter vaccine. As other makers got their products into distribution, five more polio cases developed after Cutter vaccine shots, and four cases were reported after other companies' vaccine shots.[4] Reports were then received of twenty-two cases of paralysis and two deaths in children in Idaho[1] after treatment with Salk vaccine.[5] Other simi-

*Sharpe and Dohme, Inc., was the sixth.

lar cases were reported in Georgia and Louisiana.[1] An additional alarming feature was the fact that infection was also being reported among unvaccinated persons who had come in contact with these cases.

The Cutter cases broke into the newspapers on April 27, 1955. State and local health officials blasted Washington for failing to set up stricter controls at the start. People objected that the vaccine had not been checked thoroughly before use and also that it had not been channeled freely enough where it was needed.[4]

As the news came out, public confidence in the Salk vaccine changed into doubt and then into fear.[1] There was a dramatic reduction in the number of children coming for vaccination as parents decided to wait and see what happened. Those who had accepted vaccination waited in agonized apprehension. In some areas, emergency supplies of gamma globulin were injected into children who had received Cutter material, in the hope of protecting them against infection.

The Surgeon General of the Public Health Service, Dr. Leonard A. Scheele, promptly banned further use of all Salk vaccine made by the Cutter firm, pending an investigation of the plant in Berkeley, California, by specialists.[5] Every batch of unused Cutter vaccine was recalled. California then banned the use of all antipolio vaccines until "further clarification of the present problem."

Dr. Scheele immediately counseled parents against panic.[5] He pointed out that these stricken children had not necessarily contacted polio from the vaccine. Dr. Scheele noted that the shots do not give maximum protection at once and that the victims may have been "on their way to having polio." Later the same week (May 1955) he recommended that polio vaccinations be continued, using supplies other than Cutter's.

The Cutter cases immediately raised the possibility that vaccine containing live virus had been used.[6] A narrow margin of safety existed in using the formaldehyde: an insufficient amount left some deadly live virus, while too much destroyed the virus so thoroughly that it didn't stimulate the production of antibodies. Investigators concluded that the 1954 vaccine, made in small pilot plants under laboratory conditions, apparently succeeded in staying within this margin. They speculated that the margin of control might be too slight for large-scale commercial production. In 1954, each batch of vaccine was subjected to three rigorous tests: by the manufacturer, by the Salk laboratory, and by the NIH Biologics

Control Laboratory. In 1955, however, the full testing was done only by the manufacturers, while the Federal laboratory merely made spot checks.[6]

On May 8, 1955, the U.S. Public Health Service ordered a temporary halt in release of new batches of vaccine and recommended that the entire polio vaccination program be held up pending a recheck; new testing procedures were being considered.[7,8] Government scientists[8] began a batch-by-batch recheck of vaccine production and testing of all the plants producing Salk vaccine. In mid-May, 8 million doses were released from Lilly and Parke, Davis.

The Surgeon General then discovered, to his consternation, that other firms in addition to Cutter were having difficulty in making the vaccine to specification,[1] and even more troubling were reports that polio had developed in children receiving vaccine made by Lilly and Wyeth, although these later appeared to have been coincidental cases. At any rate, the clearance program was again halted, but by May 24, the Surgeon General announced that all vaccines except for two lots from Cutter had been safe.* The PHS then established drastically revised and much stricter testing standards for Salk vaccines, to which the producers agreed.[9] The "Cutter incident," as it would become known, had ended.

Disposition

As a result of general ineptitude in handling the situation, public reaction was divided somewhat paradoxically between outrage at the shortage of the vaccine and a disinclination to use it even if it were offered.[1] This confusion was not confined to lay people; the medical profession was also alarmed. There was great resentment at the fact that complete scientific information on the Salk vaccine was not yet available in any medical publication.

In July 1955, the nation's top polio experts and health authorities revealed to the public what had gone wrong with the Salk polio vaccine program.[10] They confirmed what critics had suspected: it was very risky to jump, in a little over a year, from laboratory production of vaccine to manufacturing in tank-car volume. Many vital facts simply were not known when the leap to

---

*In June, the Surgeon General published a report which made it clear that some of the production lots of Cutter vaccine had contained live virus.[1] Clumps of virus had not been deactivated.

factory-scale production was made. During the same month, the Surgeon General announced that the Public Health Service's testing unit, the Laboratory of Biologics Control, was being overhauled and given new direction, status and staff to improve its efficiency.[10] Still another development might be interpreted as a comment on the handling of the vaccine program: Secretary Hobby handed in her resignation during July 1955. The Surgeon General and the Director of NIH also resigned.

Dr. Albert Sabin, long the foremost critic of the Salk vaccine, urged (in July 1955) that both production and inoculation be stopped until the vaccine could be made consistently safe.[11] A panel of experts agreed that one step toward consistent safety would be the elimination of the virulent Mahoney strain of virus used in Salk vaccine. The panel agreed that innoculations of Salk vaccine should continue through the summer polio season—that the probable benefits outweighed the possible hazards.

The PHS reported in November 1955, that after the testing standards had been revised earlier in the year, no indications of live virus were being found in Salk vaccines.[12] Later in the month, a panel of experts at the American Public Health Association declared that the Salk vaccine was safe and effective, and that live virus had been eliminated by a filtration technique used during production.[13]

In 1959, Dr. Albert Sabin announced the promising results of large-scale experiments with his live polio vaccine in the Soviet Union.[1] Two years later, a controversy developed in the United States concerning the long-term effectiveness of Salk vaccine and the possibility of replacing it with Dr. Sabin's vaccine.[1] Most virologists agreed that an oral vaccine would be cheaper to produce, easier to administer, and would provide longer immunity.[14] In May 1961 President Kennedy requested Congress to appropriate $1 million for the "stockpiling" of the Sabin vaccine in case of polio outbreaks in the United States. At the same time he requested continued use of the Salk vaccine.[14]

On June 30, 1958, the PHS established the Committee on Live Polio Virus Vaccine.[14] From that date until licensing of the Sabin oral (attenuated) vaccine in 1962, the committee met some fifteen times. Numerous articles were published reporting to the scientific community on progress and problems in development of the vaccine. PHS issued several interim reports and proposed standards to manufacturers, disseminated warnings about hasty production, and consulted frequently with industry to discuss the safety and efficacy of the vaccine. PHS also continued surveillance over the vaccine once distribution began.

According to one reviewer,[14] the production and distribution of the Sabin vaccine was orderly and deliberate because of the previous Salk experience, and acceptance went smoothly. In fact, however, the Salk episode was *repeated* to an astounding degree with the Sabin vaccine, according to an article in *Science*.[15] About a hundred cases of polio are believed to have resulted from an improperly attenuated Sabin vaccine (Type III).[15] One of these reportedly[15] resulted in a settlement by the manufacturer and a court judgment of over $2 million against the United States Government for negligence by the National Institute of Health's Division of Biologics Standards—the very group established to regulate vaccines as a result of the Salk episode.

After the Salk and Sabin vaccines were adopted, the number of paralytic polio cases in American fell from 14,000 in 1955 to exactly thirty-four in 1967.[16] A dread crippler had been overcome.

Comment

The development of the Salk vaccine appeared to be a long-awaited answer to polio. In this case, the publicity generated strong pressures to modify conventional manufacturing and safeguard procedures, to rush production, and to raise unrealistically high expectations in the public of immediate availability of the new vaccine. The near-panic produced by the unsafe vaccine caused much loss of confidence in the medical profession, the pharmaceutical industry and governmental regulatory agencies. The episode illustrates that the time scale for the development of physical technology cannot be altered at command, without running the risk of unsuspected complications.

REFERENCES

1. Wilson, John R., *Margin of Safety*, Doubleday and Company, Inc., Garden City, New York, 1963.
2. "Clearing Up the Polio Tangle," US News, **38**, 30 (June 3, 1955).
3. "The Polio Story—Millions and 52," *Newsweek*, **45**, 34 (May 16, 1955).
4. "Salk Vaccine at the Crossroads," *Bsns W*, 29 (May 7, 1955).
5. "No Time for Panic," *Newsweek*, **45**, 68 (May 9, 1955).

6. "Vaccine Crises," *Time*, **65**, 49 (May 9, 1955).

7. "The Truth About the Polio Scare," *US News*, **38**, 21 (May 13, 1955).

8. "Salk Off—Salk On," *Bsns W*, 31 (May 14, 1955).

9. "Near-Disaster," *Time*, **65**, 58 (June 6, 1955).

10. "Premature and Crippled," *Time*, **65**, 45 (June 20, 1955).

11. "Vaccine Safety," *Time*, **66**, 60 (July 4, 1955).

12. *New York Times*, p. 37 (November 16, 1955).

13. *New York Times*, p. 19 (November 18, 1955).

14. "Technical Information for Congress," Report to the Subcommittee on Science Research and Development of the Committee on Science and Astronautics, U.S. House of Representatives, 91st Congress, First Session, prepared by the Science Policy Research Division, Legislative Reference Service, Library of Congress, p. 309, April 1969.

15. "Division of Biologics Standards: Reaping the Whirlwind," *Science*, **180**, 162 (1973).

16. *Good Things About the U.S. Today*, by U.S. News and World Report, A Division of U.S. News and World Report, Inc., Washington, D.C., 1970.

The Thalidomide Tragedy

Abstract

Thalidomide, a drug developed in Europe for use in sleeping tablets and tranquilizers, was used in more than a dozen countries during the late 1950's. In late 1961, birth defects were attributed to the use of this drug in Germany. An application for the drug was withdrawn in the United States in March 1962 as the story created a sensation. Although the drug was never marketed in America, seventeen deformed children were born as a result of testing and importing the drug. This tragedy stimulated passage of a drug reform bill in 1962.

Background

The discovery that useful medical chemicals could be separated from coal tar in the late 1800's and the introduction of the wonder drug aspirin in Germany in 1899 ushered in an era of great growth for the pharmaceutical industry. Some of the potent new medicines that were discovered were not without danger, however. In 1938 the U.S. Congress passed the Food, Drug and Cosmetic Act—the first legislation intended* to provide that drugs would be safe when used as directed—in response particularly to the tragic loss of 108 lives that occurred when a toxic solvent was used to dissolve one of the new sulfa drugs. Spurred by the commercialization of the wonder drug penicillin in 1944, cortisone in 1949 and tranquilizers in the early 1950's, new drug development became a highly competitive and profitable business. Medical practitioners

---

*The Pure Food and Drug Act of 1906 was primarily concerned with adulteration and labeling.

140

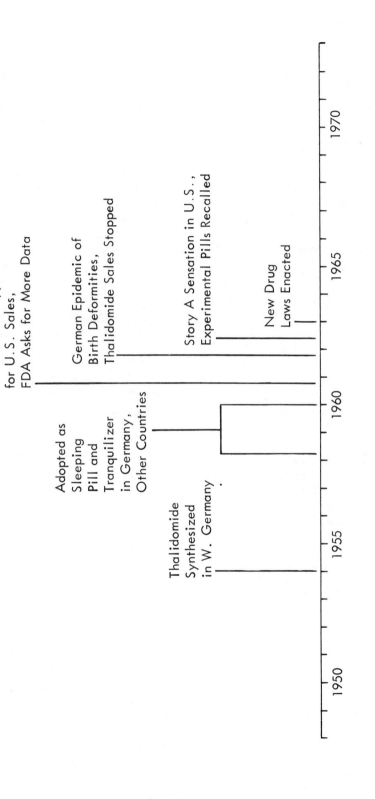

were almost overwhelmed by the amount of promotional literature supplied by the pharmaceutical companies through their brochures and their advertisements in hundreds of medical journals. In 1959 Senator Estes Kefauver (D—Tenn.) and his Subcommittee on Antitrust and Monopoly began investigative hearings on the economic and medical aspects of the drug industry; over a two-year period it accumulated over 8,600 pages of testimony and exhibits. It then accumulated an additional 4,200 pages during legislative hearings in 1961–1962 on an antitrust bill for the drug industry that also included provisions to strengthen the control of new drugs. But events had already occurred in Europe that were to make a far greater impact on the public than the eighteen volumes of these hearings.

In 1954, chemists at the firm of Chemie Gruenenthal in Stolberg, West Germany, first synthesized the drug, thalidomide [alpha-(N-phthalimido)-glutarimide]. It seemed to be an ideal sleeping tablet and tranquilizer,[1,2] and after 3 years of animal tests, thalidomide was judged so safe it was approved throughout West Germany for over-the-counter sale without prescription. Chemie Gruenenthal claimed that the drug was well suited for calming down nervous children, and was excellent for babies. Some obstetricians prescribed it to allay morning sickness in pregnant women. In addition to its wide acceptance as a sleeping compound, thalidomide was also used for grippe, neuralgia and asthma. Thalidomide had become Germany's most popular sleeping tablet and tranquilizer by 1960. Consumption also spread rapidly to other countries. Thalidomide was made or marketed, alone or in combination with other drugs, in nearly two dozen countries under fifty trade names.[1]

Key Events and Roles

On September 12, 1960, the American pharmaceutical firm, William S. Merrell Company of Cincinnati, Ohio, applied to the Food and Drug Administration for permission to market the highly successful thalidomide in the U.S. as "Kevadon," under an exclusive license with the German firm. The application, accompanied by reports showing that over three years of animal tests and human use of the drug in Europe and human tests in America* had re-

*As permitted under the law, Merrell had sent Kevadon samples to American doctors for testing.

vealed no adverse effects, appeared headed for routine processing and—Merrell hoped—quick approval. In fact, under the Food and Drug Act, the application would be automatically approved if the FDA did not raise objections within 60 days. It was assigned to Dr. Frances O. Kelsey, a new medical officer at the FDA who had previously taught medicine and pharmacology; it was her first case.

To the Merrell Company's distress, Dr. Kelsey returned the application and requested more data; she was concerned that thalidomide behaved quite differently in animals than in man (it was not a sedative in animals) and also differently from chemically similar drugs. Merrell Company officials went to Washington and complained that Kelsey was being unreasonable in view of the years of successful use, and provided further favorable results from tests by American doctors. But a letter to the editor in the *British Medical Journal* of December 1960 which came to Dr. Kelsey's attention strengthened her suspicions; it described a few incidences of peripheral neuritis—impaired nerves of the hands and feet—among long-time thalidomide users. A similar report had already appeared in Germany and the British manufacturer of the drug had added a warning about this possible side effect in its literature in August 1960,[1] i.e., even before Merrell's application had been submitted.

The Merrell Company then proposed to add a warning statement to its label. Dr. Kelsey replied that the observation of neuritis symptoms in adults (which had caused thalidomide to be made a prescription drug for the first time in Germany), would now require new data to show that thalidomide could be safely taken during pregnancy without harm to the fetus. The prospect of undertaking long-term animal and clinical studies to prove thalidomide's safety all over again appalled Merrell officials, who complained bitterly about the "stubborn bureaucrat"; but FDA officials stood firm.

Long before these tests could be completed, however, disturbing news about thalidomide came from Europe: it was linked to an outbreak of phocomelia (literally "seal limbs")—a terrible deformity in which babies are born with tiny flipper-like stumps instead of arms and hands (and instead of the legs and feet in severe cases). The outbreak had an unrecognized start in 1959, but had grown to near-epidemic proportions by 1961 and was of great concern to medical authorities. Then, almost simultaneously, Dr. Widuking Lenz in Germany and Dr. W. G. McBride in Australia determined that the mothers of several babies with phocomelia had one thing in common: they had each taken thalidomide in the first twenty to forty days of pregnancy. Lenz warned Grunenthal of his findings on November 15, and the German firm withdrew the drug from the market on November 26. A day later McBride's cabled warning arrived at the British manufacturer, Distillers Limited, and that firm withdrew its product on December 3, while awaiting further data. The Merrell Company had just started its clinical tests, which it said involved 762 "studies" and 29,413 patients* by the end of November; it promptly informed the FDA of the news, sent a letter of warning to the doctors participating in its tests, and sent investigators to Europe.

The thalidomide story was headline news in Europe during the winter but received little attention in the U.S. In February 1962 *Time* magazine[2] described the European epidemic of phocomelia. In March thalidomide was taken off the market in Canada; the Merrell Company withdrew its FDA application for the drug and quietly recalled all tablets. In April Dr. Helen B. Taussig reported at a meeting of the American College of Physicians that the European tragedy involved thousands of deformed babies; the story was picked up by the *New York Times*, but appeared only on page 37.[3] This story did prompt Representative Emanuel Celler's Committee on the Judiciary (which was holding hearings on the House version of the bill before the Kefauver subcommittee) to invite Dr. Taussig to testify. But neither her appearance on May 24, nor articles on thalidomide in May in *America*[4] magazine and in *Science*[5] were noted by the mass media.

But the thalidomide story was soon to become a national sensation: on July 15 Morton Mintz described dramatically on the front page of the *Washington Post* how one woman, Dr. Kelsey, had single handedly stood firm against great pressure and abuse to save America from the tragedy of armless and legless children. Suddenly thalidomide was big news. Dr. Kelsey was invited to testify before Senator Hubert Humphrey's Government Operations Subcommittee (which was considering coordination of drug research by government agencies) and her appearance received massive publicity. President Kennedy, who was supporting the Harris bill introduced May 3 to increase the regulation of the safety—rather than the prices—of drugs, praised Dr. Kelsey and awarded her a medal for Distinguished Civilian Service on August 4. The public now had a

*The report of subsequent congressional hearings states that 1,297 investigators received 2.5 million thalidomide tablets which were distributed to 19,822 patients including 624 pregnant women and 3,760 women of child-bearing age.[1]

recognized heroine and the news media gave her and the drug she halted wide coverage.[6-17]

The tragic extent of the thalidomide disaster in West Germany was officially confirmed in September 1962. Since 1957, when the sleeping pill-tranquilizer was first approved for over-the-counter sale, thalidomide had caused 10,000 cases of birth malformations in West Germany, according to the Public Health Ministry. Close to 1,000 cases of deformed babies were reported for Britain and untold scores of other cases occurred across Western Europe, in Japan, and South America. Coupled with the disaster itself was raging arguments over whether pregnant women who had, or may have, taken thalidomide should have abortions to prevent the possible birth of additional deformed babies. In the United States, only seventeen thalidomide-connected malformations were reported. One case which received detailed coverage by the news media was that of the Arizona housewife who traveled to Sweden to have a legal abortion performed.

Disposition

The thalidomide episode exposed gaps in medical knowledge, and set researchers all over the world to new studies of (1) how and why prenatal deformities occur, and (2) how to test drugs in the future for possible teratogenic (embryo-deforming) defects. Never before in medical history had a drug been implicated in widespread malformations in the developing human embryo. The public also wondered whether there had been bad side effects from use of other drugs. As a result of the thalidomide episode, a new Commission on Drug Safety was appointed by the Pharmaceutical Manufacturers Association in August 1962.[11]

The significance of the thalidomide incident at least for Americans, was its effect as a catalyst in stimulating reforms for the handling of drugs.[15] Some of these benefits were intangible. The honor accorded Dr. Kelsey, for example, would serve to bolster other FDA officials in resisting pressures commonly applied by drug companies to hasten clearance formalities. Physicians would be less likely to prescribe new and potentially dangerous drugs for pregnant women. Many patients would be less eager to demand the latest drug for every symptom.

The legislative aftermath of the thalidomide episode produced more tangible results.[1] On October 10, 1962, the President signed a new drug-reform bill, the Kefauver-Harris drug law. Several major provisions of which may be traced back to this case are:

1. The Department of Health, Education and Welfare was authorized to issue regulations requiring drug manufacturers to maintain facilities and controls that would assure the reliability of their products.

2. New drugs were required to be shown effective for their intended use—as well as safe—before they could be marketed.

3. The Department was authorized to withdraw clearances granted on new drugs when substantial doubt existed as to the drug's efficacy.

4. A new or established drug could be withheld from FDA approval on the basis of false or misleading labeling.

5. Even an established drug could be immediately ordered off the market by FDA if found unsafe.

6. The time allowed for FDA approval on a new drug application was lengthened from the previous provision of sixty days.

7. FDA was authorized to require the recording and prompt reporting of any adverse effects relative to the safety and effectiveness of new drugs.

8. A firm and explicit statutory basis was provided for the imposition by FDA of detailed procedures for the testing of new drugs.

9. HEW was authorized to inspect drug manufacturing facilities.

10. Drug firms were required to register with the FDA annually and were to be inspected by FDA biennially.

11. HEW was directed to establish a standard "official" name for each drug, when desirable.

12. The Federal Trade Commission was authorized to require the disclosure of ingredients of prescription drugs, their efficacy and their adverse effects in advertisements directed to physicians.

The various manufacturers of thalidomide products around the world were the subject of numerous claims for damages. In some instances the litigations were extremely lengthy and as recently as 1972 the *London Sunday Times* was banned by court order from publishing an article on the plight of Britain's thalidomide children, because the story might influence still-pending proceedings in that country. Ultimately, the English courts awarded $50 million to some 430 deformed children, to be paid by the British distributor of the drug, while a Japanese firm and the government agreed to provide $20 million compensation to 63 affected families. Claims for about 74 Canadian children were still in litigation in U.S. courts in 1973.[19]

Comment

The thalidomide tragedy in Europe and the popular press's account of how one woman, Dr. Frances Kelsey, kept it from becoming an American tragedy, changed forever the atmosphere in which drugs, food additives, and food contaminants are regulated in the United States. Whereas the carcinogenicity of such ingredients had previously been a primary concern, teratogenicity and mutagenicity were now recognized as being of equally great concern. Because of the incident, procedures for laboratory and clinical testing (particularly in records keeping) of new drugs were greatly revised and made more strict. Although thalidomide is dangerous only during early pregnancy, its use even by males or females of nonchildbearing age was totally prohibited.

That the public alarm over thalidomide was so much greater than that over the 50,000 deaths caused by the automobile per year in America is of interest. Because thalidomide can cause disfigurement—an effect abhorred by modern society far more than death—it is like cancer, extremely alarming particularly since reproduction of the species is involved.

REFERENCES

1. "Technical Information for Congress," Report to the Subcommittee on Science, Research and Development of the Committee on Science and Astronautics, U.S. House of Representatives, 91st Congress, First Session, prepared by the Science Policy Research Division, Legislative Reference Service, Library of Congress, Serial A, 357 (April 25, 1969).

2. "Sleeping Pill Nightmare," Time, 79, 86 (February 23, 1962).

3. New York Times, 37 (April 12, 1962).

4. "Harmless Sleeping Pill," America, 107, 194 (May 5, 1962).

5. "Dangerous Tranquility," Science, 136, Editorial Page (May 25, 1962).

6. "Doctor and the Drug," Newsweek, 60, 70 (July 30, 1962).

7. "The Thalidomide Disaster," Time, 80, 32 (August 10, 1962).

8. "The Drug that Left a Trail of Heartbreak," Life, 53, 24 (August 10, 1962).

9. "Tough Drug Law Coming," Bsns W, 37 (August 11, 1962).

10. "Inside Story of the Medical Tragedy," U.S. News, 53, 54 (August 13, 1962).

11. "The Thalidomide Lesson," *Science,* **137,** editorial page (August 17, 1962).

12. Lear J., "Doctor Francis Kelsey's Struggle Against Thalidomide," *Sat. R.,* **45,** 36 (September 1, 1962).

13. "10,000 Malformed Babies," *Time,* **80,** 49 (September 7, 1962).

14. "In the Backlash of Drug Scandal," *Bsns W,* 98 (September 15, 1962).

15. "Toward Safer Drugs," *Consumer Rep,* 509 (October 1962).

16. "Doctor Kelsey's Stubborn Triumph," *Good H,* **155,** 12 (November 1962).

17. Mintz, M. "Dr. Kelsey Said No," *Read. Digest* **81,** 86 (October 1962). (This article is a condensation of Mintz's original exposé in the *Washington Post.*

18. *Kansas City Times,* p. 8A (November 18, 1972).

19. *Kansas City Star,* p. 5G (March 14, 1973).

Hexa, Hexa, Hexachlorophene

Abstract

In April 1971, scientists of the Environmental Protection Agency announced that the disinfectant hexachlorophene (HCP) could cause brain damage in animals. At the time HCP was in three hundred to four hundred products ranging from baby cleansers to toothpastes to femine deodorant sprays. The announcement that these products might be dangerous, and that it could be absorbed through infant's or broken skin, raised much public concern in the latter half of 1971. The Food and Drug Administration moved slowly to restrict the use of HCP amid charges that it had been ignoring the danger signals and over the protests of the industries involved and the American Medical Association. The AMA opposed a by-prescription-only status for HCP. Staphylococcus infection outbreaks were reported in hospitals after the use of HCP was temporarily suspended at FDA suggestion. In August 1972, however, the death of thirty-five French babies was attributed to the use of a baby powder that had been accidentally formulated with an excess amount of HCP. The FDA established strict controls on HCP in September 1972, and limited to sale by prescription only all products that contained HCP as a principal ingredient.

Background

The hexachlorophene story began in 1941, when W. S. Gump first patented an antibacterial compound (known chemically as 2,2'-methylene bis(3,4,6-trichlorophenol). Hexachlorophene is chemically related to the herbicide 2,4,5-T, which was discovered about the same time (1944); both products are synthesized from one of the same chemical raw materials, 2,4,5-trichlorophenol. In 1948, the Givaudan Corporation (a Swiss-based company with a U.S.

149

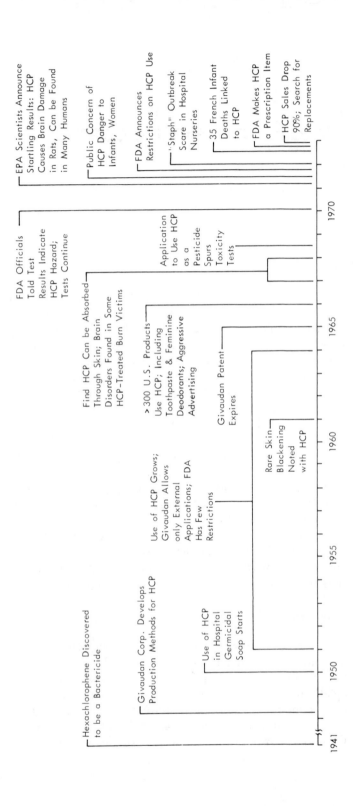

branch) developed an improved synthesis method and subsequently produced and distributed HCP on an exclusive basis until its patent expired in 1965.[1,2] Because of the known toxicity of HCP (it is also a relative of phenol), the Givaudan Corporation limited its use to topical applications and refused to sell it in preparations intended for internal use, like throat lozenges. Givaudan sold HCP only to companies that could demonstrate a safe and effective use for it in their products. HCP was used as an antibacterial agent in a variety of special soaps, deodorants, creams, and sundry cosmetics.

Starting about 1950, HCP became widely used in many hospital nurseries to bathe infants, as approximately a 3 percent solution. At some hospitals, babies were routinely bathed in HCP solutions on a daily basis to prevent skin infections.[3] One of the most popular of the liquid skin cleansers was "pHisoHex"—a bacteriostat manufactured by the Winthrop Laboratories division of Sterling Drug Company. This product (a sudsing emulsion containing 3% HCP on a weight basis) was recommended for bathing infants, cleansing burns and wounds, treating scalp infections, and for personal hygiene. The Winthrop Laboratories also marketed several other products containing HCP, such as "pHiSoAc", a cream (containing 0.3% HCP) for control of acne.[4]

Soon after the expiration of the Givaudan patents a rapid growth occurred in both the volume and diversity of HCP-containing products marketed in the United States. During this period of growth the FDA placed virtually no restrictions on the use of HCP in various products. The burgeoning use of HCP included toothpastes and mouthwashes, and a popular singing commercial proclaimed "Its got Hexa, Hexa, Hexachlorophene." By the early 1970's HCP was an ingredient of some three hundred to four hundred products, ranging from fungicides for vegetables and citrus fruits to shoe-liners, furnace filters, and after-shave lotions. In 1971, HCP represented a 4 million pound per year market;[5] Winthrop Laboratories' pHisoHex totaled $30 million per year,[6] and the highly advertised vaginal deodorants* containing HCP had grown to a $53 million annual market.[2] Although other companies were involved in HCP manufacture and sales, the Givaudan Corporation was credited with 90 to 95 percent of the sales of the chemical in the United States.[7]

*Since HCP is not effective against the bacteria chiefly responsible for vaginal odor, this appears to have been a needless use stimulated by high-pressure advertising and exploitation of modern phobias concerning body odor.[2]

Key Events and Roles

Two side effects of HCP had been noted during the 1960's although these danger signals were unrecognized by regulatory organizations. They were chloasma, a blackening of the face, reported in 1961, and burn encephalopathy, a state of coma often observed in burn patients treated with HCP, described by Dr. D. L. Larson of the Galveston Shrine Burn Institute in 1968. In 1967 it was apparently observed that HCP can enter the body through intact skin as well as through burns and wounds.[2] The heavy use of HCP thus posed the hazard that it could be absorbed through the skin and reach harmful concentrations in the blood.

The use of HCP also posed one other potential danger; the HCP might contain, as a manufacturing impurity, a group of chemicals known as chloro-dioxins, minute quantities of which can cause violent skin eruptions and acne.[2] In the mid-1960's, the dioxins were found to cause the skin disease named chloroacne that affected workers in a 2,4,5-T herbicide plant. The chloasma associated with HCP is somewhat similar. These results did not, however, cause the FDA to restrict the use of HCP in cosmetics, etc., but the nature of the toxicity of HCP was soon to be revealed to the FDA by its own scientists.

In response to a manufacturer's application to use HCP as a fungicide, toxicity tests were undertaken at the FDA Toxicology Branch in Atlanta, Georgia. Investigators Renate D. Kimbrough and Thomas B. Gaines found that rats became paralyzed following a 2-week diet which contained 500 parts per million of HCP. Examinations of the rat's brain and spinal cord revealed a "peculiar edema of the white matter resembling spongy degeneration" (the damage was reversible).[2] Later studies showed that the same brain lesions had been produced by a diet containing only 100 parts per million of HCP. These studies were submitted to FDA's Washington Office in prelininary form as early as July 1969; a paper was drafted for publication in April 1970, but encountered a seven-months' delay in getting FDA approval and was not published until August 1971.

The relevance of the data on rats to man was then studied by Robert E. Hawk and August Curley, also of the FDA Atlanta Toxicology Branch. They found that rats fed a diet containing 100 parts per million of HCP were carrying an average of 1.2 in their blood. More important, HCP levels in the blood of twelve human subjects with no unusual exposure to the chemical ranged from a minimum of 0.005 parts per million to 0.089 parts per million for one subject—the one who had made the greatest recent use of an HCP product. The latter concentration is almost a tenth of that which caused gross brain damage in the rat. This startling data was submitted to FDA for approval to publish in June 1970, but it still had not been approved six months later.[2] In addition to these results, the four Atlanta scientists studied the use by hospitals of concentrated HCP solution to wash infants. Working with personnel of the Montefiore Hospital in New York City, they found that at the time of release from the hospital, the infants had accumulated blood levels of HCP averaging 0.109 parts per million. The highest recorded level (0.646 parts per million) was more than half the blood concentration which causes brain lesions in rats.[2] It was found in a baby washed five times with a 3 percent solution of HCP.

In January 1971, the Atlanta Toxicology Branch was transferred to the newly created Environmental Protection Agency. Hawk and Curley soon received permission to present their research results at a meeting of the American Chemical Society in March 1971.[2] The paper alarmed some of the scientists present, but did not generate a great deal of national publicity and an FDA spokesman said that "We have no feeling of concern with hexachlorophene, and at this time, with the information at hand, do not plan any regulatory action."[2]

The industry, however, quickly responded: the Givaudan Corporation, the prime developer of HCP, reported in May 1971, that its own rat-feeding tests had shown no toxicity and no evidence of mutagenicity or cancer during the first four months of an eighteen-month study.[5] In addition, the FDA's Washington staff also decided quietly to make a study of HCP accumulation in human volunteers. People showering with pHisoHex accumulated between 0.1 and 0.38 parts per million HCP in their blood, and people gargling with 0.5% HCP solution for three weeks accumulated 0.06 parts per million HCP in their blood.[2] The 0.38 parts per million level was about a third of the average level that had caused brain damage in rats. These results, apparently in hand by June 1971,[2] also did not precipitate FDA action against HCP products.

In August 1971, Kimbrough and Gaines's research was published and was picked up by the newspapers, revealing to the public for the first time that HCP was a potentially dangerous substance which was being widely used by consumers. Consumer advocate groups, public officials, and concerned scientists pressed the FDA to restrict the use of HCP and to warn consumers of alleged hazards, such as neurological disturbances in infants and in animals. The FDA stated that it had almost completed a comprehensive investigation of HCP, including toxicological and metabolic studies.[8]

In November, the FDA received a report from Sterling Drug, Inc., makers of pHisoHex. They had studied five newborn monkeys washed daily for 5 minutes with pHisoHex over a period of three months. Although the monkeys showed no adverse reactions, autopsies revealed that they had brain lesions similar to those of the rats fed with HCP. Monkeys washed in the same manner without HCP were normal.[1] At the same time, the frequently-heard consumer advocate, Ralph Nader, asked the government to remove products containing HCP from the open market and to make them prescription items, because the chemical found in many deodorants, soaps and creams had caused brain damage in experimental animals. He also called for an FDA program to determine the "magnitude of the effects of HCP on humans and laws to control its use."[9]

Finally, in December 1971, the FDA, citing the findings of brain damage in baby monkeys, warned against regular bathing of infants and adults with cleansers containing 3 percent HCP.[10] In a letter to 600,000 doctors and other health professionals, the agency said that recent studies had shown that HCP can be absorbed through the skin into the bloodstream; this absorption had resulted in damage to the white matter of the brains of newborn monkeys. The agency also cited the evidence of brain damage shown in Dr. Kimbrough's study on rats involving ingestion of HCP. This action was the first attempt by FDA to regulate HCP products and was followed by an announcement that it would publish in the *Federal Register* a proposed new labeling for products containing 3 percent HCP. This new label would require conspicuous warnings about the use of such products for daily bathing of infants and adults.

The following month the FDA proposed a ban on the use of HCP in feminine hygiene sprays until it could be determined whether or not the chemical was safe and effective.[11] The FDA said it did not know whether HCP was harmful, what the human tolerance level might be, or the number of sources that a consumer may be exposed to. FDA commissioner, Charles C. Edwards, said, "Until we have such information the only prudent course is to reduce the total human exposure to HCP." The FDA also suggested a ban on the use of HCP in cosmetics (it could be used as a preservative at a level no higher than 0.1 percent), and specified that soaps and other skin cleansers containing more than 0.75 percent HCP be made available only by prescription. The FDA action also included taking a hard look at the long-term safety of other antibacterial agents.[12]

The HCP action had one undesired effect: by February 1972, the numerous hospitals that had stopped using HCP, reported 166 cases of staphylococcal infections developed in twenty-three hospital nursery units.[13] Critics of the FDA lost no time in blaming the agency's action for the outbreak of infections.[14] These developments immediately created widespread concern among the public (as well as hospital officials). A survey by the A. C. Nielson Company showed an exceptionally high consumer awareness of HCP. The report, issued in January 1972, showed that 60 percent of 2,000 random households sampled could identify HCP and 90 percent of women in households with young children could identify it.[15]

The Center for Disease Control (CDC), an agency of the Department of Health, Education, and Welfare, was the recipient of the staph outbreak reports. CDC felt that immediate action was required, and met with the FDA and the American Academy of Pediatrics (AAP) on February 2, 1972. The facts brought out at the meeting showed that the reported outbreaks were mild skin infections, and that many hospitals had no infectious outbreaks after they had discontinued bathing infants with HCP. In many cases, the hospital workers had quit washing their own hands with HCP, which was contrary to FDA recommendations, and a likely cause of the spread of infectious disease. At the conclusion of the meeting, the FDA and AAP remained opposed to routine bathing of infants with HCP products. However, the FDA did modify its warning by recommending short-term, once-daily bathing of infants with HCP solution in case of infectious outbreaks where other procedures failed.[1]

The controversy and public concern over HCP continued during the spring of 1972. In March 1972, the news media carried the story[16] that two biochemists had found that HCP interferes with the ability of brain cells to produce the energy they need to function properly, and that this may explain how the chemical could cause brain damage. In April, the American Medical Association (AMA) registered strong objections to the proposal by the FDA that products containing 3 percent HCP be sold only by prescription.[17] A spokesman for AMA said that placing HCP cleansers on prescription would create an imposition for those who need it and would not protect the newborn in hospital nurseries from its use.

In August 1972, HCP was dealt a severe blow: thirty-five French babies were killed in an episode where excess HCP was accidentally added to baby (talcum) powder during manufacture. The HCP level was 6 percent. The baby deaths were caused by encephalitis after the children developed severe skin irritations. The lesions in the infants' brains were identical to those caused by HCP in experimental animals.

In September 1972, a University of Washington pathologist, Dr. Ellsworth Alvord, Jr., had informed FDA that his studies at U.S. children's hospitals link the chemical's use to infant brain damage.[19] Dr. Alvord studied brain tissue of dead infants born since 1966 in hospitals which had used HCP. On one group of 204 babies (all dead from other causes) that had been scrubbed three times daily with the HCP cleanser, Dr. Alvord found that all of them showed brain lesions. Of the older children, fewer than 20 percent had lesions.[19]

Disposition

In September 1972, after more than a year of indecision regarding HCP's safety, the FDA announced restrictions amounting to a virtual ban on HCP.[18] The FDA stated that "there remains no doubt that HCP is a potent human neurotoxin, at high levels of use, e.g., 3 percent in emulsion and 6 percent in powder."[18] For those exposed to HCP during the last twenty years—particularly to the 3 percent solution—it appeared that the margin of safety had been frighteningly small.

Dr. Alford's data against HCP spurred a grim search. In September 1972, a national study was being organized to determine whether HCP caused brain damage in other babies who died soon after birth; the findings could conceivably link HCP to thousands of deaths or instances of brain damage. This horrifying news was widely publicized during 1972. Among some observers, the HCP affair was spreading doubts about the FDA's ability to evaluate consumer products scientifically.[19]

The FDA's new regulations on HCP on September 27, 1972 were: (1) recall of all infant powders which contained more than 0.75 percent HCP; (2) recall of all other products which contained more than 0.75 percent HCP, except those in pharmacies, where sale was by prescription only; (3) no recall of over-the-counter products containin less than 0.75 percent, HCP but further production and shipment of these products was banned; and (4) HCP content in cosmetics were restricted to a maximum of 0.1 percent when no alternative preservative was available.[20]

The following month FDA tightened its regulation on the use of HCP in cosmetic products. HCP was banned from use in products applied to mucous membranes; such as chapsticks, feminine hygiene sprays and rectal ointments.[21]

The regulations imposed by the FDA on HCP were effective in curtailing its use. The limitations set on HCP in January 1972 cut HCP sales in half during 1972, as many manufacturers began reformulating their products to eliminate HCP.[22] The strict regulations imposed in September 1972 further reduced HCP sales, and by January 1973 they were down to 10 percent of the level before the ban.[6]

The controversy over the use of HCP in hospital nurseries continues. A CDC report issued in mid-1972 showed a substantial increase in staph outbreaks since the FDA restrictions went into effect, and the FDA admits that there has been an increase in cases.[6] Their position on the bathing of infants with HCP products was the same in January 1973 as it had been a year earlier. The agency recommends dry skin care, washing with plain nonmedicated soap and tap water, or with just tap water.

The actions of the FDA effectively curtailed the indiscriminate use of HCP in consumer products, while it continues to be available when its benefits outweigh its risks, as determined by a physician. Its role in products used on infants, as an effective antibacterial against staph, is still undecided, and this question remains to be resolved in the future. Its use in consumer products has been replaced by trichlorocarbanilide and related products.

Comment

The hexachlorophene case illustrates several points. It clearly shows that a product that is used by (or on) almost all members of a segment of society has a large potential for causing public concern. When the segment is the nation's infants, the potential is tremendous, and even long-used products will be subjected to newer and more sensitive safety tests. The faddish use of HCP in hundreds of heavily advertised new products by American companies, after Givaudan Corporation lost control of its use, gives little evidence that possible toxic effects were seriously considered; it shows that the concept of technology assessment needs to be applied in the private sector as well as in the public sectors. The FDA's lethargic reaction in the face of mounting evidence that HCP was a safety hazard is puzzling; in part it may have been a result of the preceding "messy" cases they had been involved in, such as the cyclamate furor, the MSG debate, and the mercury-in-tuna scare. In part also, it may have resulted from the opposition of doctors and hospitals, who could not apparently conceive that they had been using a hazardous product on babies for twenty years; they were soon confronted with evidence that they had been relying on HCP to cover deficiencies in hospital cleanliness. The shock of thirty-five baby deaths in France was required to overcome the opposition of the medical interests and the intransigence of the FDA.

REFERENCES

1. "The Hexachlorophene Story," *FDA Papers*, p. 11 (April 1972).

2. "Hexachlorophene: FDA Temporizes on Brain-Damaging Chemical," *Science* 805 **174**, (November 19, 1971).

3. "How to Bathe Baby," *Newsweek*, **79**, 53 (February 14, 1972).

4. *Physicians Desk Reference to Pharmaceutical Specialties and Biologicals*, p. 1542, Medical Economics, Inc., Oradell, New Jersey, 1972.

5. *Chem W*, 50 **108**, (May 19, 1971).

6. "FDA Cover-Up in Staph Hassle?" *Chem W*, **112**, 14 (January 31, 1973).

7. "Hexachlorophene Action," *C&EN*, **49**, 21 (December 13, 1971).

8. "Hexachlorophene," *C&EN* (October 11, 1971).

9. *Kansas City Times*, p. 6 (November 29, 1971).

10. *Kansas City Star* (December 7, 1971).

11. "FDA's Prudence on Hexachlorophene," *Science*, **175**, 148 (January 14, 1972).

12. "Restricted Use of Hexachlorophene," *C&EN* **50**, (January 10, 1972).

13. *Kansas City Star*, p. 9 (February 2, 1972).

14. "The Staph Scare," *Time*, 69 **99**, (February 14, 1972).

15. "FDA Hexachlorophene Warning Stands Despite Staph Outbreaks," *Chem Market Rep*, p. 4 (February 7, 1972).

16. *Kansas City Star*, p. 3B (March 6, 1972).

17. *Kansas City Star*, p. 2F (April 19, 1972).

18. "Hexachlorophene Curbed," *Science*, **177**, 1175 (September 25, 1972).

19. "The FDA Bans a Suspected Killer," *Bsns W*, 24 (September 30, 1972).

20. "Hexachlorophene is Rapped by a Severe FDA Decision," *Chem Market Rep*, pp. 3 and 18 (September 25, 1972).

21. "Hexachlorophene Use is Further Hemmed In," *Chem Market Rep*, pp. 5 and 18 (November 13, 1972).

22. "Hex Ban Spurs Germ-Killer Switch," *Chem W*, 19 **111**, (October 4, 1972).

Krebiozen—Cancer Cure?

Abstract

In 1949 and 1950, the two Durovic brothers arrived in the United States with apparently about one-fourteenth of an ounce (2 g) of an unidentified white powder which they subsequently proclaimed and sold widely as a cure for cancer. The material, named Krebiozen, soon generated a controversy which was to last for fifteen years. Along the way, the controversy ensnared a previously respected scientist, Dr. Andrew Ivy; brought about the firing of the president of the University of Illinois; raised charges of "conspiracy" against the American Medical Association and of "sectarian interest" against the Catholic Church; and eventually involved the National Cancer Institute and the Food and Drug Administration in a search to identify and determine the efficacy of the drug. Ultimately, Krebiozen was generally discredited. An effort by the U.S. Government to convict its promoters of fraud in 1965 was unsuccessful, but interest in Krebiozen waned.

Background

Cancer is one of the most dreaded and least understood diseases, and the absence of effective treatment has long made desperate cancer sufferers the biggest of all targets for the sale of quack medical cures: five thousand cancer cures have been reported over the years. Hence, any claim of a cancer cure immediately raises not only the highest hopes of the public, but also focuses critical examination by medical science.

The Krebiozen story must begin with the histories of people involved, particularly of its developers, the Yugoslavian-born brothers, Stevan and Marko Durovic. Before World War II, Stevan

159

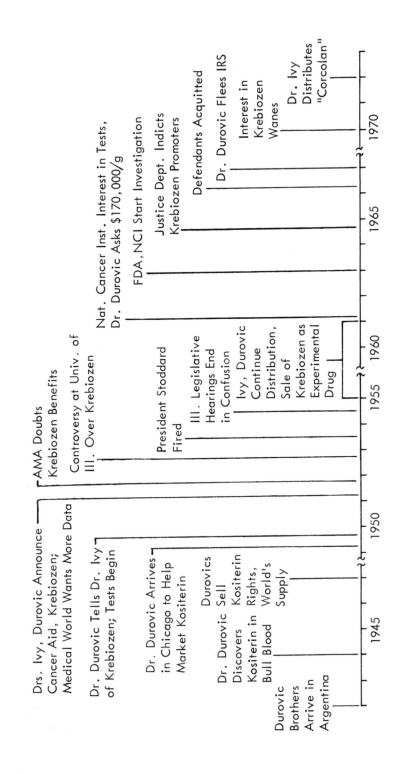

was said to have been a professor of medicine* at the University of Belgrade, while Marko was a lawyer and co-owner of a munitions plant. Early in World War II, the Durovics left for Italy and traveled on Vatican visas to Argentina.

By 1942, Dr. Durovic had gained access to the research facilities at the veterinary school of the University of Buenos Aires and at the state bacteriological laboratory. He is said to have engaged in research on the reaction to stimuli of the reticuloendothelial system in animals. In 1944, Dr. Durovic claimed to have established his own research laboratory and to have discovered a new drug which he called Kositerin.† Dr. Durovic apparently started using Kositerin for treatment of high blood pressure (hypertension) charging about $1,250 for two injections. By 1945, he claimed it not only cured 98 percent of six hundred cases of hypertension, but was also good for other cardiac problems, such as heart murmur, and a great number of other diseases including those involving virility. It could save millions of lives, he told prospective financial backers.‡

In 1946 the company, URBE, S.A., set up a Departmento Duga, which in turn set up the Durovics in a large house (Instituto Duga), which served as a laboratory, office and home, and arranged for a suburban farm where cattle could be kept for drawing blood—the source of Kositerin, Dr. Durovic revealed. Durovic was to obtain authorization from the Argentina Public Health Administration to manufacture and sell Kositerin. URBE stocked the farm with fifteen bulls which Dr. Durovic injected with a "secret substance." Two weeks or so later, Durovic drew the blood and secretly "extracted" all the Kositerin the company was ever to need. The Durovics apparently placed the Kositerin in ampules as needed and injected or sold it to patients who quietly came to the Instituto; Public Health approval had apparently not been requested. In 1947 a new firm, Duga, S.A., purchased the entire supply§ of Kositerin from the Durovics for a reported $72,000 cash and $89,000 in stock. However, Duga, S.A. retained the Durovics to assist them in marketing Kositerin and, in fact, left all of it in their possession.

*His advocates later stated that he had published extensively the results of his researches, but others have said[1] that he never published in Yugoslavia or elsewhere.

†Also spelled Cositerin in the literature.

‡According to the testimony[1] of Mr. Humberto Loretani, who became a business associate of the Durovics.

§Technically still the property of URBE, according to testimony of Mr. Loretani;[1] in addition, he claimed that "black market" Kositerin was sold in Uruguay through 1950.

At about this time, the Durovics made the acquaintance of two Chicago businessmen, R. E. Moore and Kenneth Brainard. Dr. Durovic and his associates were soon convinced that the market potential for Kositerin was greater in the United States, especially since Dr. Durovic said it might be a cure for malignant tumors. In March 1949, he went to Chicago to meet Moore and Brainard, who intended to handle U.S. distribution of Kositerin. Durovic brought a small amount of Kositerin with him, and Northwestern University was soon persuaded to test some of it on animals for hypertension. The results were apparently negative. For Dr. Durovic, Kositerin suddenly almost ceased to exist, but not so his drug-promotional efforts. Moore and Brainard soon arranged for him to meet a man whose story was to become inextricably entwined with those of the Durovics; he was Andrew C. Ivy, M.D., Ph.D.

Dr. Ivy was a man of impeccable academic and professional credentials. He was a vice-president of the University of Illinois, in charge of all their Chicago health-profession colleges, and head of the Department of Clinical Science with the title of Distinguished Professor of Physiology. He was a director of the American Cancer Society, past president of three national medical organizations, and had nearly a thousand published research papers to his credit. He had been lured from Northwestern University to the University of Illinois in 1946, and had been successful there in obtaining millions of dollars for new educational and research facilities for the Chicago campuses.

In the meeting, Dr. Durovic told Dr. Ivy he had a new anti-cancer drug and showed slides which he said indicated the effectiveness of the drug in producing regressions in carcinomas of several dogs and a cat. The new drug, which Dr. Durovic subsequently said he had obtained from horse blood in 1947–1949,* was called "Krebiozen."

Key Events and Roles

Dr. Ivy was intrigued by the potential for Durovic's new drug since it fitted nicely with theories already held by Ivy (and shared by many other researchers) on growth-controlling substances in the body. He soon agreed to assist Dr. Durovic in research on it. Dr. Durovic said that his supply of Krebiozen at the time consisted of 1,000 ampules of oil and water-oil solutions of the drug, which he

*Dr. Durovic later said this work had been supported by his rich brother, Marco.

had brought from Argentina. But apparently, he indicated more could be made, and on August 25, 1949, Krebiozen research started with a search for subjects to try it on. Dr. Ivy had already made one mistake, according to medical colleagues who testified subsequently: he did not know what the material under test was, or whether it was pure.

Nevertheless, after meager tests showed that Krebiozen was nontoxic on rats and mice (although without apparent anticancer activity), the researchers made observations on four dogs, and then quickly moved ahead to a study with human patients. Dr. Ivy was successful in persuading several doctors to try Krebiozen on their cancer patients. Dr. Durovic sent for his brother, Marco, who arrived in February 1950, bringing about 2 g (1/14 oz) of a white powder that Dr. Durovic said was the Krebiozen, but which his erstwhile associates at Duga, S.A. thought was their Kositerin. The Durovics did not give any of the Krebiozen itself to Dr. Ivy, but supplied solutions of the drug in oil as needed.

By January 1951, the results of the human cancer studies were beginning to come in, and were regarded as favorable. Then the Durovics reportedly did an astonishing thing: they proceeded to dissolve all their Krebiozen—the entire world's supply—in mineral oil and distributed it in about 200,000 ampules, with each ampule containing 0.01 milligrams of the drug. Thus, no pure Krebiozen remained for any kinds of tests or analyses.

Nevertheless, the research associates proceeded to plan for bigger things for their product. The intentions of the various members at this time were a subject of later disagreement. Moore and Brainard claimed that they made a contract with the Durovics to handle Krebiozen, and they attempted to screen Dr. Durovic from outside contacts. The Durovics felt the need for, or convinced Dr. Ivy that they needed, some favorable publicity to insure that their visitors' visas were extended by the U.S. Immigration Service. The group did not file a patent application or prepare a scientific paper for publication, but prepared a 106-page brochure which described the results of Krebiozen treatment on twenty-two cancer patients.

On March 26, 1951, Dr. Ivy presented Dr. Durovic, the brochure, and the first announcement of Krebiozen to a hundred invited doctors and other guests at a meeting in Chicago. Dr. Ivy said that Krebiozen had shown remarkable success in the treatment of human cancers, but warned that it was not the final answer to the chemotherapy of cancer, that it should not be called a "cure," and that supplies were not available for general use. He also stated that he would have preferred to make no announcement at this early stage of research, but did so because with doctors experimenting with it all over the country, rumors would get out anyway, and because of the Durovics' visa problems. In short, he said it merited careful clinical study and evaluation.

Dr. Durovic did not reveal what Krebiozen was, except that it was a white powder, a regulator substance, which he obtained from horse serum after "stimulation" of its reticuloendothelial system.* The secrecy, he explained, was to prevent world communism from learning his method. He indicated that as little as 0.01 milligrams of Krebiozen was effective against malignancies. Ampules of Krebiozen solution were apparently offered to some of the doctors present for tests on their own patients.

The doctors at the meeting deplored the secrecy of Dr. Durovic, and received with considerable skepticism† the reserved statements by their colleague, Dr. Ivy. Moore and Brainard were considerably less secretive and reserved; after Drs. Ivy and Durovic finished, they invited newsmen (whom they had thoughtfully included as the "other guests") into a nearby suite. Moore and Brainard said they were responsible for bringing Dr. Durovic's drug to America, and then provided a news release and a number of comments on Krebiozen which Marko Durovic immediately disavowed and the doctor later called sensational.[1]

The result of the March 26 meeting was a bombshell for the public. The news media carried numerous stories[2-5] on the new drug, Krebiozen, which they variously explained was Greek for "creator of biological force" or "vital substance." Although news accounts generally noted the reservations of the medical world,[6] many newspapers apparently did call Krebiozen a "cure" and hundreds of urgent telephone calls and telegrams poured in on the sponsors, the University of Illinois medical schools,[3] the American Cancer Society's Chicago office, and other medical facilities around the country. The State Department asked that samples be furnished to eight foreign countries.

The publicity of the March meeting had one immediate benefit for the Durovic brothers; Senator Paul Douglas (D-Illinois) introduced, and the Congress passed, a bill to admit them as U.S. citizens. The immense interest shown also led Dr. Ivy to set up, on

---

*Including the liver, spleen, bone marrow and lymphatic tissue. The stimulant was revealed a decade later to be the fungus *Actinomyces bovis*.

†The twenty-two patients had been observed for only a little over a year; eight had, in fact, already died; and absolutely no clinical controls had been used.

April 18, 1951, a nonprofit Krebiozen Research Foundation, which was to be dedicated to making the new drug available to all cancer patients.[1] Dr. Ivy was president. The Durovics apparently set up their own pharmaceutical concern.* (R.E. Moore and Kenneth Brainard were suddenly off the developing Krebiozen bandwagon, and sought legal assistance.) The Durovics then proposed to sell to the Krebiozen Research Foundation the 200,000 ampuled doses for $1,326,000 (according to a letter from the Foundation to the Internal Revenue Service[1]). This amount, they said, had been their production expenses in Argentina, and could be paid from future earnings of the Foundation.† The Foundation, however, could not get FDA approval to sell the drug in the absence of more scientific data; it did distribute samples to interested physicians for testing. By midsummer, Dr. Ivy expressed confidence that they would know with reasonable certainty within eight to twelve months if Krebiozen was of value as a cancer cure.[7]

The absence of scientific data was to become the plague of Dr. Ivy's life. The medical profession was generally confused by Dr. Ivy's support of Krebiozen, because clinical use of secret medicines had long been opposed by the American Medical Association. In September 1951, the respected journal, *Science*, published an invited comment to the effect that no evidence had been presented which showed anticancer activity by Krebiozen,[8] and also published a reply by Dr. Ivy in which he agreed, but again maintained that the substance merited careful clinical study.[8] The American Medical Association's Council on Pharmacy and Chemistry conducted a study of the claims for Krebiozen, and on October 27, 1951, reported (in the *JAMA*) that on the basis of data for one hundred cancer patients given this drug in six cancer clinics around the country,‡ the drug was without effect; forty-four had died, only two had shown any sign of improvement (and these were deemed temporary), and no evidence was observed of tissue changes due to the drug.[9] As a result, the Chicago Medical Society, in November, found Dr. Ivy guilty of unethical conduct and suspended him for three months for his part in the promotion of the secret drug. Dr. Ivy retorted that his conduct had not been unethical because the samples were given freely and he had not made any money on the

*Commodore A. B. Barriera, an Argentine aviation official, was apparently a financial backer.

†The similarity to the sale of Kositerin to Duga, S.A., is striking.

‡About one-third of the cases were from the Tumor Clinic, Department of Surgery, University of Illinois.

drug. These developments received considerable publicity[10-13] and tended to confuse the public.

The Stoddard Storm: Dr. Ivy's suspension from the Chicago Medical Society posed a most unusual situation for the University of Illinois. Dr. George D. Stoddard, President of the University immediately announced[14] his support for Dr. Ivy.* The University announced[15] that Dr. Ivy would remain in his position, but on December 16, 1951, Dr. Ivy received, at his own request, a leave of absence so that he could work full-time on Krebiozen.[16]

Considerable uneasiness existed within the University of Illinois faculty over the Krebiozen controversy, however, and Stoddard soon appointed an interuniversity "research validation committee" headed by Dr. W. H. Cole.† The Cole committee reported its conclusions to Stoddard on September 10, 1952; it had received no information on what Krebiozen actually was, and could find no evidence that the drug had any beneficial effect on cancer patients, despite Dr. Ivy's claims that it reduced (but did not cure) tumors and reduced pain in cancer sufferers.[17] The committee report was immediately disputed by Drs. Durovic and Ivy.[18]

The secrecy surrounding the identity of the drug was now becoming a paramount issue for the University. Stoddard, therefore, appointed Dr. R. E. Johnson‡ to head up a committee to dispel the mystery, but it also was unable to develop any new information. The Durovics would not tell how Krebiozen was made, because, they said, the University could not be trusted with the manufacturing details (Durovic subsequently claimed a great pharmaceutical company was offering him $1 million for these). The committee requested 4,000 ampules of the Krebiozen solution in mineral oil. The Durovics first said they would furnish this amount, then said they would not because attempts to extract the Krebiozen might give equivocal results, and besides, it was needed for cancer patients. Dr. Durovic finally said they would give the University 30 to 40 milligrams of the powder "from the next batch they made;" they expected to make some in Argentina in six months or so. Thus, the committee would not even get a sample from the batch already used in clinical tests. The Johnson committee reported its results on November 12, and on November 15, 1952, Stoddard announced that all Krebiozen research at the University of Illinois was ended.[19]

*Stoddard became president in 1946.
†Dr. Cole was head of the University of Illinois Department of Surgery; five universities were represented on the committee.
‡Head of the Department of Physiology and acting dean of the Graduate College.

During the latter half of November 1952, the newspapers were filled with almost daily accounts of the controversy, with statements from Dr. Stoddard and the University faculty, the Durovics, Dr. Ivy and the Krebiozen Research Foundation, the AMA, and many others around the country. On November 28, Stoddard asked the University's board of trustees to abolish the position of vice president held by Dr. Ivy and to reduce Ivy to a "distinguished professor."

Dr. Ivy immediately mounted a counterattack, with charges that he was a victim of a business conspiracy, that his academic freedom was being violated, and that he was being libeled and persecuted by Stoddard. Ivy's supporters, including several in the Illinois legislature, demanded explanations of why Stoddard would do this. Governor Adlai Stevenson, a personal friend of Stoddard's, was at a loss over what to do about the developing storm.

By January the legislature was calling for a hearing. The hearings of the Joint Legislative Committee began on March 18, 1953, continued periodically for a year (until March 24, 1954), and eventually heard over seven thousand pages of testimony, as the committee struggled to "ascertain the facts." Long before the hearings ended, however, a startling turn of events occurred.

On July 25, 1953, Dr. Stoddard was forced to resign by the University of Illinois trustees after a vote of no confidence.[20,21] The Stoddard ouster probably had several reasons,* but the Krebiozen issue was predominant. The trustees retained Dr. Ivy, although he resigned his position as vice-president.

The legislative hearings themselves were more than a three-ring circus of conflicting testimony and denials, charges and counter-charges, and even a parade of cancer patients who had been treated with Krebiozen (with unstated effect). Represented at the hearings in person, by counsel, or by statements were Dr. Ivy, the Durovics, Moore and Brainard, some Argentinians, Dr. Stoddard and numerous medical representatives. On the day the hearings opened, the Krebiozen Research Foundation published a pamphlet called "The Truth About Krebiozen" and, except for two brief written reports submitted by Stoddard, the "pro-Krebiozen" forces held the stage through the early months of the hearings. Before Dr. Stoddard, the University of Illinois and the American

---

*Including his long-standing differences with local Roman Catholic officials.[1,21] Stoddard charged that the Church took a sectarian interest in Dr. Ivy and the Durovics.[1] A leader of the anti-Stoddard group was apparently former football star Harold (Red) Grange.[21]

Medical Association testified, Stoddard had already been fired, and the news media had become weary and confused with the tangle of testimony.[1]

The legislature never could ascertain whether the University had offered the Durovics sufficient proprietary guarantees, whether medical, industrial or communist conspiracies existed, whether Ivy's academic rights has been violated, or whether Stoddard had been correct in stopping research on the dubious, secret drug. The legislature was hopelessly confused over the nature* of Krebiozen (or whether it was Kositerin) and the conflicting testimony on clinical, chemical and historical matters. On the basis of Ivy's testimony it was clear that 5,000 to 20,000 horses would have been required for the preparation of the initial batch of Krebiozen, but it was never clear where these had been kept. Even more suggestive were the revelations that the Durovics had purchased over one million glass ampules; the original 200,000† had been used, but over 262,000 others had disappeared—"broken" said Dr. Durovic. The fact that the Durovics had also purchased 300 gal. of white mineral oil (roughly enough for 1 million doses of Krebiozen) suggested to critics that 262,000 "doses" had already been sold and that these had considerably less "Krebiozen" in them than its producers recommended. Near the end of the hearings, six cancer clinics from across the country submitted affidavits on the status of patients treated with Krebiozen: seventy-three of the seventy-six treated had died.

Following the hearings, Krebiozen nearly dropped from the news for several years, although two books were published on it in 1955: *K: Krebiozen —Key to Cancer?* by Herbert Bailey, a strongly "pro" book, and Stoddard's own book.[1] Ivy filed a $360,000 libel suit against Stoddard for statements in the book. Krebiozen itself was apparently neither approved for nor banned from interstate shipment and the Krebiozen Research Foundation continued to distribute it as an experimental drug. The Durovics apparently continued to offer their product directly to cancer sufferers, and the Foundation issued another lauditory *Krebiozen Report* on its product in 1972.

The government acts: In 1961 the National Cancer Institute started gathering data on Krebiozen-treated patients because of the

*Near the end of the hearing, Dr. Ivy claimed they had finally extracted some Krebiozen from the mineral oil and analyzed it: he reportedly[1] called it a polysaccharide—a starch or cellulose.

†These were the ones said to have received all of the Krebiozen in 1951.

libel suit and to determine whether a long-sought official test of the drug was merited. An inquiry to Dr. Durovic concerning a sample of Krebiozen brought the reply that he could furnish a sample at the price of $170,000 per gram (over $77 million per pound); the government apparently said "no," and was finally given a few milligrams of the dry material and an ampule of the solution in July 1963.

Under the new drug law inspired by the Thalidomide tragedy, interstate shipments of Krebiozen would be prohibited after June 1963 unless a permit was obtained to make experimental drug studies. The Krebiozen sponsors declined to supply the required information in their application, and said that patients would have to come to Illinois for treatment. This announcement brought many protests and even pickets to the White House.

The FDA now began a full-scale investigation of Krebiozen as an effort to help the National Cancer Institute. The investigation soon developed into a campaign to reconstruct all aspects of the drug's clinical, financial and chemical history and particularly to determine the chemical identity of samples of Krebiozen which it had acquired. In late 1963 the FDA announced[22] its conclusions: Krebiozen powder was a common body substance known as creatine;* the contents of Krebiozen ampules were essentially mineral oil, with minute amounts of amyl alcohol (a solubilizing additive) and 1-methylhydantoin.† The FDA's chemical analysis was supported by the findings of the National Cancer Institute that Krebiozen did not produce any anticancer activity in man. From this time on, FDA officials considered the Krebiozen episode an open-and-shut case; if Krebiozen was merely creatine, it was an obviously fraudulent drug and the men marketing it were not erring scientists, but crooks.[22] At a September 7 news conference, the FDA announced the results of the studies on Krebiozen, and revealed that criminal charges would be brought against its sponsors —an unusual and often difficult procedure.

In November 1964, after thirteen years of controversy, the Justice Department obtained a forty-nine-count indictment charging Dr. Ivy, the Durovics and an associate with conspiracy,

---

*Chemical name and formula (αmethylguanido)acetic acid; NH_2-C(NH)-N(CH_3)-CH_2-CO-OH. It cost about 30 cents per gram.

†1-Methyl glycolyl urea, a cyclic compound, $\overline{NH\text{-}CO\text{-}N(CH_3)\text{-}CH_2\text{-}CO}$, which could be formed from creatine. A second decomposition product, creatinine was also later identified. In a December 1963 press release, the FDA said that Krebiozen ampules before 1960 contained only mineral oil.

mail fraud, mislabeling, and the making of false statements to the government. The indictment raised a new storm of controversy. The news media tried hard to explain the technical arguments: *Life* magazine even showed[23] the FDA's photo of the infrared spectra of Krebiozen and creatine hydrate, à la Dick Tracy. Other stories presented details of the Krebiozen sponsors' claims that the government scientists had erred by using improper samples or methods and by misinterpreting the results. Senator Paul H. Douglas, a long-time friend of Ivy's, inserted into the *Congressional Record* (on December 6, 18, 19 and 20, 1963) nearly forty pages of statements, technical reports and newspaper articles favoring the cause of Krebiozen.

During the long federal trial in Chicago (April 1965 to January 1966), the prosecution focused on the charge that "Krebiozen" was simply mineral oil to which traces of creatine had been added. Additional charges were that the product cost pennies to produce, that it was valueless in the treatment of any disease, and that the defendants had made at least $500,000 annually from the sale of Krebiozen by collecting from $9.50 to $95 per dose. The efficacy of the drug as a cancer cure was not specifically at issue.[24]

The details of the government's case were complex, but the underlying thesis was simple; "Krebiozen" never existed. The defense contended it had processed horses at a plant in Rockford, Illinois, and had a Chicago laboratory, and offered testimonials of satisfied Krebiozen users. The government's case apparently suffered from being too intricate for the jury to follow, and at the same time was basically too simple to be believed. The nine-month trial produced a 20,000 page record: it culminated in acquittal for all of the defendants. Immediately following the trial, the FDA and the American Medical Association issued statements emphasizing that the verdict in no way altered their scientific judgment that the alleged anticancer agent was therapeutically worthless.

Disposition

Because it had made criminal charges and lost the case, the government could take little further legal action,* barring some new provocation by the "Krebiozen" promoters such as an open

*The government did, however, eventually convict one of the *jurors* in the trial on an obstruction-of-justice charge involving circulation of a letter from a Krebiozen advocate from Pennsylvania. The Krebiozen case had already cost the federal government an estimated $1 million.

renewal of interstate shipments. The state government in Illinois had for several years been attempting to find a way to ban "Krebiozen" distribution there, and had apparently concluded that it could only be done through special legislation. The prognosis for such an effort was poor and, therefore, continued distribution was allowed to patients who lived in or came to Illinois.[22] The Internal Revenue Service, however, started to prepare charges that the Durovic brothers owed the government several hundred thousand dollars in back income taxes.[22] Dr. Stevan Durovic left almost immediately for Paris for health reasons, and has since lived in Switzerland.[25] The IRS seized about $600,000 of the brothers' assets. Marko Durovic continued his fight in court to enjoin the FDA from interference with the distribution of Krebiozen, and to obtain approval for his product.[25] But in late 1973, the U.S. Supreme Court allowed to stand a lower court ruling that blocked the marketing of the controversial drug as a treatment for cancer.[26]

    Dr. Ivy's Krebiozen Research Foundation has been dissolved. Ivy moved to Roosevelt University in Chicago in 1961 and then, in 1966, established the Ivy Cancer Research Foundation, which distributes, for use in cancer studies in Illinois, a product called Carcalon, which is said to be made like Krebiozen. Dr. Stoddard subsequently accepted important administrative positions at New York University and Long Island University. Ivy's libel suit against Stoddard was eventually dropped and the IRS claims against Marko Durovic were apparently unsuccessful. Thus ends the fifteen-year drama.[26]

Comment

    The Krebiozen affair contains a nearly unbelievable chain of events which could occur only because of the seriousness of cancer and man's fervent desires for a cure. Dr. Andrew Ivy, a man with a high standing in the academic and medical science world before his involvement with Krebiozen, became a "true-believer." His dedication to the goal that Krebiozen deserved a fair trial was to be continually frustrated, because the Durovics would never provide an adequate supply of a bona fide sample for such a test or even for analysis. His witness as an expert led to much confusion in the public, the state legislature and academic circles, in addition to his medical colleagues. His scientific objectivity was questioned and he eventually lost much of his professional stature. His repeated statements that he did not make money on Krebiozen did not address the ethical questions that concerned many medical authorities. Krebiozen, if it ever existed as a horse-serum extract (no patent application was apparently ever made), has never justified the hopes it generated in March 1951. The medical world and the government agencies found it difficult to clarify the Krebiozen affair in the face of testimonials and conflicting testimony. Reputable research continues on possible natural tumor-controlling substances in the body, and anti-cancer nostrums continue to generate controversy—for example, the current one over Laetrile. The Krebiozen case illustrates the eagerness of much of the public to grasp at rumored miracle cures for those widespread ailments for which modern medicine does not have an answer, such as cancer, schizophrenia, arthritis, obesity and sexual inadequacy.

REFERENCES

1. Stoddard, George D., 'Krebiozen'—The Great Cancer Mystery, The Beacon Press, Inc., Boston, Massachusetts, 1955.

2. New York Times, p. 31 (March 27, 1951).

3. New York Times, p. 10 (March 31, 1951).

4. "Krebiozen For Cancer?", Newsweek, 37, 42 (April 9, 1951).

5. "Earthquake in Chicago," Time, 57, 70 (April 9, 1951).

6. New York Times, p. IV-9 (April 1), p. 75 (April 15), p. 62 (April 29, 1951).

7. "Krebiozen Cancer Cure?", Sci Digest, 29, 49 (June 1951).

8. Rhoads, C. P., "Krebiozen," with reply by A. C. Ivy, Science, 114, 285 (September 14, 1951).

9. "Adverse Krebiozen Report," Sci NL, 60, 279 (November 3, 1971).

10. New York Times, p. 22 (November 14, 1951).

11. "Ivy Suspended," Newsweek, 38, 96 (November 26, 1951).

12. "The Doctor and His Ethics," Time, 58, 50 (November 26, 1951).

13. "The Case of Dr. Ivy," Sci Am, 186, 40 (January 1952).

14. New York Times, p. 14 (November 15, 1951).

15. New York Times, p. 9 (November 24, 1951).

16. New York Times, p. 51 (December 16, 1951).

17. New York Times, p. 25 (September 22, 1952).

18. New York Times, p. 32 (September 25, 1952).

19. "The Krebiozen Question," Newsweek, 40, 82 (December 8, 1952).

20. New York Times, p. 50 (July 26) and p. 21 (July 30, 1953).

21. "Storm Over Stoddard," Newsweek, 42, 52 (August 3, 1953).

22. "The Krebiozen Case: What Happened in Chicago," Science, 151, 1061 (March 4, 1966).

23. "Answer on Krebiozen: It's Useless." *Life,* **55,** 47 (October 4, 1963).

24. "Krebiozen Acquittal," *Facts on File,* p. 104 (March 17–23, 1966).

25. Private communications with Chicago office of the FDA, 1973.

26. *Kansas City Star,* p. 2 (October 15, 1973).

27. Private communications with Dr. Andrew C. Ivy, 1975.

28. The *New York Times* carried a total of 21 stories on Krebiozen during 1951–53 and 22 more stories from 1955–66, with 5 of these in 1961 and 6 more in 1965–66 during the trial.

29. "Two Grams of an Unknown." E. Keller, *Chem.* **38,** 6 (September 1965).

30. "The Krebiozen Story—Is Quackery Dead?" J. F. Holland, *J. Am. Med. Assoc.,* **200,** 125 (April 17, 1967).

DMSO—Suppressed Wonder Drug?

Abstract

During 1964 and 1965, DMSO (dimethyl sulfoxide), a versatile industrial chemical solvent of low toxicity, received widespread and increasing public attention as a potential miracle drug. National magazine articles over-enthusiastically described the use of DMSO in treating burns, arthritis, sinusitis, headaches and sprains and as a vehicle for carrying other drugs into the body right through the skin. The Food and Drug Administration became greatly alarmed that clinical studies were not being properly controlled and that laymen were even applying DMSO to themselves. When evidence developed in November of 1965 that large doses of DMSO could have adverse effects on the eyes of some experimental animals, the FDA recalled the New Drug application on DMSO and banned all clinical testing of it on humans, referring it back for basic toxicological and pharmacological studies. This action raised a great controversy that FDA was suppressing a beneficial drug, and the FDA relaxed its ban to permit clinical studies under strict controls. Since then, DMSO has been used only in veterinary medicine in the United States, but has been used to treat human patients in Mexico and several European countries. The DMSO controversy was subsequently the subject of a study by the National Academy of Science—National Research Council, which concluded that the FDA's unusually strict action was probably justified at the time, but that the substance was relatively safe and effective for certain therapeutic purposes.

Background

Dimethyl sulfoxide (DMSO), a clear, colorless liquid chemical compound at room temperature, was first synthesized in 1867 by

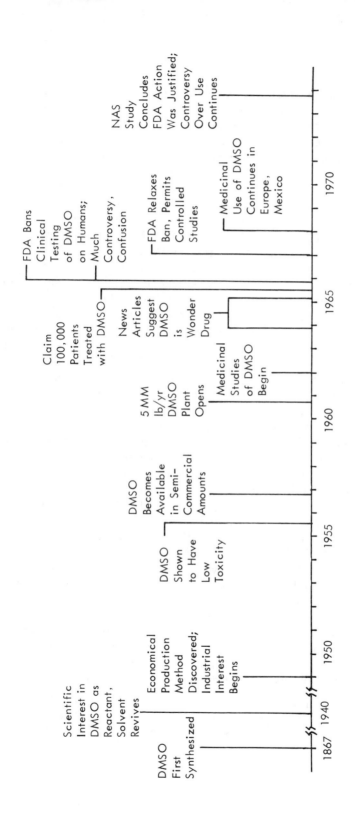

Alexander Saytzeff in Russia. One of a series of organic sulfoxides, a variety of which are found in nature,* DMSO remained a laboratory chemical of little interest for nearly seventy-five years. During the 1940's interest in the reaction chemistry and solvent capabilities of DMSO began to increase, both in the United States and in Sweden, where efforts were underway to recover useful chemicals from wood pulp wastes. By the late 1940's, an economical method of getting DMSO from spent kraft pulping liquors seemed to be available, and efforts were intensified to find industrial uses for this interesting, water-soluble, polar solvent. A number of potential solvent uses were identified (in synthetic fiber manufacture, as an antifreeze, or an absorber for gases). In 1955, studies showed that DMSO had very low toxicity—the average adult human would apparently have to drink over a quart to get a lethal dose.

The following year, Crown Zellerbach Corporation, second largest paper producer in the United States, opened a plant at Camas, Washington, to produce from wood wastes 1.5 million pounds per year of dimethyl sulfide (DMS), a compound of commercial interest itself, but also a potential raw material for DMSO. Using this source, Stepan Chemical Company in late 1956 made DMSO available for the first time in large amounts.

Industrial and academic research interest in DMSO then increased rapidly. The results indicated that DMSO had a significant toxicological and economic advantage over certain competitive synthetic fiber solvents (such as dimethyl formamide), and held promise for other industrial and agricultural uses (such as a solvent for pesticides). When Stepan suffered a bad explosion in their DMSO plant, Crown Zellerbach soon started construction of a new 10 million pounds per year DMS plant and a 5 million pounds per year DMSO plant—both based on new Swedish technology.[1] By September 1960, DMSO was available in tank car quantities at 33 cents a pound and Crown Zellerbach was inviting researchers the world over to ask for free samples.

And the researchers were making many new discoveries about this unusual solvent, not only in basic and applied chemical studies, but also in studies of natural products and biological materials. Thus DMSO was found to be a solvent for starches and to dissolve proteins without denaturing them, as many solvents did.

*Although DMSO, $(CH_3)_2SO$, is nearly odorless to most people (some are more sensitive), other sulfoxides are responsible for the characteristic odors of garlic, turnips and certain other plants. Sulfoxides are also found in small quantities in mammalian systems.

By 1962 it had been found that DMSO could be used to dissolve certain enzymes without inactivating them and was useful in cryogenic preservation of biologically active materials such as bone marrow, cell mitochondria, bovine spermatozoa, chick embryos, and even human kidney and lung cell cultures.[2] But the investigations of one medical researcher were soon to make DMSO a widely publicized and controversial substance.

Key Events and Roles

Crown Zellerbach Corporation, the sole American producer of DMSO, maintained an active search for new uses for their new product. In 1959, Robert J. Herschler, a chemist at their Chemical Products Division laboratories in Camas, began a study of DMSO as a solvent for pesticides. Not only was DMSO an excellent solvent for many hard-to-dissolve pesticides, but Hershler soon found that when the pesticidal solutions were applied to plants, the pesticide was carried inside the plant and acted systemically throughout. Further research showed that an insecticide solution in DMSO, for example, could be effectively used by injecting it directly into the tree—much like an intravenous injection in humans—and that a fungicide solution in DMSO could be taken up by the roots to protect the whole tree. Even tree wounds appeared to heal rapidly when treated by DMSO. Herschler also found, according to news accounts, that the pesticide in DMSO could be absorbed rapidly through intact human skin,* when he and a technician suffered poisoning symptoms after coming in contact with the solution.

At that time, Dr. Stanley W. Jacob, at the University of Oregon's Medical School in Portland just across the river from Camas, was investigating the freezing, storing, and transplanting of animal tissues and organs. By 1961, Herschler and Jacob had begun co-operative investigations on DMSO.

Jacob soon discovered, as had many previous investigators of DMSO, that when it is accidentally spilled on a person's skin, it quickly causes a peculiar taste in the mouth and a slight garlic-like odor on the breath.† He recognized that the DMSO had penetrated

---

*Numerous organic liquids (such as nitrobenzene) and solids (such as nicotine) were well known to be unusually hazardous because of their abilities to penetrate skin.

†Most subsequent news stories incorrectly portrayed the Jacob group as the discoverer of these phenomena. The odor arises from traces of dimethyl sulfide formed in the body.[3]

the skin and circulated in the blood, and reasoned that DMSO could have considerable medicinal value. Studies soon showed that colored dyes were rapidly carried throughout the organism when animals were swabbed or injected with DMSO solutions, or when fish were placed in them. Because of Herschler's earlier results with tree wounds, purified DMSO was next tried in the treatment of burns on animals: it was found to relieve pain quickly and to promote healing, not only in rats, but in Herschler, himself, who accidently suffered a bad chemical burn in 1962 and volunteered to be a human guinea pig. The outlook for medicinal uses for DMSO became very promising. Crown Zellerbach and the University of Oregon quickly began preparing patent applications on Herschler's and Jacob's discoveries and applied to the Food and Drug Administration for a license to conduct limited clinical testing on humans.

Jacob's early clinical trials were conducted on a variety of ailments as the opportunity arose. Because DMSO had been found to reduce surgical adhesions in animals, similar to the way cortisone does, Jacob decided to determine whether DMSO might act in the same manner as cortisone against the body's many inflammatory ailments. Topical application of DMSO to a sprained thumb, a mashed finger, and a swollen jaw was found to relieve the pain and reduce the swelling in each instance. Further experimentation indicated that DMSO helped heal cold sores, aided in clearing nasal congestion in colds and sinusitis when used as nose drops, and relieved headache pain when simply rubbed on the forehead. By October 1963, Jacob was investigating the more serious inflammatory diseases of the joints, arthritis and bursitis, and had enlisted the aid of Dr. Edward E. Rosenbaum, a clinical professor of rheumatology at the university medical school.

But already rumors were beginning to spread that a new miracle drug had been discovered: inquiries began coming to the medical school and the company. In early December the company and the university publicly announced a fifty-fifty profit-sharing arrangement on all pharmaceutical applications of DMSO, together with an indication of the potential extent of the applications. The announcement brought DMSO and Dr. Jacob to the public's attention for the first time[4-6] and set off a short speculative boom in Crown Zellerbach stock.

In February 1964, Jacob, Herschler and coworker Bischel published their first brief preliminary note[7] describing a wide range of potential medical uses for DMSO, and shortly thereafter Rosenbaum and Jacob published a preliminary communication[8] describing successful results in treating arthritic conditions with DMSO. These publications, in relatively minor medical journals, were met with considerable reserve by the medical world, but were soon to lead to sensational stories touting DMSO as a wonder drug.[9–12] The publicity not only spurred other medical researchers into the DMSO field and led[13] to the licensing of six pharmaceutical companies by Crown Zellerbach to do further testing and development on DMSO, but it generated a chill at the FDA, where the memory of thalidomide was still strong. Dr. Frances O. Kelsey, heroine of that sad episode, was placed in charge of overseeing the DMSO research. She quickly restricted the clinical DMSO studies of humans to topical applications to the skin.

But unlike the situation with most new drugs, where availability is limited and readily controlled, DMSO of a commercial grade was already available throughout the country by the gallon or tank car. Evidence soon indicated that a great many uncontrolled applications of DMSO were being made by laymen as well as by doctors, and rumors of black market DMSO circulated. By April 1965, when Rosenbaum, Herschler and Jacob presented their findings on DMSO to the American Cancer Society,[14] the *New York Times* editorially[15] urged caution in its use; and when they published their first major paper* in the *Journal of the American Medical Association*,[16] the *Journal* ran a companion editorial warning that the study had serious limitations, that claims for the drug had not been adequately proven, and that persons who medicated themselves with DMSO might be endangering their health or lives.

The public controversy over DMSO continued[17–20] until November 12, 1965, when the FDA suddenly withdrew the New Drug application and banned all clinical testing of the substance on humans, because evidence from three studies showed that animals given large doses over long periods of time developed eye trouble.[21] Shortly thereafter, the public was hearing[22] that the FDA was seizing bottles of the "wonder drug" intended for human use.

Disposition

The unusually restrictive action by the FDA, in the absence of evidence that serious adverse effects had been encountered in any

---

*The paper described results on 548 patients.

of the dozens of clinical studies,* created much controversy and brought charges of suppression of a valuable medicine. By March 1966, a Congressional subcommittee held hearings on the matter, at which FDA Commissioner Goddard and his aide, Dr. Sadusk, stated that the FDA had failed to enforce properly the laws on test of the "miracle drug" on humans.[23] A few days later the New York Academy of Sciences held a widely publicized[24-27] three-day conference on biological actions of DMSO at which seventy-nine papers were presented.[28] The gist of the conference was summarized in part by one reviewer[26]: ". . . rarely has a new drug come so quickly to the judgment of the members of the health professions with so much verifiable data from so many parts of the world, both experimentally and clinically, as to safety and efficacy."

Under this pressure, and after additional evaluation and further animal studies, the FDA lifted the total ban on clinical testing of DMSO on humans on December 22, 1966, to permit studies under strict control for conditions "where effective treatments were unavailable."[29-31] This provision still effectively limited the study of DMSO in the United States and continued to be a source of controversy among medical researchers and the public. On September 10, 1968, the FDA further relaxed its position after completion of toxicity tests on humans, but still maintains close restrictions of experimental studies or therapeutic use in the United States. DSMO has been more widely adopted for therapeutic uses by doctors in Europe and Mexico.

The initial ban by the FDA has remained a matter of curiosity and controversy. In 1973 the National Academy of Sciences — National Research Council (NAS-NRC) completed a study of DMSO and the FDA's decisions: it concluded that benefits of certain DMSO usage appeared to be established and that adverse effects of normal usage had not been established (including eye damage), but that the initial FDA action was justified on the basis of evidence available in 1965.[32,33] the FDA's decisions on DMSO have, however, inspired a recent book[34] and articles by a nationally syndicated columnist[32,33] that are highly critical of the FDA.

Comment

The rather cautious but enthusiastic early reports on the medicinal uses of DMSO were well suited to publicity in the news media,

*Estimates of the numbers of patients that had received DMSO by 1965 range from 20,000 to 100,000.

not only because man has an almost inherent interest and belief in miraculous cures and wonder drugs for his infirmities, but also because pharmaceutical and medical technology had successfully developed a number of remarkably effective drugs in recent years. But the publicity, which started even before any of the experimental results had been published in the medical literature, was apparently inspiring a sizable amount of uncontrolled experimentation that worried portions of the medical profession and the FDA. The latter was faced with the unusual problem of trying to regulate a new drug of known low toxicity that was already available in tank-car volumes as an industrial chemical—a problem vastly different from the situation with normal new drug applications.

Still shocked by the thalidomide tragedy, the FDA overreacted as soon as the first minor adverse effect of DMSO was reported, and took the unusually harsh and almost unprecedented step of banning all clinical testing of the substance on humans. During the subsequent controversy, the FDA retreated from this total ban, but has effectively prevented normal drug evaluation even to the present day. The original ban was apparently imposed because the FDA could not see a reasonable method of controlling the situation. But the ban became a symbol of FDA authority and policy, and this has made it difficult for the agency to bring itself to permit medicinal use of DMSO, even though nearly a decade of experience in the United States and in foreign countries has apparently shown that this unusual material can be used safely and effectively for certain therapeutic purposes.

REFERENCES

1. "Crown Zellerbach Readies Chemical Plant," C&EN, 38, 24 (May 23, 1960).
2. "Dimethyl Sulfoxide: Reaction Medium and Reactant," Technical Bulletin of Crown Zellerbach Corporation, Chemical Products Division, Camas, Washington, June 1962.
3. "Reduction of Dimethyl Sulfoxide to Dimethyl Sulfide in the Cat," Science, 144, 1137 (May 29, 1964).
4. "Sweet Taste of DMSO," Newsweek, 62, 68 (December 23, 1963).
5. "New Chemical Makes Sprays Work Better," Farm J, 88, 54 (February 1964).
6. New York Times, p. 77 (February 2, 1964).
7. "Dimethyl Sulfoxide (DMSO): A New Concept in Pharmacotherapy," S. W. Jacob, M. Bischel, and R. J. Herschler, Current Therap Res, 6, 134 (1964).

8. "Rosenbaum, E. E., and S. W. Jacob," *Northwest Med*, **63**, 167 & 227 (1964).

9. *New York Times*, p. 35 (March 19, 1964).

10. "New Wonder Drug: DMSO," *Life*, **57**, 37 (July 10, 1964).

11. "Great DMSO Mystery," B. Davidson, *Sat Eve Post*, **237**, 72 (July 25, 1964).

12. "DMSO: Fact and Fancy," W. B. Morse, *Amer Forest*, 70 (6 September 1964)

13. *New York Times*, p. N6 (February 28, 1965).

14. *New York Times*, p. 56 (April 1, 1965).

15. *New York Times*, p. 28 (April 3, 1965).

16. "Dimethyl Sulfoxide in Musculoskeletal Disorders," E. E. Rosenbaum, R. J. Herschler and S. W. Jacob, *J Am Med Assoc*, **192**, 309 (April 26, 1965).

17. "Too Good to be True," *Newsweek*, **65**, 109 (May 10, 1965).

18. "The Promise and Perils of the 'Miraculous DMSO'," *Good H*, **161**, 68 (August 1965).

19. *New York Times*, p. 25 (August 14, 1965).

20. "A Limited Wonder," *Time*, **86**, 82 (September 17, 1965).

21. *New York Times*, p. 49 (November 12, 1965).

22. *New York Times*, p. 24 (March 10, 1966).

23. *New York Times*, p. 22 (March 15, 1966).

24. *New York Times*, p. 42 (March 16, 1966).

25. *New York Times*, p. 31 (March 17, 1966).

26. "Meetings: Dimethyl Sulfoxide," C. E. Leake, *Science*, **152**, 1646 (June 17, 1966).

27. "Biological Actions of Dimethyl Sulfoxide," E. M. Wyer, H. Hutchins, M. Krauss, and C. D. Leake, *Ann New York Acad Sci*, **141**, 1–671 (1967).

28. *New York Times*, p. 27 (December 23, 1966).

29. *New York Times*, p. 30 (December 29, 1966).

30. "Bringing Back DMSO," *Newsweek*, **69**, 48 (January 9, 1967).

31. *Kansas City Star*, p. 34 (October 18, 1973).

32. Kilpatrick, J. J., in *Kansas City Star*, p. 34 (October 18, 1973).

33. *Chicago Daily News*, p. 10 (October 22, 1973).

34. McGrady, P., Sr., *The Persecuted Drug: The Story of DMSO*, Doubleday Company, Inc., Garden City, New York (1973).

X-ray Shoe-fitting Machine

Abstract

During the late 1940's fluoroscopic shoe-fitting machines were widely used. The hazards of these x-ray devices, particularly to children and to the shoe salesmen, were widely discussed by medical groups and in the press, but their use was not completely discontinued until the early 1960's.

Background

X-rays came into wide use for medical and other diagnostic purposes during the 1920's, and X-ray shoe-fitting machines, in which the customer could actually see the bones of his feet on a fluoroscope screen, came into use during the 1930's. Some boards of health apparently tried to have these devices banned, and a committee of scientists and representatives of government agencies drew up a safety code during World War II which urged use and exposure limits and the posting of warning signs.[1] After the war, however, the use of the machines spread rapidly throughout the United States, particularly in shoe stores or departments which specialized in children's shoes, in part because of the entertainment value of the machines. These machines became of increasing concern to public health authorities. The New York Roentgen Society and the Radiological Committee of the Medical Society of the County of New York expressed disapproval of the use of fluroscopic shoe-fitting machines, and in 1947 the Committee on Public Health Relations of the New York Academy of Medicine reported that the machines were potentially dangerous to attendants (who may be exposed to X-rays many times a day), particularly dangerous to children, and unnecessary in any event since they did not contribute materially to better fitting of shoes.[2] The

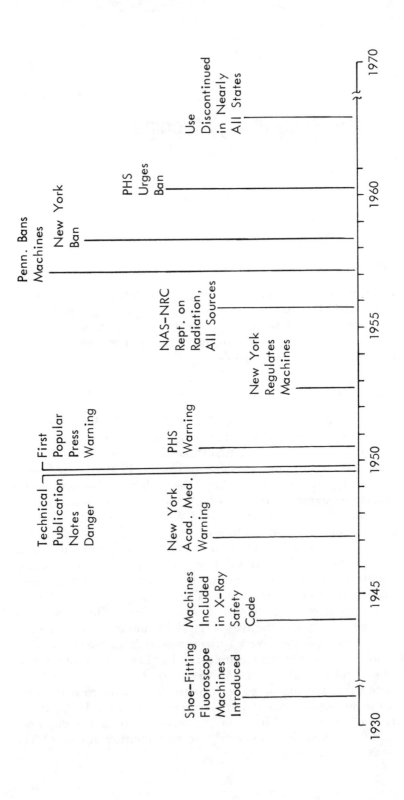

Committee noted that the results of exposure to X-rays are not always immediately apparent (biological changes may sometimes manifest themselves after the passage of years), and also noted that whereas the manufacturers of the machines claimed that exposures never lasted more than eight to ten seconds, they were an attractive hazard to children and that many exposures per day and per year were possible.[2]

The New York City Health Department, therefore, adopted in 1948 the American Standards Association[1] maximum limit of 2 roentgens per exposure and further specified limits of three exposures per day and 12 per year with the X-ray shoe-fitting machines, as well as the posting of warning signs regarding repeated exposure.[2] The New York City regulations also limited scattered X-radiation to 12.5 milliroentgens per hour in areas used by machine operators, that is, the salesmen. The more strict limits of 12 milliroentgens per minute[1] for maximum beam intensity and 5-second maximum exposure time were said[3] to be recommended in Detroit and in Massachusetts in 1948.

Key Events and Roles

Two notes in the *New England Journal of Medicine* in September 1949 were instrumental in raising the X-ray shoe-fitting machines to a national issue. In the first,[3] Dr. C. R. Williams reported that twelve typical machines in shoe stores used exposure times from five to forty-five seconds (average was about twenty seconds), and that radiation emissions were 0.5–5.8 roentgens per second with total average dose of 10–116 roentgens actually delivered to the customer's foot per exposure—five to fifty-eight times the accepted "safe" maximum.[1] Williams also reported that scatter of the X-ray beam from the port of most machines could expose customers and operators to as much as 100 milliroentgens per hour at a distance of ten feet, and that X-ray leakage through the cabinet walls ranged from 2 milliroentgens per hour to as much as two hundred milliroentgens per hour at the walls of some machines. In the second note,[4] the potentials for damage by X-rays to growing bones in children, to the skin of any customer, and to the blood-forming tissues of shoe store employees, were discussed. The dangers (particularly to children's bones) of the overuse of fluoroscopic shoe-fitting devices-or of poorly regulated devices, such as those tested by Dr. Williams, were indicated.[4] The results of these two studies were immediately publicized in *Time*[5] and in *Consumer Reports*.[2]

This publicity did not, however, trigger an immediate and effective action against the dangers presented by the fluoroscopes nor even a very productive public discussion. The Public Health Service (PHS) and the Food and Drug Administration (FDA) apparently issued a public warning on the dangers of the machines in early 1950. *Consumer Reports* commented[7] in April 1950 that it was hard to understand why public health authorities continue to permit the use of the machines in view of their known hazards and absence of real need, that all communities should outlaw their use for shoe-fitting, and that the FDA should declare them harmful and forbid their sale in interstate commerce. (In contrast, a doctor stated in *Today's Health* that the value of the X-ray shoe-fitter could not be denied, but did indicate that proper supervision was necessary.[8])

No Federal regulation of the X-ray shoe-fitting machines* was forthcoming, however, and regulatory action developed on a city-by-city and state-by-state basis.† In fact, the only Federal law which would apply even today is the "Radiation Control for Health and Safety Act of 1968" which applies to all devices which can emit *accidental* radiation. Thus, the literature of the 1950's contains periodic announcements that local regulatory bodies had set new standards for or had banned the shoe-fitting machines and numerous articles in which the hazards of the machines were discussed, especially in relation to X-ray hazards in general.[9-13] Nevertheless, approximately ten thousand X-ray shoe-fitting machines were in use in the mid-1950's according to the PHS.[12,13]

A 1956 report on radiation by the National Academy of Sciences[11-14] was a focal point in bringing a change in public opinion on X-rays in general. Thus, the New York State Health Department in late 1956[15] further tightened their 1953[16] regulations on the shoe-fitting fluoroscopes, and the Pennsylvania Health Department in 1957 and the New York City Health Department in 1958 barred the use of the machines in shoe stores.[17,18]

---

*The manufacturers of the shoe-fitting machines appear to have been almost all small companies, many of which were located in Milwaukee where the General Electric Company produced X-ray equipment for medical use.

†The Federal Government's difficulties at the time with the small producer of another product (see the AD-X2 Battery Additive case) may have had an influence here, although the U.S. Public Health Service emphasized that it lacked regulatory powers.

Disposition

In 1960, the PHS urged that all states ban or curb the use of fluoroscopic shoe-fitting machines.[19] The PHS advice seems to have been followed by most states, although the record is not at all clear. For example, in introducing legislation on radiation control, Senator Bartlett stated on July 10, 1967[20] that eighteen states still had not banned their use. In the hearings in October, one expert witness stated that the machines were still *in use* in a few states, but the Surgeon General and the director of the PHS's National Center for Radiological Health testified that they thought the machines were no longer in use or in production. Yet in May 1968, Ralph Nader stated in a popular press article[21] that "the foot floroscope machine—often used as a plaything to divert children in shoe stores—is still *used* in eighteen states" and further stated in Senate hearings[20] that "thousands of these machines, shooting radiation up to the gonadal area of these children, are still in operation." This latter statement was questioned by an American College of Radiology spokesman, and was not substantiated. A Public Health Service official indicated in 1972[22] that they were not aware of any current use of the machines, and that probably all but one or two states now had a legal ban on them.

Comment

The X-ray shoe-fitting machines were introduced at a time when Federal regulatory agencies felt either disinclined or powerless to clamp down on hazardous industrial products unless the danger was obvious and imminent. After many years of controversy, protest, and confusion over these machines and after a series of other cases of public alarm over technology, this unnecessary hazardous shoe-fitting practice was gradually discontinued. But because it was phased out quietly (on a city-by-city or state-by-state basis), confusion remained over whether the machines were still in use or not, even among witnesses before the Congress. While the X-ray shoe-fitting machine episode helped establish an atmosphere that led to more strict regulation of hazardous products, it probably did so at some small loss of confidence by the public in its regulatory agencies, in its radiological and medical technologists, and in the shoe stores that used such instruments to merchandise their products. Worth noting in this case, is that a popular new technology was glamorized for a profit at a time when little or no regulatory controls existed. These controls exist today in the X-ray area.

REFERENCES

1. "Safety Code for the Industrial Use of X-Rays," 254.1, 54. American Standards Association, New York, 1946 (quoted in Ref. 3).

2. "Fluoroscopic Shoe-Fitting: It Presents a Hazard You and Your Children should Avoid," *Consumer Rep*, **14**, 468 (October 1949).

3. Williams, C. R., "Radiation Exposures from the Use of Shoe-Fitting Fluoroscope," *New England J Med*, **241**, 335 (1949).

4. Hemplemann, L. H., "Potential Dangers in the Uncontrolled Use of Shoe-Fitting Fluoroscope," *New England J Med*, **241**, 335 (1949).

5. "Little Feet, Be Careful," *Time*, **54**, 67 (September 19, 1949).

6. *New York Times*, p. 38 (March 29, 1950).

7. "X-Rays are Not for Shoe Buyers," *Consumer Rep*, **15**, 174 (April 1950).

8. "Protecting the Feet," *Today's Health*, **29**, 13 (October 1951).

9. "X-Rays and Pregnancy," *Today's Health*, **32**, 27 (June 1954).

10. "The Choice and Fitting of Shoes," *Consumer Rep*, **19**, 486 (October 1956).

11. "Now There's a Warning About Too Much X-Ray," *US News*, **40**, 60 (June 22, 1956).

12. "Our Irradiated Children," *New Repub*, **136**, 9 (June 17, 1957).

13. "What's the Truth About Danger in X-Rays?" *Read Digest*, **72**, 25 (February 1958).

14. *New York Times*, p. 1 (June 13, 1956).

15. *New York Times*, p. 10 (December 28, 1956).

16. *New York Times*, p. 22 (May 25, 1953).

17. *New York Times*, p. 65 (January 27, 1957).

18. *New York Times*, p. 29 (January 23, 1958).

19. *New York Times*, p. 10 (August 19, 1960).

20. "Radiation Control for Health and Safety Act of 1967," Hearings Before the Committee on Commerce, U.S. Senate:

(a) S-2067, August 28–30, 1967; and

(b) S-2067, S-3211, and HR-10790, May 6, 8, 9, 13, and 15, 1968.

21. Nader, Ralph, "Wake Up, America: Unsafe X-Rays," *Ladies Home J* (May 1968).

22. Personal communication, October 1972.

Abuse of Medical and Dental X-Rays

Abstract

In 1956, the National Academy of Sciences issued a report on biological effects of atomic radiation which contained a conclusion which surprised the public. The report concludes that medical X-rays were a more important contributor to the average U.S. resident's exposure to radiation damage than the highly publicized fallout from atomic weapons testing. The report said that much of this exposure could and should be reduced, and thereby touched off a debate over medical and dental radiation practices which lasted nearly 15 years.

Background

X-rays were first discovered in 1895 by Wilhelm Roentgen and within a decade X-radiation had been tested as a medical diagnostic (combined with photographic film) and therapeutic method. The General Electric Company's invention of an efficient, heated filament, high vacuum tube X-ray source in 1913 (the Coolidge tube) catalyzed the commercial production and increased the utilization of X-ray machines. By the 1920's the medical use of the X-ray had become substantial and worldwide, but not without serious bad effects. A large number of doctors, dentists, and X-ray machine operators were suffering burns, loss of hands, and even death from frequent overexposure to X-rays; some are estimated to have received radiation doses of hundreds to thousands of roentgens per year. An increasing number of patients were also developing aftereffects of overexposure. In response,* a number of radiation pro-

---

*In addition to manmade X-rays, the discovery that cosmic rays from outer space were hitting the earth created much public concern in the 1920's.

189

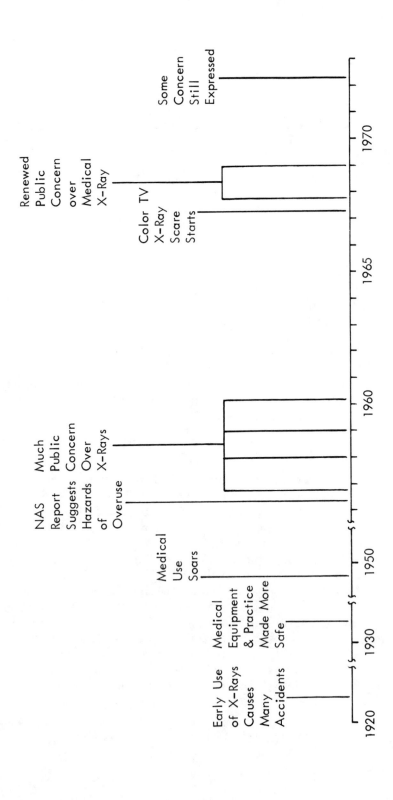

tection groups or committees were organized, including those of the American Roentgen Ray Society in 1922, the American College of Radiology in 1923, and the International Commission on Radiological Protection (ICRP) in 1928. In 1929, the predecessor of what is now called the National Council on Radiation Protection and Measurement (NCRP) was set up to speak for U.S. groups at the ICRP. The NCRP was ostensibly a nongovernmental group, although it later came to speak with almost governmental authority because of the strong involvement of the National Bureau of Standards as a collaborator with radiological societies and the X-ray industries. The efforts of these groups led to safer equipment and practices: the hazards of medical X-ray use to the worker were decreased to the point that they were no longer obvious or easily noticeable. (The NCRP in 1934 recommended the maximum permissible whole-body exposure limits of 0.1 roentgens per day or 0.5 roentgens per week for radiation workers.) The result was that as medical X-rays gained a reputation for safety, their use by doctors and dentists increased dramatically during the late thirties and thereafter.

The use of the atomic bomb on Japan by the United States near the end of World War II and the impending advent of peaceful uses of atomic energy again focused attention on radiation problems.* In 1946, the NCRP organization was enlarged (to include individuals concerned with atomic energy problems), and two industry groups (the present American National Standards Institute and the Underwriters Laboratories) moved to include radiation in their standard setting and testing programs. The discovery that postwar atomic weapons-testing was producing worldwide radioactive fallout, coupled with the clear recognition in 1948 that radiation could produce detrimental genetic effects (a concept which had been under study for some fifteen years), brought greatly increased concern over radiation in general. As a result, the NCRP decreased its acceptable radiation exposure level for workers to 0.3 rem† per week in 1949, and recommended a maximum permissible limit of exposure of 0.03 rem per week in 1952 for the population at large. The primary public and news media concern, however, was over

---

*The Atomic Energy Act of 1946 gave to the federal government, for the first time, control over certain artificial radioactive materials, but control over X-ray machines continued to reside with the states.

†By definition the roentgen applies only to X- and gamma radiation; the rem is the biological equivalent unit for all ionizing radiation, i.e., for X-rays the rem is essentially interchangeable with the roentgen.

the fallout from weapons testing—a concern which became a national election issue by 1956—and the news media gave relatively
little coverage to the dangers of X-rays,[1-4] even to children or pregnant women. (See, however, the case of the "X-Ray Shoe-Fitting
Machine").

Key Events and Roles

On June 13, 1956, the National Academy of Sciences, a con-
gressionally chartered, quasiofficial governmental advisory group,
released the results of a study by their National Research Council
entitled "The Biological Effects of Atomic Radiation."[5] The major
conclusions of the report were announced at a press conference on
June 12, and included one which surprised the public: fallout from
atomic weapons tests to date was a minor source of radiation danger
compared to the radiation Americans were receiving from medical
and dental X-rays and fluoroscopes. Furthermore, the report noted,
man's most radiation-susceptible biological system is the genetic-
inheritance mechanism*. The report recommended limits of exposure of the reproductive cells of the gonads (testicles and ovaries)
including:

1. Total exposure of any individual's gonads should not be
   more than 50 roentgens from conception up to age thirty,
   including the natural background radiation of 4.3 roentgens
   up to that age, or more than 100 roentgens up to age forty.
2. The total average exposure of the population's reproductive
   cells should not exceed 10 roentgens above the natural
   background up to age 30.
3. Each individual should have a record showing his total accumulated lifetime exposure.
4. The medical use of X-rays should be reduced as much as is
   consistent with medical necessity, because X-rays were already using up nearly one-third of the allowable average
   radiation limit.†

The press conference and report created an immediate sensation in the news media, which was supplemented by numerous

---

*In addition, the report noted that higher radiation levels (for example, 100 R in
a single exposure) could cause shortening of life, bone damage, and even radiation-
induced diseases such as cancers and leukemia.

†The average radiation of reproductive cells of Americans up to age 30 was
estimated to include 4.3 roentgens from natural background, 3.0 roentgens from
X-rays, and 0.02–0.5 roentgens from atomic weapons test fallout.

comments of agreement and disagreement by various authorities.[6-15]The public X-ray debate continued through 1957, 1958, 1959 and into 1960 before subsiding.[16-28] Included in the debate were charges that many doctors, radiologists, and dentists took far too many diagnostic X-rays, used therapeutic X-rays for many conditions (such as acne, plantar warts, eczema and superfluous hair) where other treatments were less hazardous, failed to use modern, safe equipment or proper shielding, failed to keep adequate exposure records for patients, and frequently used fluoroscopic examination when film recording would expose the patient to far less x-radiation. Much of the concern centered on the incomplete state of knowledge regarding genetic and other effects of small doses of X-rays, and on the fact that more than 200,000 medical and dental X-ray machines were being used to expose millions of Americans annually to these possibly harmful effects. (Published estimates of the number of Americans X-rayed annually increased from 3 million in 1957[21] to over 100 million in 1958;[23] one 1958 account[22] lists 100 million dental X-ray films, 54 million diagnostic X-ray films and fluoroscopies, and 15 million mass-survey chest X-rays with photofluorographic machines.) As a result of the scare, many articles noted, medical people were fearful that many people were avoiding having needed X-rays.

During the early to mid-1960's the concern for the hazards of medical X-rays essentially dropped from the public[29-31] to the technical level (where concern was also being given to radiation from other sources). Then, in 1967, the discovery that many color television sets were emitting significant amounts of X-rays (see "X-Radiation From Color TV") regenerated the entire medical X-ray controversy,[32-35] particularly as a result of the hearings in the U.S. Congress in the fall of 1967.[36,37a] These hearings continued on into 1968[37b] and generated strong claims and counterclaims between scientific experts regarding the safety of medical and dental X-ray practice* as well as further articles in the popular press.[38-41] In general, the American College of Radiology, the American Dental Association, and other spokesmen from private medical practice contended that existing training, equipment, procedures and regulations were adequate for public safety, whereas a number of independent spokesmen seriously questioned numerous aspects of these features.

*The published reports of these hearings consist of over 2,000 pages of testimony and supplementary material, of which perhaps 30 percent involved medical and dental X-ray considerations.

Disposition

One result of the 1956 NAS report and the concern which followed, was that the government concluded it should not rely entirely on private groups to set radiation standards: the Federal
Radiation Council (FRC) which contained representatives of seven
major government agencies was established in 1959. The FRC
(which was subsequently merged into the new Environmental Protection Agency in 1970) was enpowered to consult with qualified
scientists or experts and to provide guidance to the President in
matters regarding all types of radiation. A National Center for
Radiological Health was also established within the U.S. Public
Health Service in January 1967 (with a budget of $15.8 million and
staff openings for 814 persons) and subsequently assigned to FDA
in May 1971. In October 1968 the Congress passed the Radiation
Control for Health and Safety Act of 1968.

Although this act primarily pertains to unintentional radiation
exposures and does not attempt the regulation of X-ray usage by
doctors, dentists or radiologists (which was left up to the states and
professional standards), concern over possible dangers of medical
X-rays nearly disappeared as a news item during ensuing
months.[42-44] In mid-1970 the national furor which erupted over nuclear radiation standards[45-46] largely displaced the medical X-ray
from concern.

In the fall of 1970, however, the Public Health Service presented preliminary results of a national survey of X-ray exposure in
the healing arts. The PHS presentation[47] emphasized the positive:

- The population exposure to X-rays may have been less in
  1970 than in 1964, despite an increase in the number of persons X-rayed (20%) and the average number of X-rays per
  person (10 percent).
- The use of X-ray beam limitation and patient shielding had
  increased from 1964 (less than 50 percent of all examinations) to 1970 (about 67 percent of examinations).
- The X-ray examination rate declined among persons in the
  most fertile ages (fifteen to twenty-nine years).
- The use of standard X-ray film had extensively replaced
  miniature film in mass chest X-ray programs.

These results did not still all concern about medical and dental
X-rays. Since only a few states even today have licensing laws regarding X-ray technicians and only one state(California) requires

that a medical doctor be licensed to operate or supervise an X-ray machine,* the qualification of operators of medical X-ray machines continued to be of concern.[48]

Other articles[49-50] looked at the data from a different view: they emphasized that medical and dental X-rays were still too widely used (130 million Americans had 640 million exposures), by too many unqualified operators (25 percent were not qualified technicians), by too many unqualified operators (25 percent were not qualified technicians), and with too much unnecessary exposure to parts of the patient's body (in one-third of all examinations). The mobile TB X-ray program also continued to be a source of concern[51] until it was discontinued in 1972.[52]

Finally, in late 1972 a National Academy of Sciences study of radiation standards† (funded by the old FRC) noted that medical exposures were still the major source of radiation to the public: they were averaging 72 millirems per person per year and were increasing at a rate of one to four percent per year. The report further asserted that the level could be reduced substantially at little cost and no sacrifice of medical benefits.[46]

Comment

Public concern over medical and dental X-rays has continued for over a half century. This long period without reaching a satisfactory solution indicates not only the difficulty in determining what constitutes acceptable risk-benefit relationships, but also suggests that powerful interests are involved. Indeed the medical and dental communities, the instrument manufacturers, and the private standards-setting bodies constitute powerful and rather close-knit (if not frequently interlocking) groups. The governmental regulatory bodies, in deference to the "sanctity of the doctor-patient relationship," have been extremely reluctant to initiate regulatory controls or even guidance. Periodic surveys of public opinion concerning faith in medical and dental X-ray practice would be of much interest, particularly because of the news media coverage and the numbers of people having (and paying for) X-rays. To a degree, this

*Most states now require licensed practitioners for fluoroscopy according to a PHS source.

†The NCRP had recommended a maximum of 170 millirems per person per year of man-made radiation exposure excluding medical exposure and background radiation.

196

case illustrates the old maxim that familiarity breeds contempt; the medical doctors may have been aware of the dangers of X-rays at one time, but liked the convenience and aid of the pictures and gradually increased their use.

REFERENCES

1. "X-Ray May Stunt Children's Growth," *Sci Digest*, **33**, 53 (April 1953).
2. "X-Rays and Pregnancy," *Today's Health*, **32**, 27 (June 1954). This article was stimulated by a conference held by the Biology Division, Oak Ridge National Laboratory, which also led the *Journal of the American Medical Association* to warn editorially against X-raying the pelvic region during pregnancy.
3. "X-rays Can Be Dangerous," *Coronet*, **38**, 49 (August 1955).
4. "Small X-Ray Doses Damaging to Unborn," *Sci NL*, **69**, 215 (April 7, 1956).
5. "The Biological Effects of Atomic Radiation," National Academy of Sciences, National Research Council, Washington, D.C., June 13, 1956.
6. *New York Times*, p. 14 (June 14, 1956).
7. *New York Times*, p. 32, Editorial, June 14, 1956.
8. "Atomic Radiation, the r's are Coming," *Time*, **67**, 64 (June 25, 1956).
9. "Now There's a Warning About Too Much X-Ray," *US News*, **40**, 60 (June 22, 1956).
10. "What You Should Know About Danger from X-Rays," *US News*, **40**, 44 (June 29, 1956).
11. *New York Times*, p. 8, Letter (July 1, 1956).
12. *New York Times*, p. 17 (August 2, 1956); p. 5 (August 7, 1956).
13. "X-Ray Danger," *Time*, **68**, 67 (October 1, 1956).
14. *New York Times*, p. 68 (November 8, 1956); p. 30 (November 21, 1956).
15. "Fight X-Ray Scare," *Sci NL*, **70**, 338 (December 1, 1956).
16. The *New York Times Index* shows the following numbers of entries on X-ray dangers: 1957—12, 1958—12, 1959—9, 1960—7.
17. *Sci NL* carried X-ray danger articles as follows: 1957—2, 1958—6, 1959—3.
18. "Radiation; Hazard No. 1 of the Atomic Age," *Sr Schol*, **70**, 10 (March 15, 1957).
19. "X-Ray Duel," *Sat R*, **40**, 36 (June 1, 1957).
20. "Our Irradiated Children," *New Repub*, **136**, 9 (June 17, 1957).
21. "The Bad in a Good Thing," *Newsweek*, **50**, 78 (October 14, 1957).
22. "What's the Truth About Danger in X-Rays?" *Read Digest*, **72**, 25 (February 1958).

23. "Safe X-Ray?" *Newsweek*, **51**, 70 (May 5, 1958).

24. "Will X-Rays Make You Sterile?" *Sci Digest*, **43**, 66 (June 1958).

25. "Aftermath of X-Rays," *Time*, **73**, 62 (February 23, 1959).

26. "Common Sense About X-Ray Dangers," *Changing Times*, **13**, 35 (February 1959).

27. "Let's Talk Sense About X-Rays!" *Coronet*, **46**, 92 (August 1959).

28. "Will Radiation Harm Your Child," *Parents Mag*, **35**, 35 (June 1, 1960).

29. The *New York Times* carried five articles during 1961–1966 on X-ray problems.

30. *Sci NL* carried four articles during 1961–1966 on X-ray problems.

31. "Must You Fear Medical X-Rays?" *Good H*, **154**, 155 (April 1962).

32. *New York Times*, p. 16 (August 31, 1967); p. 35 (October 12, 1967).

33. Nader, Ralph, "X-Ray Exposures," *New Repub*, **157**, 11 (September 2, 1967); *Discussion*, **157**, 37 (September 30, 1967).

34. "X-Ray Excess," *Time*, **90**, 69 (September 8, 1967).

35. "Ray of Danger," *Newsweek*, **70**, 63 (September 11, 1967).

36. "Electronic Products Radiation Control," Hearing before the Subcommittee on Public Health and Welfare of the Committee on Interstate and Foreign Commerce, U.S. House of Representatives, on HR 10790, August 14, September 28, October 5, 11 and 17, 1972.

37. "Radiation Control for Health and Safety Act of 1967," Hearings before the Committee on Commerce, U.S. Senate:
    (a) on S-2067, August 28, 29 and 30, 1967, and
    (b) on S-2067, S-2311 and HR-10790, May 6, 8, 9, 13, and 15, 1968.

38. *New York Times*, p. 30 (May 18, 1968).

39. Nader, Ralph, "Wake Up America, Unsafe X-Rays," *Ladies Home J*, **85**, 126 (May 1968).

40. "How Safe are X-Rays?" *Today's Health*, **46**, 12 (June 1968).

41. "Are Doctors Using Too Many X-Rays?" *Good H*, **167**, 151 (August 1968).

42. *New York Times*, p. 78 (December 8, 1968).

43. "What You Should Know About X-Rays and Pregnancy," *Redbook*, **132**, 30 (February 1969).

44. Radiation Phobia," *Nat R*, **22**, 361 (April 7, 1970).

45. "Radiation Standards: Are the Right People Making Decisions?" *Science*, **171**, 780 (1971).

46. "Radiation Standards: The Last Word or at Least a Definitive One," *Science*, **178**, 966 (1972).

47. "Radiological Health Protection," *FDA Papers*, **5** (10), 44 (December 1971–January 1972).

48. "HEW Reveals Lack of X-Ray Control," *Prevention*, 193 (January 1971).

49. Edelson, Ed, "Peril in Too Much X-Ray," Universal Science News Service, Inc., from *Kansas City Star*, January 26, 1972.

50. "Warning: X-Rays May Be Dangerous to Your Health," *Read Digest*, **101**, 173 (August 1972).

51. "Mobile TB X-Ray Units: An Obsolete Technology Lingers," *Science*, **174**, 1114 (1971); also Discussion *ibid.*, **176**, 184 (1972).

52. "Chest X-Ray," *Science*, **177**, 562 (August 18, 1972). This letter indicates action on February 18; an Associated Press story indicated April 23 (*Kansas City Times*, April 24, 1972).

X-Radiation From Color TV

Abstract

The General Electric Company announced on May 18, 1967, a recall of about 100,000 of its large-screen (eighteen to twenty-three inch.) color television sets made between June 1966 and February 1967, because they were found to emit low levels of X-rays toward the floor. The incident was the start of a three-year flurry of excitement which eventually involved several TV brands and stimulated passage of a Radiation Control Act in 1968.

Background

Although the broadcasting of television programs in color began in 1954, only about 2 million color sets were in use in the U.S. by 1963. The big swing to color came in 1964–1965 after the CBS and ABC networks joined NBC in presenting color telecasts: 2.7 million color sets were sold in 1965 alone. About 15 million color TV sets were in use at the beginning of 1967.

The public had heard a good deal about the hazards of X-rays in the years preceding this case, particularly since the 1956 report, "The Biological Effects of Atomic Radiation," made by the National Academy of Sciences.[1] In addition, a broad bill was before the Congress in 1967 regarding products which presented undue hazards to the public. Those familiar with high-voltage vacuum tubes had, of course, long been aware of their potential for X-radiation. The Underwriters Laboratories had in 1958 set a limit of X-ray emission of 2.5 milliroentgens per hour for TV sets. The National Council on Radiation Protection and Measurements (NCRP, a nonprofit group established in 1929, having representatives from the X-ray industries, radiological societies and the National Bureau of Standards), had in 1959 set the "generally-agreed upon" safe maximum X-ray

199

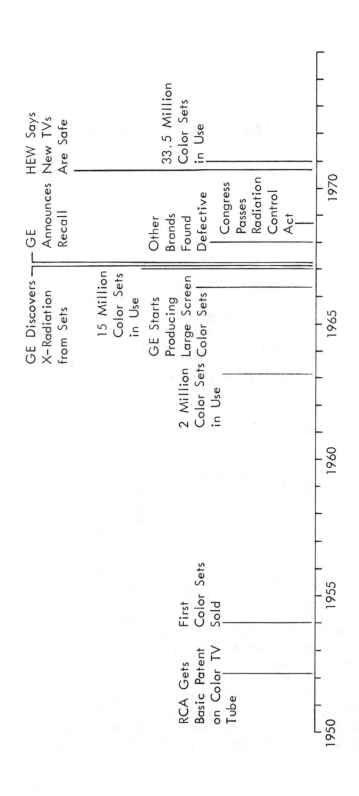

RCA Gets Basic Patent on Color TV Tube

First Color Sets Sold

2 Million Color Sets in Use

GE Discovers X-Radiation from Sets

15 Million Color Sets in Use

GE Starts Producing Large Screen Color Sets

GE Announces Recall

Other Brands Found Defective

Congress Passes Radiation Control Act

HEW Says New TVs Are Safe

33.5 Million Color Sets in Use

1950   1955   1960   1965   1970

leakage to 0.5 milliroentgens per hour at any point five centimeters from the surface of the TV set, under normal operating conditions.

The TV manufacturers were all aware that TV sets could be sources of X-rays, and that the color sets, in which the cathode ray tube operated at much higher voltage (25,000 volts) than in black-and-white sets, were potentially more hazardous. Therefore, metallic radiation shielding was a normal part of set design and radiation monitoring was a part of the quality control program. The General Electric Company (GE) began marketing color television sets in 1963. The sets were apparently made for GE by RCA until 1965 so that GE was a relatively new producer when this case opened. As the largest manufacturer of medical X-ray equipment, however, GE was certainly aware of the hazards of X-rays from television sets.

Key Events and Roles

On May 18, 1967, the GE Company and a U.S. Public Health Service spokesman announced that up to 90,000 of GE's large-screen (eighteen-in. or larger) color television sets could emit soft X-radiation, but that danger to the public was not serious and that the sets would be fixed by GE. The chronology of events leading up to this must be pieced together from subsequent testimony at Congressional hearings[2,3] rather than from press stories of the time.

In the spring of 1966, GE developed a new TV set which incorporated a special GE-designed voltage regulator tube* of a type which required external X-ray shielding. It was novelly mounted in a well in the bottom of the chassis inside the set. The Underwriters' Laboratories approved a prototype set, and production started in June 1966 at GE's Syracuse, New York plant, with the first sets reaching market about September 1. On November 21, 1966, a GE radiation safety officer's inspection detected a narrow beam (about one square inch) of X-radiation coming out of the bottom of some sets which routine quality control checks at the front, top and sides had failed to detect. The intensity of this beam was apparently greater than the NCRP limit, but did not cause immediate alarm. Company medical and radiological experts said that chances of personal injury from the sets were remote. In late November GE adopted a slightly different voltage regulator tube which apparently eliminated most of the downward emissions. In December, GE informed the New York State Department of Labor that an in-plant

---

*Also referred to as a shunt regulator tube.

radiation event had occurred, as they were required to do by state law.

In January, the New York State Department of Labor arranged for the assistance of the PHS's National Center for Radiological Health (NCRH) in measuring radiation in GE's plant after the correction program was installed. The NCRH, as a result of an August 1966 report of TV hazards and a February 1967 report of X-radiation through the face of the picture tube, had apparently already requested data from other agencies, including the manufacturers' Electronic Industries Association (EIA), and had even tested a few color sets. In January 1967, the EIA had merely replied that the industry was observing the NCRP recommendations.

In February, GE perfected external shielding for the voltage regulator tube which, combined with the new tube* and also with additional shielding for a voltage rectifier tube, "eliminated X-radiation from sets coming off the production line." At this point, GE had produced about 154,000 of the possibly defective sets: more than 43,000 of these were still in GE warehouses, and 21,000 were estimated to be in dealer stock, but 90,000 were estimated to have been sold to consumers. In addition, some 128,000 possibly defective voltage regulator tubes had already been shipped as replacement parts to dealers and to over 80,000 TV repairmen.

The factory procedures were not considered adequate for field modification work, however, and work continued at GE's Owensburg, Kentucky, plant to develop a completely new voltage regulator tube which was internally shielded for X-ray emissions. By April, these self-shield tubes were perfected and after consultation with outside medical and physical radiologists, GE started planning quietly for a field modification program.

Rumors were apparently circulating already, however, and on March 6 the NCRH had asked the EIA for specific test data on color TV sets. On April 10, a GE radiation expert informed the director of NCRH, Mr. James G. Terrill, Jr., that GE had sold color TV sets which gave excessive radiation and that GE planned a recall campaign. The NCRH requested further information from GE and also from the EIA on April 20 for all twenty-three TV-producers. On May 11, NCRH discussed the nature and scope of the problem with GE and then suggested that GE make a public announcement of the recall program, to be coordinated with a release from NCRH. Again, on May 17, NCRH, the New York State Labor Department,

*This may have been the second tube designed since the trouble started.

and the New York State and City Health Departments, all met with GE to discuss the problem. GE disclosed test data on 100 of the defective voltage regulator tubes which showed unshielded X-ray emissions at their anodes ranging from 300–4,200 milliroentgens per hour and averaging 725 milliroentgens per hour, and which indicated that emissions of many sets already in the hands of the public greatly exceeded the NCRP standards. The NCRH had apparently estimated emission levels of 10–100 milliroentgens per hour from the GE sets in early 1967, and had by April 24 found levels in the 200 to 300 range. Yet GE again indicated no plans to make a public announcement, nor apparently did the NCRH say it intended to release its statement immediately.

The story was leaked to the press and, when the *New York Times* inquired about the matter on May 18, the NCRH and GE both made public statements which were published jointly.[4] The announcements emphasized[5] that no serious public danger was involved: GE indicated that up to 90,000 sets could emit X-radiation, but the quantity emitted was far less than the 0.5 milliroentgens per hour maximum recommended by the PHS and the NCRP, and that much of it would be absorbed in the air beneath the set; NCRH said an individual would have to lie under the set in the path of the beam for forty hours to accumulate a slight radiation burn. GE announced[6] that servicemen would check and repair the sets free at the owner's convenience (although the modification would cost GE $10 to 20 a set, possibly a total of $1 million). The operation was expected to be completed by July 31. Initial press accounts of the announcement indicated that GE had discovered the problem only in February[5] rather than in the preceding November, but even the three-month delay raised questions.[6,7]

The recall campaign went more slowly than expected (the August issue of *Time*[8] said there were 9,000 unlocated sets; 1,000 were still missing on September 29,[9] but of more importance to the public, radiation levels were soon found to be substantially higher than initially indicated: *Consumer Reports* disclosed in July[6] that two GE color sets gave radiation "several times" the safe level. NCRH tests of 185 of GE's voltage regulator tubes found many gave excessive* unshielded emissions, "10–100,000 times the rate considered safe."[8] The Surgeon General, Dr. W. H. Stewart, thereupon released on July 21 a public statement regarding the GE sets concluding that "anyone who owns one should disconnect it until the

---

*Many were in the range 5 to 5,000 milliroentgens per hour; four gave emissions in the 5,000–50,000 milliroentgens per hour.

faulty tubes have been replaced." The alarm potential was greatly increased, of course, because children tend to sit on the floor very close to a TV set and would be the viewers most susceptible to genetic or bone damage by radiation.

Almost immediately, the GE recall campaign became involved in Congressional hearings, first on July 31, 1967, by the Senate Subcommittee on Commerce and Finance, and subsequently in August through October 1967, by both House and Senate hearings on radiation control for electronic products.[2a,3a] A major source of discontent* was that the original tubes required external shielding in the first place, that these posed particular threats to production line workers and repairmen, and that the recall of these tubes from the distribution channel was less effective than the field modification campaign (GE claimed less than 5,000 still extant on September 28). Despite these difficulties GE assured the hearings committees that the problem was near solution, and in September the Underwriters Laboratories adopted the lower 0.5 milliroentgens per hour emission standard of NCRP.[2a] The EIA testified on October 5 that the GE incident was "an isolated case and not representative of television set operation"[2a] and that they had on June 29 given NCRH data that tests of over 500 sets made by 20 manufacturers (excluding GE), which represented 95 percent of all color TV sales, had shown acceptably low levels of X-radiation.

Almost immediately, however, evidence began to accumulate that sets of other brands of color TV's also emitted X-radiation. In August 1967 a Pinellas County, Florida, PHS survey had suggested that other sets could also give radiation under certain conditions, and in January 1968 the *Consumer's Union* reported[10] that sets made by Admiral and by Packard Bell gave detectable X-radiation which increased above the 0.5 milliroentgens per hour level if the line voltage was raised from "normal" (120 volts) to a "normal high" of 125 volts. The radiation ranged up to four times the "safe" level, and came variously from the tops, sides and backs of the sets tested. In December the PHS had indicated that such emissions

---

*During the hearings, NCRH disclosed that a GE set operating with one particularly defective tube gave a calculated maximum exposure rate of 800,000 milliroentgens per hour at four inches below the set (normally floor level), that the "beam" had widened sufficiently to cover 30 to 50 percent of a child's head at 1 yard, and that levels of over 600 milliroentgens per hour could have been produced in a room below this set in normal home use. [The NCRH tests employed film techniques and apparently used the center (most intense) part of the beam whereas industry used a meter which averaged across the beam.]

might be an industry-wide problem.[11] The NCRH initiated (at the request of Secretary of HEW Gardner and with the cooperation of the EIA) a survey in the Washington, D.C., area among 20,000 PHS staff members' homes.[2c] The NCRH survey results, reported to the Congress[2b] and the public[12] in February, showed that many brands gave excessive radiation: 66 (nearly 6 percent of 1,124 sets) gave X-radiation which exceeded the standard and included ten brand names. Surveys in New York, Florida, and Hawaii gave similar results: 5 to 20 percent defective.[13] The NCRH even refused initially to make public the brand names,[13] but did in April under a "Freedom of Information Law" request[14] and indicated that as many as 1.4 million color sets in the country exceeded the radiation standards. The vast majority of the defective sets found in the PHS survey apparently represented no serious hazard to the health of viewers.[15]

Disposition

President Johnson's February 1968 message to Congress urged consumer protection.[17] In response particularly to the color TV issue, the Congress moved rather rapidly to pass the "Radiation Control for Health and Safety Act of 1968" to protect the public from radiation emissions from all electronic products. It was enacted on October 18, after lengthy hearings[2,3] and, according to some accounts, considerable lobbying by the Electronics Industries Association.[15,17] This legislation greatly reduced the news coverage of the color TV episode, although its lack of power for routine in-factory inspection and seizure of defective products were deplored.[17] The Bill also provided that the Secretary of HEW had until January 1, 1970, to set a legal standard for radiation from TV sets. The previously accepted figure of 0.5 milliroentgens per hour was adopted, although it was considered by *Consumer Reports*[18] as being too high and led Ralph Nader to charge that the 1968 radiation control act was insufficient.[19] Nevertheless, by late 1971, the Department of HEW was able to say that radiation emissions from new TV sets were no longer a "significant" health hazard.[20]

The recall of defective sets seems never to have been completely assured, however. In April 1969, for example, the New York PHS noted that a two-year survey of 5,000 homes showed that 20 percent of the color sets gave off "harmful" radiation and that all sets gave off some, but that little could be done except to provide free inspection.[21] The absence of an effective national policy

sparked the marketing of do-it-yourself radiation test kits, some of
which were banned in New York City as useless and possibly
misleading[15] and eventually became the subject of a warning by a
Department of HEW spokesman.[22]

In brief, the Government's action on the potentially large
number of defective sets already in use consisted of counsel by the
Surgeon General[15] and the Bureau of Radiological Health[22] to have
viewers always sit at least six feet from large-screen color sets, and
to use a "competent, well-trained serviceman" if one wished to
have his set checked. Many sets which were not defective when
sold had apparently been made defective by readjustments and
alterations of local servicemen so that the expense of recall could
not be assessed to the manufacturer. By late 1968, the manufacturers had set a goal of making sets which were "tamper-proof," and
such design is now required by law.

While the case caused considerable alarm and received rather
intense coverage by the news media, it had almost no impact on the
overall sales of color TV sets and sales continued a sharp, nearly
uniform climb throughout the 1966–1971 period.[23] Contributing to
the level of concern in this episode was the fact that a color TV
represented a large family investment,* (often exceeded only by an
automobile and the home). The owners of large-screen sets were
predominantly individuals in the upper economic classes and institutions or commercial organizations, and, therefore, exerted more
than average influence.

Comment

The great color television recall episode stemmed primarily
from deficiencies in analytical methods for X-radiation used in
quality control; it affected millions of Americans nationwide because of the mass-production capability of the technology. The efforts by the General Electric Company (and subsequently by other
TV manufacturers and the industry's trade associations) to
minimize the extent of the hazard posed and, in many eyes, to cover
up the recall and its associated problems, did much to generate
suspicion by the news media and the public. The seeming cooperation at times of the Public Health Service in some aspects of the
alleged cover-up probably decreased the public's confidence in its
regulatory agency as well as in the industry. The concern showed

*The *average* price of the 5.78 million sets sold in 1967 was $525.

by the news media and the Congress was surprisingly swift and intense, considering that no actual harmful effects could yet be attributed to the defective TVs; this concern reflected the high socioeconomic status of the sets' owners and the interests of the electronic news media.

REFERENCES

1. "The Biological Effects of Atomic Radiation," National Academy of Sciences, National Research Council, Washington, D.C., June 13, 1956.
2. "Electronic Products Radiation Control," Hearings Before the Subcommittee on Public Health and Welfare of the Committee on Interstate and Foreign Commerce, U.S. House of Representatives on HR-10790:
    (a) August 14, September 28, October 5, 11 and 17, 1972; and
    (b) Supplemental Hearings, February 1, 1968.
3. "Radiation Control for Health and Safety Act of 1967," Hearings Before the Committee on Commerce, U.S. Senate:
    (a) on S-2067, August 28–30, 1967; and
    (b) on S-2067, S-3211 and HR-10790, May 6, 8, 9, 13 and 15, 1968.
4. New York Times, p. 17 (May 19, 1967).
5. "TV That Can Bite," Bsns W, 61 (May 27, 1967).
6. "Owners of GE Color TV Sets — Take Note," Consumer Rep, 349, 32 (July 1967).
7. New York Times, p. 67 (August 1, 1967).
8. "X-Rays in the Living Room," Time, 90, 62 (August 4, 1967).
9. New York Times, p. 95 (September 29, 1967).
10. "Too Much Radiation From Two Sets," Consumer Rep, 33, 18 (January 1968).
11. New York Times, p. 94 (December 7, 1967).
12. New York Times, p. 71 (February 2, 1968).
13. "Are X-Rays Danger in Color TV?" Bsns W, 46 (March 30, 1968).
14. Washington Post, p. B2 (April 2, 1968).
15. "A Sensible Look at Radiation from TV Sets," Consumer Rep, 33, 492 (September 1968).
16. New York Times, p. 1 (February 7, 1968).
17. "TV Radiation," Consumer Rep, 34, 73 (February 1969).
18. "Some Observations on X-Radiation," Consumer Rep, 35, 23 (January 1970).
19. New York Times, p. 26 (January 5, 1970).
20. New York Times, p. 25 (September 17, 1971).
21. New York Times, p. 39 (April 8, 1971).
22. "Don't Try to Test TV Radiation at Home," Consumer Rep, 34, 560 (October 1969).
23. Merchand W, 104,(9), 17 (February 28, 1972).

Introduction of the Lampreys

Abstract

From 1939 to 1949 sea lampreys invaded the upper Great Lakes via the Canadian Welland canal system and extensively destroyed commercial fishing. In the 1960's, the double action of treatment of lamprey spawning grounds with a new chemical pesticide (TFM) and later restocking of the Great Lakes with young trout and salmon proved highly effective. By 1972, the commercial fisheries in the Great Lakes had been successfully restored, and the lamprey was under control.

Background

The sea lamprey (the scientific name, *Petromyzon marinus*, means rock-sucker) is a primitive fish closely resembling an eel. This fish is typically anadromous; that is, it hatches and spends its early life in fresh water, migrates to the ocean, where it grows to maturity and returns to fresh water to spawn.[1] During its adult life of one year, it is a parasite on fish.[2] Sea lampreys are found in streams along the Atlantic coast as far south as northern Florida.

This predator fish is a member of an odd and extremely ancient class of vertebrates, the round-mouths or Cyclostomata. This group, which is characterized by the absence of jaws, includes the lampreys and hag-fishes. Instead of jaws, the lamprey has a round, muscular, toothed sucker,[3] equipped with about 150 chitinous teeth positioned in concentric rows. The oral opening, which is located slightly below the center of the sucker, contains a toothed tongue, rough as a rasp, which can be moved back and forth.[3]

The full-grown lamprey, which ranges from twelve to twenty-four inches in length, preys on other fish by attaching itself to its victim with its sucker, rasping holes with its tongue in the body of

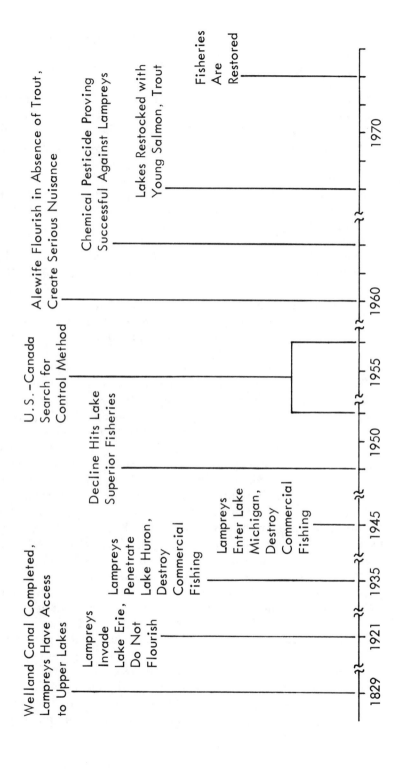

Welland Canal Completed, Lampreys Have Access to Upper Lakes

Lampreys Invade Lake Erie, Do Not Flourish

Lampreys Penetrate Lake Huron, Destroy Commercial Fishing

Lampreys Enter Lake Michigan, Destroy Commercial Fishing

Decline Hits Lake Superior Fisheries

U.S.–Canada Search for Control Method

Alewife Flourish in Absence of Trout, Create Serious Nuisance

Chemical Pesticide Proving Successful Against Lampreys

Lakes Restocked with Young Salmon, Trout

Fisheries Are Restored

1829   1921   1935   1945   1950   1955   1960   1970

the fish, and then sucking the blood and body juices.[4] Once it has gained a hold on a fish, the lamprey hangs on until it is satisfied, or the victim dies. The lamprey is capable of destroying fish much larger than itself; for example, it can kill a delicate fish, such as trout, in only four hours. After the lamprey attaches itself, the fish is slowed and becomes easy prey for other lampreys. Sometimes as many as four or five lampreys attach themselves to a single fish, bringing much quicker death to the predators' host.[2]

Elimination of the lamprey is difficult because the adult fish has no natural enemies. The lamprey's chief weakness is its breeding system; the adults swim up rivers in early spring to spawn. The young lampreys, which look like tiny worms, bury themselves in mud and lead a wormlike life, eating microorganisms.[9] After five years of this nonparasitic-type existence (when they have reached a length of about seven inches), they develop the toothy suckers and drift downstream to hunt fish in the ocean. The adult lamprey spends only 1 year in the deep water and then returns to a stream to spawn and die. In some localities, the sea lamprey has adjusted itself to spend its entire life cycle in fresh water, passing its adulthood in lakes instead of in the ocean.

The sea lamprey is an old inhabitant of the St. Lawrence River and Lake Ontario. Lampreys apparently did not create a serious problem in Lake Ontario, although the trout fishing there was normally inferior in both quality and quantity to that found in the upper lakes (Erie, Huron, Michigan and Superior). Prior to 1829 the Niagara Falls served as a natural barrier which blocked the lamprey from migrating from Lake Ontario into Lake Erie.[4] The completion of the Canadian Welland Ship Canal in 1829[1] provided a waterway for ships and, unfortunately, also for lampreys[5] between Lake Ontario and Lake Erie. In the 1950's this canal became part of the St. Lawrence Seaway, a navigation system extending from the Atlantic Ocean along the St. Lawrence River and through the Great Lakes to Duluth, Minnesota (a distance of 2,342 miles).[5]

Even after completion of the Welland Canal, lampreys did not soon gain entrance into the upper Great Lakes: they were not seen in Lake Erie until 1921, nearly a century later.[4] It is possible, of course, that sea lampreys straggled through the canal repeatedly, but were not seen or captured. The first known spawning population of sea lampreys in this area was discovered in 1932 in the Huron River, a tributary of Lake Erie. The lampreys multiplied very slowly in Lake Erie, apparently because of poor spawning conditions and never did become firmly established there.[2] Lampreys were reported in Western Lake Erie in the late 1920's and in the Huron and Clinton rivers of southeastern Michigan in the early 1930's. Still, they remained an oddity and nuisance rather than a real problem; unfortunately, neither fisherman nor scientists realized at that time the menace they presented to the upper lakes.

Key Events and Roles

By the mid-1930's the lampreys began moving from Lake Erie into Lake Huron, which had plenty of streams for spawning, deep and cold water for living, and an almost unlimited supply of food.[2,4] Their food was the same fine fish sought for sport and by commercial fishermen. By 1939 Lake Huron fisherman began to complain about scarred lake trout, and a deterioration of the quality of trout fishing. They soon expressed a grave fear that the lampreys would completely destroy the trout fishing.[1]

The lamprey then spread through the Straits of Mackinac into Lake Michigan, and starting about 1945, similar fears began to develop for the trout fishing in that lake. To some extent, the small Soo locks at Sault Ste. Marie had protected Lake Superior from the invasion of lampreys. However, during the 1940's the Soo locks were greatly enlarged to accommodate huge ore boats for the heavy mining industry in the area. This enlargement of the locks permitted a more rapid entry of lamprey. Sea lampreys were first reported in Lake Superior in 1945,[1] and soon they were also threatening the fisheries of that lake.[6]

As the result of lamprey depredations, an alarming decline in the trout-fishing industry for the Great Lakes began about 1939. The records for trout catches in Lakes Huron and Michigan (Table IV) show a marked decline in lake trout fishing.[1]

The catastrophe began in Lake Huron in 1939 and within ten years trout had all but disappeared from that lake. Aside from the difference in timing, the trend was much the same in Lake Michigan. The situation in Lake Superior was less abrupt. During the period of 1945 to 1949, the trout catch from U.S. waters of Lake Superior declined from 3,369,000 lb to 2,965,000 lb.[1]

At the prevailing market prices in May 1951, the actual loss of cash income to commercial fisherman of Lakes Huron and Michigan resulting from the decline in the catches of lake trout amounted to about $3.5 million.[1] The loss of income to wholesalers and retailers of fish, transportation companies, and other associated commercial operations adds considerably to this value. In addition,

TABLE IV

RECORDS OF TROUT CATCHES IN LAKE
HURON AND LAKE MICHIGAN

Year	Lake Huron Catch[a] (pounds)	Lake Michigan Catch (pounds)
1935	1,743,000	4,873,000
1936	1,400,000	4,763,000
1937	1,340,000	4,988,000
1938	1,270,000	4,906,000
1939	1,372,000	5,660,000
1940	940,000	6,266,000
1941	892,000	6,784,000
1942	728,000	6,484,000
1943	459,000	6,860,000
1944	363,000	6,498,000
1945	173,000	5,437,000
1946	38,000	3,974,000
1947	12,000	2,425,000
1948	4,000	1,196,000
1949	1,000	343,000

[a]Catch in United States water of Lake Huron.

this estimate excludes financial losses for fish species (such as whitefish, suckers, walleyes, northern pike and smallmouth bass) other than lake trout that were destroyed by the sea lamprey. Apparently all commercial species of fish were subject to some lamprey depredations.[1]

By the summer of 1955 these marine predators had wiped out the trout fisheries of Lakes Huron and Michigan. Whitefish suffered nearly the same fate. Steelhead trout disappeared from many tributary streams; other valuable fish which were decimated included burbots, ciscoes, smallmouth bass, northern pike, and suckers.

New trouble also appeared without warning. The lake trout and whitefish had filled important roles as predator fish in the Great Lakes ecology. When the lampreys destroyed these fish, a saltwater trash fish, the alewife, invaded the chain of lakes. The alewife, a small (6-inch) silvery fish with little commercial and no sporting value, flourished and soon became a nuisance. The alewife was capable of displacing many other fish species from the lakes,[2] by eating their eggs and young. By 1965, one fisheries official reported that "the alewife represented 90 percent of all fish, by weight, in Lake Michigan."[2] During the 1960's, periodic mass deaths of the over-abundant alewife from natural causes created massive sanitation problems along the lake shores.

In an attempt to save the fishing industry in this region, the U.S. Fish and Wildlife Service, the Great Lakes states and Canada began to carry out research and testing measures against the lampreys. A treaty for joint control action by the United States and Canada was signed on September 10, 1954.[4] The U.S. Congress lent support to these control activities by approving a lamprey eradication program in May 1956.[7]

Beginning in 1951 the U.S. Fish and Wildlife Service used electrically charged batteries to trap sea lampreys in the streams where they spawn.[8] Electromechanical barriers were installed in forty-four tributaries of Lake Superior in the United States and more in Canada. The planning called for protecting the remaining trout in Lake Superior, which would in turn provide a supply of eggs for restocking the other lakes when the sea lamprey was brought under control. Although these barriers, erected at great expense, did succeed in destroying large numbers of lampreys, they failed to bring this predator under control.[4]

Tests with mechanical weirs to prevent lampreys from migrating upstream to spawn also gave discouraging results. One major difficulty was that the weirs also prevented desirable fish from spawning.[2] A desperate search for better control methods continued.

Disposition

Starting in 1953, Dr. Vernon Applegate, a government fisheries biologist, began investigating the possibility of using a selective chemical poison which would kill the larvae of lampreys, but would not effect other fish.[10] Weeks of experiments lengthened into months, and months into years, as thousands of chemicals were tried and abandoned.

In the fall of 1955, after nearly six thousand different compounds had been tested, Dr. Applegate finally identified one compound known as TFM (3-Trifluoromethyl-4-nitrophenol), which met all the requirements.[11] A successful field trial with TFM was conducted in 1957.[6] The major advantage of this chemical treatment was that it could effectively kill infant lampreys without harming fish.[9]

TFM had been developed none too soon. By the spring of 1958 the lamprey had worked its way steadily westward, and trout fishing had all but collapsed in the eastern two-thirds of Lake Superior.[2] By the fall of 1960 all streams on both the American and Canadian sides of Lake Superior had been treated with TFM.

By the spring of 1962 encouraging results from the TFM test were in hand.[2] Annual counts at certain weirs had shown a normal total of 67,000 lampreys the previous spring. The early 1962 reports showed that the lamprey total had dropped to a mere 9,000. The chemical had done its job.

An equally important part of the lamprey story involves Russell Robertson, the superintendent of the state fish hatchery at Marquette, Michigan. Beginning in the late 1940's, he began to collect lake trout eggs from various parts of Lake Superior. At this time the lampreys were moving into the lake, and Robertson realized it would be the last opportunity to save some of these trout. Previous attempts to raise lake trout in hatcheries had met with little success, but Robertson developed methods to hatch lake trout eggs, to keep the young fish alive, and to maintain the adults.[2] As the lampreys pushed westward through Lake Superior, Robertson's collection became all that remained of many of the subspecies of trout which had lived in the lakes.

Although many observers viewed this broad collection of lake trout stock as little more than a curiosity, they suddenly became priceless,[2] following the successful TFM tests during the spring of 1958. Robertson moved ahead immediately with production of young from adults in his tanks for a lake restocking program. With the offspring of his lake trout serving as the nucleus for the entire trout rehabilitation program, he soon had a whole new "industry." Adult fish were transferred to other hatcheries, and soon millions of young trout were being produced and placed in Lake Superior.

The program of simultaneous restocking and dispersion of the TFM lampricide[3] soon proved effective. By 1961, the lamprey population growth had been halted (the population was actually reduced to 95 percent of the average for the years of 1958 to 1961), and the lake trout population had tripled in Lake Superior. Similar encouraging results were obtained in Lake Michigan in 1965. One year later, Lake Huron streams were treated with TFM and that lake was stocked with trout. By the summer of 1968, several million young fish could be released each year into the upper lakes, and natural reproduction was reestablished.

Beginning in the spring of 1966, conservationists in several states established a cooperative fisheries project designed to stock the Great Lakes with two kinds of salmon—the coho and the chinook. An important consideration was that these fish would not only attract sportsmen and benefit the commercial fisheries, but would also feed on alewives. This initial program in Lake Michigan proved to be very successful and by the late 1960's fishing and associated business operations were booming in some areas of that lake. By 1968, New York, Pennsylvania, Ohio, Indiana, Wisconsin, Minnesota and Ontario had started salmon stocking programs of their own in the Great Lakes waters.[2]

Since 1966, the rehabilitation program involving application of lampricide (TFM) and restocking with trout and salmon has accomplished a tremendous transformation in economy, growth and hopes of every community surrounding the Great Lakes.[2] By 1972, many of the Great Lakes fisheries were restored and the lampreys were under control, if not eliminated: the records for fish caught off the Split Rock River area in Lake Superior during one week in September 1972, showed that 4 percent of the catch had fresh wounds, and that about 80 percent had old scars caused by lamprey attacks.[12] All the fish in the total catch (119) had markings which identified them with fish hatcheries, indicating they had been recently restocked into the lake. The record might have been even better except that a serious reduction occurred in federal funds for lampricide treatments in the Lake Superior area.

Comment

The construction of the Welland canal permitted the sea lampreys to bypass the natural barrier of Niagara Falls and gain entry into Lake Erie, the first of the upper Great Lakes. For over a hundred years the lampreys were not a problem. But because they either became adapted or had reached more favorable spawning conditions, they then suddenly multiplied. They nearly ruined the commercial sports and fishing industries in Lakes Huron and Michigan within ten years, and made serious inroads into Lake Superior. In addition, the loss of indigenous predator fish species upset the balance of the lakes and permitted the undesirable secondary effect of explosive growth of alewife. Thanks to the development of a new chemical pesticide which controls the lampreys, and the success of a restocking program, the lampreys appear to be under control. The lesson that the technological advance (the canal) produced a biological time bomb (the lampreys) — should be remembered. And the role of Russell Robertson in preserving trout stocks should not be forgotten.

REFERENCES

1. "The Sea Lamprey in the Great Lakes," *Sci Mo*, 275 (May 1951).
2. "Defeat of the Killer Eel," *Audubon*, **60**, 40 (July–August 1968).
3. "Lampreys in the Lakes," *Sea Front*, **16**, 143 (May–June 1970).
4. "The Sea Lamprey," *Sci Am*, **192**, 36 (April 1955).
5. *The World Almanac*, Newspaper Enterprise Association, Inc., New York, New York, p. 110, 1971.
6. "A Surfeit of Lampreys," *Time*, **65**, 42 (May 9, 1955).
7. *New York Times*, p. 52 (May 22, 1956).
8. "The Sea Lamprey Defeated?", *Sci Am*, **201**, 84 (December 1959).
9. "Death for Baby Lampreys," *Time*, **66**, 84 (November 21, 1955).
10. "Science Dooms the Lampreys," *Read Digest*, **75**, 193 (July 1959).
11. "Victory on the Lakes," *Time*, **80**, 46 (July 13, 1962).
12. Private communication with personnel at the Minnesota Department of Natural Resources, Division of Game and Fish, concerning fishing records in Lake Superior near Split Rock River.

The Donora Air Pollution Episode

Abstract

An air-pollution episode in Donora, Pennsylvania, in 1948, caused twenty deaths and nearly six thousand illnesses. Although more serious episodes had occurred previously in Europe, Donora not only shocked the nation but created a public awareness that, even in the United States, air pollution can be a serious health hazard as well as a nuisance. The U.S. Public Health Service conducted a detailed investigation of the Donora episode; it concluded that two or more atmospheric pollutants and unusual weather conditions combined to produce the tragedy, and it recommended a program for pollution control and the establishment of an adverse-weather warning system. Following a similar episode in New York City, which caused an estimated two hundred deaths, a National Air Pollution Control Act was enacted in 1955. This act authorized a program of research and technical assistance for abatement of air pollution, but proved insufficient: serious episodes recurred in New York in 1963 and 1966. In December 1970, the establishment of the U.S. Environmental Protection Agency finally signaled the beginning of a major governmental effort to assure systematic abatement and control of pollution.

Background

The combustible property of coal was known even two thousand years ago, but coal was used very little as a fuel; wood was generally available, and the burning coal could produce a black, choking smoke. The antagonism to coal grew so strong in England that King Edward I (died 1307) once decreed the death penalty for anyone using it. But as good chimneys became accessible to the

217

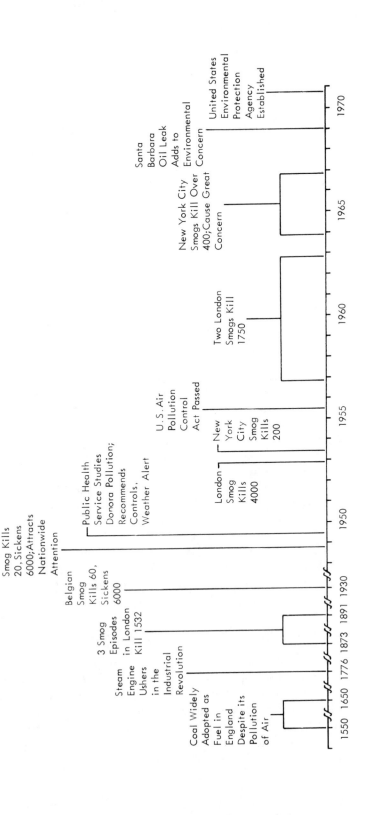

public* and as local wood supplies became depleted, the use of coal increased rapidly, particularly in England from 1550 to 1650. The invention of the blast furnace in 1735, and the steam engine in 1776 ushered in the industrial revolution which was soon to fill the skies of many European cities with smoke and other pollutants. As early as 1873, London was afflicted with an unusual epidemic of respiratory illnesses which are now attributed to air pollution. The incident was repeated in 1880 and again in 1891; a total 1,530 deaths are now attributed[1] to the three episodes. Although, "smoke-abatement societies" already existed in Europe, these three episodes (and a subsequent one in Scotland in 1909) aroused surprisingly little public attention, particularly outside of England.[2] The United States (where the first coal mine had been opened in 1745) also suffered many of the polluting side effects of the industrial revolution, but had not been visited by such tragic consequences; smoky skies and dirty waters were accepted as a nuisance and annoyance.

The first "smog" incident to attract widespread attention occurred in an industrial district along the Meuse River (between Liege and Huy) in Belgium during December 1–5, 1930.[1] Approximately six thousand people became ill, and 60 extra deaths were attributed to this smog. It was believed that sulfur trioxide pollution had combined with moisture in the air to produce sulfuric acid, which burned the lungs of its victims.

During the 1940's, Donora, Pennsylvania was known as a grimy industrial town located in a bend of the Monongahela River, twenty-eight miles south of Pittsburgh. Of its 12,300 people, who were largely of Slavic origins, two-thirds of the men and many of its women worked in the town's three principal industries—a steel plant, a wire plant, and a zinc-sulphuric acid plant—which lined the river front for three miles.[3] The steel and wire operations began in 1900, with construction of the blast furnaces, open-hearth department, and blooming mill.[4] In 1901, two looping-rod mills, a wire-drawing department, and a wire finishing department were constructed. The zinc-sulphuric acid plant was built in 1915; it also recovered cadmium and unrefined lead as by-products. Some other heavy industries in the nearby area included steel and by-product coke plants in Monessen and Clairton, and a power company and railroad yard in Elrama. Two railroads ran through the Monon-

---

*Bricks, made in coal-fire ovens, could be used to make chimneys much more cheaply than the previously used iron. The chimney increased the efficiency of combustion and conveyed the smoke outdoors.

gahela Valley in this region, one on the Donora side of the river and the other on the opposite bank, adding their streams of coal smoke to the air.

The Donora mills were major contributors, through smoke and waste-gas emissions, to heavy air pollution in the area. The main business district of Donora was adjacent to the plants, and the residential area extended to the top of the surrounding hills. At riverbank level the altitude above sea level was 760 feet, but the hills on the east bank of the river rose abruptly to a height of about 1,100 feet, and the hills on the Donora side of the river rose more gradually to a height of 1,150 feet.[4] Thus, Donora was unusually smoky even for an industrial town, since the surrounding bluffs and hills often prevented the daytime wind from dispelling plant emissions, and at night[2] the bottomland formed a drainage basin for cold, downslope winds which did little to remove pollutants. Unless the skies were high and clear, the streets were usually dim with smoke. No vegetation grew around the zinc-acid plant, and farms on the hillside had soot-covered vegetables and blackened sheep.[5] The townspeople, however, were accustomed to living under a cloud of black smoke and soot.

Weather conditions in Donora made autumn the smokiest season of the year, and October generally had the densest fogs. Gagging fogs of two or three days' duration were not uncommon, and a few had lasted four days. But near the end of October, 1948, a dense, choking fog persisted for six days. Unlike its predecessors, it was more than just another irritating experience. It was a disaster that made headlines in the newspapers across America.[3]

Key Roles and Events

The beginning of this tragic episode was well described by Berton Roueché, who based his history on eye-witness accounts:

The fog closed over Donora on the morning of Tuesday, October 26th. The weather was raw, cloudy, and dead calm, and it stayed that way as the fog piled up all that day and the next. By Thursday, it had stiffened adhesively into a motionless clot of smoke. That afternoon, it was just possible to see across the street, and, except for the stacks, the mills had vanished. The air began to have a sickening smell, almost a taste. It was the bittersweet reek of sulphur dioxide. Everyone who was out that day remarked on it, but no one was much concerned. The smell of sulphur dioxide, a scratchy gas given off by burning coal and melting ore, is a normal concomitant of any durable fog in Donora. This time, it merely seemed more penetrating than usual.[3]

By the third day of the Donora episode, a sizable fraction of the population became ill as ambient concentrations of pollutants escalated,[2] and by the fourth day, Friday, area doctors were flooded with calls from patients. The climax of the episode came on Saturday, October 30, when the atmospheric stability intensified even more: seventeen deaths were recorded on that day, and then two more on the next.[2] As the death rate mounted, alarm swept through the population. Scores of people, particularly the elderly, were complaining of extreme difficulty in breathing. The victims appeared to suffer partial paralysis of the diaphragm, and doctors found that nothing except oxygen seemed to bring relief. A state of emergency was declared, and elderly people were warned to leave the area.[6,7] The deaths occurred exclusively among elderly people (over fifty-two years of age), many of whom had a history of cardiac or respiratory disease. Nearly 43 percent of the population suffered adverse effects from the smog, although some seemed hardly to be bothered by it. In all, 17 percent of the population were moderately affected and 10 percent severely affected.[4] An emergency meeting between the city government and factory representatives was scheduled for Sunday, October 31, to consider means to curtail the emission of pollutants.[2] But fortunately, relief came to Donora on that day as rain fell to clear the air.

Donora buried its dead and the panic there soon subsided. But the impact of the air pollution episode—the first of its kind in the United States—was considerable. News of the Donora incident appeared on the front page of the *New York Times* for two days and received widespread public attention.[5-8] Governor James H. Duff of Pennsylvania called for an investigation to "ascertain the cause and prevent its recurrence."[6] Dr. William Rongann of the Donora Health Board was certain his town's tragedy was the result of industrial fumes collecting in the humid air, and said, "It's plain murder."[7] The Donora town council demanded an investigation.

The U.S. Public Health Service[4] assigned a field team to make a study of the Donora episode. After five months' intensive field work by that team, a preliminary report was filed and additional personnel were assigned by the U.S. Weather Bureau. This exhaustive study indicated that the clinical syndrome was characterized by irritation of the respiratory tract and was especially severe in elderly persons and those with known chronic cardio-respiratory disease. The data indicated that the episode was not due to an unusual air pollutant or an accidental occurence, but rather resulted from an accumulation of atmospheric pollutants during an unusually intense and prolonged period of stable air. The chronology was believed to be [2] as follows: As night approached on October 25 and the ground cooled, the high relative humidity remaining from previous storms quickly led to a saturated condition within the lower region of a cold and stagnating anticyclone. The subsequent condensation caused dense fog, particularly in the Donora River Valley, where evaporation sustained the high humidity, and a high aerosol concentration in the polluted air encouraged the formation of water droplets. The fog was held close to the ground by the stability of an elevated "inversion layer," caused by meteorological conditions known as a "temperature inversion," which occurs when air temperatures in the layer are higher than those near the ground. The cold, dense air cannot rise and mix with the atmosphere as surface air normally does.

While the weather was a factor, it could not be blamed entirely for the episode. Within the interior of the fog-bound valley, pollutants continued to be emitted. At the river-front establishments which represented the major industry of Donora, the bulk of the plant fumes and pollutant particles were released from exhaust stacks of various heights, none exceeding 40 meters.[2] Great quantities of sulfur dioxide and particulate matter were being emitted as pollutants along with significant quantities of several other contaminants. No single specific pollutant could be identified as responsible for the illnesses and deaths,* but the report concluded that the syndrome could have been produced by a combination of two or more of the contaminants. Sulfur dioxide and its oxidation products, together with particulate matter, were considered the most significant contaminants. Medical scientists suspected that improper functioning of the lung under the irritating conditions led to many heart failures, and that some delayed effects might also occur.†

The Public Health Service made ten recommendations based on the Donora study, which can be summarized briefly as follows:

---

*Autopsies on three persons who died during the smog showed acute change in the lungs characterized by capillary dilations, hemorrhage, oedema, purulent bronchitis and bronchiolitis; autopsies were not performed on most of the deceased.

†The subsequent mortality rate of those affected by the episode has been higher than that of the general population, but the overall death rate in Donora has been no higher than for other nearby towns.[3]

• Make a significant reduction, by control devices or process modifications, in the atmospheric contaminants (especially sulfur oxide, carbon monoxide, and particulate matter) emitted by the industrial plants.

• Establish a program of weather forecasts to alert the community of impending adverse weather conditions so that adequate measures can be taken to protect the populace.

Disposition

The Donora tragedy focused national attention on the growing problem of air pollution. The impact was not caused by the death toll alone, since greater mortality rates have occurred with more traditional causes , such as epidemics. The real shock came from the realization that it was apparently not abnormally high emissions of pollutants that produced the mortality and morbidity. The local sources and types of pollutants evidently had not changed substantially for decades. The episode was triggered by a combination of meteorological factors which had produced the temperature inversion and thus confined the pollutants. Unfortunately, temperature inversions can be produced in a number of ways and there is essentially nothing that can be done to prevent their occurrence. These inversions serve as an air-pollution barrier; rising exhaust gases and other air pollutants are trapped beneath the inversion layer and can, therefore, build up to high concentration. Thus, any temperature inversions that last several days could be very dangerous. The unavoidable conclusion for the nation was that unless emissions were curtailed, the frequency and duration of tragic episodes would be determined solely by meteorological factors that could not be controlled.

While the Donora incident greatly stimulated public awareness of pollution problems in the United States, remedial action was long delayed. In December 1952, London was hit by the worst killer smog ever—four thousand deaths—and in 1953, New York City suffered a serious smog episode which was blamed for nearly 200 deaths.[1] Legislative action in the Congress soon followed, and in 1955 the United States passed a National Air Pollution Control Act. It authorized the Department of Health, Education, and Welfare and the Public Health Service to sponsor research programs, including federal grants-in-aid.

But this Act was insufficient to stop the continuing pollution of the atmosphere by industry, automobiles, and other sources: tem-

TABLE V

NOTABLE SMOG INCIDENTS AND ESTIMATED NUMBER OF
EXCESS DEATHS*

Location	Year	Excess Deaths
London	1873	268
London	1880	692
London	1891	572
Glasgow	1909	592
Meuse Valley, Belgium	1930	60
Donora, Pa.	1948	20
London	1952	4000
New York	1953	200
London	1956	1000
London	1962	750
New York	1963	200–400
New York	1966	168

*Source: Reference 1.

perature inversions and smogs had become a way of life in Los
Angeles, and New York City was struck again in 1963 and 1966 with
killer smogs. By now the grim list of smog incidents had taken
thousands of lives (see Table V), and strong pressures were building up for strict nationwide controls of polluting sources. In January
1969, the Santa Barbara oil leak occurred, and proved to be the final
environmental insult needed to stimulate further legislative action.

The establishment of the U.S. Environmental Protection
Agency (EPA) in the executive branch on December 2, 1970, signaled the beginning of a major and sustained government effort to
assure systematic abatement and control of pollution.

Comment

The Donora, Pennsylvania, air-pollution episode of 1948 was a
direct result of society's indifference to the environment. The industries in the area dumped their wastes into the air and the waters
with little or no concern for the living conditions of the
citizenry—that was just the way steel, or zinc, or acid, or whatever,
was made in those days. The townspeople were dependent on the
industries for their livelihoods and were disinclined to ask for improvements, either individually or through their political institutions. But the air-pollution episode sowed the seeds of change far
beyond Donora, and eventually stirred the nation's conscience to

demand environmental reforms. Because of its small size, limited number of pollution sources, and meteorological "isolation," Donora provided a unique opportunity to study the anatomy of an air-pollution episode—and probably no incident has been studied more. Public officials and scientists involved in later, more complex, and more disastrous episodes, such as those of New York City, benefited from what was learned in Donora, and helped bring about needed environmental legislation and improved pollution-control practices. This case illustrates that not only is a crisis often needed to recognize a problem, but sometimes repeated crises are needed when we resist recognizing the facts.

REFERENCES

1. Hodges, L., *Environmental Pollution*, Holt, Rinehart and Winston, Inc., New York, New York, 1973.

2. Williamson, S. J., *Fundamentals of Air Pollution*, Addison-Wesley Publishing Company, Reading, Massachusetts, 1973.

3. Roueche, B., *Eleven Blue Men*, Little, Brown and Company, Boston, Massachusetts, 1953.

4. Schrenk, H. H., et al., "Air Pollution in Donora, Pennsylvania," *Public Health Service Bulletin*, 306 (1949).

5. "Death Over Donora," *Life*, **25**, 107–110 (November 15, 1948).

6. "Death in Donora," *Newsweek*, **32**, 25 (November 8, 1948).

7. "Death at Donora," *Time*, **52**, 25–26 (November 8, 1948).

8. "Death in Donora Dramatizes Smoke Problem," *Bsns W*, 21 (November 20, 1948).

The Torrey Canyon Disaster

Abstract

A huge sea-going tanker ran aground on March 18, 1967, sixteen miles off the southwest tip of Britain and spilled most of its cargo of 119, 328 tons of crude oil into the ocean. The winds and tide carried the foul-smelling oil along the entire coast of Cornwall causing extensive damage to beaches and death to sea birds and marine creatures. The shipowners eventually agreed to pay $7.2 million in damages. This incident caused worldwide reappraisals of potential oil pollution problems and safety precautions. In May 1967, President Lyndon B. Johnson ordered the start of an extensive study of oil pollution problems.

Background

During the 1960's, the tonnage of crude oil and its products transported by sea-going tankers was growing and ships of unprecedented cargo-carrying capacity were being built.[1] Whereas tankers of 80,000-ton capacity had previously been considered large, 200,000-ton supertankers were plying the seas; by the summer of 1967, 300,000-ton vessels were on order and the advent of the 500,000-ton jumbo tankers appeared to be fast approaching.[2]

Key Events and Roles

The *Torrey Canyon*, a supertanker of 127,000-ton displacement, flying the Liberian flag, was on charter in 1967 from the Union Oil Company of California to British Petroleum, to haul oil from the Persian Gulf to Milford Haven, Wales.[3] The crew was Italian and operated under Italy's maritime law. Union Oil had

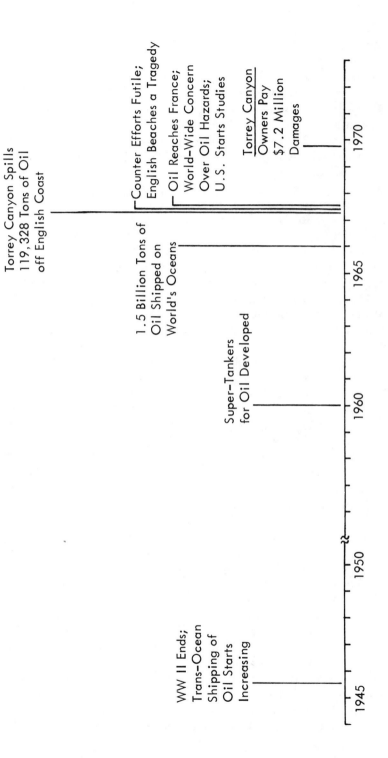

insured the vessel's hull and machinery for $16.5 million and also had $2.5 million coverage for third-party liability.[4]

On the morning of March 18, 1967, the *Torrey Canyon* was traveling near the Scilly Isles returning to Milford Haven with a cargo of 119,328 tons of Kuwait crude oil.[5] When the Scilly Isles appeared on the ship's radar screen, the chief officer realized immediately that the giant tanker had gone off course and would pass not west, but east of the isles. The captain told the chief officer not to change course and the tanker was set to pass between Seven Stones Reef and the island of St. Martins in the Scilly archipelago. The chief officer realized, too late, that the ship was dangerously close to the reef. By the time the helmsman could disengage the automatic pilot and "come hard left" under manual control, the jagged ridge of rock below the surface ripped into the bows of the 974-foot long *Torrey Canyon* and gouged a 650-foot hole in the hull. As the tanker ground onto the reef, thousands of gallons of crude oil poured from fourteen separate tanks that had been ripped open, causing what was to become the biggest and most expensive episode of marine pollution in history. The grounding occurred sixteen miles off the southwest tip of Britain,[6] an area near England's loveliest summer playground and tourist attraction, including the beaches of Cornwall and the Scilly Isles.[7] The site of the accident was outside British territorial waters, making it questionable how much jurisdiction Britain could assert.

Experts were not sure why the accident had happened.[5] The visibility that day was eight miles and *Torrey Canyon* was well within range of a lightship, three lighthouses and a major radio beacon. The ship was well-equipped and had an experienced crew; Lloyds of London had rated the ship a best risk. Yet, incredibly, the ship ran at top speed onto a well-known navigational obstacle in a channel which was six miles wide although it was normally considered out-of-bounds to large tankers.

On the morning after the wreck, a slick of oil covering sixty square miles of sea lay twenty miles off Land's End in Cornwall on the west coast of England.[5] Within hours after the wreck, Royal Navy vessels left Plymouth loaded with liquid detergent. All hopes were pinned on emulsifying the oil at sea before it had a chance to hit the beaches.

The press described the incident as one of the decade's worst disasters and Britain's newspapers called it another Battle of Britain.[6] She was the biggest ship ever to be wrecked. As consternation spread in the coastal countries of Britain, Prime Minister Wilson set up a ministerial-level emergency committee to coordinate local and national measures, and named it "Operation Canute."

Soon after the wreck, unsuccessful attempts were made, over a period of seven days, to salvage the *Torrey Canyon*, but all efforts failed to free the ship from the reef.[5] Beginning on March 28, 1971, the ship was bombed for three days with napalm, high explosives and kerosene, by British planes (at a total cost of more than $2.5 million) to split open the remaining cargo tanks and fire their contents before the oil could spill into the sea.

Because of an offshore wind, nearly a week passed before the first oil washed onto Britain's beaches.[5] Then for about nine weeks the winds and tide carried the foul-smelling oil along both the north and south coasts of Cornwall, blackening and polluting beaches and harbors all along the way—about one hundred miles of coasts were ringed with oil and sixty miles were polluted. Civilian volunteers and many soldiers and marines worked long hours at cleaning up the beaches. The first approach was to use the detergents and other clean-up techniques developed by the Navy for dealing with small spills in harbors. Firemen washed down docks with high-pressure water hoses.[8] Occasionally a beach would be completely cleaned—only to be fouled again by the next incoming tide.

For the sea birds and marine creatures of the Cornish coast, the oil pollution was a great tragedy.[3] The birds most affected were auks, razorbills and guillemots. Other forms of life that flourish in and near the sea suffered a similar fate—many limpets, crabs and shellfish were killed and fishermen were greatly concerned about damage to the commercial fishing in the area. Some oyster beds and mussel and clam farms were ruined by the oil pollution.[9] Not only were tens of thousands of birds doomed, but the fishing industry was virtually suffocated by the vast film of oil.[10]

A major new problem was then noted: the toxicity of detergents to sea life. Eventually some 2 million gallons (10,000 tons) of detergents were used to treat an estimated 13,000 tons of oil on Cornish coasts and another 0.5 million gallons were sprayed at sea. A British biologist found that the use of detergents had probably caused much more damage to shellfish than the oil; therefore, from the conservationists' point of view the use of detergents in the clean-up operations made the *Torrey Canyon* incident a double disaster.[5] In addition, some of the aromatic hydrocarbons used to dissolve the detergents and to aid in mixing with the oil probably caused much damage to wildlife also.

The *Torrey Canyon* oil eventually reached France and caused oil pollution problems.[5] Altogether, fifty of Brittany's 950 miles of beaches were affected. Twenty-five coastal villages were officially declared devastated because of damage caused by oil pollution.[9] Because of the British experience, the French shunned the use of detergents, and adverse effects on sea life were less. The French used chalk treatment to sink oil at sea and straw and other absorbents on shore.

About seven months after the tragic shipwreck, the British government received a scientific report explaining how the needless slaughter of sea birds and marine life by toxic detergents could be avoided in the event of any future oil tanker wreck.[12] This study revealed that no detergent would ever provide the answer to marine oil pollution; the researchers recommended the use of silicones.

Aside from the expense (about $2.5 million) of bombing the *Torrey Canyon*, the total bill paid by Britain for cleaning up the oil amounted to $7 million.[5] The *Torrey Canyon* itself was a total loss: it split in half during the bombing.

Disposition

The *Torrey Canyon* disaster pointed up the extent of ignorance and lack of preparedness at that time in dealing with such an emergency. Several important lessons were learned from the tragic incident. For the first time vivid evidence was available that the oil pollution problems were getting worse and that oil on water can do great harm to wildlife and recreation. There was considerable evidence that the use of detergents did more harm than the oil. Studies by British scientists, for example, showed that while oil killed 30 percent of the plankton, detergents killed up to 96 percent.

As a result of the *Torrey Canyon* incident, extensive reevaluations were started by the fall of 1967 by the big oil companies, the United States Government and international commissions.[13] Should there be special sea lanes for tankers and special offshore docking facilities with pipelines to the mainland? Should the size of tankers be limited? Can't science develop some emulsifier that will render oil harmless without destroying marine flora and fauna? There were also a host of unresolved legal questions involving owners of ships and their cargoes on international waters.

The *Torrey Canyon* tragedy pointed out a glaring loophole in the U.S. oil pollution law.[13] Under the Oil Pollution Act of 1924, offenders are liable to penalty only for grossly negligent or willful spilling, leaking, pouring, emitting or emptying of oil. Nothing much could be done to prosecute the tanker operator after most accidents. Even when gross negligence was established, the maximum fine—only recently raised to $10,000—was trifling compared to the damage that could be done. In November 1969, the *Torrey Canyon* owners agreed to pay $7.2 million in damages.[15]

On May 27, 1967, President Johnson ordered a study to be made on oil pollution.[14] The Oil Pollution and Hazardous Substances Control Act of 1968 was one result.

The *Torrey Canyon* disaster has been the subject of at least two books[16,17] and two study reports[18,19] and has had a place in many other publications in which environmental problems were discussed.[20-22]

Comment

The *Torrey Canyon* disaster illustrated dramatically the extent to which ocean shipping technology had developed a capability to inflict ecological damage and how poorly society was prepared to cope with either the danger or the regulation of its cause. That the danger was further amplified by other incidents of oil pollution has emphasized the need for control and also emphasized the dependence that modern technology has on oil and other energy sources. Recent statistics that indicate that nearly 15 percent of the world's ships (of all kinds) have some kind of collision each year indicate that serious risks of incidents similar to the *Torrey Canyon* will continue.

REFERENCES

1. "Oil Pollution," *Science*, **156**, Editorial Page (May 26, 1967).
2. "Mopping Up Oily Oceans," *Time*, **90**, 68 (July 28, 1967).
3. "The Huge Slick, Spreading With Tide and Current," **62**, *Life*, 27 (April 14, 1967).
4. "The Tanker's Messy Wake," *Bsns W*, 128 (April 1, 1967).
5. "A Tragedy of Errors," *Audubon*, **69**, 72 (November–December 1967).
6. "Operation Canute," *Time*, **89**, 28 (April 7, 1967).
7. "Battling the Blob," *Newsweek*, **69**, 44 (April 3, 1967).
8. "Britain's Great, Ghastly Ooze," *Newsweek*, **69**, 48 (April 10, 1967).
9. "Letter From Paris," *New Yorker*, **43**, 166 (April 29, 1967).
10. "The Oily Flotsam That Fouled Fair England," *Life*, **62**, 26 (April 14, 1967).

11. "Detergents at Sea," *Newsweek,* 110 (April 10, 1967).

12. "Silicones Could Have Saved Sea Life," *Sci N,* **92,** 343 (October 7, 1967).

13. "One Answer Shows Through the Oil Slick," *Audubon,* **69,** 4 (November–December 1967).

14. *New York Times,* p. 1 (May 27, 1967).

15. *New York Times,* p. 1 (November 12, 1969).

16. Cowan, E., *Oil and Water: The Torrey Canyon Disaster,* Lippincott, Philadelphia, First Edition, 1968.

17. Petrow, R., *In the Wake of the Torrey Canyon,* D. McKay Company, New York, 1968.

18. *Torrey Canyon Pollution and Marine Life,* F. E. Smith, Ed., Report of Marine Biological Association of the United Kingdom, Cambridge U.P., London, 1968.

19. "The Biological Effects of Oil Pollution on Littoral Communities," Proceedings of a Symposium at Penbroke Wales, February 17–19, 1968, I. D. Carthy and D. R. Arthur, Eds., Field Studies Council, London, 1968.

20. McCaull, Julian, "The Black Tide," *Environ,* **11,** 2 (November 1969).

21. *Oil On The Sea,* O. P. Hoult, Ed., Plenum Press, New York, 1969.

22. Jackson, Wes, *Man and the Environment,* Wm. C. Brown Company, 1971.

The Santa Barbara Oil Leak

Abstract

In January 1969, a massive oil leak developed at the Union Oil Company's drilling operations in a porous rock formation 5.5 miles offshore Santa Barbara, California. Before the well was cemented shut eleven days later, an estimated 0.5 million gallons of oil had leaked, eighty miles of beaches and harbors were fouled, and much wildlife was endangered. The blowout received nationwide news coverage (with hundreds of photographs showing blackened beaches and oil-soaked birds) and created a tremendous controversy. No sooner had the sealing of the initial leak been announced than secondary leaks started that were even more difficult to control. A presidential panel concluded that the best method to stop the leaks would be to exhaust the oil reservoir—an undertaking that could require twenty years—and recommended that drilling be resumed.

Despite the furor this recommendation raised, no agreeable alternative could be found. Union Oil—the villain in many eyes—was permitted to start limited but profitable production which, together with a cementing program for the porous rock seabed, sealed the leak by the end of 1969.

Union Oil and its partners spent an estimated $5 million in cleanup costs, and were assessed $4.5 million damages to property owners and fines of $500 each, but 342 counts of criminal pollution were dropped. After much litigation, oil operations in the area have been resumed, under more stringent controls.

Background

The first oil well in the United States was drilled in Pennsylvania in 1859; the first offshore oil well was begun off California

233

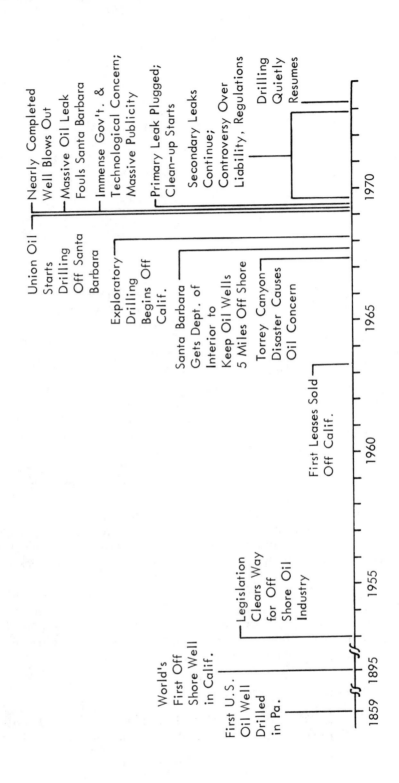

about 1896 near Summerland, a few miles south of the small city of Santa Barbara. Interest in offshore oil did not become intense, however, until after World War II.

A long argument between the states and the Federal government over the rights to offshore oil ended in 1953 when Congress passed the Outer Continental Shelf Act. Under this act the federal government left leasing and regulatory control within the "tidelands"—the three-mile territorial limit—largely to the states, while it maintained control in the adjacent "international" waters. The major California legislation covering leases and royalties to the state, the Cunningham-Shell Tidelands Act of 1955, barred drilling in certain "sanctuaries" inside the three-mile limit in order to preserve a seascape unmarred by oil rigs. By early 1969, 925 wells dotted 150,000 acres of California tidelands from Santa Barbara to Orange County.

In 1963, the first federal lease for oil off the Pacific coast was sold; California soon pressed for jurisdiction beyond the three-mile limit, contending that state beaches could be polluted by spillage from poorly regulated wells in the "federal zone." Concern was greatly heightened in March 1967 when the tanker *Torrey Canyon* ran aground off the British coast and spilled 100,000 tons of crude oil; the oil proceeded to blacken the loveliest beaches in England and was even carried to France. (See case history of the *Torrey Canyon* disaster.)* Concern was strong in Santa Barbara which had become a beautiful city of 70,000; it drew 80 percent of its livelihood from its $46 million a year tourist business and was bordered by a twenty one-mile long "sanctuary area." The Santa Barbara Channel between the coast and several islands twenty-five miles offshore was in fact one of the best sport and commercial fishing grounds in California, yielding 27 million pounds of fish in 1967.[2] Santa Barbara officials soon persuaded the Interior Department to create a two-mile "buffer zone" beyond the three-mile limit, so that no drilling would be allowed within five miles of shore. When a few oil slicks began to appear along the shoreline in 1968, Santa Barbara officials even asked Secretary of Interior Stewart Udall for a further extension of the buffer, but Mr. Udall assured the town that the federal government would keep a close eye on the drilling; the inference was that Santa Barbara had nothing to fear.

*After the *Torrey Canyon*, a "National Multi-Agency Oil and Hazardous Materials Contingency Plan" was drawn up. It was approved by President Johnson in September 1968 after the foundering of the tanker *Ocean Eagle* off San Juan, Puerto Rico (March 3, 1968).

Exploratory drilling on federal leases off California beyond the three-mile limit actually started in February 1968. By that time, 110 federal leases on seventy one tracts covering about 363,000 acres had been sold for $602.7 million. This sum included $61.4 million paid by Union Oil Company and its three partners, Mobil Oil Corporation, Gulf Oil Corporation, and Texaco, Inc., for oil rights on 5,400 acres. Union Oil Well A-21 was the fifth of five wells being drilled from Platform A, just 5.5 miles off shore, scheduled to be the first producing wells in the federal zone.[1,2]

Key Events and Roles

By the morning of January 28, 1969, Union Oil's drilling rig* had bored 3,479 feet below the ocean floor, almost to the oil reservoir believed to exist at 3,500 ft, despite some of the worst rainstorms in the history of the area. Then, as riggers pulled up the drill pipe to replace a worn bit, the drilling "mud" pumped into the well to maintain pressure apparently became inadequate for the job and the well "blew": first a twenty-foot high fountain of mud, and then a heavy mist of noxious fumes. The hole itself was successfully capped at the surface after thirteen hectic minutes.[1,2] But about 15 minutes later flammable natural gas and then oil began boiling up in the ocean—initially about two hundred yards from the drilling platform, and then so close that it drove off the drilling crew. The oil had come up the drill hole to within 700 feet of the surface, then escaped laterally through a porous stratum in the earth, and finally up to the ocean floor through a fissure. At noon, Union Oil officials notified the Coast Guard and the U.S. Geological Survey that Well A-21 was in trouble.[1,2] The Santa Barbara oil leak had begun. Union Oil, the owner of the *Torrey Canyon* (although not the actual operator at the time of its infamy) was again in the spotlight.

Officials of Union Oil were reasonably optimistic that the oil leak could be sealed within a short time by means of the technology and experience the oil companies had. When the gas temporarily cleared around Platform A, the crew went into action. But problem after problem was encountered as the crew attempted to pump mud and sea water down the well to force the oil back below the leak. At day's end, the black, sticky oil was pouring at an unknown rate in an

---

*Union Oil had contracted the actual drilling to the Canadian firm, Peter Bawden Drilling Company, the lowest bidder for the job.

ever-widening slick; an increasingly anxious and angry Santa Barbara had nightmares of suffering its own *Torrey Canyon.*

The following day, Union Oil began to spray chemical dispersants on the oil already on the ocean and made plans to start drilling a relief well near Platform A as a backup measure (possibly a two-week job) in case the plugging efforts were unsuccessful. By January 30 the winds had driven the oil to within one mile of the beautiful beaches at Carpinteria, east of Santa Barbara. Miles of logs were being strung together to form barriers across local harbors and marinas, and a long inflatable boom was being readied to protect Santa Barbara. By February 1 the oil had contaminated about 5 miles of unprotected beach.[2]

Union Oil officials were besieged by local and state officials and by the hordes of newsmen who were descending on the area. The repercussions of the leak were reaching all the way to Washington, where President Nixon had just been inaugurated and where a great deal of confusion existed because of the incomplete and conflicting information on what was happening. The man really on the spot was Secretary of Interior Walter Hickel, whose department had responsibility for oil spills* and who had taken office over the vigorous protests of conservationists just four days before the leak. They regarded Hickel as "the fox guarding the chicken coop" on environmental matters. On February 2, Hickel flew to Santa Barbara. He immediately ordered a review of the drilling and casing procedures used for all drilling operations in the channel and asked all companies to suspend operations until the review was completed.[2]

The Union Oil Company announced that it would accept full responsibility for cleanup costs and indicated that it had an armory of methods to fight the oil slick[3] while following the federal guidelines for all pollution control procedures.[2] At the Coast Guard's suggestion, Union Oil built a floating boom of telephone pole logs around the oil source, and the Coast Guard itself searched for additional booms, pumps, and the like, to help Union. But to no avail. The Federal Water Pollution Control Administration (FWPCA) and the Coast Guard had initially (although reluctantly) permitted Union on January 29 to spray chemical dispersants† on

---

*This responsibility dates back to the Oil Pollution Act of 1924, and was increased in the 1968 "Contingency Plan."

†They were aware of the toxic effects of the detergents used in Great Britain after the *Torrey Canyon* disaster.

the oil slick; but on February 1 the FWPCA ordered a halt to the use of chemicals because, it said, Union had already used more than the manufacturers recommended based on Union's estimates of the amount of oil spilled. And the estimates were going up: 15,000 gallons a day, 25,000 gallons a day, and more.[4] Union spread logs, straw and other coagulants in waters off threatened beaches and sent a huge cleaning machine into the slick; powered by a tug, the V-shaped strainer (250 feet across) skimmed the surface and deposited oil in a barge, but the operation was hampered by ocean swells.

By February 3, the sixth day of the blowout, the Union Oil crew was prepared to start the pumping of special heavy mud down the well to stop the leak. But after six hours and 3,000 barrels of mud, it was apparent that the leak was just too big: far more mud and bigger equipment to pump it would have to be brought in.[1] It would take two or three days to get things ready.

The oil continued to spread, and by February 4 it was at the boom protecting Santa Barbara harbor, which contained nearly $5 million worth of pleasure and fishing boats as well as beautiful beaches and other facilities. The next day, a ship accidentally rammed the boom and the oil came in. To make matters worse, a storm moved through the area for two days, sabotaging the containment efforts and spreading the oil on everything—beaches, boats, and wildlife. News coverage of the results was tremendous: aerial photographs showed the ominous spreading oil slick and hundreds of photos recorded blackened beaches, oil-soaked birds, and heroic wildlife rescue efforts. FWPCA relented under the pressure and again permitted use of chemicals. In Washington, at a February 6 news conference,[6] President Nixon said "far more stringent and effective regulations" would be one aim of a new Interior Department review of rules governing offshore drilling.

But in California, the Department of Interior astounded the nation by giving the oil companies clearance to resume drilling on February 5, despite the stricter regulations recommended in the government's three-day review. Then, two days later, as the leak continued, Secretary Hickel again ordered all drilling in the channel to a halt. In announcing the order, he said, "It has been increasingly clear that there is a lack of sufficient knowledge of this particular geological area. This lack leaves us no other reasonable course of action than to halt drilling until Union's well can be sealed and until the required geological knowledge is secured."[2] On that same day, however, the Union Oil crew succeeded in stemming the oil flow by injecting water and mud down the well and then sealed the well below the ocean's floor by pumping down 1,150 sacks of slurried cement. On February 8, Union Oil officials announced that the leak was plugged—eleven days after it had started, and just as the first of the news magazine stories was appearing.[3-8] It had spewed forth an estimated 0.5 million gallons of oil,[2] that covered an eight hundred-square-mile area and had contaminated forty miles of incomparable beaches.

With the emergency phase of the blowout apparently ended, the oil companies, the Federal and California agencies involved, and the afflicted citizenry breathed a sigh of relief and prepared to turn to the many unfinished problems. These included cleaning up the mess, assessing the damages and liabilities, determining the causes and necessary remedial actions to prevent a reoccurrence of such a leak, and preparing contingency plans in case a similar episode should occur. These problems would obviously require much study; a host of committees and study groups was soon established. Secretary Hickel established two Department of Interior committees, one to review and toughen the drilling regulations and another—a joint effort with the Department of Justice—to study the problem and make recommendations. In addition, Hickel suggested to President Nixon that he form a blue ribbon panel of nongovernment scientists and engineers to investigate the entire episode; on February 13 the President's Science Advisor, Dr. Lee DuBridge, announced the formation of the Presidential Panel on Oil Spills, which would consider how the federal government could best assist in the cleanup around Santa Barbara; how existing technology could be used to control oil pollution in general; how to prevent such massive pollution in the future; what oil pollution control research needed to be supported by the government.

Meanwhile two Congressional committees were holding investigations. Senator Edmund S. Muskie's Public Works Subcommittee on Air and Water Pollution held hearings in Washington that attracted fifty newsmen and representatives of all three major TV networks. This subcommittee also set February 24 for a visit to Santa Barbara. Representative John Blatnik's House Public Works Subcommittee on Rivers and Harbors, which had been holding flood control hearings in Los Angeles, had already arrived in Santa Barbara on February 14. In Sacramento, Governor Ronald Reagan ordered all state agencies to review their regulations on offshore oil operations, and state legislators considered the problem. In Santa Barbara, Carpinteria, and nearby communities, resolutions were being proposed that called for a halt to all offshore drilling in the area.

But almost before any of these studies could get started, the public was informed that a new oil slick had developed February 12 from the seabed near Union Oil's Platform A—small compared to the early blowout, but already reaching toward the shore, where weary public and private work crews were cleaning up beaches, boats, and birds. And Union Oil could offer little immediate hope for stopping this slow seepage, although they again attempted to sweep it from the sea with little success. On February 17, Secretary Hickel granted permission to the company to bleed off gas pressure from its drilling operations on Platform A and also issued a statement saying that oil companies would be "absolutely liable" for pollution from spills and leaks in the future, whether they were negligent or not. The following day he issued new guidelines for more precise and tougher regulations and on February 20 called a temporary halt to further sales of offshore leases.

Santa Barbara was by now a daily feature of the news[9-16] and was becoming a household name. Major points of concern were: how much oil had actually leaked, how great were the economic and environmental damages, who was to blame, and who was going to pay. In no case was there agreement. Claims were made that Union Oil had seriously underreported the amount of oil* and that the sport and commercial fishing and other marine life had been utterly destroyed in the area. It was also revealed that there had been strong disagreements within the Department of Interior over regulations for offshore drilling in general and over the advisability of allowing drilling in the Santa Barbara Channel at all—a decision ultimately made by former Secretary of Interior Stewart Udall. Even more vehement was the argument over whether the blowout could have been prevented if the regulations of the State of California had been followed instead of those of the Federal government. Specifically, California rules required that drillholes be lined with a casing for 1,200 feet below the ocean floor, whereas Federal regulations required only a three hundred-foot casing. But Well A-21 had been cased to a depth of only 239 feet in order to extract oil from a shallow sand formation, under a "variance" received from

---

*The rate during the eleven-day gusher was now being estimated at 5,000 barrels a day or more, instead of only 500 barrels a day as the company estimated.[9]

Interior's U.S. Geological Survey.* And while Union Oil had already agreed to pay cleanup costs, a complex legal battle was shaping up over total damages. By mid-February, a $1.3 billion suit had been filed against Union and its partners by the Santa Barbara fishing and boating industries and the owners and users of beachfront properties.[6] The State of California was preparing two suits of $500 million each against the federal government and the oil companies for damages to wildlife, the shore, and the "local economy." And in Santa Barbara, aroused citizens groups such as GOO (Get Oil Out) were demanding a halt to all oil operations in the area, even if it meant the government had to return the $603 million to the oil companies.

The President's advisory panel met, amidst charges of conflict-of-interest against several of its members and complaints that it was too secretive. In an interim report, it recommended that several steps be taken to contain and clean up the oil already spilled and that the oil pressure under Platform A be relieved to reduce new seepage through the ocean floor. On February 21, Secretary Hickel authorized Union Oil to put all five of its wells on Platform A into production. Two of these were opened and began producing oil from sands at a depth of about 500 feet. But on February 24, Well A-41 blew out as it was being opened, sending gas and oil from the ocean floor to the sea surface at an alarming rate. By February 26, the oil slick reached a length of 8.5 miles and a width of 1,000 feet; it was obvious that the nightmare of Well A-21 had returned and for a very similar cause. Well A-41, itself, was finally plugged on March 1, but the following day attempts to reactivate Well A-21 set off a new leak. Well A-21 was then quickly replugged down to the three thousand-foot level, but oil from both leaks continued to flow from the ocean floor for nearly two more weeks before decreasing to a slower seepage.

The beaches were yet to receive the worst assault of the episode.[17-32] By March 6, the oil had reached all the way to San Diego, well past President Nixon's Western White House at San Clemente. By March 18 the FWPCA again relented and permitted the use of dispersant chemicals. By March 27 the Department of Interior asked for a second presidential panel. Secretary Hickel in

---

*These variances were apparently granted at that time on an almost routine basis without even an inspection or consideration of special factors. In fact, Well A-21 had apparently never received a federal inspection before the blowout.[2]

the meantime authorized resumption on April 2 of drilling on five more of the other seventy two oil leases in the channel on the advice of the Geological Survey.

The new panel was composed and met on May 12 and 13 to consider methods of permanently stopping the leakage, which was believed to be still over one thousand gallons a day. The panel concluded[16] that the best way to guarantee that the leaks would not reoccur would be to pump all the oil out of the reservoir beneath the sea—an undertaking that could require twenty years or longer.[25] The panel's recommendation raised a furor of objections from members of the Congress, the press and the public. The American Civil Liberties Union filed suit July 10, seeking an injunction against further drilling.[16] No agreeable alternative could be found, however, and on June 2, 1960, Secretary Hickel authorized the resumption of unlimited drilling in the Santa Barbara Channel.[16]

Within two months, Union Oil had fourteen wells producing from Platforms A and B and had pumped out nearly 30 million gallons of oil. In addition, 24,855 sacks of cement were pumped down into the porous rock formations under the ocean's floor through eight shallow drillholes.[2] The combination of pumping and grouting gradually reduced the seepage rate. By the end of the year—after an estimated 1 to 3 million gallons of oil had been lost and eighty miles of beaches were despoiled—the Santa Barbara oil leak was technically at an end.

Disposition

By early 1970, nearly a year after the initial blowout, the oily mess had been largely cleaned up from the beaches and harbors,* but arguments still raged over the extent of the damage to the environment[32]. The loss of birds was obvious: 3,500 had been counted dead by April 1, 1969.[28] But several marine biologists† had been quoted during the height of the leak as saying the damage would be near-disasterous and decades might be required for recovery. As early as mid-February 1969, it was reported[11] that the University of California at Santa Barbara (UCSB) would apply to the National Science Foundation for a $730,500 grant to finance a two-year study

*Union Oil Company said it had spent nearly $5 million in this effort.[19]
†For example, Carl Hubbs, Professor Emeritus at La Jolla's Scripps Institute of Oceanography.[9,10]

of the spill's effects.* Within a month of the initial spill, the Western Oil and Gas Association made a $220,000 grant to the University of Southern California's Allan Hancock Foundation for a biological and oceanographical study of the episode.

But by midsummer 1969, ecological effects were not judged to be as bad as some feared,[23] in part because overuse of chemical dispersants had been avoided. By August 17, 1970, William D. Pecora, Director of the U. S. Geological Survey (blamed by critics for having been too lenient in overseeing the oil companies' drilling operations), stated that his agency's study, supported by University of California biologists, found very little damage to the ecology except for bird life.[33] On January 19, 1971, the USC group, headed by Dr. Dale Straughan, released their results.[33, 34] In general, they concluded that the damage to the flora and fauna from the oil was not too bad and could not be easily distinguished from the damage caused by massive rains at the same time, and that the area was recovering well. The study techniques and many of the specific conclusions were immediately challenged by conservationists and other researchers — even by one member of the project team; the oil industries' financial support of the study also diminished its credibility in some eyes. The 903-page report is, however, a valuable reference on the Santa Barbara oil leak, along with other books on the subject.[1,2,36] By late 1971, the UCSB group reported[36] the results of their studies; while pointing out damages to plant and animal life, they generally concluded that these were not permanent.

The Santa Barbara oil leak had a substantial impact on governmental regulatory and legislative bodies. On August 21, 1969, Secretary Hickel authorized tougher offshore oil drilling regulations. The regulations tightened the technical requirements on oil and gas operations on the Outer Continental Shelf and made final Hickel's February 17 "absolute liability" order, placing responsibility for the cost of oil pollution cleanup upon leaseholders, regardless of whether the pollution resulted from negligence. The final code required consideration of all environmental factors prior to leasing an oil site.[38] Within two years after the Santa Barbara blowout, 1,700 offshore wells had been drilled around the nation without mishap.[33]

The Santa Barbara blowout is considered one of the major factors leading to the rapid growth of federal interest in the environ-

*A grant to this group was subsequently funded jointly by NSF and the FWPCA.

mental quality movement and to the enactment of legislation in this area. Early in 1970, President Nixon signed into law the National Environmental Policy Act of 1969, which made protection of natural resources a matter of national policy and established the Environmental Protection Agency as of December 1970. This bill also established a three-man Council on Environmental Quality (CEQ) to review all federal activites that affect the quality of life and to report directly to the President.

The matter of whether to permit further extraction of offshore oil in the Santa Barbara area has been difficult to resolve. President Nixon asked Congress June 11, 1970, to approve legislation cancelling twenty federal oil leases in the Santa Barbara Channel.[39] The legislation would have created a 198,000-acre federal marine sanctuary between Santa Barbara and Santa Cruz Island. In April 1971, new Secretary of Interior Rogers C.B. Morton suspended drilling operations on thirty-five of the seventy-one leases in the Channel to allow Congress time to consider legislation.[40] But Union Oil and its partners went to court: a federal judge in Los Angeles ruled in June that the Secretary did not have authority to prohibit the drilling.[41] While an appeal of this ruling was being considered, the Environmental Protection Agency recommended that pumping from existing wells be increased.[42] The Department of Interior then quietly lifted the suspension on fourteen leases[43] and proposed a resumption of drilling.[44] By April 1973, limited production was underway on three tracts (including one owned by Union Oil) covering twenty square miles, but preparations were being made for renewed massive drilling by Exxon Corporation and three partners over a 290-square-mile area in the Channel.[45]

The damage suits were also not easily settled: in November 1972, the oil companies paid $4.5 million to property owners[46]; in January 1973, the oil companies were fined $500 each and 342 counts of criminal pollution were dismissed.[47]

An important outcome of the leak and the intensive study that it inspired was the realization that the seabed in many parts of the world and particularly in the Santa Barbara Channel has always been a significant natural source of oil "pollution." These constant but small and diffuse emissions do not have the environmental and esthetic impact of the massive man-made oil leak.

Comment

No incident has done as much as the Santa Barbara oil leak to raise the level of national concern over environmental pollution.

The ominous spreading oil slick; the frantic race against time to control it; the oil-coated marine life, birds, beaches, and boats; and the repeated cleanup efforts were all ideally suited to vivid pictorial presentation by the news media. The proximity of the leak to the wealthy tourist facilities at Santa Barbara, the convenience of getting to Santa Barbara from Los Angeles, and the danger of even larger potential economic damage if the oil reached Los Angeles probably did much to raise news media coverage, particularly by the Eastern-based "establishment" papers, magazines and networks. While the fears of irreparable damage to the environment expressed by many oil opponents did not materialize and subsequent study showed that natural sources of oil emissions to the sea were greater than had ever been suspected, the Santa Barbara oil leak clearly had high costs in environmental damage (particularly among birds), in economic damage, and in emotional strain on the government, the oil industry, and most important, on the public. The divisive conflicts generated over Santa Barbara have had long-lasting effects, in, for example, the acrimonious debate over the Alaska Pipeline.

In many ways, the Santa Barbara oil leak marked the beginning of the end of an American era of reckless overcomsumption of energy at the expense of the environment and irreplaceable natural resources, although it would be four years before the message was clear.

REFERENCES

1. "Santa Barbara Oil Spill," by R. W. Holmes in *Oil On The Sea; Proceedings of a Symposium*, D. P. Hoult, Ed., Plenum Press, New York, New York, 1969.

2. *Blowout, A Case Study of the Santa Barbara Oil Spill*, by Carol E. Steinhart and John Steinhart, Duxbury Press, 1972.

3. "Fighting the Oil in California's Troubled Waters," *Bsns W*, 60 (February 8, 1969).

4. "Environment: Tragedy in Oil," *Time*, **93**, 23 (February 14, 1969).

5. "A California Oil Strike Nobody Wanted," *Life*, **66**, 30 (February 14, 1969).

6. "The Big Oil Leak; A Messy Legal Residue," *Bsns W*, 30 (February 15, 1969).

7. "Great Blob," *Newsweek*, **73**, 31 (February 17, 1969).

8. "Runaway Oil Well: Will It Mean New Rules in Offshore Drilling?" *US News*, **66**, 14 (February 17, 1969).

9. "The Dead Channel," *Time*, **93**, 21 (February 21, 1969).

10. "Great Oil Slick," *Life*, **66**, 58 (February 21, 1969).

11. "Helpless Birds, Helpless Technology," *Sci N*, **95**, 183 (February 22, 1969).

12. "Another Oil Slick; Santa Barbara Subject to New Onslaught," *Newsweek*, **73**, 37 (February 24, 1969).

13. "Case of the Oily Waters," *Schol*, **94**, 21 (February 28, 1969).

14. "Maximum Feasible Leakage," by S. V. Roberts, *Commonweal*, **89**, 667 (February 28, 1969).

15. The *New York Times Index* lists 89 entries on the Santa Barbara oil leak during 1969.

16. "Santa Barbara Oil Leak," *Facts on File XXIX*, p. 452G2, July 17-23, 1969.

17. "Oil on the Waters," *Nation*, **208**, 304 (March 10, 1969).

18. "Oil Pressure," by S. V. Roberts, *New Republic*, **160**, 13 (March 15, 1969).

19. "Black But Clear Lessons From Santa Barbara," *Audubon*, **71**, 5 (March 1969).

20. "Life With the Blob," *Sports Illus*, **30**, 52 (April 21, 1969).

21. "Finding Lemonade in Santa Barbara's Oil," *Sat R*, **52**, 18 (May 10, 1969).

22. "Irridescent Gift of Death," *Life*, **66**, 22 (June 13, 1969).

23. "Not So Deadly," *Time*, **93**, 21 (June 13, 1969).

24. "GOO Story," *Newsweek*, **73**, 60 (June 16, 1969).

25. "Diary of a Disaster," *McCalls*, **97**, 58 (June 1970).

26. "Can Man Afford to Foul His Environment?" *Nat Wildlife*, **7**, 18 (June 1969).

27. "Black Tide," by J. McCaull, *Environ*, **11**, 2 (November 16, 1969).

28. "Santa Barbara; Oil in the Velvet Playground," *Ramp Mag*, **8**, 43 (November 1969).

29. "GOO Fishes In. Get Oil Out group fight companies drilling in the Santa Barbara Channel," *Newsweek*, **74**, 100 (December 8, 1969).

30. "Oil on Troubled Waters," *Time*, **95**, 46 (February 9, 1970).

31. "Spilled Oil: Growing Hazard to Coasts," *US News*, **68**, 42 (March 16, 1970).

32. "Santa Barbara Oil Pollution" hearings before the Subcommittee on Minerals, Materials, and Fuels; Committee on Interior and Insular Affairs, U.S. Senate, May 19-20, 1969, and March 13-14, 1970.

33. "Effects of the Santa Barbara Blowout," *US News*, **70**, 54 (February 8, 1971).

34. "Oil's Aftermath," *Time*, **97**, 27 (March 1, 1971).

35. *Biological and Oceanographical Survey of the Santa Barbara Channel Oil Spill 1969-70*. Volume I: "Biology and Bacteriology" by Dale Straughn; Volume II: "Physical, Chemical and Geological Studies," Ronald L. Kolpack, Ed., University of Southern California, 903 pages, 1971.

36. *Black Tide: The Santa Barbara Oil Spill and Its Consequences*, by Robert Easton, Delacorte Press, New York, New York, 1972.

37. "The Santa Barbara Oil Spill—Initial Quantities and Distribu-

tion": Part 1 by M. Foster, A. C. Charters, and M. Neashul, *Environ Pollut*, **2**, 97 (October 1971); Part 2 by M. Foster, M. Neushul, and R. Zingmark, *Ibid.*, p. 115.

38. *Facts on File XXIX*, p. 842 (December 25–31, 1969).
39. *New York Times*, p. 1 (June 12, 1970).
40. *New York Times*, p. 28 (April 22, 1971).
41. *Air and Water Pollution Report*, p. 265 (July 3, 1972).
42. *New York Times*, p. 50 (July 24, 1971).
43. *New York Times*, p. 36 (August 26, 1971).
44. *New York Times*, p. 54 (September 3, 1971).
45. *Kansas City Star*, p. 15 (April 17, 1973).
46. *New York Times*, p. 41 (November 26, 1972).
47. *New York Times*, p. 23 (January 13, 1973).

Mercury Discharges By Industry

Abstract

In March 1970, a Canadian graduate student notified officials that pickerel in Lake St. Clair (on the Canadian-American border near Detroit) contained 1 to 5 parts per million mercury, well above the 0.5 parts per million legal limit in the United States and Canada. Following Canada's immediate ban on commercial and sport fishing in the area, U.S. Food and Drug Administration officials traced a trail of contamination from the chlor-alkali plants of the Dow Chemical Company in Canada and the Wyandotte Chemical Corporation in Michigan to Lake Erie. The scope of the mercury pollution soon widened to include, first, many of the thirty other chlor-alkali plants in the United States that used mercury in their production processes, then the pulp and paper mills that used mercurial fungicides, and eventually dozens of industrial and institutional users of mercury in thirty-three states. Seventeen states acknowledged a public health menace from mercury pollution; the curtailment of fishing on streams in many states cost millions of dollars. The mercury losses by the chlor-alkali plants were nearly eliminated within a few months, but the arguments continue over the control of industrial and agricultural use of mercury compounds, how and how much this use contributed to environmental mercury levels, and what should be done about existing mercury-laden sediments and waters.

Background

Mercury (Hg) was known in ancient times: its brilliant red sulfide (vermilion) was used in decorative pigments and in cosmet-

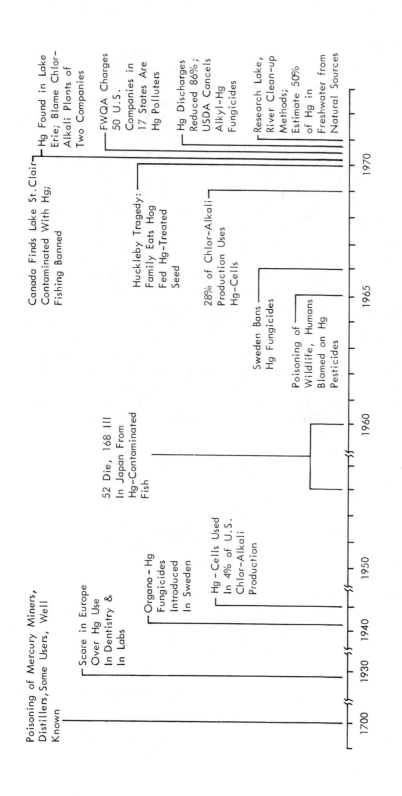

ics, and its ability to form amalgams (alloys) with silver and gold was utilized in working these metals. It was used as early as the first century A.D. as a medicinal ointment, by the year 1000 as a treatment (as $HgCl_2$) for skin rashes, and by the 1500's to help combat syphilis. More recently, calomel (Hg_2Cl_2) was used as a laxative; a number of organomercurials have been used as diuretics, antiseptics, bactericides, and fungicides; and mercurials have been given intravenously in syphilis treatment. Nevertheless, the poisonous nature of mercury has long been known. Although the illnesses of slaves in mercury mines in ancient times were of little concern, the great increase in production of mercury that started in the late 1500's (primarily for the Patio process for silver recovery) brought a wide awareness of the debilitating* and even lethal effects of mercury poisoning; the use of a mercury compound as a fungicide in felt afflicted many hat-makers and led to the phrase "mad as a hatter" in the 1700's. In the late 1920's, a mercury scare of considerable proportions developed in Europe over its use in dental fillings and in laboratory applications. Nevertheless, the industrial and medical uses of mercury continued to grow: U.S. consumption in 1969 was nearly 6 million pounds (with 232,000 pounds of the metal used in dental preparations and 55,000 pounds in pharmaceutical compounds), and total consumption since 1900 has been about 163 million pounds.[1]

Among the largest of the industrial uses developed for mercury was the mercury cathode electrolytic cell for the production of chlorine and caustic alkali (sodium hydroxide or in some cases metallic sodium). Chlor-alkali plants using this cell were built in the 1890's in the United States, but the American chlor-alkali industry largely adopted the nonmercury diaphragm cells.† In contrast, the mercury cell became predominant in Europe, particularly after Germany's World War II technology became available.[2] U.S. companies then started switching, and the proportion of chlor-alkali produced in mercury cells rose from 4 percent in 1945 to 28 percent in 1969‡ as newer and bigger mercury cells of several types were installed. Mercury cells were used in thirty-two plants by 1969.[1,3] A typical large chlor-alkali plant might use as much as 600,000

*Mercury tends to accumulate in the brain and affects the entire central nervous system.

†Both cells were developed at about the same time, following the development of the electric dynamo in about 1865.

‡By 1969 the chlor-alkali industry was a $688 million a year business, producing about 20 billion pounds a year each of chlorine and caustic.

pounds of mercury in its cells and recirculating system at one time.[4] Mercury prices rose as its use increased and reached nearly $10 a pound by 1968.

Chlor-alkali plants were not the only industrial users of mercury: its excellent electrical properties made its use in other electrical equipment nearly as large; its excellent ability to interact in organic chemical reactions led to wide usage as a catalyst; and the high degree of biological activity of its organic compounds led to wide usage as fungicides in outdoor paints, in pulp- and paper-making, and in agriculture. These fungicidal applications utilize the more toxic alkyl mercury compounds (compared to the aryl compounds used in pharmaceuticals). The first of these, Panogen®, was introduced as a seed treatment in Sweden in 1938, although it was not permitted in the United States until 1949.

The widespread industrial usage of mercury and its compounds was not without health hazards, but industrial safety practices generally minimized occupational illnesses in the United States. In 1953, however, a chemical manufacturing plant* at Minamata Bay in Japan apparently had a significant amount of its mercury chloride catalyst escape in the form of a methyl mercury compound with a wastewater discharge: the mercury was then taken up by shellfish, crabs and fish in the bay, and produced an epidemic of illnesses and death in some people who ate the fish. The cause of the mysterious malady could not be identified at the time, but as additional cases occurred over the years (possibly from further discharges), it became known as "Minamata disease." By 1960 the problem had become a major social issue, and after a determined research effort, the chain of events was finally established: the toll was already fifty-two deaths and 168 illnesses over the seven-year period.[5] The source of pollution was closed, only to have a similar, but less severe, outbreak of the disease at Niigata, Japan in 1964.

In addition, the industrial and agricultural use of mercury compounds was being blamed for the indirect poisoning of fish and birds in Sweden, and the accidental human poisonings from mercury-treated seed in Iraq (1956), Pakistan (1961), and in Guatamala (1963).[6] Sweden banned the use of the alkyl mercury fungicides in pulp and paper plants and in agriculture in 1966. In

---

*The plant, operated by Chisso Corporation, has been variously called a fertilizer plant, a chemical plant and a plastic plant in the press. The mercury loss apparently occurred from an acetaldehyde production process. The same process was used at Showa Denko Corporation at Niigata.

the United States, however, mercury fungicide production was approaching 1 million pounds a year, and little concern was apparent over mercury levels on the land or in the water. Natural levels of mercury in the earth's crust and in the oceans had been variously reported in the literature, respectively, as 0.01–0.9 parts per million and 0.03–0.7 parts per billion, but little was known about mercury levels in rivers and lakes. In fact, neither the then Federal Water Pollution Control Administration (FWPCA) nor the U.S. Geological Survey included mercury among the metals it routinely monitored in surface waters of the United States during the 1960's. The Public Health Service's Bureau of Water Hygiene did not list mercury in its toxic-substance guidelines in 1962, nor did FWPCA establish maximum permissible limits for mercury in approved public·water supplies in its 1968 criteria[7] (although it did list mercury as toxic).

The difficulty and expense of performing the analyses for trace levels of mercury was part of the reason that mercury levels were unknown at that time. In addition, the methods used did not differentiate between the inorganic or aryl mercury forms and the far more toxic methyl or ethyl mercury compounds. The absence of data, in turn, probably led to classification of mercury with other "safe" elements. But newer, quicker and more sensitive analytical methods (such as atomic absorption and neutron activation analysis) were coming into practice, and the recognition of DDT as a global pollutant was creating many environmental champions who questioned previous assumptions.

In particular, concern about mercury had developed in Sweden and in 1966 authorities there called an international symposium on mercury pollution; it was attended by five official American delegates (including one from the President's Office of Science and Technology) but did little to stir the complacency of U.S. officialdom.[6] The following year two Swedish scientists determined that inorganic mercury could be methylated by anaerobic bacteria in underwater muds and a third one, Dr. Carl Rosen, came to the U.S. and worked with Dr. J. M. Wood and F. S. Kennedy at the University of Illinois to identify the specific bacteria. By April 1968, they had submitted their results for publication in the prestigious journal, *Science*, but their paper was declined; it was resubmitted to the equally prestigious British journal, *Nature*, and was published in October.[6] The popular journal, *Environment*, then picked up the issue and summarized and interpreted the meaning of the Minamata, Swedish, and methylation data in its May 1969 issue.[6] At this point, the Food and Drug Administration reacted and established the first regulation on mercury in food or drink: the maximum level of mercury permitted in fish or shellfish was set in July 1969* at 0.5 parts per million.

And in New Mexico, six obscure farmers started feeding leftover mercury-treated seed grain from a local granary to their hogs in August 1969. Thus the stage was set for the great mercury scare.

Key Events and Roles

In May 1968, Norwegian-born Norvald Fimreite embarked on a doctoral research project in zoology at the University of Western Ontario in Canada. Mindful of the Scandinavian experience with environmental mercury problems, he began a study of mercury concentration in food chains: hawks that were fed steadily on chicken livers contaminated with 3 parts per million mercury developed levels of up to 18 parts per million mercury.[8] Fimreite then turned to pike—a predator fish—and collected samples from Lake St. Clair on the Canadian-American border between Lakes Erie and Huron.

At about the same time, the tragedy of the Huckleby family in New Mexico became a national sensation when it was described on the February 17, 1970, Huntley-Brinkley network television newscast and in the press. Huckleby had butchered for his family's use one of the hogs that had been fed mercury-treated seed grain. The tragic result was that the Huckleby children all suffered severe brain damage with various degrees of permanent crippling, blindness, and other disabilities.*

Fimreite was finding that pickerel from Lake St. Clair contained mercury at levels as high as 5 parts per million (and average 1.36 parts per million) in the methyl form. On March 19 he sent a letter describing his results to the Canadian Wildlife Service and a day later released the letter to the press. Within a week, the Canadian Government had confirmed the results and imposed a ban on the sale and export of fish from the St. Clair River and Lake—just before the season opened.

The ban sent a shock across the border to U.S. Food and Drug

---

*The World Health Organization had suggested a limit of 0.05 parts per million mercury in the total weekly diet, but Swedish scientists felt that about 0.2 parts per million mercury was almost universally distributed in "nature."[6]

*Sadly enough, the Hucklebys continued to eat the contaminated pork, even after several other hogs became sick or died from the mercury.

Administration officials, who quickly verified the high mercury
levels of fish in Lake St. Clair, and who also found contamination
downstream in western Lake Erie. Michigan and Ohio authorities
banned commercial fishing in the area by the end of April, and
Ontario had banned both commercial and sport fishing. These developments also surprised officials of the Dow Chemical Company
and the Wyandotte Chemical Corporation. Authorities pointed to
Dow's chlor-alkali plant on the St. Clair River at Sarnia, Ontario,
and to Wyandotte's plant on the Detroit River in the United States
as the sources of mercury pollution. Government officials were astounded to learn that Dow's plant had formerly been losing up to
two hundred pounds a day of the expensive mercury, although they
had in late 1969 started design and housekeeping changes which
had cut the loss to twenty pounds a day.[9] Wyandotte was losing
about 10 pounds a day.[8] Both companies agreed to step-up mercury
containment programs—Wyandotte consenting to install a temporary $1 million brine processing unit while designing permanent
facilities.

But the authorities were concerned and confused over the mercury which had already escaped; and, of course, the commercial
fishing industry, the sports fishing interests (when fishermen were
warned not to eat their catch) and related industries were alarmed
and vocal. Michigan officials, who had first resisted a ban on sports
fishing, decided caution was in order, and Governor Milliken ordered a ban on all fishing in the Lake St. Clair area. Ohio's Governor Rhodes rescinded the ban on fishing in Lake Erie on April 23,
but on April 28, Ohio filed a federal suit seeking damages from Dow
and Wyandotte for polluting the lake. A major question was, "How
did the mercury, which was discharged in inorganic (metallic or
salt) form, get converted to the methyl form?" Dow spokesmen,
including President Herbert D. Doan, maintained that no one knew
until very recently that inorganic mercury could be methylated in
biological systems. Subsequently, critics pointed out that Swedish
conferences had discussed evidence of this in 1966, and Dr. John
M. Wood of the University of Illinois had demonstrated methylation reactions in 1968.[1] An even bigger question was, "Has mercury
pollution occurred elsewhere in the United States?" The answer
was soon to come.

Before FWQA could pinpoint any other industries that were
discharging mercury, a list of the users was required. A list of the
consumption of mercury in major uses in the United States was
prepared by the Bureau of Mines[1] as shown in Table VI.

TABLE VI

CONSUMPTION OF MERCURY

Purpose	1969 Consumption (thousands of lbs)
Chlor-alkali production	1,575
Electrical apparatus	1,417
Paints	739
Industrial control and instruments	531
Dental preparations	232
Catalysts	225
Agriculture	204
General laboratory use	155
Pharmaceuticals	55
Pulp- and paper-making	42
Other uses	736
Total	5,911

An extensive investigation of 129 companies throughout the Spring of 1970 revealed that fifty of them were discharging mercury.[10] The industries which bore most of the blame were the thirty-two mercury cell chlor-alkali plants[11] and the pulp and paper mills.

On July 25, 1970, Attorney General Mitchell authorized the Justice Department to file suits under the 1899 Refuse Act against five chemical firms and three papermakers operating ten plants across the United States over alleged mercury pollution. The companies charged, the locations of offending plants, and the amounts of mercury discharged (pounds a day, based on Department of Interior data) were: Diamond Shamrock at Delaware City (11.5) and Muscle Shoals, Alabama (6.5–8.6); Olin-Matheson, Augusta, Georgia (8.7–12.9), and Niagara Falls, New York (26.6); Allied, Solvay, New York (4.4); International Minerals and Chemical Company, Chlor-Alkali Division, Orrington, Maine (2.6); Pennwalt Chemical Company, Calvert City, Kentucky (1.54); Oxford Paper Company, Rumford, Maine (26.2); Weyerhauser Company, Longview, Washington (15.1); and Georgia-Pacific Company, Bellingham, Washington.[12]

Under this pressure, most companies acted quickly to reduce or eliminate mercury discharges by using better in-plant housekeeping and by installing various control measures. A report issued by Secretary of the Interior Walter J. Hickel in September 1970 showed that the fifty industrial plants previously discharging mer-

cury had reduced their discharges into waterways from about 287
pounds daily in July to forty pounds daily in September, a reduction of 86 percent.[13]

Several states then enacted legislation to control future mercury contamination of waterways. Massachusetts, Florida, Vermont,
Arizona, Texas, Oklahoma, and others restricted mercury discharges and, in some cases, its use. Illinois limited plant discharges
to 5 pounds a year and effluent concentrations to 0.5 parts per
billion.[1]

Disposition

In March 1971, the U.S. Supreme Court declined to hear the
mercury pollution case that the state of Ohio had filed against Dow
and Wyandotte Chemical Companies for polluting Lake Erie. They
ruled, in an 8 to 1 decision, that they did not have the time or
expertise to try antipollution lawsuits that had not made their way
through the lower courts. Justice John M. Harlan wrote, in a majority opinion, that if the way was cleared for states or citizens to come
directly to the high court to challenge pollution, it would mean "a
serious drain on the resources of this court."[14]

Five of the eight lawsuits filed by the Justice Department earlier were dropped when the companies agreed to discharge less
than one-half pound of mercury per day on a weekly average.[10] The
other three suits were still pending.

A report, "Hazards of Mercury," was released by the Department of Health, Education, and Welfare and the Environmental
Protection Agency in January 1971. The report recommended that
commercial uses of mercury be curbed, and that the United States
begin an immediate, intensive program to decontaminate waterways containing mercury deposits. It also estimated that even with
complete elimination of mercury losses by industry, existing deposits in the waterways would yield highly toxic methylated mercury to waters over a period of decades.

The chlor-alkali plants were particularly cooperative in lower-
ing their mercury discharges according to government stipulations.
Many of them helped lead the way in developing abatement programs and conducting research on mercury's environmental impact.
Dow Chemical Company began a multimillion dollar project to
convert its Sarnia, Ontario plant from the mercury cell process to
the diaphragm cell process, which uses no mercury, and shut down
its only mercury cell plant in the U.S. at Plaquemine, Louisina.[16] By

the middle of 1971, thirty of the original thirty-two mercury cell chlor-alkali plants were still operating, but their losses of mercury in wastewater had been reduced on an industry average to 2 percent of the January 1, 1970 level, according to the Chlorine Institute.[11]

The problem of restoring mercury-contaminated lakes was far greater. Intensive studies in Sweden and the U.S. recommended four different methods, but each had its drawbacks. The four methods and their negative aspects were: (1) covering the mercury-rich sediments with inert clay or other absorbing materials (this process would be both expensive and inefficient); (2) dredging the lakes to remove the sediment (also expensive and impractical in large lakes); (3) adding iron sulfide or hydrogen sulfide to form mercury sulfide, which does not methylate (addition of these chemicals may kill the fish); and (4) adding nitrogenous compounds to raise the alkalinity and change the methylation process (this method would spread the pollutant over a larger area).[17]

To date, no feasible method has been found to remove the existing mercury deposits in the U.S. waterways. The intensive effort by both industry and government to reduce the mercury pollution of water in 1970 and 1971 substantially reduced the mercury content of the nation's waterways. However, even at the height of the industrial dumping problem in early 1970, the normal geological loss of mercury from rocks matched the losses of U.S. industry.[17] The existing mercury deposits will add to the natural background level for decades, and those bodies of water already contaminated will remain so for many years. In Minamata Bay, over four hundred tons of mercury are estimated to contaminate bottom sediments and no clean-up efforts are underway. In fact, fishing from the Bay was never officially prohibited until 1973—seventeen years after the cause of the illnesses was first suspected. During that time a total of sixty-five deaths and an additional 332 known illnesses have occurred from mercury poisoning there.[18] On March 20, 1973 the Japanese manufacturing firm was ordered to pay 937 million yen to victims of the disease after a four-year court case.

Comment

The mercury pollution episode illustrates very well the causes of many of our environmental problems: process-control waste-disposal methods that were imperfect but tolerable when American industries (and cities) were small and relatively isolated are no longer acceptable when the amounts of materials handled become so large and the sources so numerous that a natural system's ability to dissipate damage is overloaded. In addition, it illustrates the fact that little-known pathways of chemical and biological systems in nature may become exceedingly important when the natural conditions are radically altered. The explanation of the companies involved in this episode—that no one knew inorganic mercury, even in the amounts being discharged, could be dangerous to the environment—is of particular interest. One wonders whether these companies may have had chemists and biologists who, if asked, would have suggested caution or studies before discharging large amounts of it.* But the company managements or the plant designers and operators apparently never asked. Those scientists within or out of the companies who might have helped were unaware of mercury losses (proprietary operating data), and had little reason to consider the consequences. No mechanism operated that raised the right questions in the minds of the right people.

The discovery of the mercury pollution problem, coming on the heels of the widely publicized Huckleby tragedy and in the midst of the DDT debate, received much publicity and caused a strong emotional reaction on the part of the public, public officials, and the industry. Remedial action was promptly undertaken but could not be complete. The emotional atmosphere that was generated made difficult a reliable assessment of any damage that low levels of mercury might cause, and set the stage for the great tuna-fish scare which followed.

In addition, the realization of the ease with which inorganic mercury could be converted in nature to the more biologically dangerous methyl and ethyl mercury forms, soon grew into widespread concern over the little-known movements and effects in the environment of a host of other heavy metals.

*Chemists know that organomercury compounds are easy to make.

REFERENCES

1. "Mercury: Anatomy of a Pollution Problem," *C&EN*, **49**, 22 (July 5, 1971).

2. Sconce, J. S., Editor, *Chlorine, Its Manufacture, Properties and Use*, American Chemical Society Monograph Series, Reinhold Publishing Company, New York, pp. 127-F (1962).

3. "Tighter Limits on Mercury Discharge?" *Chem W*, **107**, 35 (July 22, 1970).

4. Parker, J. T., "Facts Behind the Mercury Menace," *Pop Sci*, **197**, 62 (December 1970).

5. "Mercury in the Environment," *Science*, **171**, 788 (February 26, 1971).

6. Montague, P., and K. Montague, "Mercury: How Much Are We Eating?" *Sat R*, **54**, 50 (February 6, 1971).

7. "Water Quality Criteria," Report of the National Advisory Committee to the Secretary of Interior, Federal Water Pollution Control Administration, April 1, 1968.

8. "Mercury's Turn as Villain," *Chem Eng*, 84–86 (July 27, 1970).

9. "Mercury Mars the Catch," *Chem W*, **106**, 16 (April 8, 1970).

10. "Mercury Polluters are Under Attack from Representative Reuss," *Oil Paint & Drug Rep*, 4 (October 26, 1970).

11. "Chlorine Production Drops as Demand Lags," *C&EN*, **49**, 10–11 (November 8, 1971).

12. "Blast at Industry," *C&EN*, **48**, 14 (August 3, 1970).

13. "Mercury Dischargers Still Feel Prod, Despite Efforts to Clean Waterways," *Oil Paint & Drug Rep*, pp. 5 and 22 (September 21, 1970).

14. "Pollution: Attack From All Sides," *C&EN*, **49**, 9–10 (March 29, 1971).

15. "Mercury: Potentially Grave," *C&EN*, **49**, 10 (January 25, 1971).

16. "Chlorine-Caustic Soda," *C&EN*, 9 (November 8, 1971).

17. "Mercury in the Environment," *Environ Sci & Technol*, **4**, 890–892 (November 1970).

18. Huddle, N., and M. Reich, "Minamata Trial Results," *Ecologist*, **3**, (6) 201 (June 1973).

The Mercury-in-Tuna Scare

Abstract

In December 1970, a chemist at a New York university reported that deep-sea tuna fish in cans off the grocery shelf contained mercury levels higher than the 0.5 parts per million maximum set by the Food and Drug Administration. This news closely followed the uproar early in 1970 over mercury pollution of rivers and lakes by some American industries, and a widely publicized case of mercury poisoning via contaminated pork. On December 15 the FDA announced that nearly 1 million cans of tuna were being recalled and estimated that 23 percent of all tuna packed that year was contaminated, while insisting that all of it was still safe to eat because the standard had a "large margin of safety." In addition, the FDA soon recalled nearly all the swordfish on the market for the same reason. These developments caused a great scare for fish eaters, and seafood sales plummeted. They also aroused fears that man had polluted the entire ocean.

Further testing showed that only 3.6 percent of the tuna was actually above the safety limit, and on February 5, 1971, the FDA announced that all tuna on the market was untainted. The source of the mercury was controversial: some suggested industrial wastes, and others the use of marine paints treated with mercury fungicides on fishing boats.Subsequent tests soon indicated that tuna have always had mercury levels near or even above the FDA safety limit and that the levels in swordfish are frequently above the limit. The controversy over whether these levels constituted a hazard for fish eaters continued to rage.

Background

In early 1970 the American people had two traumatic experiences involving the hazards of mercury compounds,[1] as described

260

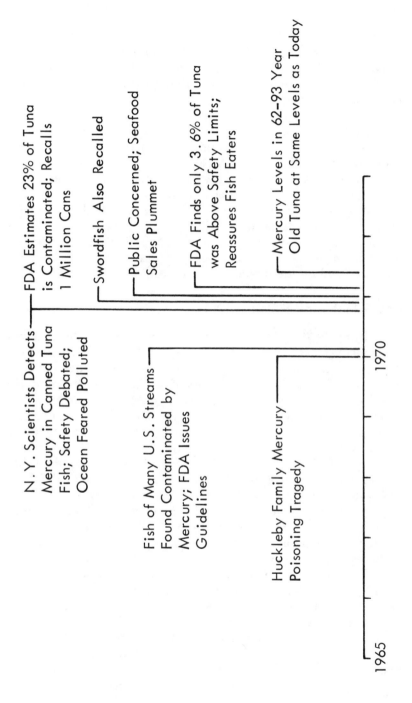

N.Y. Scientists Detects Mercury in Canned Tuna Fish; Safety Debated; Ocean Feared Polluted

FDA Estimates 23% of Tuna is Contaminated; Recalls 1 Million Cans

Swordfish Also Recalled

Public Concerned; Seafood Sales Plummet

FDA Finds only 3.6% of Tuna was Above Safety Limits; Reassures Fish Eaters

Mercury Levels in 62-93 Year Old Tuna at Same Levels as Today

Fish of Many U.S. Streams Found Contaminated by Mercury; FDA Issues Guidelines

Huckleby Family Mercury Poisoning Tragedy

1970

1965

in detail in the case history, "Mercury Discharges by Industry." In
the one, the children of the Huckleby family in New Mexico suffered brain damage from eating a hog which had been fed grain
treated with a mercury fungicide; the tragedy was widely publicized on national television and in the press. In the other, mercury
discharges, initially from chlor-alkali manufacturing plants and ultimately from dozens of industrial plants, were found to be polluting streams and lakes in two-thirds of the states. The mercury was
showing up in unacceptable levels in fish; commercial and even
sport fishing was banned in many localities across the country. The
results were the loss of millions of dollars to fishing and related
industries, a great amount of publicity over industrial and governmental mercury clean-up plans, and a public awareness of environmental pollution problems and of the scientist's ability to detect smaller and smaller traces of contaminants. The nation was
thus prepared for the mercury-in-tuna scare.

Key Events and Roles

In July 1970, the Canadian government seized some contaminated whale meat in the northern reaches of Hudson Bay, some two
thousand miles from the nearest known industrial source of waste
mercury and a long way, also, from any significant agricultural use
of mercury. Although this event was mentioned in at least one
popular publication,[2] its significance was not widely recognized—
even by U.S. Governmental authorities—in the tumult which was
then existing over industrial mercury discharges. In late September,
the National Marine Fisheries Laboratory in Seattle reported that
seals caught off the coast of Washington contained up to 172 parts
per million of mercury in their livers. This discovery—said to be
the first report of high mercury levels in a marine organism[3]—was
in marked contrast to results of previous FDA studies that found
negligible contamination of marine organisms near coastal pulp
mills that were discharging mercury. This discovery also caused
alarm because seal livers (from seals obtained off Alaska's Alleutian
Islands) were being used to make a non-prescription vitamin pill
that was sold in many health food stores. On September 30, the
FDA announced the recall of 25,000 of the pills, some of which
contained as much as 30 parts per million mercury, and a month
later the popular press carried stories,[4-6] that the government was
studying Aleuts who had been eating seals.

Scientific studies on river and lake fish had indicated that larger fish higher on the food chain would contain, through biomagnification, a greater concentration of the mercury than smaller fish. Realizing this, Professor Bruce McDuffie, a chemist at the State University of New York at Binghamton, took a can of tuna from his pantry shelf and analyzed the contents for mercury: the tuna contained 0.75 parts per million, well above the 0.5 parts per million limit the FDA had set in 1969. On December 4, 1970, McDuffie reported his findings.[1] The Food and Drug Administration soon confirmed the presence of high mercury levels in tuna. On December 8, the FDA's announcement that preliminary tests had shown that Grand Union brand tuna contained unsafe levels of mercury and that the company had agreed to withdraw its product from store shelves, was carried deep in the newspapers.[7, 8] Almost immediately, however, the story grew to front-page importance as more brands were implicated; FDA Commissioner Charles C. Edwards, Jr., announced on December 15 that almost 1 million cans of tuna were being recalled from market.[9-11] The FDA, he said, had found that in 138 samples tested, the mercury levels averaged 0.37 ppm, and ran as high as 1.12 parts per million. The mercury was found in tuna from all eight major commercial tuna fishing areas, and the FDA estimated that 23 percent of the 900 million cans of tuna packaged in 1970 were contaminated. The Commissioner insisted, however, that the tuna being removed was "absolutely safe" to eat, because the 0.5 parts per million limit had a substantial margin of safety, and because Americans didn't rely on fish as a staple of their diets anyway.

The FDA's announcement caused a sensation during late December, reinforced by the FDA's recall on Christmas Eve of nearly every brand of frozen swordfish, a 25 million pounds-per-year market. The FDA said that swordfish samples contained from 0.18–2.4 parts per million mercury with 89 percent of it over the 0.5 parts per million limit, but also stressed that swordfish was still fit to eat.[12] At the same time, however, FDA Deputy Commissioner James Grant said that regular consumption of swordfish would be an unnecessary and avoidable risk. The public, particularly expectant mothers and those on a high-fish diet, were concerned about the mercury hazard, and confused by the FDA's reassurances while issuing recalls. Consumer crusader Ralph Nader said the incident was a resounding indictment of the FDA.[11] Environmental crusader Dr. Barry Commoner complained to delegates at the annual meeting of the prestigious American Association for the Advancement of Science in Washington, D.C. during Christmas week that the FDA

appeared to be "minimizing the hazard of mercury." He contended
that the "safety margin" on mercury was inadequate, and urged that
the FDA lower the 0.5 parts per million limit.[13] In the December 15
announcement, Dr. V. O. Wodicka, Director of FDA's Foods
Bureau, had emphasized that scientists did not know how the mercury had got into the food chain,[10] and a report the following week
that tainted tuna was being found in Britain reinforced some
ecologists' contention that man had polluted the oceans.[11] On the
other hand, the National Cannery Association was reported[10] to be
speculating that the tuna might be exported to Europe and Asia.

The mercury-in-tuna scare continued into early 1971 and inspired a spate of popular magazine articles,[1, 14-28] as well as a CBS
documentary. Spot checks around the country showed that tuna
sales in grocery stores dropped drastically, following the initial
FDA announcement. A New York Times survey of eleven major
cities in January found consumers confused and concerned, and the
seafood industry suffering adverse economic effects.[29] On January
14, Professor McDuffie reported that analysis of blood, hair, and
urine samples from sixty-two Binghamton, New York, residents indicated that people who ate contaminated fish had five times as
much mercury in their systems as those who did not.[30] The following day Weight Watchers International, Inc. revised its weight-
reducing diet to minimize an emphasis on eating tuna and other
fish.[31] Many weight-watchers apparently sought out their already
busy doctors for a mercury analysis.

The FDA had continued testing tuna samples for mercury during January, but even before their tests were completed, serious
questions were being raised about the "normal" level of mercury in
fish. As early as January 6, 6, the New York State Environmental
Conservation Department reported that preserved fish, caught in
state waters as long ago as forty-three years, contained twice as
much mercury as the fish now being banned; the story made page
one news.[32] Some observers held that fish may always have been
slightly "contaminated" with mercury, and that it was "highly irra-
tional" to conclude that this level was dangerous.[33] Others noted
that references to rich mercury levels in fish* had been made as
early as the 1930's.[34] Still others speculated that the tuna had become contaminated after it was caught by coming into contact with
mercury fungicide-treated paints or timbers on fishing boats.[18]

By January 8, 1971, the FDA announced that its survey of
American canned tuna was half completed, and that the contamina-

---

*One source listed about 0.2 ppm as the maximum in fish.[34]

tion was less than feared[35] On February 4, Commissioner Edwards announced[36] that final results showed that only 3.6 percent of 166 million pounds of domestic and imported tuna tested exceeded the FDA limit—rather than the 23 percent estimated in December—and that all tainted stocks had been removed from the market.

Disposition

The FDA had found that larger varieties of tuna contained more mercury than smaller varieties and said that tuna could remain a staple of American diets if only the smaller ones were packed.[35] Fortunately, the most popular tuna was the white-meated albacore (about forty-five lbs) followed by the progressively darker yellow fin (weighing about 150 pounds), skipjack (thirteen pounds), big eye (235 lbs), and blue fin (250 pounds).[1] The FDA's earlier, erroneous conclusion that 23 percent of the tuna was contaminated was based on analysis of yellow fins.* By February, the tuna industry was reporting that the 1971 catch from the fishing grounds near Central America did not contain unsafe levels of mercury.[37] The Tuna Research Foundation, an association of California tuna canners, launched a vigorous† public relations and information campaign to overcome the adverse publicity and revive tuna sales from their 1971 first-quarter collapse.[38] The campaign even included a press release that the Apollo 14 astronauts were taking tuna fish sandwiches to the moon, implying that NASA considered tuna safe. It was generally successful; sales rebounded, and were good for 1971 and 1972.

The swordfish story had a different ending, however. Swordfish must normally run from 100 to 500 pounds each in order to provide an adequate number of marketable "swordfish steaks"; in the February 5 announcement clearing tuna, the FDA said that 87 percent of the swordfish samples had mercury levels above the safety limit.[36] The FDA urged the swordfish industry to withhold all shipments until they could be tested, but here a second problem arose; unlike the tuna industry which was concentrated among a few major producers, the swordfish industry was composed of dozens of small, scattered operators.‡ As the testing continued, nearly

---

*All samples of big eye and blue fin were over the 0.5 parts per million limit.

†The campaign reportedly had printing, mailing, and associated costs of $65,000, plus the staff-time costs of a public relations firm.

‡Of the 26 million pounds of swordfish consumed in the U.S. annually, about 22 million pounds were actually imported.

95 percent of the 853 samples of swordfish failed to pass, and 832,000 pounds were seized—some with mercury levels of 1.5 parts per million.[26] The FDA concluded that it could not control the industry under these circumstances, and on May 6, 1971 issued what amounted to a virtual ban on swordfish: a well-publicized official warning to Americans to stop eating it because of mercury contamination.[26,39] If a further blow was needed to kill the swordfish industry, it was not long in coming: testimony before a Senate Commerce subcommittee on May 20 that a New York woman had suffered mercury poisoning from subsisting on a swordfish diet for nineteen months was widely publicized.[27,40] The diagnosis was almost immediately disputed[41] and the New York City Health Department subsequently reported that a five-month study showed no differences in mercury level between habitual fish eaters and people on normal diets,[42] but by then the swordfish case was closed.

The question of how the mercury got into wide-ranging ocean fish has apparently been satisfactorily answered: the levels discovered in tuna and swordfish in 1970–71 have apparently always been there. Museum samples of tuna sixty-two and ninety-three years old and a swordfish twenty-five years old had mercury levels in the same range as those caught today[43] and further studies on deep ocean, ancient, and even prehistoric fish continue to find mercury levels above the FDA's 0.5 parts per million standard (critics have suggested that these levels may be the results—or the artifacts —of preservation conditions).

The question of whether mercury levels slightly above the FDA limit pose any hazard to fish eaters on a normal diet has not been answered. Mercury is known to be cumulative in human tissues upon continuous consumption (but it is also excreted, with a half-life of about seventy days*), and symptoms of poisoning appear when blood levels of mercury reach about 0.2 parts per million in humans. The 0.5 parts per million limit on mercury in foods was believed to provide a "safety factor" of 10, although FDA usually allowed a safety factor of 100; the latter would have placed the limit at 0.05 parts per million, which would be below the 0.2 parts per million that Swedish scientists had called the "universal level in nature" and would have ruled out a great many foods. In fact, Sweden has set its standard of mercury in food at 1 parts per million so that nearly all the fish in its coastal waters would not be ruled inedible.

*One-half of the amount present would be excreted in seventy days.

Comment

Because it followed earlier mercury episodes, the detection of mercury in tuna fish became a great source of public concern and confusion. The FDA's initial efforts in trying to handle the problem did not inspire confidence, and were complicated by the several self-appointed spokesmen and the intense news coverage. The case became important because a numerical value of 0.5 parts per million had been established as a maximum safety limit on fish at a time when it was not realized that ocean fish at the top of the food chain frequently built up mercury levels of this magnitude, and when trace mercury analyses were rare. But when the number had been set and was being demonstrably violated by the tuna, the FDA felt obliged to order a recall, even if no danger to humans who ate such fish had ever been observed. Subsequent information revealed that tuna could meet the FDA standard if the smaller "low-mercury" types were used, but the swordfish industry in the United States was wiped out. While fish have probably always tended to accumulate mercury, and man has probably consumed fish for tens of thousands of years—evolving a tolerance to whatever mercury levels were normally present in his diet—it has been only very recently that his technology has made the very large deep ocean fish (with the high mercury levels) available.[44]

REFERENCES

1. "Mercury: How Much Are We Eating?" P. Montague and K. Montague, Sat R, 54, 50 (February 6, 971).
2. "A Grim Pursuit of Quicksilver," Bsns W, 42 (July 18, 1970).
3. "Ubiquitous Mercury," Sci N, 98, 366 (November 7, 1970).
4. New York Times, p. 1 (October 6, 1970).
5. New York Times, p. V16 (November 1, 1970).
6. New York Times, p. 26 (November 6, 1970).
7. New York Times, p. 51 (December 8, 1970).
8. Wall St J, p. 17 (December 8, 1970).
9. Wall St J, p. 1 (December 14, 1970).
10. New York Times, p. 1 (December 16, 1970). Also, a p. 94 story on mercury poisoning.
11. "The Tainted Tuna," Newsweek, 76, 42 (December 28, 1970).
12. New York Times, p. 1 (December 24, 1970).
13. New York Times, p. 14 (December 29, 1970).
14. "Meddlesome Mercury," Sci N, 99, 7 (January 2, 1971).
15. "Weight Watchers, Beware!" Newsweek, 77, 91 (January 25, 1971).

16. "Mercury: No Need for Hysteria," *Life*, **70**, 32 (January 29, 1971).

17. "Metallic Menaces in the Environment," *Fortune*, **83**, 110 (January 1971).

18. "Mercury: More Than a Fish Tale," *Chem W*, **108**, 57 (January 27, 1971).

19. "Tuna on Toast, Please," *Newsweek*, **77**, 70 (February 15, 1971).

20. "Mercury in Fish: A Slippery Question," *Sr Schol*, **97**, 7 (February 8, 1971).

21. "Mad Hatter Disease: Mercury Poisoning," *Sci Digest*, **69**, 61 (March 1971).

22. "Mercury: A Grave Threat to Human Life," *Consumer Bul*, **54**, 32 (March 1971).

23. "Tuna Fish and Swordfish: Are They Safe to Eat?" *Read Digest*, **98**, 93 (April 1971).

24. "No-Nonsense Report on Mercury in Fish," *Redbook*, **137**, 80 (May 1971).

25. "Truth About Mercury," *Field and Stream*, **76**, 14 (May 1971).

26. "Good-bye to Swordfish," *Newsweek*, **77**, 93D (May 17, 1971).

27. "The Mercury Menace," *Newsweek*, **77**, 47 (May 31, 1971).

28. "Catch Is, Should You Eat It" *Sports Illus*, **35**, 46 (July 12, 1971).

29. *New York Times*, p. 61 (January 10, 1971).

30. *New York Times*, p. 74 (January 14, 1971).

31. *New York Times*, p. 16 (January 15, 1971).

32. *New York Times*, p. 1 (January 6, 1971).

33. *New York Times*, letters from Professor L. J. Goldwater, p. IV 14 (January 17, 1971).

34. "Mercury in Fish," letter by D. M. Wicken, *C&EN*, **49**, 7 (April 19, 1971).

35. *New York Times*, p. 15 (January 8, 1971).

36. *New York Times*, p. 1 (February 5, 1971).

37. *New York Times*, p. 73 (February 3, 1971).

38. "Combatting Alleged Danger of a Product," *Public Relations News*, Case Study No. 1347, New York, New York, 1972.

39. *New York Times*, p. 1 (May 7, 1971).

40. *New York Times*, p. 1 (May 21, 1971).

41. *New York Times*, letter from Dr. L. J. Goldwater, p. 40 (June 2, 1971).

42. *New York Times*, p. 20 (June 23, 1971).

43. "Mercury Concentrations in Museum Specimens of Tuna and Swordfish," G. E. Miller, P. M. Grant, R. Kishore, F. J. Steinkinger, F. S. Rowland, and V. P. Gaines, *Science*, **175**, 1121 (March 10, 1972).

44. Tuna were first canned commercially in 1903, the sardine slightly earlier.

The Rise and Fall of DDT

Abstract

DDT emerged from World War II to ready public acceptance as the insecticidal atomic bomb, but in March 1949 DDT was charged with contributing to a mysterious "X-disease" in cattle and "virus X" in humans. These subsequently disproved charges, coupled with increasing evidence that the insecticide could have undesirable side effects, caused a considerable public scare—the concern heightened by disillusioning indications that flies and mosquitoes were becoming DDT-resistant. DDT survived this scare, however, and remained for nearly twenty years the most widely used pesticide in the world, reaching in the United States alone a production of 180 million pounds in 1963. Nonetheless, DDT remained the focus of controversy which peaked in 1969 over the side effects of pesticides. The result was a near-total ban of its use in the United States in 1972.

Background

From ancient times man has investigated plants animals, and minerals for their value in warding off the attacks of obnoxious or dangerous insects. Of the approximately 1 million species of insects, over nine hundred are known to be obnoxious to man, his animals, or his crops. Synthetic inorganic and organic chemical pesticides came into significant use after about 1850, and several chlorinated organics were utilized or patented in the early 1900's.*

*Carbon tetrachloride, a fumigant for nursery stock (1908); paradichlorobenzene, a fumigant for clothes moths (1913); pentachlorophenol, a wood preservative (1936) and herbicide (1940); and chloranil, a fungicide for crops (1937). Chlorinated naphthalene used as insecticides (1936) were patented as preservatives for wood, paper, and textiles (1909) and were widely used on electrical cables. Chlorinated biphenyls, introduced as industrial fluids (1929), had little pesticidal use.

269

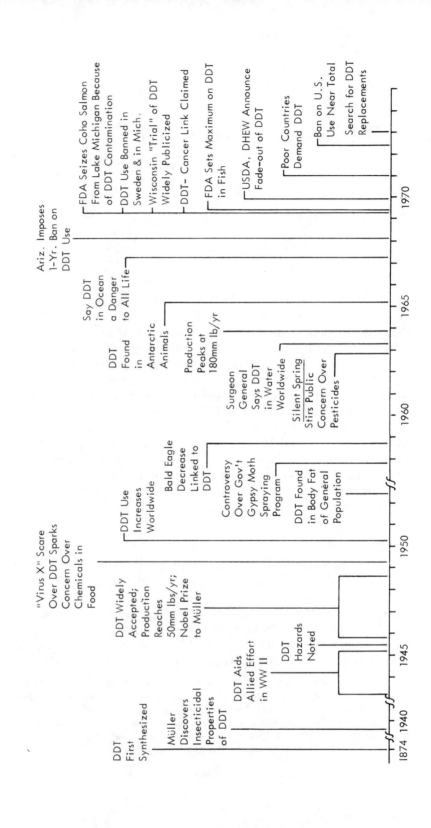

The total sales of synthetic organic pesticides, however, were only 287,150 pounds in the United States in 1940 compared to a production of over 1 billion pounds in 1970.

In 1939 Paul Müller, a chemist for the Swiss company, J. R. Geigy A-G, discovered the remarkable insecticidal properties of DDT* during a search for new mothproofing chemicals. In 1948, Muller was awarded a Nobel Prize for his discovery. That this substance was a veritable time bomb for mankind was hardly suspected: it had first been synthesized in 1874 in Germany. The outbreak of World War II was to have a profound effect on the DDT story, however, for not only did the threat of insect-borne diseases to military troops increase the demand for insecticides, but pyrethrums, the major insecticide, was in reduced supply because Japanese sources were lost to the Allies after 1941.

In September 1942, Geigy sent DDT samples to the United States. It was quickly tested for its insecticidal properties and acute toxicity in the United States and in England: it was an effective "contact" insecticide and far less toxic to humans than nicotine or lead arsenate. The Army gave DDT tentative approval in June 1943 to be used as a body louse powder. By 1943, low-cost production methods had been developed, and by the end of the war eight U.S. companies were making DDT.

The successes of DDT in controlling malaria, typhus, and other insect-borne diseases during the war were spectacular and became well known to the public from reports of returning troops (whose bodies, hair and clothing had received liberal dustings of DDT) and from news stories.[1] According to McIntire's review,[2] "Countless articles appeared in newspapers and magazines, playing up DDT's accomplishments and speculating on its possibilities; it seemed to many that a new age would commence with the release of DDT for civilian use at war's end, an age in which DDT would eliminate once and for all the insect pests that plagued mankind." Hence when DDT was approved for civilian use in August 1945, the large production capacity was quickly matched by an eager buying public. And the use of DDT was virtually unfettered by governmental regulations at the time.

The result was an unprecedented acceptance of a new pesticide for use in households, agriculture, public health, and government spraying programs, as well as the development of new and

*The letters are taken from the chemical name, dichlorodiphenyl trichloroethane, for the primary ingredient. More correctly named 2, 2-bis (p-chlorophenyl) trichloroethane, it is usually called p,p'-DDT.

ingenious methods of applying it, although the U.S. Army and the U.S. Public Health Service (PHS) issued a joint statement in 1945 urging caution in the use of DDT. By 1948, DDT was being produced by some sixteen companies in the United States at a rate of nearly 50 million pounds and was grossing $15 million per year. Furthermore, the post-war successes of DDT and other wartime pesticides, such as the insecticide, benzene hexachloride,* the herbicide, 2,4-D, and the rodenticide, sodium fluoroacetate,† spurred intensive research on other chlorinated organics and on other kinds of chemical pesticides. Thus, a pesticide industry was generated for the first time. In addition, the nation's food and fiber supply, and public health measures became increasingly dependent on pesticide use under the recommendations and approval of the U.S. Department of Agriculture (USDA), state extension services, U.S. Forest Service, and U.S. Public Health Service (PHS). (The latter even told the householder how to formulate his own DDT mixtures cheaply.[2]) DDT was thus elevated to the same pedestal by the public as penicillin, the new miracle drug that flowed also out of World War II research.

While public idolatry of DDT was developing, a number of warning signs appeared. For example, undesirable effects on non-target species were noted as early as 1944, when DDT was found to kill not only the intended coddling moth, but also predator lady beetles—an event which allowed a population explosion of red mites[3]—and when toxicity to fish and frogs was noted.[4] By 1945, ingested DDT was known to be stored in the body fat[5] and to appear in the milk[5,6] of laboratory animals. The latter result raised similar suspicions for cows feeding on DDT-sprayed pastures.[7,8] These hazards were published not only in technical journals, but numerous warnings and precautions appeared in the scientific news journals[9-11] and in the public press[12-15] in 1944–46. Ornithologists, entomologists, and other students of nature were particularly concerned: *Audubon* magazine was a favorite forum for pleas of caution.[16]

But "the voices of numerous biologists, entomologists and toxicologists expressing doubts of its widespread application were simply drowned in the uproar of public excitement in anticipation of the new insect-free age, to say nothing of the quieter but more influential voices of the chemical industry and agriculture."[2]

*Discovered to be an insecticide in France and England in 1941–42, although first synthesized in 1825.
†Also known as 1080.

In response to the increasing use of economic poisons and these expressions of concern for safety, the United States Congress passed the "Insecticide, Fungicide and Rodenticide Act of 1947." This act, which replaced the "Insecticide Act of 1910," brought the safety aspects of pesticide usage under federal control by requiring pre-marketing registration of each use, package labeling, and reports of transfer of pesticides. It passed with very little discussion or opposition and apparently gave a great feeling of assurance to the lawmakers that they had handled the pesticide problem.

Key Events and Roles

In January, March, and April of 1949, a New York physician, Dr. Morton S. Biskind, published three articles in which he claimed he had found direct relationships between DDT poisonings and a mysterious "X-disease" in cattle,[17] and a "virus X" (and neuropsychiatric disorders[18]) in humans.[19] These alarming reports were widely publicized by the newspapers[20,21] and in the news magazines[22-34] throughout 1949. Dr. Biskind's controversial conclusions created much concern in the public, which tended to link "X-disease" with the rising incidence of the dreaded crippler, polio.

The proposed relationship of DDT with "X-disease" was later generally discredited, as was apparently "virus X" itself, but the scare, coupled with evidence that flies[32] and mosquitoes[33,34] were becoming resistant to DDT, had a sobering effect on the public. It focused the attention of the manufacturers and government agencies on potential hazards of pesticides, and brought about some changes in use patterns. For example, the USDA recommended that farmers use the less toxic methoxychlor around dairy cattle instead of DDT. Furthermore, public awareness increased in the general problem of chemicals in foods. That awareness led to Congressional hearings in 1951* and 1952, and in 1954 to the so-called "Miller Amendment" to the Food, Drug, and Cosmetic Act of 1938, which established control of pesticide residue levels on farm crops under the USDA and the Department of Health, Education and Welfare,† and provided for advisory committees from the National

*During the hearings, a link between pesticides and cancer was also suggested—a question still unresolved; the Salk polio vaccine development in 1953–55 ended the controversy over polio.

†The USDA was to set permissible residue levels; the FDA to see that they were met.

Academy of Sciences. These hearings also led in 1958 to the so-called "Delaney Clause" regarding carcinogens in food. (See "The Cyclamate Affair.")

In addition, the "virus X" scare apparently solidified the opposition of a great number of critics of DDT. These critics included individuals of remarkably diverse backgrounds and interests, ranging from entomological, marine, and avian scientists, naturalists, and "organic" gardeners, to opponents of big business, big agriculture, big government, big technology and big science. But primarily, all were opponents of the increasing wholesale use and abuse of pesticides, which now numbered in the hundreds and were reaching a total production in the hundreds of millions of pounds per year. And although many of the new pesticides—such as, parathion—were far more toxic than DDT, it was the latter with its ideal, headline-length name and atomic-bomb reputation* that the public and the news reporters thought of when pesticides were mentioned.

Despite the controversy over the immediate and long-range hazards of DDT, its use after the 1949 scare increased rapidly as it was adopted worldwide for public health and agricultural purposes. In fact, as shown in Figure 3, the amounts of DDT exported from the United States surpassed the amounts consumed domestically by the mid-1950's, as many Americans switched to more effective—although more expensive—new insecticides. In addition, DDT production plants were operating or being designed in Western Europe, in India, and apparently in Eastern Europe, Russia, and China.

New evidence of the side effects of pesticides, and of DDT in particular, was reported from time to time during the 1950's and occasionally reached the public's attention, as is shown graphically in Figure 4 by the fluctuating number of stories in the news magazines and in one newspaper. Of particular interest was the discovery before 1952 that small levels of DDT had been found in human body fat—first of one man,[35] then of the general population[36,37] (presumed to result from eating vegetables or meat with residues)—and even in samples of human milk.[36] The persistence of DDT in soil had been well documented by 1954.[38] In 1957–58, much controversy† arose over the federal government's

---

*By the early 1950's the atomic bomb had also fallen into considerable disfavor, and atomic weapons testing had become a controversial issue.

†A suit to prevent the government's spraying was unsuccessful and was appealed eventually to the U.S. Supreme Court, which refused to review the case in 1960.

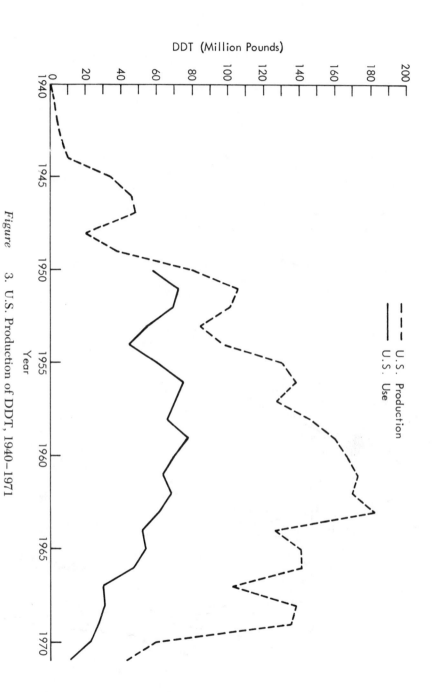

Figure 3. U.S. Production of DDT, 1940–1971

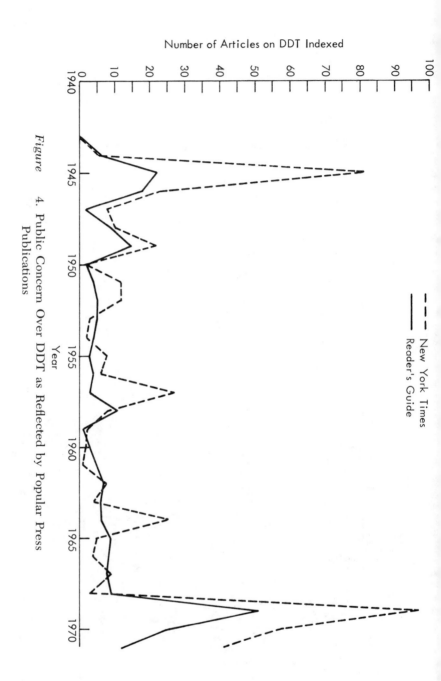

Number of Articles on DDT Indexed

- - - - New York Times
———— Reader's Guide

Year

Figure 4. Public Concern Over DDT as Reflected by Popular Press
Publications

program for widespread aerial spraying of DDT for control of the gypsy moth that was attacking New England forests.[39] In addition, the DDT spraying programs of many local communities to combat Dutch elm disease was also causing controversy;[40] earthworms tended to concentrate the DDT, and to poison birds which fed on them. *Audubon* magazine carried an article[41] which suggested that the decline of the bald eagle might be a result of DDT-caused sterility.* And in 1960, the first of what was to develop into a stream of papers on the biomagnification of a chlorinated hydrocarbon pesticide in food chains was published.[42] It was based on studies since 1957 at Clearlake, California. By now DDT had reached what manufacturers call "its maturity." U.S. consumption was declining and the number of manufacturers had dropped to six. But DDT's hour in the spotlight was yet to come.

In June 1959, *Reader's Digest* carried an article[43] strongly critical of the way pesticides were being used. And in November of that year, a panic over pesticides in cranberries developed (see "The Great Cranberry Scare"). In addition, the controversy over the USDA's fire ant eradication program in the South had grown to the point that the National Academy of Science—National Research Council, was called upon in mid-1960 to make a study.† By 1961, President John F. Kennedy and Congressman John V. Lindsay (R-N.Y.) and other elected officials were expressing their curiosity about pesticides. Then, in early 1962, the thalidomide tragedy hit the headlines and focused attention on the unexpected side effects of chemicals. (See "The Thalidomide Tragedy.") The timing could not have been more perfect for the appearance of one of the most remarkable books of our time: *Silent Spring*.[44]

Rachel Carson, marine biologist, author of the best-selling *The Sea Around Us* and other popular science books, had spent four years writing the book[45] (even then she was afflicted with cancer[45]). *Silent Spring*, published in September 1962 (after serialization in the prestigious *New Yorker* magazine) opened with a melodramatic allegorical account of a town that was completely destroyed and lifeless because of a "snow" of pesticides. This was followed by chapter after chapter of almost every adverse observation‡ on pes-

---

*Speculation over the causes for the decline of bald eagles dates back to 1921, and had been attributed in 1943 to the loss of timbered areas.

†The study committee developed such internal conflicts that the compromise report was not completed until 1963. The pesticide involved was Mirex.

‡These were often anecdotal accounts of uncontrolled medical cases or other incidents that the author gave without references.

ticides that had been reported, told in an alarming tone (with almost no mention of any benefits of pesticides). This book had a simplicity and style that made it easily readible for those without scientific training and included a long bibliography, although references were not cited in the text.* The center of attention was the chlorinated hydrocarbons and, in particular, DDT.

Silent Spring received countless reviews, quickly became a Book-of-the-Month Club selection, and leaped onto the best-seller list. Although Miss Carson did not call for a complete ban on pesticides—she did call for more care, more research—the tone of the book and chapter titles such as, "Elixirs of Death" led many persons, friends and foes alike, to that conclusion. And the friends and foes were immediate, legion, and so strongly polarized that an impartial review or opinion of it was a rarity. The pesticide and agriculture industries overreacted so strongly against the book that their protests lost effect. The biologists, who since World War II had been jousting the physical scientists and engineers for bigger research budgets, applauded vigorously. The public was probably confused and concerned, although the numbers of publications in the popular press on DDT itself, apparently† increased very little, as Figure 4 indicates, after publication of *Silent Spring.*

Nevertheless, the book had a great influence on public opinion and on the activities of governmental officials.[45] In early April 1963, the television program *CBS Reports* carried "The Silent Spring of Rachel Carson" and the Senate Subcommittee on Reorganization, under Senators Hubert H. Humphrey (D—Minn.) and Abraham A. Ribicoff (D—Conn.), started prolonged hearings on the roles of federal agencies in pesticide usage. Secretary of Interior Stewart L. Udall became one of the first governmental figures to speak against existing pesticide use patterns. In May, the President's Science Advisory Committee, Life Sciences Panel, issued the report of its eight-month study on pesticide use. It was a critical evaluation of both the hazards and benefits, and recommended the elimination of

---

*Although Dr. Morton Biskind is one of seventeen specialists whose help Miss Carson acknowledged, his name appears neither in the text nor the index and no reference is made to the "virus X" matter; five of his pre-DDT publications are listed as sources, but not his extensive 1953 review of health aspects of pesticides.[46] Neither is he mentioned in *Since Silent Spring.*[45]

†The indexes to the popular press publications are sufficiently inconsistent and variable with time that the effect was probably greater than indicated; the media's increasing awareness of the whole range of pesticides in use tended to place DDT in context.

persistent pesticides as a long-range goal. In the fall of 1963, a massive fish kill on the lower Mississippi River—the largest of a series there since about 1959—was being traced to industrial discharges of the chlorinated insecticide, endrin,† at Memphis, Tennessee, and appeared to confirm Miss Carson in a dramatic way.

But more important, *Silent Spring* had a great nucleating effect on environmental researchers. Almost immediately, DDT was observed to be persistent in water and muds,[47] and to be widely dispersed in our nation's waters[48] and air.[49] At this point, significantly, much of the research began to focus not on the effects of sudden massive exposures to DDT (from accidental spraying and the like), but on the effects of small levels of DDT on targets far removed from the point of application—on global effects. (Analytical methods were becoming breathtakingly sensitive—from parts per million to parts per billion and even parts per trillion.) Thus, when it was reported that the U.S. population now carried an average 8 parts per million DDT in its body fat[50], the counterclaim that intense occupational exposures to DDT had produced few adverse impacts[51] was of little real interest: most people simply didn't want *any* DDT in their bodies. And when DDT's supporters recited their litany of lives it had saved, critics retorted that the world was already overpopulated. Interest shifted to such places as Antarctica, where penguins and seals were found to contain DDT[52,53] (analysis of snowfall showed no DDT; the animals were apparently bringing it to Antarctica);[54] the oceans, where fish and birds were found to contain DDT;[55–57] fisheries, where newly hatched fry were dying because of DDT in the eggs;[58] the haunts of the birds of prey and fisheaters, who were said to be cracking DDT-thinned eggshells instead of hatching them;[59–61] and later again the oceans, where a report[62] that DDT reduced photosynthesis produced a scare that the world's oxygen supply would soon be depleted.†

These results alone were enough to generate a strong movement to ban DDT, but in addition, five years after *Silent Spring*, pesticide production in the United States had grown to nearly 1 billion pounds a year ($1 billion sales); pesticides were being used on 200 million acres, and not one pesticide had been banned. Furthermore, a controversy was raging over military use of the her-

---

*Endrin is one of the most toxic of all pesticides to fish, in large measure because it is so rapidly and extensively accumulated by them from water that contains only traces.

†The likelihood of this was soon down-graded,[63,64] but was widely publicized by the critic of overpopulation, Dr. Paul Ehrlich.[65]

bicide 2,4,5-T in the unpopular Vietnam War and another was underway on agricultural fertilizer use. But the symbol of it all was DDT and the pressures were growing irrestible.

Disposition

In January 1968, Arizona imposed an experimental, one-year ban on the use of DDT. In New York and Michigan a new group, known as the Environmental Defense Fund, succeeded, by means of the publicity surrounding its unsuccessful court suits, in stopping the use of DDT in local mosquito and Dutch elm disease control programs. In December 1968, the battle moved to Wisconsin, the scene of much early furor over water fluoridation (as discussed in "The Flouridation Controversy"), a great deal of recent protest over the war and related social issues, and a current protest over a far-flung military communications transmitter (see "Project Sanguine—Giant Underground Transmitter"). In Wisconsin, the setting was not actually a court of law, but an "administrative hearing procedure," by Wisconsin's Department of Natural Resources. At this "trial" of DDT, its opponents presented three weeks of widely quoted testimony; not until after a three-month recess did the defense testify.

Before that, however, the coho salmon of Lake Michigan were reported to be contaminated with DDT* in March 1969. The coho had only recently been introduced there, after the native trout had been decimated by the sea lamprey (see "Introduction of the Lampreys") and was already the foundation of a thriving sport fishing and tourist business (and hopefully of a commercial fishing industry). The news caused tremendous local and national anger at agricultural pesticides. In March, also, word leaked out[66] that a study of a host of chemical compounds, sponsored by the National Cancer Institute (NCI) had found several that caused tumors. The list included DDT. The retraction of a proposed prepublication presentation of the results raised charges of a coverup. The paper, described as a preliminary note, was published in June,[67] and while it did not call the tumors† cancerous, an impression that it did was widely

*FDA seized about 35,000 pounds of coho said to contain 13–19 parts per million DDT.

†Evidence for tumorgenicity of DDT was based on seventy-two susceptible mice fed high doses (140 parts per million in feed) for eighty-one weeks; 43 percent developed tumors. The negative controls averaged sixteen tumors, but ranged from 3 percent for two groups of females to 32 percent for one group of males.

assumed, especially by those who could recall frequently stated (and quoted[44]) beliefs of NCI's Dr. W. C. Hueper a decade earlier that DDT was a carcinogen.

As a result of the contaminated coho and the NCI results, Secretary of Health, Education, and Welfare Robert H. Finch appointed a "Commission on Pesticides and Their Relationship to Environmental Health," and the FDA imposed a 5 parts per million limit of DDT in fish in April 1969. In addition, Michigan outlawed the intrastate sale of DDT (coincidentally Sweden imposed a two-year ban on DDT). The major U.S. use of DDT was on Southern cotton, and about 70 percent of production was exported; such actions in Northern states were not considered disastrous by DDT producers. But in May 1969, the National Research Council's "Committee on Persistent Pesticides" presented a report, under a USDA contract, that recommended additional and more effective steps to reduce the needless or inadvertent release of persistent pesticides to the environment. By this time, DDT's defenders in the Wisconsin affair (organized as a "Task Force" of the National Agricultural Chemicals Association) were preparing to present their case, but, the decision to ban the use of DDT in the United States was all but assured. Even so, it was the testimony of the Swedish Dr. Göran Löforth (brought in by the EDF), based on previously known data,[36,50,68] that DDT was even in mother's milk and the inferences that DDT had effects on human fertility that made an impression on the news media.[69,70]

On July 10 the USDA suspended the use of DDT and eight other persistent pesticides in its programs; later that month, Wisconsin and California announced DDT bans. In October the four remaining DDT producers in the United States presented their full case for DDT at the State of Washington hearings.[71] By that time also, substantial evidence had developed that the analytical methods for chlorinated hydrocarbon pesticides, including DDT, were subject to interference by the hitherto unsuspected existence in the global environment of substantial amounts of polychlorinated biphenyls (PCB's),* industrial compounds chemically resembling the organo-chlorine pesticides.[72-75]

On November 13, 1969, Secretary Finch's announcement that essentially all† uses of DDT would be phased out within two years

---

*Other substances can also interfere at low DDT levels; soil samples sealed since 1910 appeared to have DDT according to one method of analysis. About half of the "DDT" originally reported in coho was apparently PCB's.

†A few exceptions were to be made, primarily for public-health purposes.

made front-page news.[76] The FAO quickly announced that developing countries should continue controlled use of DDT and other organo-chlorine insecticides, and this also made page one.[77] The government's drive was temporarily slowed by the legal appeals in January 1970 of six pesticide manufacturers, but following an announcement from Russia in May that DDT production would cease there,[78] the U.S. Court of Appeals ordered the USDA to suspend DDT use within thirty days.[79]

The phase-out of DDT continued into 1971, although an FDA statement that DDT causes mutations in rats, in addition to tumors, caused additional concern in August 1970.[80] The underdeveloped countries of the world continued to maintain that DDT was indispensable.* On March 18, 1971, William Ruckelshaus, Administrator of the U.S. Environmental Protection Agency (EPA) announced, to the dismay of many, that all uses of DDT would not be suspended immediately, pending a one-year review; a particular concern was that more toxic insecticides, such as parathion, would replace DDT. In June 1971, a NAS-NRC study urged an end to discharge of DDT and other chlorinated hydrocarbons into the ocean.[81] A year later, on June 14, 1972, the EPA announced the cancellation of nearly all remaining uses† of DDT.[82]

Comment

The story of DDT is rather like that of World War I as depicted in the play, "Oh, What A Lovely War": it was impossible, but it happened. Introduced under the aura of a miracle, DDT was rapidly and widely accepted by the public at the urging of officials in many agencies of government, despite numerous warning signs. But in an age where man could not yet think of a global environment, now-obvious considerations and safety tests were not made, and DDT was liberally sprinkled over the world. The 1949 "virus X" episode caused a considerable temporary scare and a sobering effect, but at the same time it anesthetized the manufacturers of pesticides, those in pertinent government agencies, and the informed public. When subsequent spokesmen voiced more substantially-based criticisms, the pesticide advocates tended to dismiss them as scare-mongers.

*In Ceylon, a half million cases of malaria were reported in 1968 following the cessation of DDT use; malaria had been nearly eradicated in the early 1950's.

†In early 1973, the USDA was said to be initiating a $2 million retraining program to teach cotton growers how to use the more toxic DDT substitutes safely.

The evacative *Silent Spring* may have generated far more public concern than the facts warranted, but a public shock was the intent of the author. The book undoubtedly changed the course of pesticide history. The 1960's were remarkable for the polarity that existed between scientifically-trained spokesmen among the pro- and anti-DDT forces; objectivity often gave way to emotion. The uncertainty in analyzing environmental samples for one trace contaminant among many potential trace contaminants was not sufficiently appreciated initially, and the accuracy of possibly 75 percent of the earlier analytical results on DDT is suspect—the results on which the emotions were developed. But the discovery of DDT in corners of the earth far from its point of use, and in human tissues—especially in mother's milk—were key factors in its demise, even if the full significance of these data are uncertain.

The carcinogenicity of DDT may never be proved;* the reported counts of peregrine falcons, ospreys, and pelicans will be argued by historians; and DDT levels in the world will probably decrease faster than its critics will ever admit. But recent experiments on the synergistic effects of environmental pollutants† have shown that the deemphasis on DDT use was a good decision. Thus, emission of large amounts of any persistent global pollutant is intolerable: one simply cannot predict how it will interact with every other substance. This does not mean, however, that DDT cannot be made available for special applications for man's benefit.

*Recent reports have even claimed DDT exposure reduces cancer incidence.

†For example, DDT applied alone at the rate of 0.3 μg per fly caused only 38 percent mortality, but 100 percent mortality when 10 μg of a PCB was added. The same PCB alone at this level caused no fatalities.[83] Adverse synergistic health effects on humans are accumulating for a number of pollutants.

REFERENCES

1. The *New York Times Index* shows five entries for DDT in 1944, 81 in 1945, and 23 in 1946.

2. "Spoiled by Success" by Greg McIntire, *Environ,* **14,** 14 (1972).

3. "Laboratory and Field Tests of DDT for Control of the Coddling Moth" by L. F. Steiner, C. H. Arnold and S. A. Summerland, *J Econ Entomol,* **37,** 157 (February 1944).

4. "Toxicity of Dichlorodiphenyl trichloroethane (DDT) to Goldfish and Frogs," *Science,* **100,** 477 (November 24, 1944).

5. "Accumulation of DDT in the Body Fat and Its Appearance in the Milk of Dogs," *Science,* **102,** 177 (August 17, 1945).

6. "Transmission of the Toxicity of DDT through the Milk of White Rats and Goats," *Science,* **102,** 647 (December 21, 1945).

7. *New York Times,* Section IV, p. 9 (March 10, 1946).

8. "DDT Spraying of Pastures can Make Milk Poisonous," *Sci NL,* **49,** 5 (October 26, 1946).

9. "DDT Future Production Requirements Studies," *Chem Ind,* **55**(1), 114 (July 1944); its toxocity to beneficial insects may limit its agricultural uses—Report to Government.

10. "Insect War May Backfire; Chemicals May Harm as Well as Help," *Sci NL,* **46,** 90 (August 5, 1944).

11. "DDT; Insect Killer that can be Either Boon or Menace," (with editorial comment) *Nature,* **38,** 120, 145 (March 1945).

12. *Time,* **44,** 66 (August 7, 1944), **45,** 91 (April 16, 1945), **46,** 90 (October 22, 1945, **46,** 88 (December 10, 1945).

13. *New York Times,* p. 23, (August 9, 1945).

14. "DDT and the Balance of Nature," V. B. Wigglesworth, *Atlan,* **176,** 107 (December 1945).

15. See also 1945–1946 articles in *Read Digest, Bet Home & Gard, House Beaut, Ladies Home J, Newsweek, Colliers,* and *Pop Mech.*

16. *Audubon,* **47,** 90, 113, 311, 245 (1945); *Ibid.,* **48,** 116, 243 (1946); *Ibid.,* **50,** 167 (1948).

17. "DDT Poisoning and X-Disease in Cattle," M. S. Biskind, *J Am Vet Med Assoc,* **114,** 20 (20 January 1949).

18. "DDT Poisoning—A New Syndrome with Neuropsychiatric Manifestations," M. S. Biskind and I. Bieber, *Am J Psycho,* **3,** 261 (April 1949).

19. "DDT Poisoning and the Elusive 'Virus X': A New Cause for Gastro-Enteritis," M. S. Biskind, *Am J Dig Dis,* **16,** 79 (March 1949).

20. Albert Deutsch in *New York Post,* March 30 and April 1, 1949.

21. The *New York Times Index* shows 22 entries for DDT in 1949 compared to a total of only 12 for 1948 and 1950.

22. "Worse than Insects?" *Time* **53,** 70 (April 11, 1949).

23. "DDT Scare," *Bsns W,* 24 (April 16, 1949).

24. "DDT Scare," *Newsweek,* **33,** 54 (April 18, 1949).

25. "DDT Aftermath," *Bsns W,* 26 (May 14, 1949).

26. "Bug Killers Can Kill You," *Colliers,* **123,** 23 (June 25, 1949).

27. "Danger to Health in DDT," *Consumers Res Bul,* **23,** 20 (June 1949).

28. "DDT, Too Much Spraying, Too Little Government Action," *Consumer Rep,* **14,** 276 (June 1949).

29. "DDT: Ally or Enemy," *Sci Illus,* **4,** 7 (July 1949).

30. "Possible Hazards from the Use of DDT," *Am J Publ Health,* **29,** 925 (July 1949).

31. "DDT Danger Refuted," *Sci Digest,* **26,** 47 (July 1949).

32. "Development of a Strain of Houseflies Resistant to DDT," *Science,* **107,** 276 (March 12, 1948).

33. "Mosquitoes Resist DDT," *Sci NL,* **56,** 355 (December 3, 1949).

34. "DDT Down, 2,4-D Up," *Time,* **54,** 52 (December 5, 1949).

35. "A Case of DDT Storage in Human Body Fat," D. E. Howell, *Proc Okla Acad Sci*, **29**, 31 (1948).

36. "Occurrence of DDT in Human Fat and Milk," E. P. Laug, F. M. Kunze, S. C. Pickett, *AMA Arch Ind Hyg Occ Med*, **3**, 245 (1951).

37. "Examination of Human Fat for the Presence of DDT," G. W. Pearce, A. M. Mattson, and W. J. Hays, Jr., *Science*, **116**, 254 (1952).

38. "A Survey of DDT Accumulation in Soils in Relation to Different Crops," J. M. Ginsburg and J. P. Reed, *J Econ Entomol*, **47**, 31, 467 (1954).

39. The *New York Times Index* has 27 entries for 1957, the increase largely reflecting the gypsy moth spray controversy.

40. "Notes on Some Ecological Effects of DDT Sprayed on Elms," R. J. Barker, *J Wildlife Manag*, **22**, 269 (1958).

41. "U.S. is Losing Its Bald Eagles; Sterility Suspected, DDT Cited," *Audubon*, **60**, 275 (November 1958).

42. "Inimical Effects on Wildlife of Periodic DDT Applications in Clear Lake," E. G. Hunt and A. I. Bischoff, *Calif Fish Game*, **46**, 91 (1960).

43. "Backfire in the War Against Insects," by R. S. Strother, *Read Digest* **74** (June 1959).

44. *Silent Spring*, by Rachel Carson, Houghton-Mifflin Co., Boston, Massachusetts, 368 pp. (1962).

45. *Since Silent Spring*, by Frank Graham, Jr., Houghton-Mifflin Co., Boston, Massachusetts, 333 pp. (1970).

46. "Public Health Aspects of the New Insecticides," by M. S. Biskind, *Am J Dig Dis*, **20**(1), 331 (1953).

47. "Persistence of DDT and Its Metabolites in a Farm Pond," W. R. Bridges, B. J. Kallman, and A. K. Andrews, *Trans Am Fish Soc*, **92**, 421 (1963).

48. "DDT and Dieldrin in Rivers: A Report of the National Water Quality Network," by A. W. Breidenbach and J. J. Lichtenberg, *Science*, **141**, 811 (1963).

49. "Airborne Particulates in Pittsburgh; Association with p,p'-DDT," P. Antommaria, M. Corn, and L. DeMaio, *Science*, **150**, 1476 (1965).

50. "DDT Storage in the U.S. Population," G. E. Quinby, W. J. Hayes, Jr., J. F. Armstrong and W. F. Durham, *J Am Med Assoc*, **191**, 175 (1965).

51. "Men with Intense Occupational Exposure to DDT: A Clinical and Chemical Survey," E. R. Laws, Jr., A. Curley and F. J. Biros, *Arch Environ Health*, **15**, 766 (1967).

52. "DDT Residues in Adelic Penguins and a Crabeater Seal From Antarctica: Ecological Implications," W. J. L. Sladen, C. M. Menzies, W. L. Reichel, *Nature*, **210**, 670 (1966).

53. "Pesticides in the Antarctic," J. L. George and E. H. Frear, *J Appl Ecol*, **3** (Suppl), 155 (1966).

54. "DDT in Antarctic Snow," T. J. Peterly, *Nature*, **224**, 620 (1969).

55. "DDT Residues in an East Coast Estuary: A Case of Biological Concentration of a Persistent Insecticide," G. M. Woodwell, C. F. Wurster, Jr., and P. A. Isaacson, *Science*, **156**, 821 (1967).

56. "DDT Residues in Pacific Sea Birds: A Persistent Insecticide in

Marine Food Chains," R. W. Risebrough, D. B. Menzel, D. J. Marsten, Jr., and H. S. Olcott, *Nature,* **216,** 589 (1967).

57. "DDT Residues and Declining Population in the Bermuda Petrel," C. F. Wurster and D. B. Wingate, *Science,* **159,** 979 (1968).

58. "Reproduction in Brook Trout (*Salvelinus fontinalis*) Fed Sublethal Concentrations of DDT," K. J. Macek, *J Fish Res Bd Canada,* **25,** 1787 (1968).

59. "Decrease in Eggshell Weight in Certain Birds of Prey," D. A. Ratcliff, *Nature,* **215,** 208 (1967).

60. "Chlorinated Hydrocarbons and Eggshell Changes in Raptorial and Fish-Eating Birds," J. J. Hickey and D. W. Anderson, *Science,* **162,** 272 (1968).

61. "Measurements of Brown Pelican Eggshells from Florida and South Carolina," L. J. Blus, *Bioscience,* **20,** 867 (1970).

62. "DDT Reduces Photosynthesis by Marine Phytoplankton," C. F. Wurster, *Science,* **159,** 1474 (1968).

63. "Man's Oxygen Reserves," W. S. Broecker, *Science,* **168,** 32 (June 26, 1970).

64. "Atmospheric Oxygen in 1967–1970," L. Machta and E. Hughes, *Science,* **168,** 29 (June 26, 1970).

65. "Eco-Catastrophe," P. Ehrlich, *Ramp Mag,* **8,** 24 (1969).

66. "New Storm Brewing Over DDT," *Bsns W,* 32 (March 8, 1969).

67. "Bioassay of Pesticides and Industrial Chemicals for Tumorgenicity in Mice: A Preliminary Note," J. R. M. Innes and twelve others, *J Nat'l Cancer Inst,* **42**(6), 1101 (1969).

68. "Chlorinated Hydrocarbon Insecticides in Plasma and Milk of Pregnant and Lactating Women," A. Curley and R. Kimbrough, *Arch Environ Health,* **18,** 156 (1969).

69. *New York Times,* (May 6, 1969).

70. "DDT in Mother's Milk," *Sat R,* **53,** 58 (May 2, 1970).

71. "DDT: Selected Statements From State of Washington DDT Hearings and Other Related Papers," by Max Sobelman, published by DDT Producers of the United States, December 1970, Library of Congress No. 77-12457 MARC; Class No. QH 545.P456.

72. "Polychlorinated Biphenyls in the Global Ecosystem," R. W. Risebrough, P. Reiche, D. B. Peakall, S. G. Herman, and M. N. Kirven, *Nature,* **220,** 1098 (1968).

73. "The Need for Confirmation," editorial by M. S. Schecter, *Pesticide Monitoring J* **1,** 1 (June 1968).

74. "Polychlorobiphenyls (PCB's) and Their Interference with Pesticide Residue Analysis," I. M. Reynolds, *Bull Environ Contam Toxicol,* **4,** 128 (1969).

75. "Current Progress in the Determination of the Polychlorinated Biphenyls," R. W. Risebrough, P. Reiche, and H. S. Olcott, *Bull Environ Contam Toxicol,* **4,** 192 (1969).

76. *New York Times,* p. 1 (November 13, 1969).

77. *New York Times*, p. 1 (November 29, 1969).

78. *New York Times*, p. 6 (May 14, 1970).

79. *New York Times*, p. 8 (May 29, 1970).

80. *New York Times*, p. 18 (August 4, 1970).

81. *New·York Times*, p. 1 (March 19, 1971).

82. *New York Times*, p. 1 (June 17, 1971).

83. "PCB's and Interactions with Insecticides," by E. P. Lichtenstein, *Environ Health Perspectives*, **1**, 151 (April 1972).

The Asbestos Health Threat

Abstract

In the early 1960's, information linking public exposure to airborne asbestos particles with a rare form of cancer, and possibly also with the asbestosis and cancer of the lung long known to occur in many heavily exposed asbestos workers. The degree to which the public was being exposed to asbestos and the latent effects of this exposure were widely discussed. Between 1970 and 1972, local government agencies and the Environmental Protection Agency began instituting emission control regulations and in 1972, the Occupation Safety and Health Administration of Department of Labor set stringent regulations for controlling asbestos concentrations at work sites.

Background

Asbestos is a generic term applied to several naturally occurring, fibrous hydrated mineral silicates which are incombustible in air and resistant to corrosion. Chrysotile ($3MgO \bullet 2SiO_2 \bullet 2H_2O$), a tubular form of serpentine, is the type most widely used in the United States,[1] and accounts for about 95% of the world's production. Historical records indicate that asbestos has been known for more than two thousand years, but the modern asbestos industry began in Italy only in about 1866.[2] Asbestos has since become virtually essential to industrialized countries, with annual world production exceeding 3 million tons by 1962.

As early as the first century, the adverse biological effects of asbestos were observed by the Greek historian, Strabo, and by the Roman naturalist, Pliny the Elder.[3] They both mentioned the "sickness of the lungs," which we now know as asbestosis (a non-

288

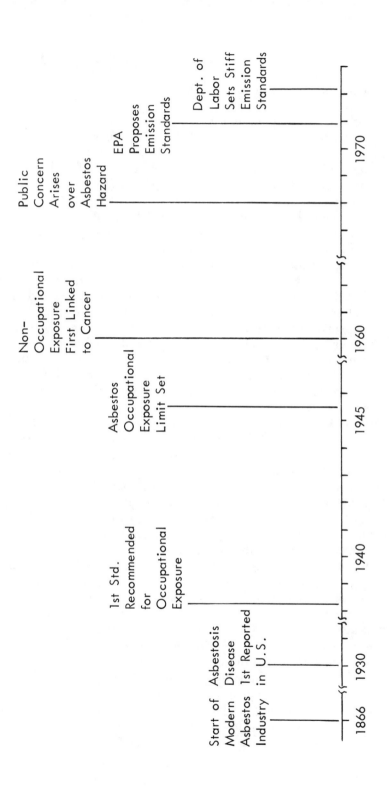

malignant pneumoconiosis), in slaves who wove asbestos into cloth.

With the increasing use of asbestos materials in the last quarter of the nineteenth century, and increasing reports of asbestos-related disease, a concern developed over the role of these minerals as factors in human disease. The first recorded case of an asbestos-related disease was reported in England in 1906,[1] and the first use of the term "asbestosis" and a detailed description of the "curious bodies" seen in lung tissues was given in 1927, with an indication that these bodies originated from asbestos fibers. Cases were first reported in the United States in 1930.[1]

Various studies[1] have since shown that many of the workers exposed to asbestos dust develop asbestosis if the dust concentration is high or the duration of their exposure is long. As early as 1918, American and Canadian insurance companies would not insure asbestos workers because of the working conditions. In 1947, the frequency of bronchogenic cancer was indicated to be greater among persons engaged in the manufacture of asbestos products than among the general male population.[1] When inhaled into the lungs, asbestos is indestructible, and it continues to exist there for a lifetime.

In the United States the use of asbestos reached nearly 1 million tons per year by 1972—eight times the amount used just thirty years ago—even though domestic production meets only fifteen percent of U.S. needs.[1] The major use of asbestos in the United States, accounting for about 75% of the total, is in the construction industry. Asbestos cement products are the largest market, including pipe (45 percent of the total asbestos used), shingles, sheets, and electrical panels. Asphalt and vinyl floor tiles, roofing felts, asbestos papers and millboards, and heat insulation also contain asbestos. The use of sprayed asbestos to fireproof steel skeletons of buildings has considerably displaced the use of cement, but was said to constitute less than five percent of the asbestos market in 1970.[4] Other construction uses included putties, molding compounds, calks, and fillers. Outside of the construction industry, asbestos is used in textiles such as safety clothing, fireproof curtains, gaskets, conveyor-belts, ironingboard covers, draperies, rugs, and filters for beverages and pharmaceutical solutions. Asbestos is used as the primary constituent in brake linings and clutch facings for motor vehicles and other commercial and industrial equipment.[5] In all, over three thousand asbestos products are said to be in use.[6]

About 92 percent of all asbestos used by the construction in-

THE ASBESTOS HEALTH THREAT

dustry is firmly bonded, that is, the asbestos is "locked in" in such products as roofing felts and shingles. The remaining 8 percent is in a friable or powder form in insulation materials, asbestos cement powders, and similar products and generates most of the dangerous airborne fibers encountered in construction work. An estimated 50,000 workers are involved in the manufacture of asbestos-containing products. About 40,000 field insulation workers in the United States are directly exposed to asbestos dust.[1]

The first standard for controlling occupational exposure to asbestos dust was recommended[1] in 1938, following a study of employees in four asbestos textile plants, where massive exposures occurred. A tentative limit of 5 million particles per cubic foot was recommended and was maintained by the American Conference of Governmental Industrial Hygienists (ACGIH) during the period 1946–1970.

Despite the increasing evidence of the occupational health dangers of asbestos, its use increased rapidly after World War II.

Key Events and Roles

Studies conducted in a South African mining community and published in 1960 indicated that thirty-three cases of mesothelioma, a rare and incurable form of cancer of the lining of the lung or the abdominal cavity, could be correlated with asbestos exposures.[7] Most importantly, many of the exposures had occurred as much as two decades earlier and sixteen of the cases were not asbestos workers. Thus, exposure could be hazardous even for people who did not come in contact with asbestos on the job; it could affect, for example, children playing in asbestos mine tailings.

Further studies in South Africa and autopsy screening studies in the United States soon showed that "asbestos* bodies" were to be found in the lungs of as much as 20 percent or even 50 percent of the general public. These results stimulated the New York Academy of Sciences to sponsor an international conference on the biological effects of asbestos in 1964, which, in turn, initiated public awareness of the problem.[8,9]

While air was assumed to contain normally small background levels of asbestos and other particles from natural sources, the suspicion that numerous industrial uses could be adding a patent car-

*Also called ferruginous bodies, since inhalation of other insoluble materials cause similar-appearing bodies and the identification of the core fiber or particle is difficult.

cinogen to the air generated scientific concern over asbestos use and its effects. To add to the uncertainty, however, the analytical methods which were used in the asbestos industry to count dust particle levels in worker areas were not applicable to measuring the degree of ambient air pollution caused by asbestos. Not until 1968 was the first rough estimate made: 600–6,000 asbestos particles per cubic meter of urban air[10] in the vicinity of heavily traveled streets.†

The explosion of scientific and technical interest in the asbestos problem is well illustrated by a 1969 literature review[10] of asbestos air pollution: two-thirds of the over 240 references were published during the years 1965–1968, and many of the pre-1960 papers dealt with asbestos production technology or the old occupational asbestosis problem. Of particular concern was the evidence of a latency period of twenty or thirty years after asbestos exposure before the effect might become evident.

A leader in arousing public concern over asbestos appears to have been Irving J. Selikoff, M.D., head of the Environmental Medicine Science Laboratory, Mount Sinai School of Medicine in New York City, and past president of the New York Academy of Science. He had become interested in health effects of asbestos in the 1950's and had published results of his studies as early as 1964.[11] His estimates of the health dangers of asbestos pollution have been widely publicized.[12-14] In late 1968, before a House subcommittee on labor, Dr. Selikoff testified that of the insulation workers whom he had studied in the New York-New Jersey area, 20 percent died of lung cancer and nearly 50 percent by some type of cancer, and Dr. William H. Stewart, former Surgeon General, estimated that more than 3.5 million construction workers who were not working directly with asbestos were nevertheless being exposed. A 1966 estimate had placed one-fourth of the general population in possible danger from asbestos dust.[10]

The concern over asbestos pollution mounted in 1970 and 1971. The *New York Times Index* lists only three articles on asbestos during 1964–1969, but twenty articles in 1970 and 1971.

A July 1970 *Today's Health* article[15] summed up many of the scare aspects of asbestos, particularly the thoughts of Dr. Selikoff, although noting that medical opinions on the dangers were not unanimous. The concern over asbestos in the air sparked concern over other sources of asbestos danger. Cases of asbestos powders

†The street-level estimate apparently reflects the suggestion by South Africa's Dr. J. T. Thompson that automobile brake- and clutch-linings might be a major cause of asbestos pollution.

used in do-it-yourself jewelry kits, children's cosmetic kits and kindergartner's "playdough" were cited in 1970.[15]

Still more alarming information on the extent of the danger posed was a report published in 1972[16] (although first presented at a seminar of the Food and Drug Administration in October 1969),[16] that significant asbestos contamination had been found in samples of seventeen widely used parenteral drugs, such as solutions used for intravenous, intramuscular and intraperitoneal therapy. The asbestos filters used by the pharmaceutical industry to purify the medicinal solutions had actually *added* a contamination of their own to the filtered liquid.

Since asbestos filters are also used in food-processing industries, it was postulated[16] that certain beverages (for example, beer) may also be contaminated with asbestos. In fact, studies[17] in Canada (which produces over 40 percent of the world's asbestos) found 2 to 4 million asbestos fibers (1–15 micrometers long) per liter in the tapwater of three major cities, and also found asbestos fibers in Canadian, American, and European wines, beers and soft drinks.

The dangers of ingested asbestos are unclear, although Selikoff reported increased stomach cancer in asbestos workers in 1964.[11] A 1971 report[18] presented epidemiological evidence that, among Japanese men, asbestos-contaminated talc, which is used to dust rice, is a carcinogen or co-carcinogen which causes a high incidence of stomach cancer, but these conclusions have been questioned.[19]

Disposition

In April 1970, New York City became the first governmental unit in the nation to attempt the control of pollution from aerial spraying of asbestos. These regulations required that tarpaulins be used to cover areas where asbestos was being sprayed, that waste asbestos be vacuumed and swept into closed containers, and that the containers be buried.[20] The building contractors did not meet the regulations and, under pressure from the city, voluntarily stopped asbestos spraying. In June 1970, Mayor Lindsay ordered that all spraying of asbestos be halted in New York City.[21]

In early 1971, the new Environmental Protection Agency (EPA) identified asbestos as a dangerous air pollutant (along with mercury and beryllium) which required control,[22] and at a December 3 press conference (and in the *Federal Register* of December 7), the Agency *proposed* emission standards[23] which called for use of air filters for mining, milling, manufacturing, and fabrication processes, for the elimination of all visible emissions from these processes, and for the banning of spraying of asbestos fireproofing and insulation except where spraying is enclosed and emissions are contained. The EPA did not propose numerical emissions standards, because of the absence of accepted test methods. The EPA said that public hearings on the proposed standards would be held and final standards would be set within six months thereafter.

In the meantime a number of local agencies have acted to control asbestos pollution sources. A stringent pollution control code was passed by the New York City Council in July 1971, to abolish asbestos spraying six months after the code was signed into law.[24]

By January 1972, Illinois banned asbestos spraying and even required that workers clean themselves of asbestos fibers before leaving the job site.[25] Boston, Chicago and Philadelphia had also banned the spraying of asbestos fireproofing material altogether.[13]

In June 1972, the Occupational Safety and Health Administration of the U.S. Department of Labor set stiff regulations for controlling asbestos concentrations at work sites.[13] The new regulations called for even more stringent control standards in 1976, with the delay intended to give industry a chance to adjust.

An article on December 11, 1972, noted that EPA would have to defend in court its failure to set emission levels for three hazardous air pollutants—asbestos, mercury, and beryllium.[26] The Environmental Defense Fund (EDF) and Ralph Nader's Health Research Group had filed a citizen's suit against EPA—charging that final standards were 6 months overdue. Companies were reported to be reluctant to plan for and purchase control equipment until standards were set.

The asbestos threat has become of special concern to two communities: Manville, New Jersey and Duluth, Minnesota. Manville is the home of the world's largest asbestos processing plant. Recent studies have concluded that sixty-two of the town's approximately 15,000 residents have died of mesothelioma since 1963 and Dr. Maxwell Borrow has estimated that four thousand residents have asbestosis. The community's economic life is threatened now as the plant's operator, Johns Manville Corporation, indicates it cannot afford the control technology required to reduce asbestos levels to those demanded by the labor unions and Federal government. A shutdown of asbestos operations would mean job losses for eight hundred of the plant's two thousand workers and an annual payroll loss of $10 million for the community. Duluth, Minnesota, has fears that the city's water supply is being contaminated with asbestos in the taconite tailings being dumped in it by Reserve Mining Company (see Taconite Pollution of Lake Superior). The discovery that Duluth's water supply contains asbestos fibers has stimulated an examination of the water supplies nationwide. The results: some asbestos-like fibers appear to be present in many of them, possibly from natural sources, but possibly from industrial activities.[27]

Comment

The incombustibility, corrosion resistance, and other techno-economic properties of asbestos led to such widespread use that it has become virtually essential in industrialized countries. Although its health hazards to those working with it were known and more or less accepted as a risk, the revelation that it might be developing into an airborne source of an incurable cancer for the general public was frightening. Even though the analytical methods for measuring the asbestos particle levels and their effects on the populace are poorly developed, the risks were decidedly too high to forestall an attempt to control the technology. The asbestos episode probably increased the public's suspicion of American industry and technology. The case illustrates the potential that widely-quoted scientific experts have in raising the national consciousness about technological problems. Undoubtedly, they will demand and get greater control of this potential health hazard.

On the other hand, the discovery of the presence of small amounts of asbestos fibers in drinking water has the potential of causing considerable unwarranted public concern; the toxicity of abestos by oral ingestion is probably very small. Insufficient data are available to confirm this supposition, however, because of the long incubation times observed with some of asbestos's other effects.

REFERENCES

1. "Criteria for a Recommended Standard—Occupational Exposure to Asbestos," U.S. Department of Health, Education and Welfare, Public Health Service, Health Services and Mental Health Administration, National Institute for Occupational Safety and Health, 1972.
2. "Asbestos," *Encyclopedia of Chemical Technology*, second edition, Vol. 2, p. 734, Interscience Publishers, 1967.

3. "Asbestos and Talac—Inescapable Hazards," *Prevention*, 209 (May 1972).

4. "Asbestos Builds for Steady Growth," *Chem W*, **106**, 113 (June 17, 1970).

5. "National Invention of Sources and Emissions—Cadmium, Nickel and Asbestos, 1968; Section III, Asbestos," W. E. Davis and Associates, under Contract No. CPA 22-69-131; February 1970, National Technical Information Service PB 192 252.

6. "Asbestos Controls Given," *A/W PR*, 492 (December 6, 1971).

7. "Asbestos—the Need For and Feasibility of Air Pollution Controls," Committee on Biological Effects of Air Pollutants, National Research Council, National Academy of Science, Washington, D.C., 1971.

8. "Dangerous Dust?" *Sci Am*, **211**, 64 (December 1964).

9. "Asbestos Workers Live Longer But Get Cancer," *Sci NL*, **86**, 297 (November 7, 1964).

10. "Preliminary Air Pollution Survey of Asbestos," Litton Systems, Inc., under Contract No. PH 22-68-25. U.S. Department of Health, Education and Welfare, Public Health Service, National Air Pollution Control Administration APTD 69-27, October 1969.

11. Selikoff, I. J., et al., "Asbestos Exposure in Neoplasia," *J Am Med Assoc*, **188**, 22 (1964).

12. "Danger of Lung Cancer in Asbestos Workers," *Sci NL*, **86**, 276 (October 31, 1964).

13. *Wall Street J* (midwest edition), p. 1 (June 8, 1972).

14. *Kansas City Times*, p. 18A (October 5, 1972).

15. "A Cancer Threat Greater than Cigarettes?" *Today's Health*, **48**, 30 (July 1970).

16. "Asbestos Contamination of Parenteral Drugs," *Science*, **177**, 171 (July 14, 1972).

17. "Asbestos, Asbestos Everywhere," *Chem*, **45**, 23 (May 1972).

18. "Talc-Treated Rice and Japanese Stomach Cancer," *Science*, **173**, 1141 (September 17, 1971).

19. *Science*, **175**, 474 (February 4, 1972).

20. *New York Times*, p. 26 (April 10, 1970).

21. *New York Times*, p. 44 (June 16, 1970).

22. *New York Times*, p. 82 (April 1, 1971).

23. *A/W PR*, 9, 491 (December 6, 1971).

24. *New York Times*, p. 70 (July 14, 1971).

25. *A/W PR*, 45 (January 31, 1972).

26. *C&EN*, **50**, 5 (December 11, 1972).

27. *C&EN*, **51**, 18 (December 10, 1973).

Taconite Pollution of Lake Superior

Abstract

In 1956, Reserve Mining Company, a subsidiary of Republic and Armco Steel Corporation, began operation of a huge taconite (iron ore) beneficiation plant at Silver Bay, Minnesota, on the shore of Lake Superior. The plant discharged some 67,000 tons a day of taconite tailings into the lake. Within a few years the discharging slurry had formed a long delta extending into the lake, had discolored the water at least eighteen miles away, and had crossed the state boundary into the waters of Wisconsin. A heated controversy developed that pitted environmentally-oriented citizens groups, fishermen, and eventually agencies of state and federal governments against the company. As the case moved into the state courts in 1969, residents of Silver Bay, union officials and state authorities worried over the company's claim that alternative waste disposal methods were so expensive that the plant might have to close if it lost the case. In 1972, the Environmental Protection Agency began lengthy federal court proceedings against the company to halt the pollution, and in addition, an international commission of the U.S. and Canada held hearings on the problem. In late 1973, the case reached a fever pitch when evidence developed that the discharge was contaminating the drinking waters of nearby cities such as Duluth with asbestos-like fibers that might produce gastrointestinal cancer. Prosecution of the case continues.

Background

Lake Superior is not only the largest fresh-water lake in the world, but is also the coldest, deepest, and cleanest of the five Great Lakes. The 350-mile long lake is bounded by Michigan and Wisconsin on the south, Minnesota along the northwest shore and

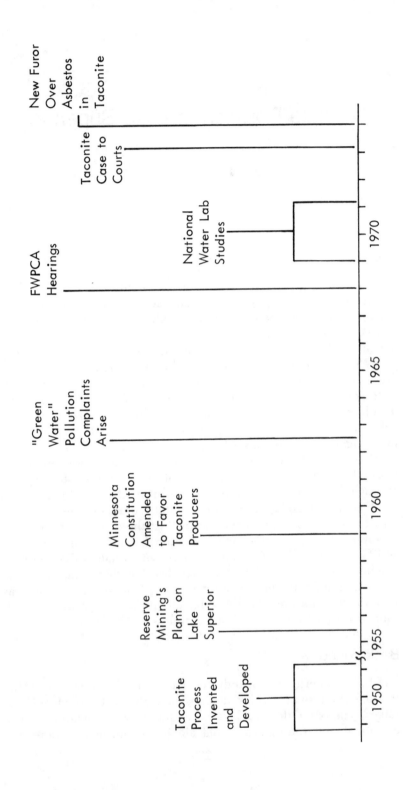

Taconite Process Invented and Developed

Reserve Mining's Plant on Lake Superior

Minnesota Constitution Amended to Favor Taconite Producers

"Green Water" Pollution Complaints Arise

FWPCA Hearings

National Water Lab Studies

Taconite Case to Courts

New Furor Over Asbestos in Taconite

1950   1955   1960   1965   1970

Canada on the north. The lake is situated in generally rugged, scenic terrain. The land rises sharply from near the shore in most places around the lake to a 400–800-foot ridge, and then on the west to more mountainous heights. The poor soil and cold climate severely limit agriculture in the area and the economy has traditionally been based on fishing, logging and mining. Compared to the lower lakes, few people live around the shores of Lake Superior, with the major concentration (~150,000) being in the bistate Duluth-Superior area at the western tip of the lake.

The mining activities have centered in the enormous iron ore deposits of the Mesabi Range in Minnesota. Beginning in the 1890's, vast quantities of high grade hematite (70 percent iron oxide) were removed from huge open pit mines (that themselves became tourist attractions) to feed the steel mills of the east. But by the late 1930's, the hematite deposits were rapidly being depleted, and forecasts for the economy of the area were gloomy. The Mesabi still contained immense quantities of iron, but most of it was in the form of taconite, a low-grade, greenish ore (25–30 percent iron oxide) that was spread widely through the rock formation in a 600-foot thick layer. The problem was the separation of the iron ore particles from the tightly adhering rock. In 1939, Reserve Mining Company was formed (a jointly-owned subsidiary of Republic Steel Corporation and Armco Steel Corporation) and began to search for a practical ore beneficiation process. In the ensuing years, E. W. Davis and coworkers developed a successful process in which the taconite was crushed and washed in a series of steps to give a fine gray powder from which the iron-containing particles were recovered electromagnetically. The resulting iron-laden dust was then mixed with a binder and fired in a kiln to give marble-sized pellets which could be shipped East.

The process required, however, a great quantity of water for the washing steps (as much as 10,000 gallons per ton of finished pellets) and also a place to dispose of the waste water. Therefore Reserve decided to build its beneficiation plant right on the shore of Lake Superior; the site selected was a Silver Bay, a small community sixty miles northeast of Duluth and forty-five miles from the mine at Babbitt, from which the rock would be hauled by rail. In 1947, agencies of the State of Minnesota happily gave Reserve permission to draw water from the lake, but public hearings were held on the proposed discharge of up to 130,000 gallons per of wastewater. Reserve officials spoke convincingly of the harmless sand-or gravel-like nature of the solids in their washings and received a permit. The Corps of Engineers (which guarded against navigation obstructions) also gave its discharge permit the following year; construction was soon underway on a $350 million plant that would produce—even twenty years later—15 percent of the nation's iron ore.

Key Events and Roles

The construction and the opening of the taconite processing plant at Silver Bay in late 1955 provided an immediate boast to this economically depressed area of Minnesota. In the early years of the plant's operation, the state readily approved several increases in the wastewater discharge rate and in 1959 even passed a special ammendment to its constitution to give the infant taconite industry a favorable tax break. Within a decade, fourteen other taconite beneficiation plants were in operation in Minnesota and Michigan's Upper Peninsula. All of these, however, had an important difference from Reserve's plant: their wastewaters were sent to settling lagoons and then recycled—they had no discharge to public waters.

In contrast, Reserve's plant, since it began operating at full capacity in 1956, was discharging 67,000 tons a day of taconite tailings into the cold clear waters of Lake Superior. Although Reserve officials had previously indicated that the solids would sink in the lake's great depths, it soon became obvious that such was not entirely the case. By the mid-1960's the results were ominous: the discharge had formed a delta that extended out into the lake, and complaints of turbid, polluted water were being voiced by fishermen and conservationists, including some from Wisconsin. The passage of the Federal Water Quality Act in 1965 focused new interest on the matter.

The pollution of the lower lakes was already of great concern, and indeed Lake Erie had already been pronounced "dead" by some environmentalists. In early 1967, a scientist reported[1] that Lake Erie would require about twenty years to flush itself clean even after pollution was stopped, and that Lake Superior would require over five hundred years. The story was picked up by *Time* magazine.[2] The Corps of Engineers became concerned and told Reserve that its permit to discharge would be reviewed—for the first time since 1960. The Corps asked its parent agency, the Department of Interior, for assistance. Secretary Udall formed a taconite study group, with members from the Federal Water Pollution

Control Administration, the U.S. Geological Survey, the Bureau of Mines, Commercial Fisheries, and Sport Fisheries and Wildlife, under the direction of Interior's Charles H. Stoddard.

The study group concluded that Reserve was polluting the lake and should be required to stop discharging its taconite tailings within 3 years by switching to on-land disposal. The "Stoddard Report," as it became known, was issued in December 1968, but only to a few closely involved organizations and individuals, including Reserve Mining. It was never made public; Reserve—either through Representative John A. Blatnik (D-Minnesota) of the Duluth Silver Bay district or otherwise—apparently pressured the Department of Interior to withhold it. In fact, Assistant Secretary Max N. Edwards stated that the report contained "inaccurracies" and was "unapproved." In contrast, FWPCA Commissioner Joseph G. Moore, Jr., publicly defended the unavailable report, and FWPCA issued a separate report on its contribution. Because of the change of administrations in January 1969 (from Democrat to Republican) all three men (and also Stoddard) were soon to leave the government. But before he left (and just before the Santa Barbara Oil Leak) Udall scheduled public hearings in Duluth on pollution abatement in the Lake Superior basin for May 13–15, 1969.

The hearings (or conference as it was called) drew a large, conservation-oriented audience, including representatives of the newly formed 1,600-member Save Lake Superior Association, the National Wildlife Federation and the Sierra Club, and Senator Gaylord Nelson. While the hearings ostensibly involved the entire lake, the evidence presented and the discussions that followed centered on the taconite problem. Because of FWPCA's statutory limitations, however, the emphasis was on whether the pollutants either crossed state boundaries or were a "health and welfare" threat, rather than on the question of whether pollution existed. The "Stoddard Report" continued to cause controversy; while Edward's successor at Interior, Carl L. Klein, insisted that it was officially nonexistent, Representative Blatnik inferred that it merely called for keeping the situation "under surveillance," and Stoddard complained that his report had been "revised." The hearings and related studies developed the following general picture.[3–6]

Reserve's taconite tailings were contributing more solids to Lake Superior at one point in twelve days than did all of its natural U.S. tributaries in a year. The taconite tailings have a range of sizes. About 45 percent of the solids discharged since 1957, had deposited off shore to build up a large delta that protruded into the lake.* The smaller particles (40 percent are less than 44 micrometers in diameter; 9 percent less than 4 micrometers and ~3 percent less than 2 micrometers†) were carried farther into the lake as billowing gray clouds; some particle settled out on the lake bottom as far as ten miles off shore and fifteen miles southwest (down current) of the plant. As the concentration of solids in the water decreased the lake surface took on a characteristic "green water" appearance (either from reflection of the sunlight from the particles or from partial dissolution of them). Turbid "green water" (that is, with tailings visibly present) had been observed eighteen miles southwest of Silver Bay, and clear "green water" had been reported off the Wisconsin shore. The chemical composition of the tailings had also been partially determined: a projection showed that the plant discharged each day approximately 314 tons of manganese, twenty-six tons of phosphorous, three tons each of chromium and lead, two tons of copper and one ton each of nickel and zinc.

Environmentalists felt that the addition of a pollutant of this nature to a rather delicate ecosystem such as Lake Superior—which has a relatively small number of plant and animal species and quite limited shallow breeding areas—was clearly dangerous. The taconite also worried the commercial and sport fishermen, who were just recovering from the lamprey invasion of the 1950's and were concerned about DDT pollution of the recently introduced coho salmon.

The three-day hearings recessed amid public demands that the taconite discharges stop. Reserve Mining immediately issued a press release stating that it had not been shown that the taconite tailings had crossed state boundaries or were a health hazard. In addition, Reserve claimed that switching to on-land disposal of its wastes would not only cost $250 million (compared to an estimate of $7 million in the "Stoddard Report"), but would, in fact, create new conservation problems. Reserve hinted that if it were forced to alter its disposal method, it might have to close its mining and processing operations. The fear of losing 3,100 local jobs (with a $34 million payroll) soon created union labor support for the company's position.

But by the time the conference resumed in executive session on September 30, new data were in hand.[3] Dr. Donald I. Mount, of

---

*The deposit was variously reported in the press as a half-mile peninsula[4] and a 9 sq mile delta.[6]

†One micron equals about 0.00004 in.

the National Water Quality Laboratory in Duluth, reported that taconite tailings were not only present in Wisconsin's waters, but that they were killing important organisms (including fish foods) in the lake.[6] In addition, taconite tailings were shown to be present in the drinking water supplies of Beaver Bay, Two Harbors and Duluth (based on studies using cummingtonite as a tracer*), although the health effects of this presence were still uncertain.

The conference had little choice but to conclude[3] that the tailings had adverse ecological effects on a part of the lake, and that presumptive evidence existed that they endangered the health and welfare of persons in Wisconsin. The conference committee took little action, however. It recommended, that FWPCA and the states keep the discharge "under continuing surveillance" and report to the conferees at six-month intervals. Reserve Mining was requested to restudy the engineering and economics of alternative disposal methods and submit a report and timetable for action six months after the conference report was published by the Secretary of Interior (which occurred in January 1970). The conservationists were appalled at the lack of action, whereas Edward M. Furness, President of Reserve Mining expressed disappointment at the conference's conclusions and mild recommendations.[6]

Disposition

In late 1969 Minnesota had filed a suit against Reserve Mining claiming that their discharge violated water quality standards for turbidity and suspended solids. Reserve filed and won a counter suit claiming the state's standards were too stringent and that harmful pollution of the lake had not been proved. The state had appealed and a long legal struggle ensued.

The creation of the Environmental Protection Agency in December 1970 added a new dimension to the struggle. The EPA quickly commissioned an independent, six-month study of alternative disposal methods, and on April 28, 1971, gave Reserve a 180-day notice to stop polluting interstate waters.[7] The study developed five alternative methods for consideration.[8] Reserve raised many objections to each and proposed to pipe the waters to the bottom of the lake instead. EPA Administrator, William Ruckelshaus, concluded that further discussion was pointless. Through Attorney General John N. Mitchell of the Department of Justice,

*Cummingtonite is a silicious mineral of composition $Mg_4Fe_{2.5}Mn_{0.2}Ca_{0.4}Si_{7.9}Al_{0.1}O_{22}(OH_2)$ that is present in taconite.

EPA brought suit against Reserve on February 17, 1972, demanding that it stop violating the 1965 Water Quality Act.[9,10] Then, in a puzzling change, the Justice Department amended its suit[11] on May 4 to claim that Reserve's discharge constituted a "continuing public nuisance" that should be halted under provisions of a seldom used and poorly developed federal "common law." The Justice Department claimed that this "new legal ground" would be very useful in pollution suits in general and gave the government a better chance to win the case. The result was, however, to slow down prosecution of the taconite case.

Reserve immediately asked the court to delay federal action pending conclusion of its case in the Minnesota Supreme Court. Reserve did agree[12] with the Minnesota Air and Water Pollution Control Agency to undertake a 2-½ year $3 million program to reduce air pollution at its Silver Bay plant, but requested a variance on water quality regulations. Even while the case dragged in the courts, the taconite tailings problem received the attention of the International Joint Commission of the U.S. and Canada at hearings in Duluth on water quality of the Great Lakes.[13]

The taconite problem took on an entirely new dimension in June 1973, when EPA announced[14] that the drinking waters of Duluth contained asbestos fibers—which are widely feared as carcinogens—in amounts up to 100 billion fibers per liter. EPA warned that young children should not drink the water until further studies could be made; bottled water soon disappeared from store shelves in Duluth. The evidence that Reserve Mining's taconite tailings constituted a direct health threat to 150,000 persons stirred action in several quarters. On the regulatory front Russell E. Train, Head of the Council on Environmental Quality, was appointed to coordinate government action. (The appointment itself was controversial in that environmentalists considered this a move by Representative Blatnik to add another layer of bureaucracy and thus slow EPA's urge for quick action.[15]) The Public Health Service and the Food and Drug Administration (FDA) were immediately faced with the decision of whether to ban the use of the drinking waters of Duluth (and adjacent communities) for either local consumption or for consumption on ships and airplanes in interstate commerce. Before a decision was reached, however, evidence began to develop that water supplies of many other cities around the nation (particularly San Francisco) also contained asbestos-like particles.[16] In fact, an estimated 800,000 miles of water pipes made with asbestos in them were already in use. The regulatory implications were astounding. But the actual health effect of these asbestos-like particles in water were difficult to determine; even 5 months after the Duluth discovery, EPA's cancer research team could not prove that they had caused any harmful effects.[17] Meanwhile, the Corps of Engineers puzzled over how to comply with a court order to filter the Duluth water supply.

On the judicial front, EPA brought a new suit against Reserve Mining (and its parent companies) in August 1973, charging that its wastes were a health threat. An 8-½ month trial before Judge Miles W. Lord that developed 1,621 exhibits and 18,000 pages of testimony of the U.S. District Court, came to a stormy head in early 1974. Judge Lord first ordered Reserve to develop an onshore disposal plan. Then amid rumors that Reserve had exerted undue political influence among Minnesota's Democratic administration and legislators and in the Republican administration in Washington, Judge Lord ordered Reserve Mining and its two parent companies to produce records of political contributions* and meetings with government officials.[19] Finally on April 20, Judge Lord ordered the plant to close immediately on the grounds that it was a substantial danger to public health.[20,21] Two days later, however, a three-judge panel of the eighth U.S. Court of Appeals in St. Louis, Missouri, granted a temporary stay of the closure order,[22] and then on June 4 granted a motion for a seventy-day stay while a cleanup plan was developed by the company. The Appeals Court ruled that Lord's closure ruling was not based on adequate proof that health was endangered, and required that the decision be made on the basis of conventional air and water pollution grounds.[23,24]

The U.S. Justice Department, over the objections of EPA, accepted the Appeals Court ruling.[25] But when Reserve's proposed alternative plan consisted of switching the discharge to a disposal basin on nearby Palisade Creek, Judge Lord immediately ruled it unacceptable[26,27] and appeared to be moving toward again ordering the plant to close. The outcome of the case is thus still uncertain, a decade after protests from the public arose.

Comment

The taconite pollution issue has been primarily a local one (although potential international implications are obvious). It has developed because an industrial company constructed a complex

---

*The two parent companies were not named as defendants in the initial government suit of February 1972. Both Presidents Willis Boyer of Republic and William Verity of Armco were reported as long time backers of President Nixon and important Republican party fund raisers.

plant without being aware of the need for or allowing sufficiently for the proper disposal of huge amounts of waste by-products it produced. The company was not acting much differently than companies had since the beginning of the industrial revolution, but times were rapidly changing and the public was gaining an awareness of environmental threats. The result has been the development of considerable animosity, both locally and probably nationally, toward the mining and the iron and steel industries. Minnesota's constitutional amendment to favor and encourage the taconite industry, passed prior to the controversy, appeared incongruous subsequently, but the conflict between economic benefits and environmental or social costs is one that must be faced with increasing frequency by our society.

REFERENCES

1. Rainey, R. H., "Natural Displacement of Pollution from the Great Lakes," *Science* **155**, 1242 (March 10, 1967).

2. "Salvaging the Lakes," *Time*, **89**, 60 (April 21, 1967).

3. "An Appraisal of Water Pollution in the Lake Superior Basin," Federal Water Pollution Control Administration, U.S. Department of the Interior, April 1969 (revised January 1970).

4. Hill, G., "Lake Superior, Private Dump," *Nation*, **208**, 795 (June 23, 1969).

5. "Effects of Taconite on Lake Superior," National Water Quality Laboratory, Federal Water Pollution Control Administration, U.S. Department of the Interior (April 1970).

6. Laycock, G., "Call it Lake Inferior," *Audubon*, **72**, 48 (May 1970).

7. *A/W PR*, p. 193 (May 10, 1971).

8. *A/W PR*, p. 452 (November 8, 1971).

9. *A/W PR*, p. 36 (January 24, 1972).

10. "Test on Taconite," *Time*, **99**, 44 (March 13, 1972).

11. *A/W PR*, p. 193 (May 15, 1972).

12. *A/W PR*, p. 517 (December 25, 1972).

13. *Duluth Herald*, p. 1 (December 7 and December 8, 1972).

14. *A/W PR*, p. 258 (June 25, 1973).

15. *A/W PR*, p. 281 (July 16, 1973).

16. *Kansas City Star*, p. 5D (September 12, 1972).

17. "Data Sparse in Asbestos Scare," *Medical World News*, **14**, 104 (October 12, 1973).

18. *A/W PR*, p. 195 (May 20, 1974).

19. *Kansas City Star*, p. 4B (April 10, 1974).

20. *Kansas City Star*, p. 1 (April 22, 1974).

21. *Kansas City Star*, p. 1 (April 28, 1974).

22. *Kansas City Star*, p. 16 (April 23, 1974).
23. *A/W PR*, p. 225 (June 10, 1974).
24. *Kansas City Star*, p. 10A (June 15, 1974).
25. *A/W PR*, p. 53 (June 15, 1974).
26. *Kansas City Star*, p. 11 (August 5, 1974).
27. "Protesting Pollution," *Chem. W.*, **115**, 14 (August 14, 1974).

Foaming Detergents

Abstract

Some synthetic detergents, introduced following World War II, contained alkyl benzene sulfonate (ABS) which was only about fifty percent biodegradable, compared to soap's nearly complete biodegradability. The ABS in drain waters caused foaming which became a noticeable problem in many waterways by 1961. The public equated water pollution with the visible detergent foam, although it actually represented only about 10 percent of the total contamination of the water. The detergent industry responded to the foam problem by switching to biodegradable detergents before the end of 1965.

Background

Synthetic detergents were introduced into the United States consumer market following World War II. They quickly replaced soap, since they were so effective in cleaning laundry and dishes and gave visible evidence to the user of their activity: a mountain of foam.

The surfactant used in the detergents was a branched-chain alkyl benzene sulfonate (ABS),* a by-product of petroleum cracking. It was cheap to produce, safe to the user and highly effective in its cleansing properties. However, it was only about 50 percent biodegradable—that is, only half was decomposed to CO_2 and water by bacteria of the water and soil—compared to soap's 100 percent biodegradability. The problem was that bacteria could readily degrade the ABS molecule only to the point of branching: further degradation was very slow. Consequently, the ABS re-

*Also noted in the literature as DDBS (for dodecylbenzene sulfonate)

308

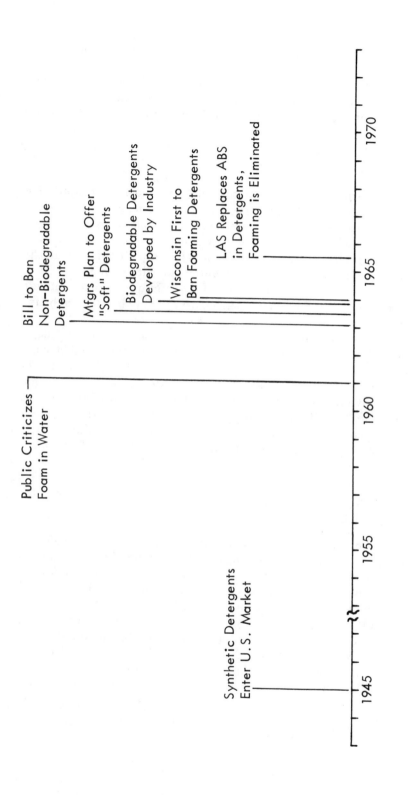

mained active after going down the drain and caused foaming in surface and underground waterways.[1] Nevertheless the sales of ABS were remarkable and reached 500 million pounds a year by 1963.[2]

Key Events and Roles

The foam became a noticeable problem around 1961. Some rivers were accumulating large amounts of foam which made the areas unsightly and on occasion even made river nagivation difficult. Some communities had foamy tap water, which caused considerable public apprehension. The public equated water pollution with the visible detergent foam, although in reality detergents contributed only about 10 percent of the total contamination of the water.[3] Even dead fish in rivers were blamed on the foam by the public.[4]

By 1963, detergent foam was the subject of much debate and some legislation. In December, Wisconsin became the first state to ban the use of "nondegradable" detergents.[5] The Wisconsin State Board of Health had reported the existence of a wall of foam thirty-five-feet wide, three hundred-feet long and fifteen-feet high on the Mississippi River.[6] Representative Henry Reuss (D-Wisconsin) introduced a bill (HR 2105) to ban the import or interstate transportation of detergents that took too long to decompose.[1] Many other states were considering similar bans and legislation.[7]

Disposition

The detergent industry was quick to respond to the foam problem. They announced that they would switch to biodegradable detergents by the end of 1965, if not sooner. Research soon produced a linear alkyl sulfonate (LAS)—that is, one which did not have the branched-chain molecule. It was readily destroyed by bacterial action and was a suitable cleaning substitute for ABS, but was more costly to produce (the C_{12} alkyl was used). Nevertheless, by mid-1965 the detergent industry* had switched to the "soft" detergents, at a reported cost of about $150 million.[6]

Although ABS was found to be harmless to humans in the quan-

*While the ABS was discontinued in the United States, it has not been dropped worldwide. In fact a new plant capable of producing 15,000 metric tons a year of dodecylbenzene will open in Thailand in 1973.[8]

tities present in waterways, the foam problem it caused initiated the intensive investigation of water pollution that followed.[4] "If all that foam resulted from only 10 percent of the total water contamination, what about the 90 percent that couldn't be so easily detected?" was the question raised in the minds of the public.

Comment

The foaming detergents controversy began the modern era of concern over water pollution. That the petrochemical industry and the detergent manufacturers were able to switch to biodegradable detergents so quickly and completely probably generated a false feeling that pollution problems could be solved easily and at relatively low cost.

REFERENCES

1. "Lather Over Detergents," *Bsns W,* 104 (February 16, 1963).
2. "New Battle of Suds," *Bsns W,* 126 (May 23, 1964).
3. "Less Suds For the Nation's Rivers," *US News,* **59,** 16 (July 12, 1965).
4. "Detergents in Water—Another View," *US News,* **55,** 16 (August 1963).
5. *New York Times,* p. 34 (December 19, 1963).
6. "Water Pollution," *Science,* **152,** Editorial Page (May 20, 1966).
7. "Just How Safe Is Your Drinking Water?", *US News,* **55,** 74 (July 15, 1963).
8. "Thai Petrochemical Complex Nears Reality," *C&EN,* **51,** 11 (January 22, 1973).

Enzyme Detergents

Abstract

In 1967 the detergent manufacturers developed several new home-use laundry products which contained enzymes. Many of these products were presoaks for clothes, but some were sold as regular laundry detergents. All were aggressively advertised during 1968–1971: the enzymes were touted to remove stain that normal detergents could not. By 1970, sales of enzyme detergent products were approaching $1 billion. Much controversy was developing, however, over the safety and effectiveness of these products: consumer groups and the Food and Drug Administration were soon investigating alleged health hazards (dermatitis, asthma, cancer) and the Federal Trade Commission was investigating alleged false advertising. Because of adverse public opinion and decreasing sales of enzyme products, many manufacturers had dropped the enzymes during 1971, before a National Academy of Sciences—National Research Council study found no evidence which indicated that enzymes posed a significant health hazard.

Background

Dry cleaners in the United States have used enzyme-based spot-cleaning products for many years. Enzymes are organic catalysts that break down or digest organic matter (such as proteins, starch, or fats) by chemical means. Used in laundry products, the appropriate enzymes will remove difficult stains such as blood, grass, mustard, lipstick, and chocolate.[1] In 1967, the detergent industry developed enzyme detergents for home use. After market testing the new enzyme detergents, products such as Drive, Tide XK, and Gain came onto the consumer market late in 1968.[2] The products were formulated into two different types. One type was

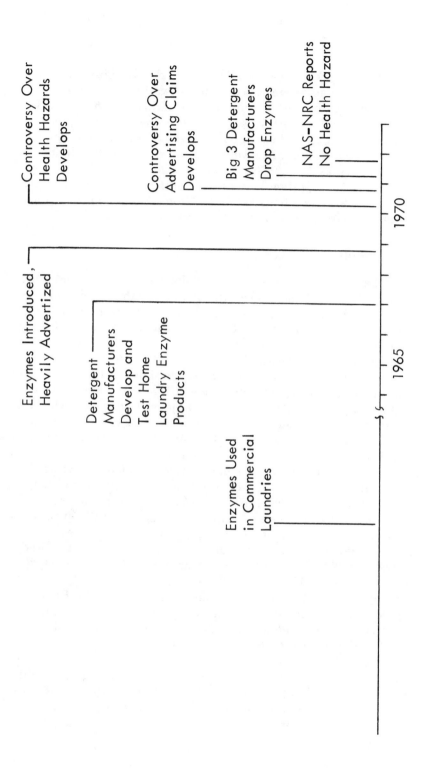

Enzymes Introduced, Heavily Advertized

Controversy Over Health Hazards Develops

Controversy Over Advertising Claims Develops

Detergent Manufacturers Develop and Test Home Laundry Enzyme Products

Big 3 Detergent Manufacturers Drop Enzymes

NAS–NRC Reports No Health Hazard

Enzymes Used in Commercial Laundries

1965

1970

called a presoak and was applied to laundry for thirty minutes or overnight to digest stains prior to normal washing. The second type was used directly in the washer in place of regular detergents.[1] The new detergents were heavily advertised as "new miracle" cleaners; the advertising relied heavily on endorsements from television personalities who were also involved at the time in antipollution activities, such as Arthur Godfrey and Eddie Albert.[2] The enzymes captured a large portion of the detergent market. By 1970, 60 percent of the $1.5 billion market in the United States for presoaks and detergents contained enzymes.[3]

Key Events and Roles

The new detergents came under attack in 1969–1970 from two fronts: the possible health hazards involved with enzymes and the cleaning effectiveness which the manufacturers were claiming in their advertisements. The FTC, FDA, and HEW began making their own investigations late in 1969.

The matter of possible health hazards of enzymes actually began in mid-1969 with a British report on skin irritation caused by enzyme dust in manufacturing plants. Allegations that enzyme detergents were harmful were denied by the big soapmakers who had run extensive skin patch tests prior to marketing the products, and had found no ill effects.[3] The AMA issued a two-page release stating that "Enzyme containing detergents are as safe and mild for consumer use as comparable nonenzyme products."[4] During the spring and summer of 1970, consumer advocate groups and Ralph Nader called on the FTC to ban the sale of enzyme detergents.[5] They stated that enzymes contributed to respiratory ailments, skin allergies, and also that some detergents contained excessive amounts of arsenic.[6] The Soap and Detergent Association offered, in October 1970, to finance a test program to determine the health effects.[7] In late 1970, a Rockefeller University study attributed respiratory diseases to enzymes and in February 1971, the FDA announced a 6-month scientific study would be conducted by the National Academy of Sciences—National Research Council of alleged health hazards of detergent enzymes.[8]

The matter of effectiveness involves the nature of enzymes: since enzymes are numerous, each enzyme performs a specific function by digesting only certain types of organic matter. It was soon obvious to users that the presoaks and detergents did not remove all stains, as the ads claimed, and on September 24, 1970, the FTC charged the three big manufacturers (Colgate-Palmolive, Lever Bros., and Procter and Gamble) with false advertising.[9,10] On March 4, 1971, the three companies agreed to accept the FTC suggestions for advertising and packaging their enzymes. Future ads and labels listed the types of stains the product could and could not be reasonably expected to remove.[11]

Disposition

The controversy had caused enzyme detergent sales to decline. The manufacturers, consequently, removed enzymes from most of their products during 1971.[12] On November 19, 1971, the NAS-NRC reported that their study revealed no evidence of major health hazards from enzyme detergents.[13] Some detergents still contain enzymes today, but the number is drastically reduced from the 1970 peak when virtually all detergents contained some enzymes. The advertising for enzyme detergents is much subdued today compared to their 1968–1971 heyday.

Comment

The enzyme home-laundry detergents flashed on the scene, under the impetus of an outlandish advertising campaign, at a time when such products were receiving increasingly close scrutiny by consumer advocates. The suggested health hazards of the enzyme detergents have not been verified, yet such products were so precipitously downgraded by the industry that the public was probably left very confused. The case illustrates the development of a public concern over technology that could and should have been avoided by more responsible actions.

REFERENCES

1. "Enzyme-Active Laundry Products," *Consumer Bul*, **51**, 4 (October 1968).
2. "Battle of Omaha Beach," *Newsweek*, **89**, 89 (October 7, 1968).
3. "Enzymes in Hot Water," *Time*, **95**, 86 (February 16, 1970).
4. *Wall Street J*, p. 1 (April 9, 1970).
5. "Nader Charges Hazards," *C&EN*, 14 (June 22, 1970).

6. "Action on Enzyme Detergents," *Consumer Bul*, **53**, 29 (October 1970).

7. *New York Times*, p. 6 (October 9, 1970).

8. *New York Times*, p. 24 (February 2, 1971).

9. *Wall Street J*, p. 2 (September 24, 1970).

10. *New York Times*, p. 54 (September 24, 1970).

11. *New York Times*, p. 19 (March 4, 1971).

12. *Wall Street J*, p. 12 (February 16, 1971).

13. *New York Times*, p. 53 (November 19, 1971).

Nitrilotriacetic Acid (NTA) in Detergents

Abstract

Nitrilotriacetic acid (NTA) was introduced into laundry detergents in 1970 as a replacement for phosphates, which were accused of contributing to the eutrophication of waterways. It proved to be a good substitute in performance, but became a subject of controversy because of its potential health hazards. In December of 1970, it was withdrawn from detergents by the manufacturers.

Background

In 1967, public concern grew over the eutrophication of the nation's waterways, particularly Lake Erie. Phosphates were accused of being a contributer to the growing problem. By 1969, the use of phosphates in detergents had become an issue of national debate and controversy. Federal, state, and local governments were introducing legislation to ban phosphates in detergents early in 1970. With the realization that the major constituent in their detergents might soon be banned, the detergent manufacturers sought a suitable substitute.

Key Events and Roles

Research and development and the screening of possible existing substitutes pointed to the chemical nitrilotriacetic acid (NTA) as one of the best available products. NTA was inexpensive and was already being manufactured by Monsanto and W. R. Grace. Tests indicated that it was as good as the phosphate builders in detergents in its performance. Early in 1970, the detergent makers began substituting small amounts of NTA for phosphates in some of their products. NTA was not considered to be a total substitute,

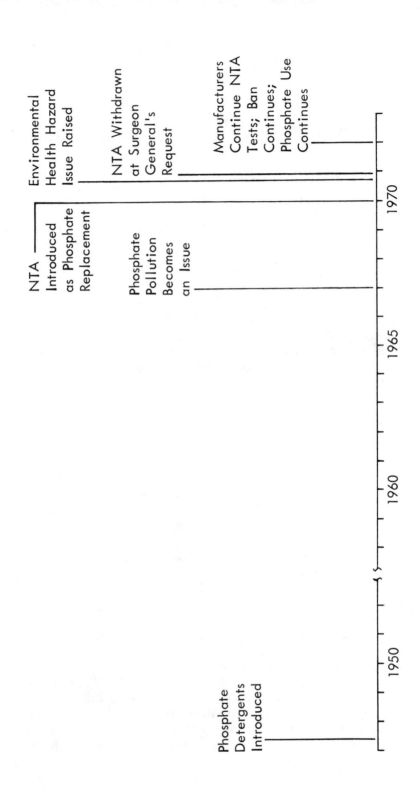

since it did not have all of the properties of phosphates. The detergents would merely have a reduced phosphate content in most cases. Nevertheless, NTA producers were planning to boost their capacities from about 135 million pounds per year in 1970 by an additional 400 million pounds per year.[1]

Throughout 1970 the detergent manufacturers replaced part of the phosphates in their detergents with NTA. There was skepticism on the part of a few ecologists and others over the question of NTA's possible health hazards.

In May 1970, Dr. Samuel S. Epstein submitted a report to the Senate Public Works Committee stating that NTA will not significantly reduce pollution and may be a health hazard in some cases. He urged that more research be conducted on the chelating properties of NTA before wide substitution of it was made.[2,3]

The National Institute of Environmental Health Sciences undertook a study to determine if any toxic or adverse health effects were possible with the use of NTA. The study concluded that although NTA itself was not toxic, its chelating properties cause it to combine with heavy metals such as mercury and cadmium found in water and soil. When the chelate compounds are taken internally, the toxicities of the metals are increased. Their ability to cross the placental barrier in rats was apparently enhanced since the fetal mortality rate increased when NTA was combined with cadmium chloride and methyl mercury. Without the NTA, the cadmium chloride and methyl mercury were not lethal at the doses given.[4]

When the evidence of the study was revealed to Surgeon General Jesse L. Steinfeld, he and EPA Administrator William D. Ruckelshaus called the representatives of the detergent makers to Washington, D.C., on December 17, 1970. When presented with the report and the Surgeon General's suggestion that the use of the additive be discontinued, the manufacturers agreed to halt use of NTA.

Monsanto, W. R. Grace, and Procter and Gamble were not convinced that NTA was harmful in the amounts that would appear in the environment. They conducted their own independent research to determine how harmful NTA was in an effort to refute the government's test results. Extensive testing was conducted ($25 million by W. R. Grace alone) throughout 1971, and the tests were reviewed by HEW. In September 1971, the Surgeon General decided to continue the ban on NTA, although the tests showed that it was not carcinogenic. He felt that the data were still too limited and the need for further research still existed.

Disposition

Since December 1970, when the detergent manufacturers voluntarily halted the use of NTA, the ban imposed upon its use by the Surgeon General has remained in effect. A committee of non-government scientists appointed by HEW in 1970 to review the information on the health effects of NTA reported in 1972 that the levels encountered in daily use are not toxic to man, but that the mutagenicity and carcinogenicity of NTA are still unknown.[5] Monsanto Company scientists have recently reported[6] that NTA in the parts per million concentration range is degraded completely over several days' time. Approval for the use of NTA will not be given by the Surgeon General until it can be clearly shown that NTA has no long-range adverse health effects on man.

Comment

The phosphate detergents were introduced at a time when long-range impacts on the environment were seldom considered. Even if possible hazards of the phosphates had been exposed then, public opinion would probably not have permitted a government ban of such a popular consumer product without clear proof of the danger. But the environmentalists' pressure to "get the phosphates out" had grown so strong by 1969 that the discovery of NTA was widely hailed as the solution to a serious national problem. On this high note, NTA was introduced by the detergent manufacturers. The suggestion, however, that NTA might be more hazardous in the environment than the phosphates it was replacing sent a shock through the environmentalists, the pertinent government agencies, and the public. The reactions of the government agencies reflect an increasing awareness of the need to look for unforeseen consequences of our technology. In this case the technology, NTA, can still be adopted at a later date when its limitations are better defined. In general, society should recognize that every newly-discovered technology does not have to be adopted immediately—that the most rapid growth is not always desirable.

REFERENCES

1. "NTA Moves Into Canada," *Chem W*, **107**, 26 (August 19, 1970).
2. *New York Times*, p. 38 (November 13, 1970).
3. "NTA," *Environment*, **12**(7), 3 (September 1970).

4. "A Solution Becomes a Problem," *Sci N*, **98**, 475 (December 26, 1970).

5. *C&EN*, **50**, 7 (May 15, 1972).

6. Warren, C. B., and E. J. Malic, "Biodegradation of Nitrilotriacetic Acid and Related Imino and Amino Acids in River Water," *Science*, **176**, 277 (April 21, 1972).

Saltville—An Ecological Bankruptcy?

Abstract

In July 1970, the Olin-Mathieson Chemical Company started closing its manufacturing plant operations in Saltville, Virginia (population 2,500), because of the cost of reducing its waste effluents to meet new water pollution standards. The impact on Saltville was tremendous since it had no other major industry. With a large percentage of its workers unemployed and a drastic cut in its tax revenues, Saltville was struggling for existence in 1973. Olin has assisted Saltville residents in many ways and is contributing $600,000 to the town.

Background

The town of Saltville, Virginia, takes its name from the nearby salt deposits. The deposits were used as a supply for a salt-making industry which started in 1788 and continued until 1864 when the Union army destroyed the salt works. In July 1895, the Mathieson Alkali Works began production in Saltville. The Mathieson firm later established installations in other parts of the country, but the Saltville facility remained the mother plant.[1] For most of the first half of the century, Saltville fit the pattern of a typical company town with most of the land, homes, stores, and utilities owned and operated by the company. Even the hospital, school system, and police department were financed by the plant. After World War II the company began to take a less paternalistic role in the community. Company houses were sold to the workers who lived in them and land was sold to workers to build their own homes. The company stores were phased out and the utilities were turned over to the town, but the community remained economically and psychologically dependent upon the chemical works. Mathieson

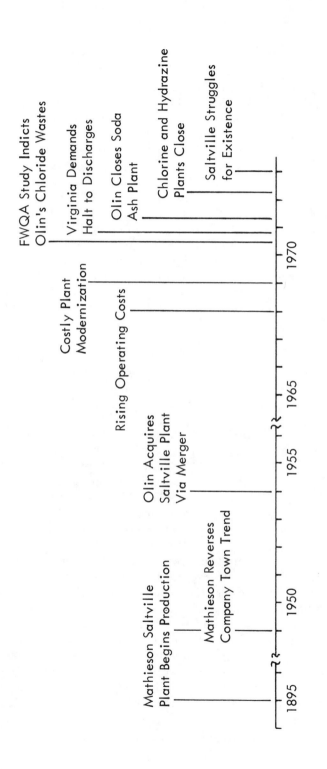

merged with Olin Industries in 1954, but the merger had little effect on the nature of the Saltville operation or on the workers.[2]

The Saltville works had been expanded through the years until it included a one thousand tons per day soda ash plant, a 290 tons per day mercury cell chloralkali plant, a 250 tons per day dry ice plant, a large government-owned hydrazine plant, and a rubber-coating facility.[2] At its peak in the early 1960's, 1,500 men were employed at Olin's Saltville site, but much of the installation was old and obsolete and the 1960's saw a decline in the number of men employed. In 1968, Olin had spent several millions to modernize the plant and increase efficiency, but the modernization had not worked out: costs had risen 35 percent, but efficiency had not.

During the 1960's, the Saltville Industrial Development Corporation had been formed under the auspices of the Chamber of Commerce in an effort to make Saltville less of a one-company town. One of its primary concerns was to bring in sources of employment for the women. They were successful in getting a garment factory which employed about 225 women. The only other industry in the town was U.S. Gypsum, employing about two hundred men.[1]

Key Events and Roles

The original and major product of the plant was soda ash (ammonia process soda) which resulted in a large amount of waste sodium and calcium chlorides being discharged into the local river.[1] These chlorides were not toxic, but they greatly increased the hardness of the water downstream. The Federal Water Quality Administration made a study of the effects of the chlorides in the stream and the findings were released in June 1970. The study revealed that the hardness created by this pollution cost $2 million annually to other water users downstream: other plants and installations had to add expensive water treatment facilities before they could use the river water and homeowners had to purchase water softeners and extra detergents.[1]

The State of Virginia had enacted laws establishing strict water quality standards and after the FWQA report the state authorities told Olin to stop discharging chlorides. In July 1970, the company announced that it was phasing out its soda ash plant because there was no other economically feasible solution. Much of the plant, which still employed about 800 men in early 1970 (and especially the ammonia process operations) was known to be economically marginal. The water pollution cleanup order was recognized as "the straw that broke the camel's back." Olin claimed to have spent $2 million on pollution control research. By early 1971, Saltville was the subject of stories in the national press.[3]

The first layoff, affecting 415 men, occurred in July 1971, when the soda ash operations were shut down and smaller layoffs continued through that year.

The water quality study had also found serious mercury pollution present in fish downstream from Saltville. Federal officials said the mercury came from the chlorine plant, and ordered Olin to stop mercury discharges, so in February of 1972, that plant was also shut down and an additional 150 workers were laid off. The government-owned hydrazine plant was still being operated by Olin, but its continued operation depended upon renewal of a government contract. Then, in the summer of 1972, the Air Force cancelled the contract and the hydrazine plant was also closed. The total number of men employed by Olin had decreased to only about one hundred, a drop of seven hundred since 1970.

The impact on the town was tremendous; most of the people had lived with the plant for generations with sons inheriting their fathers jobs at the plant. Saltville had no other major industry to take up the slack and many of the workers were unskilled and most were over forty years old. Many of the technical and managerial workers decided to relocate, but only about ten hourly-wage workers chose to take advantage of the company's offer of relocation. With most of its workers unemployed and a drastic cut in its tax revenues the outlook for Saltville was not good. With the Olin plant closing, the Saltville area lost a $6 million a year payroll, and the county lost almost $192,000 in annual property taxes.[4]

Disposition

Olin did much to ease the shock, such as offering to relocate any former employees, giving generous severance pay, and making early retirement available to many of the older employees. In 1971, Olin gave a building, used to house the plant's computers, to the town as a job preparedness center along with a $50,000 grant to run the center so workers could pass a high shoool equivalence exam and learn a skill. Some of the former employees found jobs in nearby towns and others commuted eighty miles a day to work, but in 1972 unemployment was still high and the future was uncertain for many.

There was little resentment against Olin or the state authorities for the plant closing, although some workers felt that the authorities tried to do too much too soon. The State of Virginia increased the unemployment compensation payments from twenty-six weeks to thirty-nine weeks to help alleviate the economic hardship of those who lost their jobs because of the environmental cleanup. The Virginia Highlands Community College helped to develop the program at the Saltville Job Preparedness Center for training the laid-off workers and the head of the center was provided by VHCC.

The two small remaining industries (the garment factory and the U.S. Gypsum plant) were insufficient to keep Saltville going when Olin left. The Saltville Industrial Development Corporation began looking for additional industry. The development corporation had little success in getting new business to come to Saltville because there was little land except Olin's, suitable for industrial development. Olin, which had donated the land for the garment factory site and also an adjoining ten-acre tract as an additional industrial site, gave Saltville almost all its real property holdings in the town in January 1973. Included were 3,500 acres of land and several industrial buildings.[5] Olin is also transferring its extensive mineral rights to the town so Saltville can try to attract another mining company. To compensate for the loss of tax revenue and aid in developing the area, Olin is paying the town $600,000 over a three-year period.[6]

Comment

The closing of the chemical manufacturing plants at Saltville, Virginia, under pressure for environmental improvement had tremendous adverse economic effects on the community despite considerable effort by the manufacturer. Basically, the problem is one of different time scales; while popular opinion and regulatory decisions that demand rapid remedies can quickly develop, the time scale of social systems normally tends to be of a decade or several decades. In the case of Saltville, a local symbiotic social-technological system that had grown over a seventy-five-year period was suddenly severed when the existing technology failed to meet society's changing standard: the local social system faced a difficult adjustment when the technology abruptly ceased to exist. The Saltville experience and others like it may spark the development of an "ecological bankruptcy syndrome" among industry and labor.

REFERENCES

1. "Saltville: Caught in the Cleanup!", *Tenessee Valley Perspective*, **2**, (4), 4 (Summer 1972).

2. "Saltville Rallies After Olin Shutdown," *C&EN*, **50**, 7 (March 13, 1972).

3. "End of a Company Town," *Life*, **70**, 36 (March 26, 1971).

4. *Kansas City Star*, p. 26 (August 29, 1972).

5. "Olin Gives Property, Cash to Saltville," *C&EN*, **51**, 4 (January 8, 1973).

6 "Olin Christmas Gift," *Chem W*, **112**, 17 (January 10, 1973).

Truman Reservoir Controversy

Abstract

In 1964, the Army Corps of Engineers began construction of a $294 million multipurpose dam and reservoir on the Osage River near Warsaw, Missouri (100 miles southeast of Kansas City).

In March 1972, after over $100 million had been spent or let in contracts, the work on the dam itself was 25 percent completed, and 50 percent of the land had been acquired, environmentalist groups filed a lawsuit seeking to halt all work and prevent completion of the reservoir. This suit claimed damage to the ecology and contended that the Army Engineers were not in compliance with the new National Environmental Policy Act of 1969. The U.S. District Court ruled (September 1972) that further construction work under certain restrictions would be allowed pending the preparation and assessment of an environmental impact statement. In late 1973 the environmental impacts were judged acceptable, and in mid-1974 Congress authorized funds toward completing the project.

Background

The Harry S. Truman Dam and Reservoir (then called the Kaysinger Bluff Dam) was authorized by Congress in 1954. The U.S. Army Corps of Engineers would construct a dam and a small (twelve thousand-acre) lake on the Osage River at Warsaw, Missouri, about one hundred miles southeast of Kansas City, and immediately above the privately owned* Lake of the Ozarks on the same river.[1, 2] There had been strong local opposition to the project at the time of its authorization, and to gain support and acceptance

*Union Electric Company of St. Louis is the owner but the lake is publicly utilized.

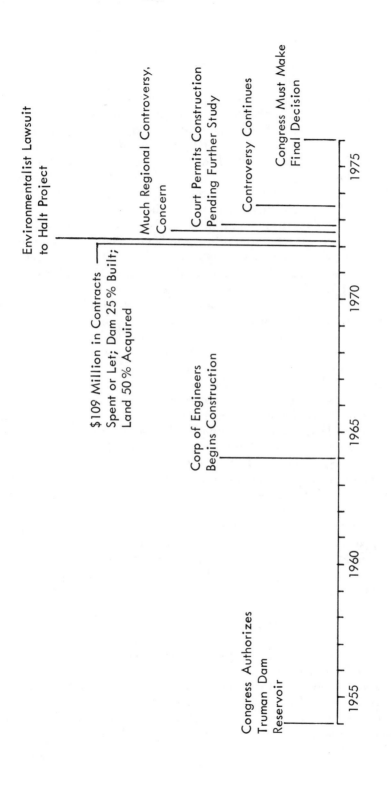

Environmentalist Lawsuit to Halt Project

$109 Million in Contracts Spent or Let; Dam 25% Built; Land 50% Acquired

Much Regional Controversy, Concern

Court Permits Construction Pending Further Study

Controversy Continues

Congress Must Make Final Decision

Corp of Engineers Begins Construction

Congress Authorizes Truman Dam Reservoir

1955 1960 1965 1970 1975

for the project, it was redesigned to include a hydroelectric power plant and a larger recreation lake—55,600 acres at normal level and up to 209,000 acres at flood stage. Resistance to the revised project gradually dwindled. Family farming is hard and, excepting in fertile river bottoms, often financially unrewarding in much of this region. Many rural and small town residents welcomed the prospects of new jobs provided by a resort boom and industrial growth.

The project would produce the largest federal reservoir in Missouri or Kansas and would be nearly the size of the Lake of the Ozarks. The dam itself would be some five thousand feet long and 126 feet high, and the reservoir would have 958 miles of shoreline. The hydroelectric generators would produce 160,000 kilowatts of power. The total estimated cost was about $179 million. The benefits to be derived from the project included flood control, hydroelectric power, recreational and economic facilities, and regional development. The Corps' projected annual benefits from the project were $12.3 million, of which $5.8 million were from flood control. The project became western Missouri's hope for reversing the loss of population to the cities and of building a new industry (tourism) to supply new jobs and income.

Because of difficulties in obtaining a Congressional appropriation, construction work on the project was delayed ten years, until late in 1964.[1] Even then funding for the project limped along with the inevitable effect of slowing down construction work. By early 1972 the estimated cost had risen to $294 million, but $64 million worth of work had been completed and $45 million was under contract. About 50 percent of the land had been acquired, and 100,000 acres of farms and the little town of Fairfield had been vacated. Overall, some $130 million had been invested in the project, construction was about one-quarter completed, and the Corps felt confident that it could meet its revised completion date of 1979. But neither the Corps nor the Truman Dam was without its opponents, and an enormous local controversy was about to erupt.

Key Events and Roles

On March 6, 1972, the Environmental Defense Fund (EDF of East Setauket, New York, and Washington, D.C.), the Missouri Chapter of the Wildlife Society, and seven residents of the Warsaw, Missouri, area filed suit in U.S. District Court in Kansas City to bring the Truman Reservoir project to a complete halt.[1,2] The plaintiffs were seeking an injunction to stop work because the Army engineers had failed to file an environmental impact statement (EIS) with the government as required under the National Environmental Policy Act of 1969, and because the project threatened "irreparable environmental damage."

The major complaints leveled at the Corps of Engineers by the plaintiffs were that the Truman Reservoir would:

- Cause the extinction of the rare paddlefish in the area.
- Cause deterioration of the Schell-Osage Wildlife area.
- Jeopardize the emerging striped bass fishing in the streams and cause an undesirable increase in carp and other nongame fish.
- Inundate nationally significant archaeological and paleontological sites.
- Force many families to move from farm and community homes.
- Fail to provide the benefits claimed by the Corps of Engineers.

The ancient and rare paddlefish, sometimes known as the spoonbill or duckbill, is represented by only two living species, *Polyodon spathula* of the Mississippi Valley and *Psephurus gladius* of the Yangtze River in China. According to some authorities, the largest concentration of paddlefish in the United States is located in the project area; they claim that the lake would flood the spawning ground and that paddlefish will not spawn in a lake. The paddlefish reach fifty to sixty pounds in size and are sought not only by local fishermen, but also—because of their unusual, human-like blood chemistry—by some medical researchers. The plaintiffs also charged that the reservoir could jeopardize the striped bass fishing in the streams. They stated that because the Truman Reservoir would be shallower and warmer than other Missouri reservoirs, it was likely the new reservoir would attract a large population of undesirable fish such as carp and other nongame species.

The complaint alleged that holding the reservoir at conservation pool would interfere with periodic flooding in the vicinity of the Schell-Osage Wildlife Area. Claims were made that without such periodic inundation, the timber in the bottomlands would be harmed and a serious reduction would occur in the available natural food supply (such as acorns) for waterfowl (including Canada geese, snow geese, mallards and many species of ducks).

The complaint pointed out that the Truman Reservoir will also inundate paleontological and archaeological sites in the area that were found during the 1960's to be of national significance. The land near Warsaw was once home to musk oxen and mastodons and their 25,000-year-old skeletons can be found at several sites in the area; and the Rogers Shelter archaeological site was inhabited by man some 10,000 years ago.

The plaintiffs also charged that because of the flooding of over 200,000 acres in seven counties to create the reservoir, families would be forced to move from 330 farm homes and 570 nonfarm homes.

The lawsuit called the flood control benefits claimed by the Army engineers "faulty." The complaint alleged that only 9 percent of the flood control benefits were for the Osage basin, while more than seventy percent were for the Mississippi River. A spokesman for the plaintiffs said the situation was "a case of flood control overkill."

The Corps of Engineers immediately asserted that it had been proceeding with proper care and deliberation, and that it was not required to have prepared and filed an environmental impact statement, because the project was authorized and started many years before the enactment of the recent federal environmental legislation. But even as the Corps was issuing this reply, it revealed in Washington to the Senate Appropriations Subcommittee that it had just reanalyzed the economics of the hydropower plant. The Corps had concluded that although the power plant might not pay for itself within fifty years (as required by law for such a project), it would be more expensive (a multimillion dollar loss) to turn back at this point than to complete the hydropower plant.[3]

Announcement of the lawsuit opened a floodgate of controversy. Newspapers in Missouri,[1-6] if not the national press[7,8] carried front page stories on it and on the court proceedings[9-13] which were to continue for two years.

Most local residents appeared to be strongly in favor of completing the project, and Missouri politicians came out strongly in support of the project. Senator Stuart Symington (D-Missouri) stated that any halt in the construction of the Truman Dam and Reservoir would be unwise. He noted that a slowdown or halt in the work, even a temporary one, would have a serious adverse effect on the economy of the area. Senator Thomas F. Eagleton also announced support. Much of the opposition to the EDF lawsuit was based on the opinion that action against the reservoir was taken too late. Many people contended that since construction on the project began six years ago, and had progressed for so long, it should be completed. People were concerned and anxious to do something (such as signing petitions and making contributions toward legal defense) to get the dam completed. Representatives from several communities in the area formed the Truman Reservoir Citizens Defense Fund.

Resolutions by the Henry County Court and the City of Osceola emphasized the crippling losses the area would suffer in jobs and money if the project were stopped at this stage. In petitioning for continuance of the Truman Dam construction, the Henry County Court said the county had lost thousands of tax dollars from land purchased by the government—which could only be returned through building and commercialization which would come with completion of the dam. The court estimated that economic loss in the area would be millions of dollars if the dam were not completed. A petition passed by the Council of the City of Osceola stated[10] that halting the project "would completely destroy our city." They noted that some 125 homes and twenty-seven businesses had been acquired by the government; that the city had $990,000 in unfinished contracts related to the Truman project; that the city had lost thousands of dollars in assessed valuation through government acquisitions and needed future development to recoup this loss.

An important aspect of the problem was that the Army claimed it could not get appropriations fast enough to buy out the landowners who were ready to sell. Property not yet acquired deteriorates because maintenance and improvements seem pointless, and the owners live in unhappy suspended animation. To some observers, it appeared that to stop work on the project would do the worst kind of harm to the very people the lawsuit was nominally supposed to help—those who had waited so impatiently for it to be completed.

Disposition

Judge John W. Oliver of the U.S. District Court heard arguments on the case in May 1972. The judge was presented with widely divergent views on the impacts of the projects. He soon urged that the attorneys agree on "factual data" concerning effects on fish and wildlife so that they could get to the economic feasibility of the project. The judge then ruled that EDF had proved its point on environmental damage with its first five witnesses and that testimony from more than one hundred other scheduled witnesses should merely be entered in the record. Testimony on the economic effects showed that governmental agencies were not in agreement; the Bureau of Sport Fisheries and the Bureau of Outdoor Recreation had estimated annual water use benefits of only $1.6 million, nearly $2 million less than the Corps estimate. Testimony also conflicted over the relative merits of proposed alternatives to the project—no lake, a lake of reduced size, a "dry lake" (that is, a controlled emergency flood plain), or a dam without the power plant and associated water needs. In addition, the judge had to weigh the effects of alternative restrictions that might be placed on construction work while an EIS was prepared: the Corps estimated that a complete shutdown for nine months would cost $22.5 million.

On September 13, 1972, Judge Oliver denied EDF's petition for an injunction to halt all construction work, but did, however, enjoin the engineers from continuing with certain facets of the project that could adversely affect the environment.[13] This work was halted pending the preparation and filing of an environmental impact statement by July 1, 1973. Other facets of work were allowed to continue without interruption. The opinion noted that the ultimate decision on whether the dam would be completed rested with Congress and added that no substantial environmental impact would occur until and unless the dam was actually closed some time in 1976, thus allowing time to review the EIS.

The EDF immediately challenged Judge Oliver's ruling in higher courts. The U.S. Circuit Court of Appeals declined to stop work on the project until it had heard oral arguments on the appeal of the March 1972 decision which were already scheduled for February 1973. The U.S. Supreme Court declined in December to issue a temporary injunction against work, pending further appeals, and the contesting parties settled back to await the EIS.[15]

The Corps of Engineers had actually started work on an EIS when the controversy erupted; by the time the court decision was announced it had prepared and begun circulating a three-volume draft EIS for comment by other federal and state agencies.[16] Buoyed by a larger-than-usual $27.5 million appropriation for the project from Congress in 1973, the Corps filed its EIS on schedule. The EDF immediately challenged the adequacy of the EIS, claiming that it was vague and in some places based on incomplete data. Following court hearings on July 23 to 25, Judge Oliver ordered that the Corps supplement its EIS and be more specific about the economics of the hydropower plant, the impacts on people, and the alternatives available.[17]

The Corps of Engineers issued its final EIS on the project on September 22, 1973.[18] On November 8, Judge Oliver denied EDF's petition to halt the project.[19] In accepting the EIS, the judge noted that it demonstrated that a conflict existed in economic theory for such a project, and that the Congress and other decision makers were now so advised. The U.S. Court of Appeals upheld Judge Oliver's final decision in early 1974, and the EDF terminated its litigation. On June 3, 1974, the House Appropriations Committee allocated $43 million to the project and efforts were under way to rush the Truman Dam and Reservoir to completion.[20,21] The estimated total cost had by now risen to $332 million.

For its part, the EDF claimed that although it lost the final decision, its legal action had produced an EIS which had been lacking, had opened up the Corps of Engineers' decision-making process to public scrutiny, and had caused the Corps to give more careful consideration to costs of all kinds on other projects. EDF noted that it had similarly stimulated the preparation of EIS's on projects ranging from the small Gillham Dam on the Cossatot River of Arkansas to the giant Trans-Alaska Pipeline project, and that environmentalists' litigation on the Cross-Florida Canal had shown that sufficient adverse effects might occur that President Nixon, on the recommendation of the Council on Environmental Quality, had terminated the whole project.

Comment

The Truman Dam and Reservoir controversy is one of several in which a previously approved technological development required reevaluation in the light of changed social values and regulations. The new requirement by law of environmental impact statements created much confusion in governmental agencies as well as much opportunity for direct citizen action. The results of the Truman Reservoir study indicate that gains in some values will inevitably be accompanied by certain losses in others. Society must learn to balance these gains and losses, not only for present users, but for succeeding generations.

REFERENCES

1. *Kansas City Times*, p. 1A (March 6, 1972).
2. *Kansas City Times*, p. 1 (March 9, 1972).
3. *Kansas City Star*, p. 3 (March 10, 1972).

4. *Kansas City Star,* Editorial (March 14, 1972).
5. *Kansas City Star,* p. 3 (March 20, 1972).
6. *Kansas City Star,* p. 10S (March 26, 1972).
7. *New York Times,* p. 56 (March 23, 1972).
8. *A/W PR,* p. 125 (March 27, 1972).
9. *Kansas City Times,* p. 3 (May 23, 1972).
10. *Kansas City Times,* p. 3 (May 24, 1972).
11. *Kansas City Times,* p. 48 (May 26, 1972).
12. *Kansas City Star,* p. 8D (July 21, 1972).
13. *Kansas City Times,* p. 12 (July 31, 1972).
14. *Kansas City Star,* p. 1 (September 13, 1972).
15. *Kansas City Star,* p. 8 (December 19, 1972).
16. *Kansas City Times,* p. 16C (September 16, 1972).
17. *Kansas City Star,* p. 1 (August 23, 1973).
18. *Kansas City Star,* p. 15A (October 7. 1973).
19. *Kansas City Times,* p. 1 (November 4, 1973).
20. *Kansas City Times,* p. 1 (June 4, 1974).
21. *Kansas City Star,* p. 1 (June 25, 1974).

Plutonium Plant Safety at Rocky Flats

Abstract

The Dow Chemical Company produces radioactive plutonium for nuclear weapons in an Atomic Energy Commission plant at Rocky Flats, Colorado, five miles west of the Denver suburbs. This defense plant suffered a disastrous fire in May 1969. Although apparently very little plutonium was actually lost as a result, the fire focused much attention on the operation of the Rocky Flats plant and raised much controversy over whether it was normally emitting contamination which endangered the health and safety of nearby residents. During 1971 and 1972, this plant became one of the AEC's most bitterly criticized facilities. Critics charged that the plant's management has been negligent from an environmental standpoint and irresponsible toward its worker population. The AEC and Dow Chemical dispute these charges. The controversy and public concern continue.

Background

Plutonium triggers for nuclear weapons are produced in an Atomic Energy Commission (AEC) plant located at Rocky Flats, Colorado, about twenty-one miles from the downtown area of Denver. In 1951, an isolated mesa at the foot of the Rockies had seemed to be a fitting location for the top secret AEC facility, but by 1970, Jefferson County, in which the Rocky Flats plant is located, was the thirteenth-fastest-growing county in the nation and the Denver suburbs were within five miles of the plant. Denver itself had grown from a medium-sized railroad hub during World War II into a full fledged industrial metropolis with an increasingly pollution-conscious citizenry.

The Rocky Flats plant is the only facility in the country man-

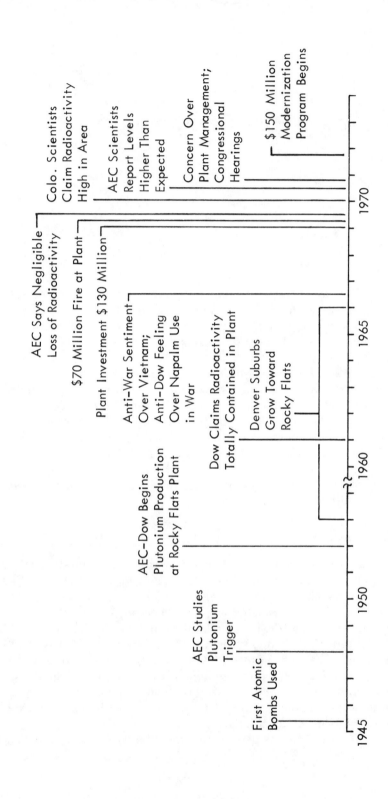

First Atomic Bombs Used

AEC Studies Plutonium Trigger

AEC-Dow Begins Plutonium Production at Rocky Flats Plant

Dow Claims Radioactivity Totally Contained in Plant

Denver Suburbs Grow Toward Rocky Flats

Anti-War Sentiment Over Vietnam; Anti-Dow Feeling Over Napalm Use in War

Plant Investment $130 Million

$70 Million Fire at Plant

AEC Says Negligible Loss of Radioactivity

Colo. Scientists Claim Radioactivity High in Area

AEC Scientists Report Levels Higher Than Expected

Concern Over Plant Management; Congressional Hearings

$150 Million Modernization Program Begins

1945   1950   1960   1965   1970

ufacturing plutonium, and is considered essential to national defense. Government investment in Rocky Flats totaled about $130 million between 1952 (when the plant opened) and 1969.[1] The plant employed about 3,700 workers in 1971.

The radioactive plutonium is a proven carcinogen in dogs and a highly dangerous potential carcinogen in man.[1] The plutonium plant tries to achieve "total containment," which amounts to keeping all radioactive materials within plant confines and completely accounted for. The Rocky Flats management has claimed that the plant has been successful in achieving its containment goals: they claimed in 1961[1] that only 0.038 curie of plutonium had escaped off-site during the entire history of the plant. This was a cumulative pollution figure for all of the building's stack effluent measurements.

The production of nuclear war materials had become a sensitive issue during the late 1960's because of the developing anti-war sentiment over Vietnam and the arms race, and because of the fears of many people over the contamination of the environment with radioactive poisons. In addition, the Dow Chemical Company, which actually operated the plant for the AEC, was the subject of much criticism because it had become the most widely known supplier of napalm to the government for the Vietnam war.

Key Events and Roles

The Rocky Flats plant suffered a disastrous $70 million fire on May 11, 1969; this incident was reported to be the largest American industrial accident.[1] The principal danger from such a fire was that a cloud of smoke containing plutonium oxide particles might escape and blow away, possibly to contaminate soils, water supplies, and buildings, and to be inhaled by people. In a 1969 report on the fire, the AEC stated[1] that "there is no evidence that plutonium was carried beyond the plant boundaries." The report noted, however, that one air filter had suffered a "minor failure" and that the roofs of two buildings had "slight exterior contamination."

After the report, a Rocky Flats public relations press release contained this reassuring statement: "Since the 1969 fire, the special investigations by the AEC, close surveillance by state health officials and a team from the Joint Committee on Atomic Energy, and a hearing by the Colorado Industrial Commission have reaffirmed the safety of Rocky Flats operations."

The report and press release were not the end of the incident, however, but only the beginning of what was to be a long controversy. The subject of the controversy soon enlarged from the direct effects of the fire to a searching examination of the entire management and operational practices in the plant.

Under the sponsorship of the Colorado Committee for Environmental Information (CCEI, a group of Colorado scientists), Edward A. Martell, a nuclear chemist at the National Center for Atmospheric Research, led a group of scientists in an independent survey near the Rocky Flats facility. He concluded in a January 1970 report that there was "a curie or more" of plutonium off-site;[1] the report concluded that the "health and safety of the people of Denver were being threatened." A second report issued one month later stated that the offsite plutonium ranged from "curies to tens of curies." The investigating group concluded that there was between one hundred and one thousand times more plutonium in the local environment near Rocky Flats than there would have been if good containment practices had been continually maintained.[2] Plutonium was found in soil located far from the plant boundaries and in the direction of the prevailing winds, which blow toward the Denver suburbs.

The Colorado Department of Health, however, concluded in May 1970, after an investigation, that although there was some plutonium in offsite soils, "No public health hazard existed during the time encompassed by this report."[1]

On August 1, 1970, two scientists of the AEC's Health and Safety Laboratory (who had been assigned in February to make an official study) reported that the plutonium 239 beyond the Rocky Flats plants could in fact be as great as 5.8 curies.

The Dow Chemical Company then suggested two other possible sources of contamination in addition to the 1969 fire.[1] One source was a fire which had occurred in 1957 at Rocky Flats. The other source was a two-acre open storage area onsite, where drums of oil, laced with tiny plutonium particles, had stood rusting and leaking their contents for several years. This storage area became the prime suspect. Dow had already removed the leaky drums and later laid down an asphalt pad over the surface where they stood to prevent plutonium still in the soil from being blown away. Following the CCEI report, however, they added more asphalt. These leaky oil drums reflected badly on Dow's stated policy of total containment: they had been permitted to stand out in the open for several years, and Dow's monitor systems apparently had not indicated anything unusual during that time.

Management's position was that "the difference between 1, 6, or 15 curies was not significant from a public health viewpoint." The Director of Research and Ecology stated that the amounts of plutonium found spread over the plains around Rocky Flats were well below permissible international standards. This rebuttal ignored the fact that the standards for plutonium have long been a subject of concern and controversy among scientists. A particularly controversial issue has been the sixteen-nanocurie-per-lung permissible burden agreed to by the National Committee on Radiation Protection and Measurements and the International Commission on Radiation Protection in 1959.

The scientific issue was the question of how nonhomogeneous bits of plutonium, emitting very intense but localized alpha radiation (so-called "hot particles") behave in body tissue. The standard for the lung was based on a model in which the inhaled plutonium behaved like an aerosol and irradiated the lung walls uniformly. The "hot particles," however, were said[1] to behave differently. Some AEC scientists, including A. H. Long of Argonne National Laboratory and Donald Geesaman of the Lawrence Berkeley Laboratory, raised questions about the standard.[1] They theorized that a "hot particle" emits much more intense radiation over a smaller tissue area* than the uniformly diffused particles, and hence, that it involves much greater risk. One AEC plutonium scientist noted that when the standards were first set, the experts had agreed that further study was needed, but in fact very little had been done in the twenty years since. Edward Martell of CCEI believed adamantly that the standards, as they applied to the "hot particles" he had found in the Denver area soils, were far too lax. Some critics of the standards have suggested that they must be revised downward by a factor of 50 to 100. The Rocky Flats management indicated that these standards would be so low that the plutonium plant might have to be closed.

In October 1970, a Congressional committee started hearings on the Rocky Flats controversy. Major General E. B. Gilles, AEC's Assistant General Manager for Military Applications, gave some unusually frank testimony which the Denver press gleefully reported in January 1971. Gilles said that Rocky Flats' buildings were crowded, its equipment corroded, and that the plant suffered the adverse effects of aging and had a number of serious deficiencies

*The theory seems to imply that the particle is absorbed on the membrane surfaces.

which required corrective action. He also told the Subcommittee on Public Works of the House Appropriations Committee, "If we were to have a fire in these buildings similar to the previous one, we would have a very severe impact on public health and safety."[1]

General Gilles said he was purposely frank because the AEC could not afford to lose as much as a few months to get the fire and safety reparations work going. But in 1971, when the time came for continuation of these funds, the Joint Committee on Atomic Energy withheld $12.8 million of AEC's fire and safety money from the $15 million request for Rocky Flats. The reason given by the committee was that "preliminary design work" on Rocky Flats was dragging behind schedule by several months.[1]

Disposition

Two years after the 1969 fire, the Rocky Flats plant was still being criticized as a health hazard to employees and the surrounding communities.[1] The prospects were that Rocky Flats would remain a focus of attention and mistrust as the Denver suburbs grew up around the plant site, and that pressures to relocate or automate the plant would increase.

On December 6, 1971, the AEC announced plans for a $150 million modernization program at the Rocky Flats plant to reduce the risk of workers' exposure to plutonium.[3] In 1973, the Rocky Flats plant again received unfavorable publicity when traces of radioactive tritium (hydrogen) were found in the water supply of nearby Broomfield; the tritium had apparently been received unsuspectingly in a shipment of scrap plutonium from the AEC's Livermore Laboratories in California. The tritium levels did not appear to constitute a health hazard.[4]

Comment

The Rocky Flats controversy apparently arose because of a combination of factors including: a widespread and long-existing fear of radioactivity; the AEC's secrecy of the plant operation; the antagonism toward war materials such as nuclear devices and napalm, which also was made by Dow, the plant manager; a laxity in maintenance and operating procedures at the aging plant by the plant management and the AEC; and the encroachment of the suburbs. The public concern in this instance has probably been beneficial in effecting remedial action at the plant, but the controversy may arise again in the future because of the plant's proximity to Denver.

REFERENCES

1. "Rocky Flats: Credibility Gap Widens on Plutonium Plant Safety," *Science,* **174,** 569 (November 5, 1971).
2. "Colorado Environmentalists: Scientists Battle AEC and the Army," *Science,* **168,** 1324 (June 12, 1970).
3. "Renovating Rocky Flats," *Science,* **174,** 1312 (December 24, 1971).
4. "Unwanted Water Additive," *Chem. W.,* p. 16 (October 10, 1973).

Storage of Radioactive Wastes in Kansas Salt Mines

Abstract

Early in 1970, the Atomic Energy Commission (AEC) decided to use an abandoned salt mine near Lyons, Kansas, as a repository for radioactive wastes from around the country. Many Kansans were against the move; they considered it unsafe and favored legislation to stop the project. In May 1972, the AEC announced that it had decided on an alternative to the underground waste repository in Kansas; the new plan called for design and construction of surface storage facilities for commercial atomic wastes.

Background

Since the beginning of the atomic age in 1942, the Atomic Energy Commission has had the problem of safely disposing of its lethal radioactive nuclear wastes.[1,2] The magnitude of this disposal problem grew significantly as more nuclear facilities were built and posed a major problem where nuclear power plants were developed. The recent proliferation of civilian commercial plants has created an urgent need for a new nuclear waste dump.[3]

By the early 1950's, the AEC began seeking permanent repository sites for these wastes. In 1955, under support of the AEC, the National Research Council of the National Academy of Sciences (NAS) completed a survey of possible repositories[3] for nuclear wastes which must "cook" for about 500,000 years before the radioactivity of the longest-lived plutonium isotopes is completely spent. The NAS report identified deep salt formations as the best storage location because of their seismic stability, compressive strength, ability to conduct heat, high melting point, and nuclear

344

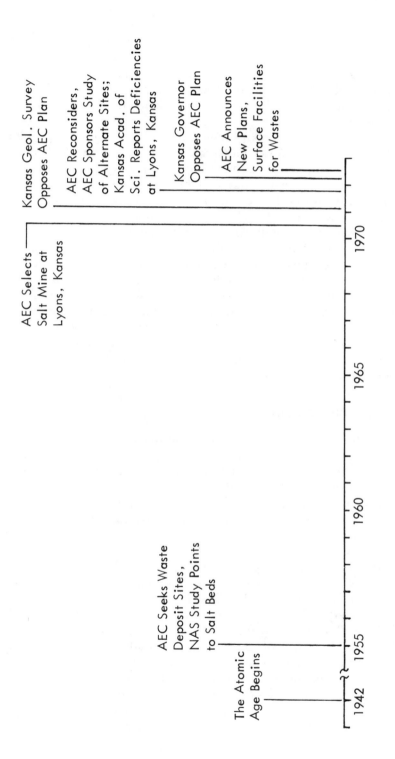

shielding property, which is similar to that of concrete. In particular
a Lyons, Kansas, salt field, three hundred feet thick and one
thousand feet below the surface, had emerged from studies as a
choice site.

On the basis of the NAS report, the AEC developed a plan for
radioactive wastes in salt formations. According to the AEC plan,
canisters of hot, solidified, high-level radioactive wastes ranging
from the size of firewood logs to eighteen feet long and two feet in
diameter, would be brought to the salt mine in railroad cars and
lowered down shafts into large rooms carved in the salt.[3] Then
drivers operating heavily shielded, motorized vehicles would use
remote controlled hoists to set the canisters into holes drilled about
twenty-two feet apart in the floor of the mine. After each vault had
received its complement of canisters, it would be filled in with salt.
It was believed that the pressure of the salt and the heat of the
cylinders (600° to 900°F) would generate sufficient plastic action to
move the salt in and around the containers to form a seal. The
steel-covered ceramic canisters were expected to disintegrate
within a period of six months to ten years, leaving the salt to hold
the lethal wastes in place.

Key Events and Roles

In June 1970, the AEC announced that, on the recommendation of two studies[3] (one made in about 1955 and a supplementary
study in 1970) by the National Academy of Sciences, it had tenta-
tively selected an abandoned salt mine* near the central Kansas
community of Lyons, as the nation's first underground nuclear
waste disposal site.[2-4] This waste dump would be a permanent
repository for solid, high-level radioactive matter imported from
commercial nuclear power plants, and for low-level (alpha radiation) wastes from the AEC's Rocky Flats, Colorado, plutonium
refining plant. This facility to be named the Lyons Nuclear Park,
would be designed to hold all of the nondefense atomic refuse
accumulated in the United States by the end of the century—a total
of about 770,000 cubic ft, or 38,000 tons.[3] An AEC spokesman
added that further geological and safety studies would be made,
and that three years would be required to prepare the facility.

The announcement caused immediate consternation in Kansas.
In August 1970, the Kansas State Geological Survey, headed by Dr.

*Owned by the Carey Salt Company.

William W. Hambleton of the University of Kansas, began studies of the surface geology, groundwater hydrology, and subsurface geology of a nine-square mile area around Lyons, Kansas, where the proposed AEC waste repository would be located.[2] The researchers obtained core samples from two deep drillings from opposite corners of the proposed 1,000-acre site. They also drilled some forty shallower holes.

The Kansas Sierra Club acted in February 1971 to make public a state geological report to Governor Robert B. Docking which was sharply critical of the AEC plans.[2,5] The report accused the AEC of giving insufficient consideration to several serious problems: possible migration of waste containers through the salt; possible thermal expansion of the salt which might cause overlying layers of rock to crack and allow water to seep in; and possible surprises caused by unforeseen radioactive interactions.[3] The report also criticized the AEC transportation plans as "completely inadequate" and pointed out that no emergency retrieval plan for the wastes existed. The report concluded that the AEC had "exhibited remarkably little interest" in the heat-flow problem and in studies of radiation damage.[2]

One of the more vocal critics of the AEC's plans was Representative Joe Skubitz (R-Kansas).[2] An exchange of letters with AEC Chairman Glenn T. Seaborg, failed to satisfy Skubitz' objections, and he sent a letter to Governor Docking citing the "paucity and unsureness of facts by those who were scientifically best informed." Skubitz urged officials to oppose "making Kansas an atomic garbage dump."

Dr. Seaborg responded to these charges with a point-by-point rebuttal[2] and stated that the Commission had no intention of burying wastes in the facility until all the pertinent data were available to support the safety of the operation.

The AEC position was presented at a hearing held by the Joint Committee on Atomic Energy in March 1971. All necessary data had been accumulated, the AEC said, and the only way to confirm the safety of the installation was through actual work on the site.[3] Statistics were cited to prove that the Kansas people had nothing to worry about hydrologically, geologically, thermally, or radioactively. The AEC asked the Committee to authorize a $3.5 million appropriation (for fiscal 1972) to begin work on the planned repository; this initial fund would cover the purchase of the site and architectural and engineering services.[2,3] The total cost was estimated at $25 million.

The five thousand residents of Lyons (who had been reassured at public meetings with AEC officials and stood to gain 200 permanent jobs manning the dump) were not apparently perturbed by the prospect of living in the country's nuclear waste capital.[3]

In June 1971, the AEC released its report on the environmental impact of the proposed demonstration project in Kansas for storage of the wastes.[6] Based on a draft of this report, officials of the Environmental Protection Agency (EPA), questioned several aspects: the long-term integrity of the salt formation; its ability to contain the radioactive material and prevent contamination of surrounding earth or of groundwater; contingency plans in case the radioactive material had to be removed; and procedures and responsibility for periodic evaluations and long-term environmental studies.

The AEC reported in October 1971[7] that it was considering alternate sites for the national radioactive waste repository. The AEC had found that the huge vault at Lyons might be in danger of flooding with water from nearby exploration wells or a salt mine using water extraction. Geologists had consistently maintained that the atomic burial area must be water-free;[8] if not, the salt might dissolve and allow radioactivity from the nuclear waste to move to the surface. The AEC reassessed its $25 million project after Governor Docking and nearly the entire Kansas delegation to Congress leveled charges that the AEC had been lax in setting up safety precautions for the proposed facility.[8]

The Attorney General of Kansas was even seeking in October 1971, to intervene in licensing hearings for two nuclear power plants outside the state, on the ground that waste from the plants would wind up in Kansas.[8]

Representative William Roy (D-Kansas) called on the Atomic Energy Commission in October 1971 to consider sites outside Kansas for its proposed radioactive wastes repository.[9] During the same month, the AEC announced that it had asked the Kansas Geological Survey to make recommendations on possible alternative sites in Kansas.[9]

In a report made public on October 30, 1971[10] the Kansas Academy of Science stated there were deficiences in the old Carey Salt Company mine making its safety doubtful for use as a national nuclear waste repository. The report concluded that thorough exploration for better sites was necessary, and that the search for suitable sites should not be limited to Kansas, nor even to the continental United States. The report observed that "overall consideration for the long-term safety and welfare of humanity must override the desire for short-term economic gains."

Governor Docking announced plans in March 1972[11] to ask the Kansas Legislature for approval of a resolution supporting his office in opposing the AEC plan to construct the Lyons atomic waste dump, and recommending that the AEC look beyond Kansas and the United States for a nuclear waste storage site.

Disposition

In May 1972, the AEC announced that it had decided on alternative action to an underground waste repository in Kansas.[12] The agency stated that it planned to design and build surface storage facilities for commercial atomic wastes. Surface storage facilities are in use at various AEC plants around the country. AEC Chairman James R. Schlesinger said the agency was surveying a variety of options for long-term storage of radioactive wastes and would proceed vigorously with a proven method—engineered surface storage—so that the public and industry would have complete confidence in the availability of adequate and safe facilities for handling wastes for the foreseeable future. The location of the facility was yet to be determined.

The AEC plans for a pilot project nuclear waste repository proposed for the 1979–1980 period were unsettled in January 1973. Dr. Frank Pittman, director of the AEC's waste management program, indicated that the Lyons, Kansas, area would not be used for this project and that three other sites in Kansas and one in southeast New Mexico (near Carlsbad) were being considered. Dr. Pittman stressed that "no decision would be made for a year and possibly two" on where the pilot plant would be located. He also suggested that the chosen pilot site would not necessarily be used for the final large-scale depository.[13]

Comment

The controversy over the storage of radioactive wastes in the Kansas salt mines at Lyons reflects an increasing problem of credibility which the AEC and several other agencies of government are experiencing. The inability of the AEC to prove its case in the face of critical opposition further justified, in many minds, this lack of credibility. On the other hand, the AEC could not possibly do what some people asked; that is, prove that nothing would go wrong at the burial site in a hundred or a thousand years. The controversy was never quantified in terms of future risks versus future benefits. The problem of radioactive waste disposal will continue to plague our society for many years.

REFERENCES

1. "Deposing of the Waste," *Sci N*, **97**, 312 (March 29, 1970).

2. "The Kansas Geologists and the AEC," *Sci N*, **99**, 161 (March 7, 1971).

3. "Nuclear Waste: Kansans Riled by AEC Plans for Atom Dump," *Science*, **172**, 249 (April 16, 1971).

4. *New York Times*, p. 11 (June 18, 1970).

5. *New York Times*, p. 27 (February 17, 1971).

6. "Radioactive Wastes—EPA and AEC," *Sci N*, **99**, 416 (June 19, 1971).

7. "AEC Seeks Alternate Sites for Radioactive Waste Storage," *A/W PR* (October 11, 1971).

8. "The AEC Has Trouble Salting Away Waste," *Bsns W*, 30 (October 16, 1971).

9. *Kansas City Times* (October 2, 1971).

10. *Kansas City Times*, p. 4B (October 30, 1971).

11. *Kansas City Star*, p. 30 (March 16, 1972).

12. "AEC Decides to Store Wastes on Surface," *Sci N*, **101**, 342 (May 27, 1972).

13. *Kansas City Star*, p. 7 (January 29, 1973).

Amchitka Underground Nuclear Test

Abstract

Amchitka Island, a small isolated island in the Aleutians off the coast of Alaska, was the site selected by the Atomic Energy Commission for its largest underground nuclear test shot (about five megatons) in the Spartan Missile Program. Controversy, both domestic and international, surrounded the need for the test. Despite fears of earthquakes and radiation leakage, the H-bomb was exploded on November 6, 1971. No apparent damage was done, and the AEC decided no further tests were required.

Background

In 1965, the Atomic Energy Commission (AEC) selected Amchitka Island as the site for Project Long Shot. This project provided for detonating a relatively small, eighty-kiloton nuclear device 2,300 feet underground. Its purpose was to provide nuclear detection information so that the U.S. Government's position could be strengthened in the nuclear test-ban treaty being negotiated with the Soviet Union. Since the Russians were against international inspections, the United States had to know whether or not nuclear tests could be detected seismographically, and distinguished from earthquakes. Amchitka Island was ideal for this type of test because it is in an earthquake-prone area.[1]

Naturalists and conservationists opposed the test since the near-extinct sea otter inhabited the area. Precautions were taken by the AEC to save the otters and the test was made on October 30, 1965.[2] No otters were harmed and the test was a success.

In 1967, the AEC decided that because they had geological data from Project Long Shot and the island is isolated, Amchitka would be a good place to conduct a new test program—the mul-

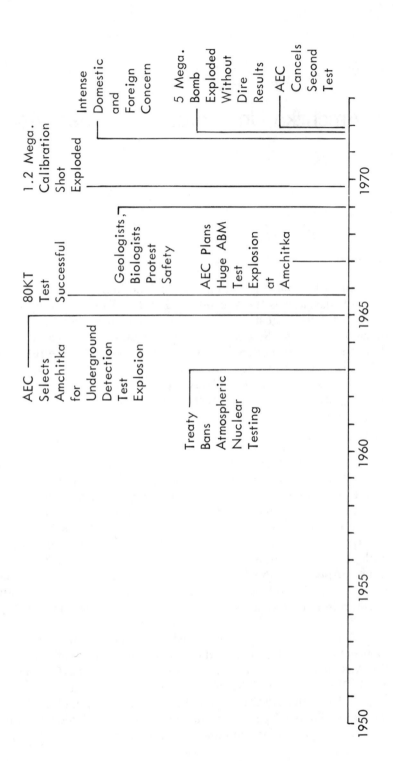

timegaton Spartan missile warhead to be used in the Safeguard Ballistic Missile Defense System. The previously-used test sites in Nevada and New Mexico were deemed unsuitable for detonations above one-megaton yield because of the direct ground shock effect on nearby inhabitants and tall buildings:[3] The Spartan warhead would yield up to five megatons, the largest weapon to be tested underground in U.S. history.

Key Roles and Events

The Amchitka test site was prepared by drilling three holes deep into the island's breccia rock foundation. The rock is so hard that up to a year was needed to drill each hole. The test program was divided into two parts. First, a one-megaton device was to be exploded as a "calibration shot" to determine if any adverse effects appeared before detonating the larger devices. The two remaining holes were to be used for two tests of the 5-megaton Spartan warheads themselves.

The one-megaton "calibration shot," designated Milrow, was scheduled to be tested in October 1969. The blast would occur four thousand ft below the island's surface and would be measured by numerous seismographic devices and monitored for radiation leakage.[4] Geologists and biologists in the U.S. protested that the scheduled test could result in earthquakes, tsunamis (earthquake-induced tidal waves) and destruction of wildlife. Japan and Canada filed protests with the United States, holding this country responsible for any damage caused by the test. The Senate Foreign Relations Committee held public hearings on the test. The AEC admitted earthquakes were a possible result of the test, but claimed that chances were small. The apparent need for the information to be gained from the tests as a vital part of the U.S. defense system won the debate. On October 2, 1969, the Milrow device, 1.2 megatons, was exploded. The AEC reported no waves or earthquakes, and found the site suitable for the larger Spartan warheads.

The AEC took two years to prepare for the first, five-megaton test, code-named Cannikin. They announced in April 1971 that the test would be conducted later that year. Again, Japan and Canada filed protests with the United States against the test, and opposition was strong at home, particularly from Alaska and Hawaii.

At the request of Alaska's Governor William A. Eagan, the EPA held public hearings in Alaska from May 16–19. The only opinions expressed in favor of the test came from AEC employees. On the other side, Jeremy J. Stone, Director of the Federation of American Scientists, challenged the necessity of the test. He said that the ABM program was changed in 1969 so that the Spartan warhead was now unnecessary and would be made obsolete by smaller warheads on the new Sprint interceptors and improved Spartans. In addition, he said "there is little doubt that the Spartan warhead will detonate; and much can be known about the warhead's effectiveness through paper and pencil calculations."[5]

A lawsuit was filed in Washington's U.S. District Court against the AEC on July 9 by the Committee for Nuclear Responsibility, a citizen's group, which claimed that the test would do irreparable harm to the environment; it asked the court to stop the test. Specifically, the suit declared that (1) the blast might cause earthquakes, contaminate the ocean, and leak poisonous debris into the air, (2) the debris could travel outside the United States, which would violate the Nuclear Test Ban Treaty of 1963, and (3) the AEC was in violation of the 1969 National Environmental Policy Act by not filing an adequate environmental impact statement on the test as required.[6]

Led by Senator Mike Gravel (Alaska) and Senator Daniel K. Inouye (Hawaii) amendments to delay or cancel the test were proposed in both the House and Senate. On July 17, the House defeated a bill to delete AEC funds for the Amchitka test, 70-27.[7] The Senate, on July 21, rejected an amendment to defer the test and approved funding of the test, 57-37.[8] However, the Senate did approve a move to forbid the Cannikin test until after March 31, 1972, unless direct approval was given by President Nixon.[9]

In August, a top-secret report was given to the Administration, which was reported to contain the recommendations of seven government agencies on the feasibility of Cannikin. News sources in Washington, D.C., although they did not see the documents, were able to determine that only the AEC and Department of Defense favored the test. The State Department was said to favor postponement until after the Strategic Arms Limitation Talks with Russia; the EPA and Council on Environmental Quality feared consequences to the environment from the test; and the Office of Science and Technology thought the test was of marginal value, since the new safeguard system did not use "heavy" weapons. A thirty-three-member Congressional delegation headed by Representative Patsy Mink (Hawaii) petitioned the U.S. District Court in Washington for public disclosure of the documents prior to Cannikin.[10]

The U.S. District Court ruled late in August against both the Committee for Nuclear Responsibility and the Congressional delegation in their attempts to halt the test.[10] An appeal by both to the U.S. Court of Appeals did not alter the decisions. However, the Court did rule that the AEC had to give the environmentalists the documents describing the potential environmental hazards of the tests. The Committee for Nuclear Responsibility appealed its case to the U.S. Supreme Court and hoped for a hearing.

The only remaining prerequisite for the AEC to conduct the test, which was scheduled for the first week in November, was President Nixon's approval. Pressure was directed at him from the Department of Defense and the AEC as to the necessity of conducting the test while, on the other hand, the possibility of both domestic and international ill will from a faulty test existed. His decision came on October 28, 1971, when he signed his approval for Cannikin.[11]

The first week of November was a week of protest. Canadian demonstrators blocked bridges into Michigan for several hours, stoned United States consulates, and threatened United States companies. Canadian newspapers were full of articles and cartoons denouncing the test. Peace groups demonstrated in front of the White House. One hundred members of the Canadian Parliament sent a letter of protest to President Nixon, and thirty United States Senators sent an eleventh-hour telegram to him urging him to call off the test. The Japanese Government once again denounced the test.[12]

The final dramatic event occurred on November 6, 1971, the day of detonation. On less than a day's notice, the U.S. Supreme Court convened in an unusual Saturday session with only one and a half hours to decide on the Committee for Nuclear Responsibility's lawsuit to halt the test. An hour after meeting the Court denied the injunction on a vote of 4–3.[12]

The five-megaton H-bomb was detonated six thousand feet below ground in a cavern fifty-two feet in diameter. The shock was measured at a magnitude of seven on the Richter Scale, and observers on Amchitka felt only a slight tremor.[12] The shock wave was registered in Japan also, and they filed an official protest with the United States. However, there were no earthquakes, no tidal waves, and the last shock wave was measured thirty-eight hours after the test.[13]

Disposition

The day after the test, the AEC announced that there was no reason for a second test and abandoned Amchitka as a future test site.[14] Later in November, they reported that the test was a success, and that no radiation entered the environment.[15] Canada also reported no radioactive fallout.[16]

None of the most dire results predicted by the Cannikin opponents occurred, although Alaska's Fish and Game Commission reported scattered deaths among the state's wildlife,[17] and AEC scientists reported 900–1,100 sea otters were killed.[18]

The lawsuit brought against the Administration by the 33-member Congressional delegation (to obtain the secret information pertaining to Cannikin) went to the Supreme Court. In March 1972, the Court agreed to hear the suit.[19]

Comment

The furor that was raised over the proposed Amchitka nuclear test was primarily a reflection of the loss of confidence and credibility in the Atomic Energy Commission and the Department of Defense by a portion of the American public during the 1950's and 1960's, and secondarily a reflection of the increased public concern for the environment during the late 1960's. That the test was conducted without dire consequences did not necessarily indicate that the public's attention had increased the AEC's precautions. Neither did the AEC's success in conducting the test have much effect in restoring the public's confidence in that agency.

REFERENCES

1. *Life*, **59**, 151 (October 15, 1965).
2. *New York Times*, p. 1 (October 30, 1965).
3. *Bsns W*, 140 (September 16, 1967).
4. *Science*, **165**, 773 (August 22, 1969).
5. *Science*, **172**, 1220 (June 18, 1971).
6. *Time*, **98**, 41 (July 19, 1971).
7. *New York Times*, p. 8 (July 17, 1971).
8. *New York Times*, p. 13 (July 21, 1971).
9. *New York Times*, p. 18 (August 1, 1971).
10. *Science*, **173**, 1004 (September 10, 1971).
11. *New York Times*, p. 1 (October 28, 1971).
12. *Time*, **98**, 15 (November 15, 1971).
13. *Time*, **98**, 84 (November 22, 1971).

14. *New York Times*, p. 1 (November 8, 1971).
15. *New York Times*, p. 11 (November 20, 1971).
16. *New York Times*, p. 96 (December 2, 1971).
17. *New York Times*, p. 75 (November 21, 1971).
18. *New York Times*, p. 49 (December 12, 1971).
19. *New York Times*, p. 1 (March 7, 1972).

The Dugway Sheep Kill Incident

Abstract

In March 1968, over six thousand sheep grazing on ranges near the U.S. Army Dugway Proving Ground in Utah sickened or died almost simultaneously of an unknown cause. Investigations revealed that the sheep had been killed by chemical warfare nerve gas accidentally released during tests made at the Army base. After much controversy the Army compensated the sheep-owners for their losses and initiated a program for tighter safety measures for testing gases and other chemical agents at Dugway.

Background

The Dugway Proving Grounds, a 1,500-square-mile military reservation near Tooele, Utah, has been used for many years as the U.S. Army's chief site for field testing of chemical and biological weapons.[1] The base is located in an isolated area—thinly populated Tooele County—and many of its 1 million acres are spread across the barren desert of western Utah. The Great Salt Desert covers the western half of the county and the eastern half contains a series of small mountain ranges—Cedar, Stansbury, Onaqui and Oquirrh. The Proving Grounds is at the edge of the desert and is about eighty miles southwest of Salt Lake City.[2]

The Dugway base has served the U.S. Army Chemical Corps as a testing station for chemical agents and weapons since about 1942.[2] In 1953, its activities were broadened to begin the first major testing of biological munitions.[3] The staffing of the base in 1968 included about one thousand civilians and six hundred military personnel. Most of the employees' families have quarters in the town of Dugway (population three thousand) located on the base near the southern tip of the Cedar Mountains. Sheep ranches are located in

358

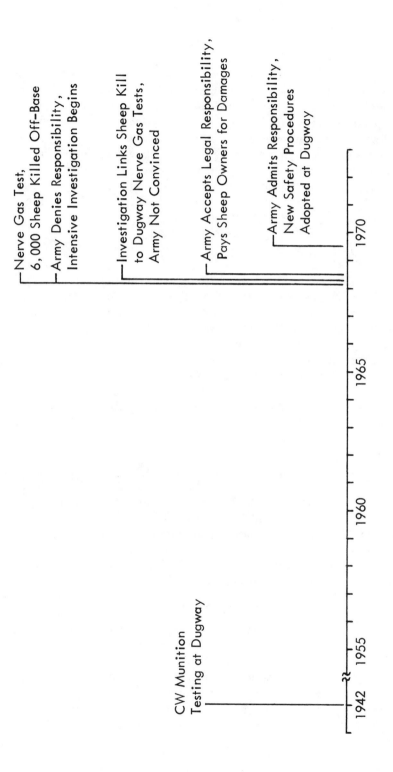

CW Munition
Testing at Dugway

Nerve Gas Test,
6,000 Sheep Killed Off-Base

Army Denies Responsibility,
Intensive Investigation Begins

Investigation Links Sheep Kill
to Dugway Nerve Gas Tests,
Army Not Convinced

Army Accepts Legal Responsibility,
Pays Sheep Owners for Damages

Army Admits Responsibility,
New Safety Procedures
Adopted at Dugway

1942 1955   1960   1965   1970

nearby valleys immediately to the west of the Proving Grounds. The nearest substantial settlement to the Proving Grounds is at Tooele (about thirty miles northeast of Dugway), a town of 9,000 on the eastern side of Rush Valley.[2] U.S. Highway 40 crosses the desert some thirty miles to the north of Dugway.

Security is tight around the Dugway Proving Grounds. Restrictions are designed to keep out unauthorized persons and to keep in the chemical and biological weapons and agents tested at the base.[2] The Army had, over the years, revealed very little of the kinds of activities they were conducting at Dugway.

Key Events and Roles

Beginning on March 14, 1968, over 6,000 sheep grazing in the Skull Valley pastures located about thirty miles east of the Dugway Army base were suddenly killed or sickened by a mysterious ailment which attacked their nervous system.[4] Many sheepherders in the pastures near Dugway discovered their sheep were either dead or in their death throes. The affected herds were located in a crescent extending generally east and northeast from Dugway. The symptoms they found in the animals still living were droopiness, wry necks, twisted spines and general muscular weakness.[1] The course of illness in sheep that died ranged from twenty-four hours to several weeks, and many of the ailing sheep were killed to end their suffering.

Veterinarians called to the scene had never before observed symptoms like these and were unable to diagnose the illness or help the sick animals. The veterinarians notified Dr. Lynn James, nutritionist, and Dr. Kent van Kampen, veterinary pathologist, with the USDA's poisonous plant research center at Logan, Utah. Their diagnosis was that a chemical intoxicant was affecting the nervous systems of the sheep ; they recommended contacting Dugway for information about recent tests at the Army Proving Grounds. Dugway scientists who traveled to the site shortly thereafter offered no suggestions for the diagnosis of the illness or for therapy.[2]

The widely publicized incident created much public concern and suspicions were immediately aroused that the sheep had been killed by some lethal substance originating at the Dugway test site.

There was much speculation by the public and by government officials concerning causes of the sheep deaths. Suggestions included a biological agent, an infectious disease, a poisonous plant, a toxic chemical from Dugway, or even a pesticide sprayed in the sheep grazing area. For a short time, rumors that a dread disease (such as Venezuelan Equine Encephalomyelitis) might have been accidently loosed in the area created a public scare.

The massive sheep kill attracted a swarm of investigators to Utah. By the end of March 1968 nine agencies were participating in the investigation: Dugway Proving Grounds and Edgewood Arsenal from the U.S. Army, the Department of Health and The Department of Agriculture from the State of Utah, the National Animal Disease Laboratory and the Poisonous Plant Laboratory from the U.S. Department of Agriculture, the Department of Ecology and Epizoology from the University of Utah, the Department of Animal Science from the University of California at Davis, and the National Communicable Disease Center of the U.S. Public Health Service.²

The Army's initial reaction to the news of the sheep deaths was to deny that Dugway had been doing any testing that could have caused the incident. However, this position was abandoned when the office of Senator Frank F. Moss (D-Utah) revealed on March 21 that three separate nerve agent operations had been conducted at Dugway on March 13, the day before sheep herders first noticed their sheep were ill.⁴ The Army had supplied this information to Senator Moss and apparently intended it to be "for official use only." But the senator's press secretary found no restrictive marking on the document and promptly released it to Utah newsmen.

Two of these nerve agent operations—a demonstration firing of 155-millimeter shells containing nerve gases, and a disposal activity involving the burning of about 160 gallons of a persistent nerve agent were ruled out as possible sources of the substance which killed the sheep. The third operation, however, in which a high-speed aircraft dispensed 320 gallons of a persistent and lethal nerve agent (VX)* in the form of liquid droplets from two pressurized spray tanks remained highly suspect.

Army officers† stated in March 1968, at an informal briefing for the Utah Congressional delegation, that the purpose of the aircraft spraying operation was to test a complete disseminating system, includingnerve agent, spray tanks and high-speed aircraft.⁴ The nerve agent itself had been released hundreds of times before at Dugway and, therefore, did not require further testing.

The Army was reluctant to discuss whether anything went

---

*VX is an extremely toxic organophosphate compound; the lethal dose for humans and animals is very small.

†Including Brigadier General William W. Stone, Jr., of the Army Material Command, who headed the Army's investigation.

wrong during the test, but three sources* who participated in the investigation, confirmed that there was, indeed, a malfunction. This malfunction apparently resulted in the lethal agent (VX) being released at a much higher altitude than specified for the test.[4]

Full details of how the malfunction occurred are not available in the open literature, but a version has been pieced together.[4] Army test plans called for the plane to dispense the nerve agent† from two tanks at an altitude of 150 feet above ground, then pull up and jettison the empty spray tanks.[1,2,4] In the March 13 test, one of the spray tanks apparently failed to empty itself over the grid target (apparently because of the malfunction of a release valve), and the agent was, therefore, accidentally dispensed after the plane pulled up.

Colonel Watts, the former Dugway commander, has been quoted as saying the lethal agent was released at a maximum altitude of about 1,500 feet.[4] Thus, the nerve agent was clearly within striking range of the ridge of the lower-lying Cedar Mountains, which stood (at an altitude of about 1,200 feet) between the sheep and the test site. General Austin W. Betts, chief of research and development for the Army, estimated that somewhat less than twenty pounds of the VX remained airborne after the test.[4]

At the time of the test, the winds at altitudes below 2,300 feet were generally from the south-southwest. Two hours later, however, the wind shifted and blew from the west. According to the Army, the "wind could have carried any very small particles of VX remaining airborne over the areas in Skull Valley and Rush Valley where sheep were later affected." Scattered showers developed during the early evening after the test (which occurred at 5:30 p.m.) and it was speculated that this rain could have washed the airborne VX out of the air and deposited it on vegetation and on the ground. The snow which fell the following mornings may have also carried the agent with it.

Investigators were initially puzzled that sheep seemed to be the only animals affected.[4] Cattle and horses, intermingled among the sheep, showed no signs of illness, though tests did indicate a slightly depressed level of cholinesterase in the blood—an indica-

---

*D. A. Osguthorpe, a veterinarian who acted as consultant to the Utah Department of Agriculture, G. D. Carlyle Thompson, director of the Utah State Division of Health, and Surgeon General Steward.

†The agent was dyed red so ground personnel could check its dispersion. A third ingredient was a diluent which gave the final preparation a consistency similar to motor oil.

tion the animals were exposed to an organophosphate compound, a class which includes the lethal nerve agents. Humans and dogs in the same area also seemed to be unaffected. Further investigation revealed that sheep are more susceptible to the VX agent than many animals and that the sheep had greater access to contaminated food than most animals.

The Army stated that the sheep were apparently affected by eating contaminated vegetation, and feeding experiments by the Agriculture Department's Poisonous Plant Research Laboratory in Logan, Utah, gave results which strongly supported this theory. When forage from the affected areas of Skull Valley was fed to healthy sheep, they showed a significant depression of cholinesterase activity (a sign of possible nerve agent exposure); some sheep developed symptoms identical to those observed in the sick Skull Valley sheep. In contrast, sheep placed in affected areas, but muzzled and fed only hay and water from outside sources, showed no signs of toxicity.[4]

Some investigators believed the sheep were sickened by licking contaminated snow, and since one laboratory identified traces of nerve agent in snow water, this remained a possible source of the poison.

During the investigation, toxic plants, pesticides and viruses were eliminated one by one as possible causes.[2] Laboratory tests were made on samples of incompletely digested food taken from the rumen and stomachs of dead sheep, as well as sheep tissue and blood samples. Lowered blood cholinesterase was found in sheep and to a lesser extent in cattle and horses.[2]

Several days after the illness first appeared, sheep were treated with atropine, an antidote for organophosphate poisoning. Sheep failed to respond, except to massive doses, and even then the recovery was only temporary.[2] Dugway personnel then began to feed VX to healthy sheep on an experimental basis; these sheep responded better than the range sheep to atropine treatment.

Army personnel explained at a hearing in the office of Senator Wallace F. Bennett of Utah on March 25, that prior to this experiment with feeding the VX agent there had been "meager information concerning its effects on sheep." Previous tests of the agent had been with laboratory animals.[2]

Finally, in late April, convincing evidence about the cause of the sheep kill came from chemical tests conducted by the National Communicable Disease Center (NCDC) in Atlanta, a branch of the Public Health Service. Chemists at NCDC used gas chromatography, infrared spectroscopy, and mass spectrometry in attempts to find traces of nerve agent in the dead sheep or in Skull Valley environmental samples such as grass and snow. In a report to Utah's health director, dated April 29, 1968, NCDC stated unequivocally that their chemists had found traces of the nerve agent VX. The report summary stated:[2,4]

- Skull Valley water and forage as well as blood and liver from ill sheep showed an agent which proved to be identical in chemical composition to a sample of the test agent (VX) supplied by Dugway.
- Rumen samples from ill sheep showed the same instrument response as the authentic test agent furnished in water by Dugway, that is, the hydrolysis products of the test agent.
- Gas chromatograms of the test agent, hay, and water extracts showed similar scans indicating identity of agents under study.
- Mass spectrometry of the test agent, hay and water isolates proved beyond doubt that these responses are in fact identical and can only be attributed to the same chemical.

The cause of the sheep kill had thus been established to the satisfaction of almost everyone but the Army, which continued to call the findings "inconclusive."[2] Army scientists initially analyzed several hundred samples of water, soil, snow, vegetation and wool from Skull and Rush valleys and found no evidence of VX. Only after very large samples of vegetation were analyzed did the Army conclude that the agent might "possibly" be present.[4] According to the Army, several thousand samples of environmental materials from a 100-square-mile area were analyzed by various agencies in the effort to determine what killed the sheep.

In addition to tests identifying the nerve agent itself, there was considerable evidence that the sheep were poisoned by an organic phosphate compound. These compounds found in nerve agents, some noxious plants, and many common pesticides interfere with the action of the enzyme cholinesterase at nerve endings, and a depressed level of cholinesterase is, therefore, considered a rather specific indication that an organic phosphate is involved. Numerous investigators reported a severe depression of cholinesterase in the blood of the affected sheep. Moreover, the Agriculture Department reported finding a cholinesterase depressing substance in snow collected from the area of the sheep kill. Since the investigation turned up no evidence that death was caused by poisonous plants or pesticides, the most probable cause among the organic phosphates appeared to be the nerve agent tested at Dugway.[4]

Additional evidence implicating the nerve agent came from feeding experiments at Dugway. Healthy sheep developed essentially the same symptoms as the sick sheep in Skull Valley when fed small doses of VX.[4]

The Army was still not convinced, however, that VX was the only cause of the sheep deaths. On November 25, the Army stated: "Although minute quantities of the agent were detected off-post, the results from these investigations have not provided conclusive eivdence that nerve agent, by itself, caused sickness or death in the sheep. The answer is still unknown and may never be determined. The evidence suggests a combination of factors or effects."[4]

The massive sheep kill also raised questions about the potential danger to human life in the areas surrounding Dugway.[4] It appeared that only good fortune had prevented the deaths of civilians in the area where the sheep were killed. For example, the Army acknowledged that it was conceivable that a strong wind might pick up some of the nerve agent from the soil and carry it to travelers on U.S. 40 about thirty miles north of the Dugway test site. Moreover, the town of Tooele is located only twenty miles north of pastures where sheep were affected by the nerve gas.[1]

Disposition

During the summer of 1968, the Army accepted legal responsibility by paying $376,685 to one rancher for the loss of 6,249 sheep (4,372 dead, 1,877 sickened) as well as a lesser amount to some Indians who lost a small number of sheep.[4] According to an Army letter to members of Congress, such compensation is proper, under the Military Claims Act, "where the Army's activities contributed to the loss." Despite this legal admission of involvement in the incident, Brigadier General John G. Appal, who had immediate command over Dugway at the time of the sheep kill, was quoted on December 6, 1968, as denying that the nerve agent caused the sheep deaths.[4]

After seventeen months of steadfast denial, the U.S. Army finally admitted responsibility for the sheep kill incident.[1] The admission came during a Congressional hearing in Washington during the summer of 1969: The Army admitted that nerve gas being tested at the Dugway Proving Grounds had killed more than 6,000 sheep in Skull Valley, Utah, in March 1968.

Starting in July 1969, a high-level advisory committee, headed by the Surgeon General, reviewed the safety of chemical testing at Dugway. All of the committee's recommendations for improved practices have been adopted by the Army. One very significant regulation required Dugway, for the first time, to establish a monitoring system to detect the entry of chemicals into the environment outside the Army base.

In 1969, several anti-CBW reviews and legislative measures were proposed by opponents of chemical and biological warfare programs. Most observers noted that this movement could be traced back to the sheep kill incident near the Dugway Proving Grounds.[6]

Comment

The public reaction to the Dugway sheep kill is probably well illustrated by a story which circulated thereafter; "The Army said, 'we didn't do it and we won't do it again.'" The sheep kill and subsequent events had a strong adverse effect on the credibility of the Army and probably on government agencies in general. It may have been a factor in the presidential decision to downgrade chemical and biological warfare.

REFERENCES

1. "How the Sheep Died in Skull Valley," *Farm J*, **93**, 34B (September 1969).

2. "The Wind From Dugway," *Environ*, **11**, 2 (January-February 1969).

3. Hersh, Seymour, M., *Chemical and Biological Warfare*, The Bobbs-Merrill Company, Indianapolis, Ind., 1968.

4. "Nerve Gas: Dugway Accident Linked to Utah Sheep Kill," *Science*, **162**, 1460 (December 27, 1968).

5. "6,000 Sheep Stricken Near CBW Center," *Science*, **159**, 1442 (March 29, 1968).

6. "CBW Secrecy Veil Pierced," *Chem Eng*, **76**, 64 (August 25, 1969).

The Nerve Gas Disposal Controversy

Abstract

In early 1969, the American public learned from a Pentagon news leak that the Army was planning to transport 27,000 tons of old chemical warfare munitions across country and dump them in the Atlantic Ocean. This revelation, in the wake of the Dugway Sheep Kill, in which six thousand sheep fell victim to an error in an Army nerve gas test, caused much public concern. Controversy centered not only on the danger of rail and ship transport of the lethal nerve agents, but also on possible deleterious environmental impacts that the gas could have on the ocean. No safe and rapid alternative disposal method was found, however, and a fearful country watched as the Army proceeded with its plan. The transport was achieved without incident and the dumping of the nerve gas weapons in the Atlantic was executed on August 19, 1970, without deleterious effects. As a result of the disposal controversy, legislation was passed in 1971 that banned future ocean dumping of any type of chemical or biological weapons.

Background

The use of toxic chemicals in warfare is not unique to the twentieth century: they were probably used before the advent of recorded history. As early as 431 B.C. the Spartans and Athenians made suffocating gases by burning pitch and sulfur. Greek fire, a highly combustible mixture of chemicals invented in 670 A.D., was used through the Middle Ages. America also saw early chemical warfare in incendiary arrows used by the Indians. In the Civil War in 1862[1] the Union Army suggested the development of artillery

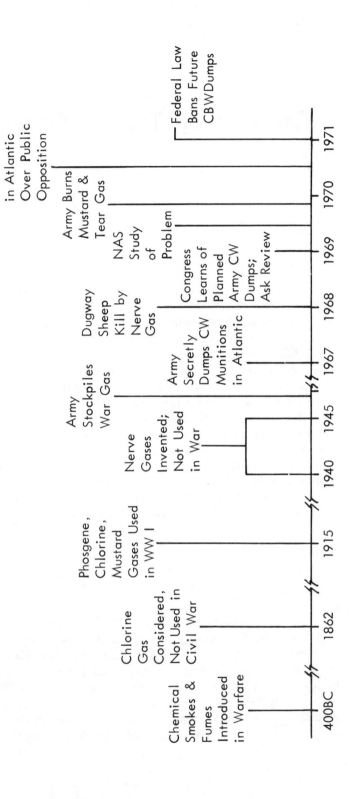

Chemical Smokes & Fumes Introduced in Warfare — 400BC

Chlorine Gas Considered, Not Used in Civil War — 1862

Phosgene, Chlorine, Mustard Gases Used in WW I — 1915

Nerve Gases Invented; Not Used in War — 1940

Army Stockpiles War Gas — 1945

Army Secretly Dumps CW Munitions in Atlantic — 1967

Dugway Sheep Kill by Nerve Gas — 1968

Congress Learns of Planned Army CW Dumps; Ask Review — 1968

NAS Study of Problem — 1969

Army Burns Mustard & Tear Gas — 1970

Army Dumps GB Rockets in Atlantic Over Public Opposition — 1971

Federal Law Bans Future CBW Dumps — 1971

shells containing chlorine, a recently discovered poison gas.* The government did not approve, however.

Modern chemical warfare has its roots early in World War I, when hundreds of chemical agents were synthesized and tested by Germany, England, and France. The Germans were the early leaders: on April 22, 1915, they dispersed chlorine gas from a cylinder in a highly successful attack and in 1917, they used the new mustard gas in an artillery bombardment. Most of the chemical agents tested in World War I, however, were eliminated as impractical. By the end of the war the only agents of significance were the military gases phosgene and mustard and the riot-control tear gases.

Immediately prior to World War II, insecticide research in Germany led to the discovery of a new type of poison, the organo-phosphorus-cholinesterase inhibitors or so-called "nerve poisons." The "nerve agents" are a group of highly toxic chemical compounds, generally organic esters of substituted phosphoric acid,[2] many of which are modern insecticides, but a few of which are military agents. The nerve agents are at least one order of magnitude more lethal than the previously known chemical agents. Sarin (GB), for example, is so lethal that 30 millionths of an ounce, inhaled or touched on the skin, can kill a 185-pound man.[3] Nerve agents are rapid acting (causing death in minutes as opposed to hours by other agents), attacking a vital enzyme, cholinesterase, so that the nerve signals to the muscles are disrupted. The "G" agents are all liquid under ordinary atmospheric conditions but possess a relatively high volatility permitting rapid dissemination in vapor form. In pure form, they are generally colorless, nearly odorless, and readily absorbable through the lungs, eyes, skin, and intestinal tract. Even a brief exposure may be fatal, with death occurring in one to ten minutes or delayed up to one to two hours depending on agent concentration. During World War II, Germany produced 10,000 tons of Tabun (GA) and pilot-plant quantities of Sarin (GB), but they were never used in the war: apparently Germany was convinced that the Allies also had nerve agents in addition to the older agents. Because of the Allies' policy of retaliation-use only, gas warfare was not employed during World War II.

Following World War II, the victors continued research and manufacture of chemical agents as the tensions of the cold war replaced those of the hot war. In the United States, carefully de-

---

*Chlorine was first prepared in 1774, but was not correctly identified and named until 1810–15. It was not produced commercially in the United States until 1892.

signed production facilities and storage and handling procedures controlled by strict safety standards were developed. By the 1950's the United States had developed a significant production capacity for the G-type chemical warfare nerve agents. In addition, an even more toxic type of nerve agent was discovered in England: the "V" agents (such as VX). These are also generally colorless and odorless liquids, but they do not evaporate rapidly at normal temperature and are thus even better suited to military use than G agents. By the early 1960's a capacity also existed in the United States for manufacturing V-type nerve agents.

The U.S. nerve agents themselves were incorporated in a variety of weapons, such as grenades, mortar shells, rockets, and artillery shells. Most of the agent weapons contained an explosive charge designed to disseminate the chemical and were stored in one thousand-pound bomb clusters, each containing seventy-six "bomblets" comprised of 2.6-pounds of GB and 0.5-pounds of tetryl burster (a high explosive).[4] The weapons were stockpiled in a few carefully guarded military facilities as the cold war continued.

But the stockpiles of nerve gas weapons were not completely a source of security and confidence: as time passed the older weapons became increasingly less reliable and some began to develop leaks as corrosion occurred. By the mid-1960's these aging weapons were posing a serious disposal problem for the Army's experts: they decided that the best bet was to dispose of them in the ocean.

In 1967–68, the Army secretly dumped three shipments of old nerve gas munitions into the ocean in a program known as CHASE (the acronym for "Cut Holes and Sink 'Em").[1,5] Three railroad shipments of nerve gas in containers (two of them from Anniston, Alabama) were loaded on surplus Navy ships, towed some 250 miles off the New Jersey coast and sunk. More than 21,000 rockets (M-55), each armed with an explosive and GB nerve gas, were disposed of by this method. In order to prevent the rockets from leaking their lethal gas during shipment, the Army first embedded them in concrete inside steel vaults or "coffins." The dumping was conducted without incident and no evidence of underwater damage was apparent.

The nerve agents were vividly brought to the public's attention in the late 1960's by two incidents: the famous 1968 Dugway sheep kill and a 1969 nerve gas "accident" in Okinawa. The sheep kill occurred near a major chemical and biological warfare testing area, Dugway Proving Grounds in Utah: an error in the testing of a nerve gas resulted in the death of over 6,400 sheep.[6] About a year later, twenty three American soldiers and one civilian were hospitalized at the U.S. Kadena airbase in Okinawa after accidental nerve gas "exposure."[7] These two incidents and the revelation that the Army had already secretly dumped one batch of nerve munitions in the ocean, set the stage for drastic public outrage upon learning of Army plans for a bigger disposal operation.

Key Events and Roles

Early in February 1969, Representative Richard D. McCarthy (D-New York) and his family were watching an NBC-TV documentary on chemical and biological warfare. Shocked and horrified by what she saw, Mrs. McCarthy sent her children from the room and asked her husband what he knew about the issue. His reply was not much. Consequently, McCarthy asked for a Pentagon briefing on CBW. When presented in March to nineteen congressmen and senators, the briefing did not offer much information but instead ended up pleading for money and public understanding. A month later, on April 21, in a major Congressional floor speech, McCarthy pointed out many discrepancies between the government's statements and actions. Instead of limiting chemical agents, the military had a stockpile of 100 million lethal doses of nerve gas in storage.

Then, in early May 1969, McCarthy learned from a Pentagon leak of an Army plan to dump large amounts of nerve munitions in the ocean. He promptly wrote Defense Secretary Melvin Laird and Secretary of Transportation John Volpe asking for a review of the plan's safety, consequences, and alternatives. McCarthy pointed out two recent cases of poison gas leaks in Kansas City and St. Louis, both of which involved transport of tanks of phosgene gas. The Army replied that the safest and cheapest way of disposing of old nerve munitions was ocean dumping. This was the fate they planned for 27,000 tons of nerve munitions located in four areas: Rocky Mountain Arsenal, Colorado; Anniston Army Depot, Alabama; Blue Grass Ordnance Depot, Kentucky; and Edgewood Arsenal, Maryland. The munitions would be encased in concrete for safety and then moved by rail to Earle, New Jersey. The "coffins" would then be loaded on four gutted Liberty ships, towed 250 miles out to sea (later changed to 145 miles), and sunk beyond the continental shelf in 7,200 feet of water,[4,8] where the concrete and metal would be slowly decomposed by the sea water. The exposed gases would then be quickly degraded.

The Army's explanation and reassurances did little to calm Congressional fears of a horrible accident. To help settle the controversy, the Army asked the National Academy of Sciences (NAS) to study the situation for feasible alternate disposal plans. In early July, the NAS submitted its report. In the case of the 12,643 tons of mustard gas, NAS suggested that instead of dumping it, the Army burn it. The Air Force's GB nerve gas bombs posed a far greater problem, however. These posed not only a grave hazard during transport (for example, by derailment or a sniper's bullet), but also in a possibly massive sympathetic detonation when the containers hit ocean bottom at an estimated speed of seventy miles per hour. Even if the transport and dump were executed without accident, what would be the ultimate effect of leaking gas on the marine environment? On the other hand, the bombs were not equipped with an explosive deactivation device that would enable easy removal of the GB. In fact, large numbers of these munitions had already been embedded in concrete "coffins." The NAS felt they were not qualified demolition experts and they could only suggest that the Army reevaluate alternatives such as land disassembly for the nerve agents. The NAS did approve ocean-dumping of two minor chemical munitions problems: 2,325 one-ton steel containers once filled with an unspecified poison, but now filled with contaminated water, and eighty-six fifty-five-gallon drums filled with concrete in which were embedded cannisters each containing eighty pounds of tear gas. In these cases, onsite demolition problems appeared to outweigh possible ocean contamination.[9]

An Army-appointed committee then suggested an alternative disposal plan for the GB rockets—destruction in an underground nuclear explosion. In late September 1969 the Atomic Energy Commission (AEC) submitted its reply to the Army on the feasibility of the proposed program (termed "Operation HARPIN"). Destruction of GB rocket vaults would require explosion of a nuclear device of about 100-kiloton yield. If the Yucca Flats Nevada Test Site were used, "HARPIN" would not only cost $3.5 million, but the August 1970 disposal deadline set by the Army would interfere with other AEC test site activities. For these stated reasons and probably more importantly the fear of adverse public reaction, the AEC refused in October 1969[5] to carry out "Operation HARPIN." The Army was left to fend for itself.

By spring of 1970, the rockets were becoming very unstable

and the need for disposal was becoming critical. The Army actively renewed its ocean-dumping plan and in July submitted an environmental impact statement on the plan to the Council on Environmental Quality as required by new legislation. The statement minimized the possibilities of major environmental impacts, but there were few field studies of past dumping for reference. Although the nerve gases were known to be decomposed by water, there appeared to be insufficient information on the rate at which they would be hydrolyzed (and thus detoxified)[5] by sea water at deep ocean pressures. However, the Army claimed that the gases would soon be destroyed and that there was little commercial value to fish found at the depth where the vaults would be deposited, anyway. The CEQ concluded that the potential hazards to the ocean environment were minimal, at least compared to the hazards of doing nothing—some of the nerve agents were being stored at Rocky Mountain Arsenal under the flight path of planes using the Denver airport. The CEQ did not oppose the Army's plan and Operation CHASE was soon on again.

During the summer of 1970 the papers were filled with news of the disposal program. Dozens of pictures showed the coffins as they waited in row after row to be loaded or were being placed aboard trains. Despite protests by cities that lay along the transportation route, a fight by the Environmental Defense Fund, and temporary injunctions, the Army began transporting their problem child to New Jersey. Daily news stories described the progress of the potentially catastrophic cargoes across country, aboard ship, and finally on August 19, 1970, to the very depths of the ocean.

Disposition

The entire nerve gas disposal operation—transport, ship loading, and burial at sea—was conducted without mishap. Ocean monitoring for nerve agents detected no leaks or explosions of the coffins after dumping and no evidence of a leak has yet been reported. In addition, the Army instituted an incineration program for disposal of its old mustard gas stocks. But the aftershock of the nerve gas disposal controversy continued. In September 1970, Representative Emilio Q. Daddario (D-Connecticut) introduced legislation that would give the CEQ jurisdiction to approve future ocean dumping of military materials.[5] In February 1971, Secretary Melvin Laird banned further ocean dumping of gas weapons[10] and in

November 1971 the Senate approved a bill prohibiting all future ocean dumping of chemical or biological warfare agents.[11]

Comment

The controversy over disposal of the nerve and mustard gases illustrates problems which can arise when new technologies are adopted with little anticipation or forethought about the possible future. If disarmament methods for the GB M-55 rockets had been considered when they were manufactured, the disassembly and detoxification would have been simplified. If the Army had not embedded the rockets in concrete, other options for disposal may have been open. Because of these oversights, the previous well-publicized accidents with nerve gas, and the military's penchant for secrecy, the public had little confidence in the military's judgment to transport the lethal cargoes across country and to dispose of them safely. Indeed, because of the news media coverage, the public almost expected a catastrophic accident before the disposal was completed. The operation was carried out with great care and safety, however, perhaps because of the controversy. The controversy certainly led to the ban on further ocean dumping, although its effectiveness in the nerve gas case has not been challenged.

References

1. "Chemical Warfare" in *Encyclopedia Britannica*, **5**, 353 (1960 edition).
2. "Nerve Gas 'Scare': A Look at the Facts," *US News*, **69**, 32 (August 17, 1970).
3. "Army is at Sea Over Nerve Gas," *Business Week*, 72, (June 7, 1969).
4. "Nerve Gas: Too Hot to Handle," *Sci N*, **95**, 499–500 (May 24, 1969).
5. "Nerve Gas Disposal: How the AEC Refused to Take Army Off the Hook," *Science*, **169**, 1296–98 (September 25, 1970).
6. "Germs and Gas as Weapons," *New Repub*, **160**, 13–16 (June 7, 1969).
7. "The Weapons Nobody Wants," *Time*, **95**, 53 (June 8, 1970).
8. *Kansas City Times*, p. 8c (May 8, 1969).
9. "NAS Suggests Modifications," *Sci N*, **96**, 26 (July 12, 1969).

four miles deep. The wires could then reflect radio signal fre-

10. *New York Times*, p. 26 (February 25, 1971).
11. *New York Times*, p. 26 (November 25, 1971).

Project West Ford—Orbital Belt of Needles

Abstract

Project West Ford was a United States Air Force program started in 1959. It involved an attempt to establish an invulnerable worldwide military communications system by orbiting a belt of tiny copper wires, capable of reflecting radio signals back to earth, at an altitude of two thousand miles. Astronomers around the world expressed concern that the orbiting needle belt, five miles wide and twenty-five miles deep, would interfere with radio waves from the stars and with optical astronomy. An international controversy surrounded the program, and it was halted in 1963 after two test shots.

Background

Project West Ford was a government-sponsored program proposed by a physicist from the Massachusetts Institute of Technology.[1] This experimental project was designed to construct a global communications network by ringing the earth with a belt of thin copper wires. Each wire was to be 0.7 inches long, and 0.001 inches in diameter (about a third the diameter of a human hair). About 350 million wires were to be placed in orbit around the earth, 2,000 miles up, in a band five miles wide and twenty-four miles deep. The wires could then reflect radio signal frequencies in the eight thousand-megacycle range back to earth.[2] The Lincoln Laboratories of MIT, West Ford, Massachusetts, conducted the project under the guidance of the Department of Defense.[3]

The idea originated in 1959 as a possible alternative to orbiting numerous expensive communications satellites. MIT proposed an

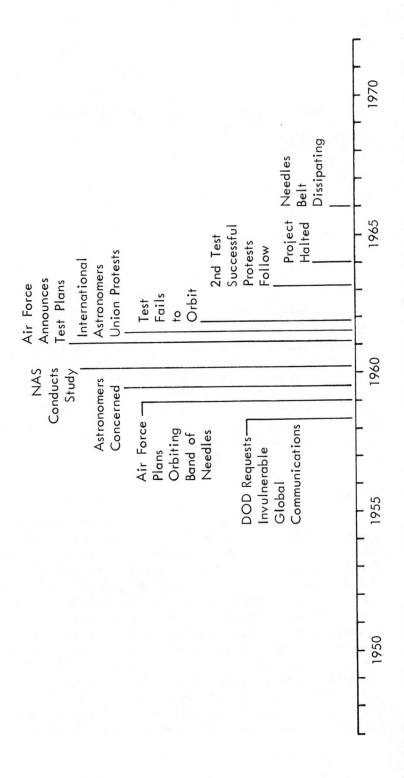

experimental test for the system and the Air Force accepted. Experimental work began at the Lincoln Laboratories that year on Project Needles (its original designation), but controversy over the program developed shortly thereafter.[4]

Key Events and Roles

Astronomers throughout the world expressed their concern when informed of the project. The major objection to the program came from radio astronomers, who feared that the orbiting needle belt would interfere with radio waves from the stars. In addition, the band of copper wires might interfere with optical astronomy as well.[5]

In order to resolve the differences between the Government and the astronomers, the Space Science Board of the National Academy of Sciences conducted a study from the fall of 1959 to June 1960. They concluded that the project would have no harmful effects on any branch of science, but warned that a large-scale operational system, put into effect if the pilot experiment was successful, would be a cause for concern.[3]

The National Aeronautics and Space Council announced that the upcoming test launch was experimental and not an operational system. They assured the astronomers that the experimental test would be completely evaluated before attempting any operational system, and that interested scientists would have a chance to review and comment on the test.

The Air Force announced that the test would be held in the latter part of 1961. At the International Astronomers Union Meeting held in August of that year, a resolution was passed by one thousand members representing over thirty countries. The resolution, proposed by the Belgians, asked the Air Force to delay its launching of the needles into space until scientific research could establish that the oribiting belt would not impair astronomical observations, and the expected lifetime of the orbiting needles could be determined.[6] MIT reported that the expected lifetime of the needles in space was 3–5 years and the President's Science Advisory Committee further assured the astronomers that no harmful effects would be produced by the experiment.

On October 21, 1963, the needles were launched "piggyback" on the Midas IV satellite. The satellite was successfully orbited, but the seventy-five-pound package of needles, imbedded in a

seventeen-inch long naphthalene cylinder, were lost.[7] The expected orbiting belt did not form and the package could not be
located with radar. It was later concluded that the needle package
was not given the proper spin when ejected, and the needles did
not disperse.[8]

The Air Force tried it again on May 12, 1963, and succeeded.
This time the launch was not announced ahead of time so that
another failure would go unnoticed. The needles formed a belt ten
miles deep and ten miles wide which encircled the earth.[9] The
USAF termed the experiment a success later in May after conducting several tests.[10] Astronomers once again called for a halt to
further experiments after five telescopes spotted the orbiting needles in space.[11]

Disposition

No additional experiments were conducted and an operational
system did not evolve from Project West Ford. In the light of international controversy surrounding the program over the issue of the
right of any nation to clutter up space the United States unilaterally
halted further testing.

In 1966, Dr. I. I. Shapiro reported that half of the needles
orbited in 1963 had reentered the atmosphere and the remainder,
held in clumps by electrical charges, should reenter within two
years.[12] The proposed band of needles gave way in communications
planning to the orbiting satellites it was initially supposed to replace.

Interpretation

Project West Ford, unlike Project Sanguine*, posed little
hazard to human health or to the terrestrial environment. It did,
however, pose a potential nuisance to a small group of elite scientists. This well-organized group was able to articulate an effective
(and even international) protest which finally brought the program
to a halt. The protest resulted in the utilization of alternative
methods that did not pose the same interference problems to accomplish the same communications goals. To many persons, Project West Ford was merely another governmental and military
"grand plan" which they could just as easily do without.

*See "Project Sanguine—Giant Underground Transmitter."

REFERENCES

1. "With Whiskers to Feel," *Newsweek*, **57**, 61 (January 23, 1961).
2. "Orbiting 'Needles' Belt," *Sci NL*, **80**, 134 (August 26, 1961).
3. "Needles," *Science*, **134**, 460 (August 18, 1961).
4. "Bouncing Radio Waves Off Orbiting Belt of Wires," *Bsns W*, 71 (October 7, 1961).
5. "Worrisome Wires," *Sci Am*, **203**, 80 (October 1961).
6. "Pondering the Riddles of the Universe," *Bsns W*, 90 (September 2, 1961).
7. "Whisk Went the Whiskers," *Newsweek*, **58**, 67 (December 11, 1961).
8. "The Needles Again," *Science*, **136**, 247 (April 20, 1962).
9. "Russians Blast Away at U.S. Project That Put Copper Needles Into Orbit," *Bsns W*, 98 (June 15, 1963).
10. *New York Times*, p. 6 (May 25, 1963).
11. "Needles Bar Space View," *Sci NL*, **84**, 163 (September 14, 1963).
12. *New York Times*, p. 38 (December 17, 1966).

Project Sanguine—Giant Underground Transmitter

Abstract

In February 1969, Wisconsinites first learned that the United States Navy had for nearly ten years been secretly planning the installation in their state of a 21,000 square mile gridwork of underground electrical cables as a global communications transmitter. Project Sanguine, as it was called, created a storm of protests from local citizens, conservationists, and some antiwar activists, and raised critical scientific question of its technical feasibility and budgetary questions of its estimated $1.5 billion cost. By 1972, the Navy had reduced the size of Sanguine and had apparently dropped the Wisconsin site in favor of a Texas location if indeed Sanguine was to exist at all.

Background

In the late 1950's, the U.S. Department of Defense (DoD) agencies became interested in global communications networks, utilizing orbiting satellites or other means. (See "Project West Ford—Orbital Belt of Needles.") The Navy had one unusual need: an invulnerable, jam-free transmitting system to give at least one-way communication to deep-running nuclear-powered submarines carrying Polaris or Poseidon missiles, to instruct the submarines to retaliate to a nuclear attack on the United States. Conventional communications systems at the time (using high-power transmission from massive aerial structures) were vulnerable to physical attack, were susceptible to jamming (particularly by aerial nuclear explosions which ionized the atmosphere between the transmitter

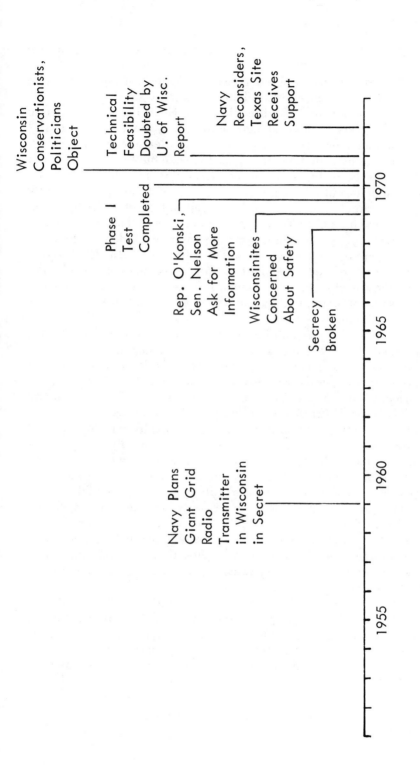

and receiver) and suffered considerable loss of signal strength when the electromagnetic radiation entered water.

To circumvent these difficulties, the Navy, in 1959, decided to develop a very long wavelength or extremely low frequency (ELF) communications system in which the electromagnetic energy (wave) was launched into the space between the earth's surface and the ionosphere, which begins at a height of about fifty miles. The ELF signal would circle the earth, but part of it would be constantly leaked (radiated) back to earth. The advantage of an ELF system over conventional transmitting frequencies was that signals below 100 cycles per second (or hertz) can penetrate the depths of the sea without appreciable attenuation. The difficulty with an ELF system was that it required a transmitting antenna comparable to or longer than the wavelength of the signal. For the frequency of interest to the Navy, 45 hertz, the wavelength would be 6,660 kilometers or over 4,100 miles.

The Navy's program for an ELF communications system was designated Project Sanguine. As initially envisaged by the Navy,[1-3] the antenna would consist of 6,000 miles of cables laid in an array of interconnected grids, with each cable separated by six miles from adjacent parallel cables; the entire grid would be 150 miles long by 140 miles wide (21,000 square miles). The cables would be buried six feet underground and would be electrically powered by 240 separate transmitter stations to provide survivability against sabotage or attack. Electrical power requirements would be high, not only because of loss of signal strength in the water, to the ionosphere, in the earth covering the cables, and in the half of the signal radiated downward, but more importantly because of direct electrical losses from earth conductivity (the cables would carry 14,000 volts), the grounding of each cable at its end points,* and the use of a modulated signal.† The Navy apparently estimated[2] the power requirements at thirty megawatts, about that of a community of perhaps 15,000 people.

In selecting a site for the transmitter grid, the Navy wanted an isolated area where the power requirements could be minimized and a location near the middle of the North American land mass for defensive purposes. Northern Wisconsin appeared ideal:

The Laurentian Shield extended under the area; that ancient

---

*A technical requirement.

†The message would be impressed as a small electric current on a larger carrier current.

dry-rock formation would reflect much of the downward half of the signal and would minimize surface conductivity losses.

• The area had no mountains to deflect the signal or to make installation difficult.

• The area was sparsely populated without large industrial or urban areas; the Chequamegon National Forest constituted much of the area of interest.

By 1963, the Navy had decided that research indicated that the ELF system was feasible and Project Sanguine was planned in three phases.

Phase I would consist of four years of preliminary research projects, culminating in the installation and test at Clam Lake, Wisconsin, of two fourteen-mile-long aboveground antennas, which were to be powered by a transmitter at the point where they crossed. Phase I eventually cost $27.5 million.

Phase II would consist of an underground grid test.

Phase III would consist of the installation of the grid network under most of the northern one-third of Wisconsin. (The land over the grid would be reusable.) The original schedule called for completion of Sanguine by 1976, and the total estimated cost (based on a somewhat reduced level of effort in 1968–1969) was about $1.25 billion.[1]

Key Events and Roles

Project Sanguine was classified "secret" by the Navy, and for nearly ten years the public was unaware of its existence. In 1967, preliminary research by Radio Corporation of America in North Carolina had found, in a pilot test run, that a surface ELF transmitter could cause nearby electrical disturbances (such as making telephones ring and railroad crossing lights blink, disrupting TV, and charging metal fences),[1,4] but this appeared to be a problem which could be worked out. Concern also existed over potential radiation hazards (sterility had reportedly occurred in some military operators during the early days of radar). By 1968, as preparations started for the Clam Lake test, papers in Milwaukee, Ashland, and other Wisconsin cities carried stories of the unusual activities, and rumors of a mysterious project circulated. In October 1968, *Electronic News* revealed[5] that electronics manufacturers were seeking contracts from the navy on a billion dollar Project Sanguine. The first official word on Sanguine to reach the public, however, appears to have come[1,6] in early 1969, when Alvin E.

O'Konski, U.S. Representative from Northern Wisconsin and a long-time member of the House Committee on Armed Services, revealed to some of his constituents that the project was about to bestow an economic windfall on their area[1] (an estimated ten thousand jobs[4]).

Not only was this the first word that the public heard of Sanguine, but it was also the first that Wisconsin's Senator Gaylord Nelson had heard.* Nelson immediately asked and received from the Pentagon a briefing on the classified project in his home state. He was not only surprised at the scope of the project, but was incensed when his staff noted that slides of clippings from eight related stories from Wisconsin newspapers (including the *Milwaukee Journal*) and the *Electronic News*[1,7,8] had all been stamped "Secret" by the Navy. The Department of Defense quickly stepped in to halt an incipient uproar by announcing that published news stories could not be classified.

Publicity over the news story classification flap prompted the Navy to disclose the first technical details† of the program to the general public.[4,9] The Navy also disclosed[4,7] that it had awarded a $1.6 million contract to RCA to devise methods of preventing interference with power lines and fences, and a $700,000 contract to Bell Laboratories to find ways to suppress telephone interference, and that a $175,000 contract was being awarded (Hazelton Laboratories was the recipient) for a study of what problems, if any, the ELF radiation might present for humans, animals, or plant life in the vicinity of the grid. In addition, to allay the concerns of local citizens, the Navy started showing locally in early 1969 a twenty-five-minute color film about the project, which pointed up the economic benefits, and planned a July public briefing in Ashland, Wisconsin, at the invitation of local political leaders.

The briefing turned out to be a disaster from the Navy's viewpoint: about half of the attendees were conservationists who had far more questions on the potential damage of the electromagnetic field to human health, animal and plant life, migratory birds and even earthworms, than on the technical details of the giant project. The Navy spokesman could not answer the environmental and conservation questions (the on-going Hazelton Labs project was the only related research) and, because plans were incomplete could not satisfactorily answer technical questions pertaining to power

---

*He was Governor of Wisconsin at the inception of Project Sanguine.

†The size of the gridwork was reduced to 100 miles long and 70 miles wide at this time.[4]

requirements ("Would an atomic power plant be required on Lake Superior for the network?"), the miles of wire to be buried (erosion was a concern), or the electric shock potential of fences, guardrails, and dogs chains.

The net result of the briefing (other than the transfer of the Navy spokesman) was that much of the local citizenry, the conservationists, influential papers such as the *Milwaukee Journal,* and a few antiwar activists* were solidified in their opposition to Project Sanguine. As the fall test date for Phase I neared, local concern grew—a State Committee to Stop Sanguine was formed†—and the national attention increased.[10-15] Secretary of Defense, Melvin Laird (formerly a Congressman from Wisconsin) was told of Wisconsinites' feelings and was reportedly blamed for Sanguine.[14] Wisconsin Senator William Proxmire, and Representatives Henry Reuss and Robert Kastenmeier began to lobby against Sanguine.[14] In addition, estimates of the power requirements for Sanguine had gone up sharply to eight hundred megawatts‡ probably requiring a nuclear power plant of its own—and cost estimates were reaching $1.5 billion.[14]

Disposition

In the face of the mounting pressure the Navy regrouped. On October 24, 1969, Deputy Secretary David Packard was said[12] to be reviewing Sanguine, and on November 7 the Navy announced a "research breakthrough" of more efficient methods of ELF communications that eliminated the need for the huge grid system in Wisconsin and could permit a smaller and lower power transmitter elsewhere.§ The Navy said it would continue research for six to twelve months on the potential new designs, and that only $6–$7 million[16] would be spent during the year.‖

This announcement tended to reduce the concern, but did not by any means eliminate it. The publications of the results of the Hazelton Labs at the end of 1969 did little to relieve uncertainty over the potential hazards to man, animals, or wildlife.

*The University of Wisconsin was a center of violent protest during this period.

†Kent Shifferd, professor of history at Northlands College, Ashland, was chairman.

‡Other estimates as low as 2.4 megawatts for a smaller Sanguine system also appeared.[15]

§Such as in isolated areas of Washington, Oregon, California, Idaho, Texas, Georgia, North or South Carolina, and even New York.[16]

‖Secretary of the Navy John Chafee was also quoted[1] as having Defense approval for expenditures of $20 million on Sanguine.

Project Sanguine virtually dropped from the national news during 1970, but independent studies were underway. In May 1971, Senator Nelson described Project Sanguine to the Senate and introduced results of analyses by University of Kansas and University of Wisconsin scientists, both of which questioned the technical feasibility of the system.[2,3] The published[2] results of the Wisconsin group's study[2] concluded that Sanguine was not technically feasible: it would be too slow and would require enormous power input to send reliable messages in a reasonable time. The report estimated that three nuclear power plants would be needed to transmit the simple message "Fire Missiles" in a one-minute period; and it would be easily jammed by lower-power surface transmitters. At Senator Nelson's recommendation, an *ad hoc* committee of the National Academy of Sciences—National Research Council also studied Project Sanguine, including apparently the Navy's still classified documentation.[3] The NAS-NRC panel split: the majority concluded that Sanguine would work substantially as the Navy anticipated, but added some qualifying recommendations for needed research. The conflicts over the technical feasibility have been reviewed.[3,19]

As a result of the attack from environmentalists and others, the Department of Defense was urged by Wisconsin Representative Les Aspin to drop Sanguine in favor of an alternative such as an airborne very low frequency (VLF) system[17] and Senator Nelson moved in September 1971 to cut $5.5 million in research and development funds from the Sanguine budget, noting that Sanguine had already cost taxpayers $50 million.[18] In addition, he was concerned that the Navy still had over $2 million earmarked for investigation of ultra-hardened sites in which the cables would be buried in tunnels several thousand feet deep.[18]

The Navy again fell back and regrouped. In July 1972, the Navy indicated it was considering a 1,600 square mile grid transmitter—one which it says would require only the amount of power needed for a community of three thousand persons.[19] The Navy asked Congress for another $12 million in 1973 and estimated the total cost of the reduced system as $0.75 billion. While Wisconsin was not completely ruled out, a central Texas site (avidly supported by Texas Senator Lloyd Bentsen) appeared favored by the Navy until citizen outcries there again required a reevaluation. By early 1974, the Navy was considering a site in the Upper Peninsula of Michigan—again in the face of expressed anxiety from local citizen's groups and legislators.[20] The future of Sanguine is most uncertain.

Comment

The intense public concern that the revelation of Project Sanguine generated resulted from a combination of circumstances including: the belated discovery that it had long been planned in secret; the immense size and cost of the installation; the weariness over the Vietnam War and the distrust of the military and the federal government that the war stimulated (especially among activist groups in Wisconsin); and the rising conservationist tide. The Navy's response to the incipient protest (staging an all too transparent public relations effort instead of furnishing answers to the technical questions asked) solidified the opposition and furthered the public's distrust. The technology was sufficiently complex that neither it nor its potential side effects were well understood by the public, nor apparently by the Navy.

REFERENCES

1. Laycock, G., "Not All is Sanguine in Wisconsin," *Audubon*, **72**, 104 (January 1970).
2. McClintock, M., P. Rissman, and A. Scott, "Talking to Ourselves," *Environ*, **13**(7), 16 (September 1971).
3. Wait, J. R., "Project Sanguine," *Science*, **178**, 272 (October 20, 1972); with correspondence, *Science*, **180**, 1300 (June 22, 1973).
4. "Wisconsin Gets Wired For Sound," *Bsns W*, 116 (April 26, 1969).
5. "Six Seek Billion $ Navy Pack," *Elect N*, **13**, 1 (October 14, 1968).
6. *New York Times*, p. 10 (February 12, 1969).
7. Connolly, R., "Navy Moving on Project Sanguine," *Elect N*, **13**, 34 (February 17, 1969).
8. *New York Times*, p. 37 (February 13, 1969).
9. Hamilton, A., "Wiring Wisconsin," *New Repub*, **160**, 8 (March 22, 1969).
10. "Project Sanguine Alarms Wisconsinites," *Ind Res*, **11**, 51 (September 1969).
11. *New York Times*, p. 49 (October 14, 1969).
12. *New York Times*, p. 26 (October 24, 1969).
13. *New York Times* (November 7, 1969).
14. Gruchow, N., "Project Sanguine Short-Circuited," *Science*, **166**, 850 (November 14, 1969).

15. "Still Sanguine," *New Repub,* **161,** 13 (November 29, 1969).
16. "Cut Project Sanguine Price," *Elect N,* **14,** 26 (November 10, 1969).
17. "Scrub Project Sanguine: Use Alternatives, DoD Urged," *Elect N* **16,** 26 (August 9, 1971).
18. "Senator Nelson Asks R&D Funding Cut for Sanguine," *Elect N,* **16,** 6 (September 20, 1971).
19. Crossley, M., Universal Science News, Inc., in *Kansas City Star,* p. 4D (July 26, 1972).
20. *Kansas City Times,* p. 6D (March 13, 1974).

Project Able—The Space Orbital Mirror

Abstract

In July 1966, the National Aeronautics and Space Administration revealed that, with the Department of Defense, it was conducting feasibility studies for an orbiting space mirror. In Project Able, as it was called, the mirror would reflect sunlight to illuminate a circular area of 220 miles diameter, on the dark side of the earth, to nearly twice the brightness of a full moon. The mirror would be controlled from the ground and could be used for military and civilian uses. News of the project generated much concern by astronomers. After a study of the potential effects and costs by a National Academy of Sciences committee, Project Able was dropped.

Background

The origins of the concept of a space orbital mirror are not entirely clear.[1-4] During the early and mid-1960's, the Department of Defense had encountered numerous problems because of the clandestine guerilla nature of the Vietnam War and the extensive use of night transportation of men and supplies southward by the North. The military had developed over the years a number of techniques for night illumination of terrain. For Vietnam the Department of Defense had developed a system whereby an airplane at 12,000 feet could use a spotlight to illuminate a two-mile diameter spot on the ground, four times brighter than could the full moon. During those same years the National Aeronautics and Space Administration was developing its capabilities for putting men and equipment into orbit around the earth and for the moon voyages, but at the same time NASA had an extensive program designed to identify and publicize civilian benefits of the space program.

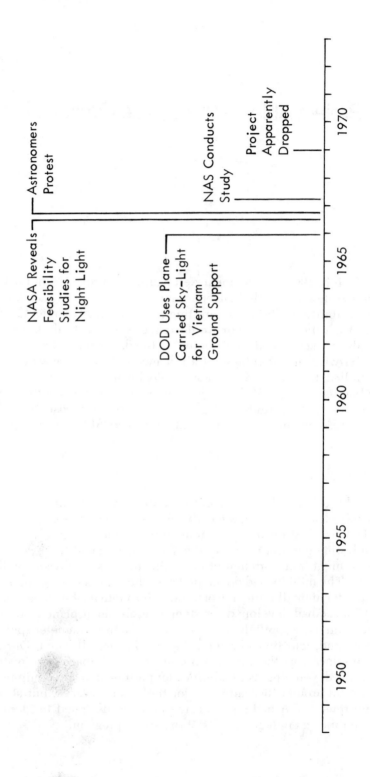

Somewhere in the mid-1960's the idea was conceived to put one or more big lights in earth orbit to illuminate selected parts of the darkened areas of the earth at night. Whether military or civilian interests were actually primary is uncertain, but the Department of Defense and NASA apparently started planning quietly for a gigantic space orbital mirror which could reflect sunlight to illuminate spots on the back side of the earth on command at night. Thus was born Project Able.

The DOD-NASA concept of Project Able called for a satellite constructed of aluminized mylar film which would be inflated in space into a round flat mirror two thousand feet in diameter. It would move in a synchronous orbit 22,300 miles above the earth and would be capable of illuminating an area 220 miles in diameter with a light intensity 1.7 times as bright as the full moon. Controlled from the ground, this mirror could reflect light to selected points on the night side of the earth and three such mirrors could provide continuous around-the-world capability.

Key Events and Roles

The first public news on Project Able appears to have been carried in a July 1966 Associated Press story out of Seattle, where the Boeing Company, a major aerospace contractor, was located. The AP story received limited attention by the newspapers. One of the newspapers which did carry the story was the *Courant* of Hartford, Connecticut, where a keen sensitivity existed because a local controversy had erupted over the installation of new city street lights. Astronomer Edgar Everhart (physics professor at the University of Connecticut; discoverer of one comet and co-discoverer of another) had contended that they were too bright and cast a sky glow which interfered with celestial observations. The first adverse public reaction to Project Able is indicated in the *Courant's* headline, "But Who Needs Sun at Night."

On August 15, the *Courant* carried a second story, based on information from Cape Kennedy, Florida, that G. T. Schjeldahl Company had received a $40,000 contract to develop a satellite package which would unfold in orbit to give a two thousand-foot diameter reflective dish which could illuminate half the breadth of Florida at night. The dispatch stated that such satellites could be used for exposing enemy positions, for search and rescue operations, for security purposes, as navigation beacons, and for radio astronomy experiments.

Professor Everhart was greatly concerned about the possible interference of the space orbiting mirror on astronomy. Prompted by Everhart, *Sky and Telescope* magazine inquired of NASA and learned that contracts of about $125,000 each for feasibility studies under Phase A of Project Able had been awarded to Boeing and Westinghouse Electric Company by NASA and DOD. In October 1966, *Sky and Telescope* published an editorial (under the *Courant* headline above) which reviewed these developments and carried Professor Everhart's objections.[1]

The news of Project Able raised concern among many other astronomers and also among other scientists. Thus, in late 1966, the House Committee on Science and Astronautics held a hearing on Project Able. A NASA spokesman described the technical details of Project Able and exhibited a drawing which showed the space orbital mirror lighting up the Washington, D. C.-New York City area. NASA also revealed that five companies* were now evaluating the mirror concept, under contracts totaling $490,000, and in addition that the synchronous earth orbit program, of which the mirror proposal was a part, was expected to become operational by mid-1970. In general, NASA played down the potential military applications and emphasized the potential civilian benefits of the night light.

In contrast, *Time* magazine, in the first major popular press story on Project Able, illustrated the concept with a drawing which showed the mirror lighting up a section of Vietnam.[2] In January 1967, both *Time*[2] and *Science*[3] carried stories discussing the pros and cons of the proposed space orbital mirror. Some of the benefits suggested by NASA included:

- Potential use in Vietnam to limit night infiltration;
- Aid to search and rescue activities and recovery operations;
- The lighting of blacked-out cities;
- Increased lighting during the winter night in polar latitudes;
- A visual navigation beacon; and
- A very large radio antenna for radio astronomy after suitable modification.

Objections to the project by Everhart and others[1-3] included:

- That it would hamper astronomy in adjacent areas: the United States did not have a right to hamper astronomers in other countries.
- That launching the first one might lead to a proliferation of orbiting mirrors which would impair astronomical observations over much of the earth's surface.

* Grumann Aircraft and Goodyear Aerospace were the other two.

• That it could swing out of control and reflect sunlight indiscriminately; the mirror could not be repaired in space or brought down.

• That it would disturb the delicate circadian rhythms that control many life processes in plants and animals.

• That mirrors would soon be damanded for nonessential purposes, such as plowing fields at night or lighting lakes for resorts.

• That NASA had no business in military operations; putting even one mirror in orbit would violate the recent treaty banning the use of space for warfare purposes.

In the face of these objections, NASA tried to reassure the astronomers by telling them that a reflecting satellite of the type contemplated would not interfere with astronomy. NASA also insisted that there had been no decision to orbit such a solar mirror, and no judgment had been made that it would be a worthwhile project.

Disposition

Because of the increasing concern among the public and astronomers, the National Academy of Science's Space Science Board commissioned its Committee on Potential Contamination and Interference from Satellites to begin a study of Project Able to ascertain any harmful effects on science. This committee had been created to study Project West Ford.* Charles Tounes, chairman of the NASA advisory committee, said they would rely on the NAS committee's recommendations and deferred any further study until that committee reported its findings.

The NAS committee chairman, John W. Findlay of the National Radio Astronomy Observatory, was reported[4] in April 1967, as saying that the project was too impractical and too expensive. He added that the orbiting mirror would require a Saturn V booster and there wasn't really any demand for the mirror.† Project Able was not officially killed at that time, but no further work was planned and it was apparently dropped quietly by NASA and the Department of Defense.

*See "Project West Ford—Orbital Belt of Needles."

†There was apparently never the firm commitment to Able that there was to West Ford or Project Sanguine. (See "Project Sanguine—Giant Underground Transmitter.")

Comment

The controversy over Project Able is one of several in which governmental projects have been resented by the public or groups within the general public. In this instance, the furor raised by the astronomers would probably have been small compared to that raised by a much larger public if the orbiting mirrors had actually become operational. Yet the technology may be adopted in a future age.

REFERENCES

1. "But Who Needs Sun at Night?", Editorial, *Sky and Tel*, **34**(4), 183 (October 1966).
2. "Mirrors Are Coming," *Time*, **89,** 56 (January 13, 1967).
3. Nelson, B., "Reflecting Satellite," *Science*, **155,** 304 (January 20, 1967).
4. "A Crack in 'Able,' " *Sci N*, **91,** 304 (April 1, 1967).

The Chemical Mace

Abstract

Chemical Mace, a spray developed in 1965 for riot control and for coping with unruly suspects, was adopted by many police departments. Mace was used frequently during the disturbances of the late 1960's. Much controversy developed in 1967–1968 over Chemical Mace, when reports indicated that it can cause permanent injury. The Surgeon General urged caution in its use, and a few police departments stopped using it. The Mace controversy peaked after its use during the tumultuous Democratic National Convention in Chicago in August 1968. In Senate hearings on the safety of Mace and similar products in May 1969, a statement that the U.S. Army had rejected Mace as an antiriot agent caused much concern. The Medical Committee for Human Rights recommended that these sprays be properly labeled and controlled, and asked the Public Health Service to distribute treatment instructions for hospitals and emergency clinics. The Food and Drug Administration, on the other hand, took the position that this lachrymating agent did not constitute a serious hazard to public health and safety. Mace is still being used by police, but presumably with caution. Civilian versions of Mace or similar products can be purchased by the public.

Background

In the opening months of World War I, Germany and then France experimented with the use of irritating or lachrymating (tear-producing) gases as weapons. On April 22, 1915, the Germans used poisonous chlorine gas for the first time in warfare with such success that it shocked both the Allies (who suffered five thousand casualties) and the Germans (who were unprepared to take advan-

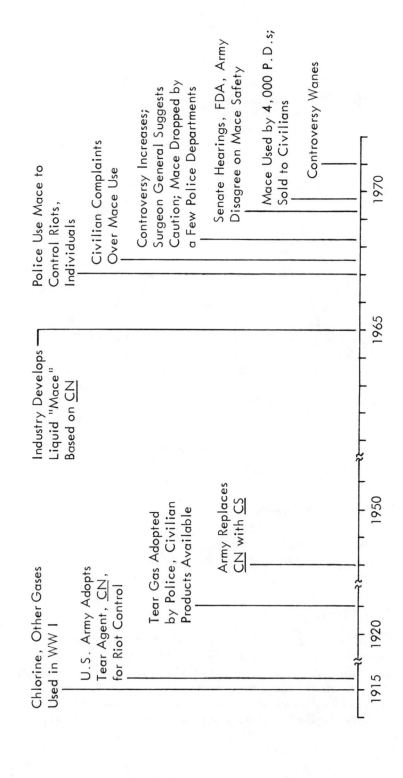

Chlorine, Other Gases Used in WW I

U.S. Army Adopts Tear Agent, $\underline{CN}$, for Riot Control

Tear Gas Adopted by Police, Civilian Products Available

Army Replaces $\underline{CN}$ with $\underline{CS}$

Industry Develops Liquid "Mace" Based on $\underline{CN}$

Police Use Mace to Control Riots, Individuals

Civilian Complaints Over Mace Use

Controversy Increases; Surgeon General Suggests Caution; Mace Dropped by a Few Police Departments

Senate Hearings, FDA, Army Disagree on Mace Safety

Mace Used by 4,000 P.D.s; Sold to Civilians

Controversy Wanes

1915   1920   1950   1965   1970

tage of their opportunity to move ahead). This initial success with a war gas spurred not only a research effort to find protective measures, but also a widespread search for different and better agents. It also left a strong revulsion, particularly in the Allied countries, against gas warfare. This revulsion lasted long after chlorine gas was replaced with less objectionable gases (that is, gases which are relatively painless and do not have lifelong after-effects).

During World War I the U.S. Army also investigated a number of chemicals as agents for different purposes, and as the war closed the Army adopted the chemical, α-chloroacetophenone* as its standard tear gas. The code name was "agent CN." The compound was actually a solid, not a gas, at room temperature, but could be effectively dispersed as a fine smoke or fog by a grenade or by heat. The Army also developed liquid formulations of the chemical† which could produce a fog or aerosol, but CN itself was the standard agent for both training purposes and riot control—it has little warfare value—until the adoption of the new agent CS (o-chlorobenzylmalononitrile) in the late 1940's.

Tear gas was also adopted by police during the 1920's and 1930's and civilian products (such as, tear gas fountain pens) became available, but these were basically the same as the Army's CN devices and apparently created little concern among the law-abiding public. During the late 1950's and early 1960's, however, tear gas was used with increasing frequency by police and the National Guard to quell civil disorders arising from racial, civil rights, or campus conflicts, and finally from anti-Vietnam war sentiment. Frequently, the individuals against whom the tear gas and other methods of control were used were well-educated and articulate (compared to the earlier recipients in labor troubles) and both they and the disorders were extensively covered in the news. In addition, the increasing use of tear gas in Vietnam itself raised questions of violations of the gas warfare conventions and added to the anti-gas sentiment.

Key Events and Roles

During the mid-1960's the Smith and Wesson Chemical Company developed a new formulation of α-chloroacetophenone (the Army's CN agent) in which the tear agent was dispensed from a

---

*Also called phenyl chloromethylketone and phenacylchloride.

†For example, with chloroform to give agent CNC; with chloroform-chloropicrin mixture (CNS); or with benzene-carbon tetrachloride (CNB).

pressurized can as a stream of heavy droplets which could hit a target individual 8 to 20 feet away. Thus, it did not create a fog or aerosol, reduce visibility, or contaminate the local area as did tear gas. Smith and Wesson, through their General Ordnance Equipment Company, announced their invention—Chemical Mace—in trade publications in mid-1966[1] and started selling it to local police units for use in controlling either riots or unruly individuals. By August 1967, the merits of Chemical Mace were being described to the general public in the newspapers.[2]

Almost immediately, however, the National Association for the Advancement of Colored People (NAACP) moved to halt the further adoption of Mace by the police[3] on the basis of objections from civilians who had been "Maced". In New York the police were reported to be testing Mace in September 1967[4] and it found wide acceptance in police departments around the country by the end of the year.[5] It was used on campus disturbances for the first time in 1967.[6] By February 1968 the Crime Council in New York was urging its adoption[7] and it was used on garbage workers during a strike in Memphis.[6] In March it was used at a concert[8] and at a civil rights march led by Dr. Martin Luther King.[9]

By this time, a considerable controversy was developing over the safety of the concentrated stream of irritant dispensed with Mace: a reporter who had been hit in the face with Mace during demonstrations at Ohio State University the previous fall was suing for $300,000; a Negro was said to have suffered partial depigmentation from Mace; and an ophthalmologist said that it produced corneal scarring in the eye of a rabbit.[6] In April the Food and Drug Administration (FDA) acknowledged that it had not made any tests on the long-range effects of Mace.[10]

In view of the growing controversy the Surgeon General of the Public Health Service (PHS), Dr. William Steward, sent a letter to state and local health authorities warning of the dangers of prolonged irritation by the Mace liquid, but indicated that prompt treatment of the exposed individual would prevent permanent damage.[6] A spokesman for General Ordnance countered that their Mace device had already delivered about 25,000 doses, with no evidence of damage worse than a sunburn.[6] From March to early June, Mace was much in the news.[6,11-14] As a result of the concern over Mace's safety, some police departments—Cleveland, Kansas City, Los Angeles, and San Francisco—said they would stop using it.[6] Others, however, said they would continue to use it, and in July the New York Police were equipped with Mace.[15]

The controversy over Mace might have waned, however, had it not been for the tumultuous Democratic National Convention held at Chicago in August. Although the Chicago police had been required to carry Mace in March 1968, [16] it was prominently announced on August 24 that the National Guard at the convention would not have Mace.[17] Three days later Mace was being used and its use added to the furor.[18] The role of Mace in the convention violence was subsequently included in a National Commission study,[19] and in May of 1969, a Senate Commerce subcommittee decided to investigate Mace and similar chemical sprays.

The widely publicized[20-23] hearing before Senator Frank E. Moss's subcommittee did much to reveal the controversy over the regulation of sprays of the Mace type and the extent to which they were already being used. The Medical Committee for Human Rights, a group which had provided assistance to incapacitated demonstrators at military facilities, had been asking that the chemical sprays be properly labeled and controlled and that the PHS distribute treatment instructions for such injuries to hospital emergency rooms. As the hearings opened, Senator Abraham Ribicoff charged that the Surgeon General had failed to warn of the dangers of Mace and, with an Army spokesman, revealed that the Army had evaluated it a year earlier, but had rejected its use in riot control because it could cause eye damage in test animals.* Dr. Herbert L. Ley, Jr., Commissioner of the FDA, then testified that they had "absolutely no report of permanent human injury resulting from the use of Mace formulations"[22] and that the sprays do not constitute such a hazard that they should be banned from interstate commerce.[21] The FDA had apparently been unaware of the Army's evaluation. Ley added that the sprays could be banned if they were found to be hazardous, but that they were, in fact less hazardous than the policeman's gun or nightstick.[21] The committee was immediately presented with case histories by Professor Joseph A. Page of the Georgetown University Law School of three individuals who had been sprayed and suffered such injuries as corneal damage, detached retina, and lingering lung inflammation. The committee was thus faced with conflicting testimony.

The committee also heard testimony by spokesmen for the chemical spray manufacturers (who of course contended that their products were safe) that the sprays were already in use by over

___

*The Army noted it has adopted CS for which the effects disappeared in ten to fifteen minutes. The news sources apparently did not note that the Army had used CN itself for thirty years or more.

4,000 police departments across the country.[22] The manufacturers' contention that this use was "under carefully prescribed procedures" was somewhat weakened when the papers the same day carried pictures of police helicopters dispersing gas over demonstrators and bystanders (several hundred in all) in the Peoples Park demonstration in Berkeley, California.[23]

The scope of the "Mace problem" took on a new dimension when it was revealed during the hearings that civilian versions were already on the market. William G. Gunn, president of Smith and Wesson, noted that six different sprays were on the market in forty-one states and that some of these, not Mace, were available over-the-counter to the public—even in Washington, D.C. One of these, a product made in West Germany was said to contain mustard gas, a blistering agent developed in World War I. Another, a product called "Preventor" made by Defensive Instruments, Inc., delivered a 4 to 6 foot mist rather than a liquid stream. Senator Moss noted that the mist could be blown back in the face of the user and called the FDA "derelict" in not requiring a warning on the label of this product, of which 250,000 units had already been sold. In addition, just before the hearings had opened, a front page news report noted that Mace had been used in committing a robbery,[24] further indicating that the chemical was out of control already.

At the close of the hearings the Federal Trade Commission commented that it, like the FDA, had been unaware of the Army's evaluation of Mace and that it was looking into possible deceptive advertising of some chemical sprays.[21] Senator Moss said he planned to introduce legislation to control the distribution and use of such weapons.[22]

Disposition

The use of Chemical Mace and similar sprays by police departments has continued, but apparently with more caution and after better training of the officers in most departments. Its use is not highly controversial today. The civilian products are still available, but are presumably under better control (or at least better labeled) than when first introduced.

Comment

The controversy over the police adoption of the nonlethal weapon, Chemical Mace, illustrates the point that law enforcement agencies are only loosely regulated in what new technologies they employ; they apparently get much of their information from manufacturers' agents, and are not closely attuned to the adverse effects of these technologies on the public. The controversy itself, however, primarily arose because of the use of weapons with a strong warfare connotation against an unusually well-educated and articulate group of demonstrators, at a time of increasing concern for civil rights.

REFERENCES

1. Coates, Joseph F., *Nonlethal Weapons For Use by Law Enforcement Officers*, Institute for Defense Analysis, November 1967. (Clearinghouse for Federal Scientific and Technical Information No. AD 661, 041).

2. *New York Times*, p. 17 (August 3, 1967).

3. *New York Times*, p. 27 (August 28, 1967).

4. *New York Times*, pp. iv–6 (September 17, 1967).

5. *New York Times*, p. 41 (January 18, 1968).

6. "Mace Questions," *Time*, **91**, 52 (May 17, 1968).

7. *New York Times*, p. 26 (February 19, 1968).

8. *New York Times*, p. 32 (March 24, 1968).

9. *New York Times*, p. 1 (March 29, 1968).

10. *New York Times*, p. 10 (April 17, 1968).

11. The number of *New York Times* articles on mace were: March—5, April—3, May—7, and June—1, 1968.

12. "Mace in the Face," *New Repub*, **158**, 14 (April 13, 1968).

13. "Mace's Secret Formula," *New Repub*, **158**, 8 (May 11, 1968).

14. "Case of Mace," *Newsweek*, **71**, 79 (June 8, 1968).

15. *New York Times*, p. 40 (July 28) and p. 30 (July 30, 1968).

16. *New York Times*, p. 53 (March 10, 1968).

17. *New York Times*, p. 1 (August 24, 1968).

18. *New York Times*, p. 29 (August 27), p. 36 (August 28), p. 1 (August 29), and p. 12 (August 30, 1968).

19. *New York Times*, p. 3 (December 2, 1968).

20. *New York Times*, p. 10 (May 18), p. 95 (May 20), p. 29 (May 22), p. 25 (May 28), and p. 47 (June 3, 1969).

21. "Antiriot Sprays: Are They Safe or Hazardous?", *US News*, **66**, 10 (June 2, 1969).

22. "Counterattack on Mace," *Newsweek*, **73**, 69 (June 16, 1969).

23. "Mace in the Face," *Commonweal*, **90**, 402 (June 27, 1969).

24. *New York Times*, p. 1 (May 13, 1969).

The Bronze Horse—A Technological Definition of Art?

Abstract

A bronze statuette of a horse, estimated to have been made around 470 B.C., had been a prize exhibit of the Metropolitan Museum of Art for over forty years. From 1961 to 1967 tests on the statue were made which indicated that it had been made by a process not invented until a few hundred years ago. The statue was designated a fake, and the Met began reauthenticating most of its collection of ancient art. New scientific tests on the bronze horse may now have proven that it is, in fact, as ancient as first believed. After eleven years of testing with various new techniques, the controversy continues.

Background

One of the prizes of the Metropolitan Museum of Art's collection of ancient Greek art, a fifteen-inch bronze statue of a horse, was purchased from a Paris dealer in 1923. The dealer said that it had been recovered from a shipwreck discovered in 1908 near Tunisia.[1] The statue was hailed as "the quintessence of the ancient Greek spirit," and dated at about 470 B.C.[2] The bronze was used by art historians to define the style of that period and by professors acquainting their students with the art of ancient Greece.[2-4] Pictures of the popular exhibit appeared in numerous museum publications, art books and the *Encyclopedia Britannica* as an excellent illustration of ancient Greek art. Thousands of plaster casts were made of the horse, including one for the 1936 Olympics,[1] and the replicas were for sale in New York for $75 each in 1967.

Forgeries of art were, of course, long known and the forgery of

402

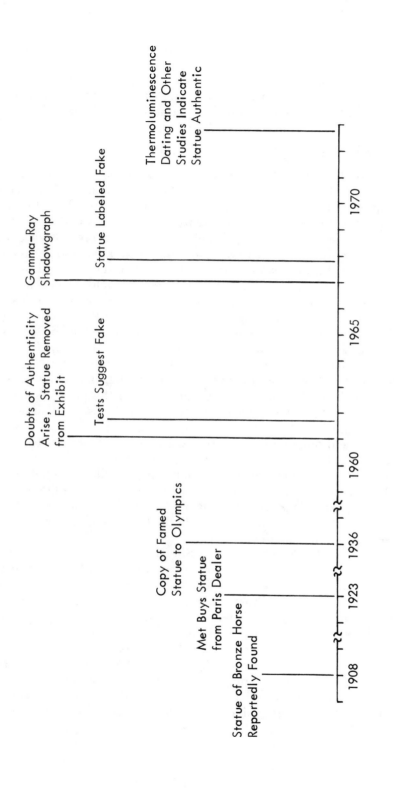

Statue of Bronze Horse Reportedly Found

Met Buys Statue from Paris Dealer

Copy of Famed Statue to Olympics

Doubts of Authenticity Arise, Statue Removed from Exhibit

Tests Suggest Fake

Gamma-Ray Shadowgraph

Statue Labeled Fake

Thermoluminescence Dating and Other Studies Indicate Statue Authentic

1908   1923   1936   1960   1965   1970

Egyptian and other antiquities had been practiced for about 125 years.[5] In 1961, the Met had had an "Etruscan warrior" statuette declared a forgery after it had owned it for thirty years.[5]

Key Events and Roles

In 1961, the Metropolitan's Vice Director, Joseph V. Noble, noticed a faint line running from the top of the horse's mane to the top of its nose and continuing less obviously to circle the entire body.[3] He suspected that the line was a mold mark, but the methods used in ancient Greece did not leave mold marks. Instead, the ancients used the "lost wax process." In this technique, a wax model had a cast poured around it, and then the cast and model were fired to burn out the wax and harden the mold. The cavity left where the wax had been is a perfect, one-piece mold, with openings for the molten bronze to be poured in and the air to escape. The result is a solid one-piece statue with no encircling mold marking. In the newer, sandcasting method for bronze, developed in the fourteenth century and perfected in the nineteenth century, a mold is cast in two sections around the original model, then removed and reassembled; the bronze is then added. During this process some of the bronze can seep between the sections and must be filed down afterwards.

In addition to the seams, the statue had a tiny hole in its head. Dietrich von Bothmer, the Met's curator of Greek and Roman art, noted after a visit to Greece that the life-size marble horse statues of the Acropolis had holes in the head which the Greeks fitted with anti-pigeon spikes. A small statue intended for indoor display would not need this, but the hole in the bronze could be a forger's mistake of including every detail.[3]

Soon after Noble suspected the possibility of a forgery, the statue was quietly removed from display. Two fairly simple tests were quickly made. One revealed that the horse's specific gravity was too low for solid bronze, but just about right if the horse had a sand core held in place by wire and tacks, as might have been used in Paris about 1920.[2] The other test, with magnets, indicated the presence of points of iron and therefore the likelihood of a sandcast piece.[3] However, these results were not sufficient proof. Ordinary X-rays could not penetrate the bronze to reveal the internal structure, but in September 1967, a new technique, the gamma-ray shadowgraph (developed to inspect the thick steel hulls of nuclear submarines) was used. When the film was deve₁₂ped, Noble said, the sand core with its iron wire and points was clearly visible. Noble announced that the horse was a fake in December 1967.

Disposition

The announcement of the forgery had a considerable impact on the art history world: it is exceedingly rare that a work of art which has served as a standard for an era is proven a fake. The director of the Metropolitan, Thomas Hoving, said the museum would start authenticating every exhibit they had and he stated that it was the duty of the museum to reveal its fakes. A careful program of scientific testing and artistic evaluation was conducted on the museum's collection which resulted in the reattribution of some of the works.

This was not, however, the end of the bronze horse story: in 1972 a new verdict was given. Four experts at the Metropolitan, two conservators and two chemists, declared the piece was, after all, a genuine work of antiquity. This conclusion was based on further, three dimensional gamma-ray study and the use of another new technique: thermoluminescence dating. The latter, performed on the ceramic core by the Space Physics Laboratory of Washington University, St. Louis, provided evidence that the statue was between two thousand and four thousand years old.[4] Additional investigations indicated that the methods of manufacture, the composition of the metal and core, and the nature of the corrosion were all consistent with the statue being authentic. The horse had been coated with a dark green and black wax early in this century for preservation and appearance: when the wax was removed the seam disappeared.[1] The "seams" had apparently been formed by the many reproduction casts made of the horse. A study of several other ancient bronzes showed that they also had ceramic cores which had been inside the wax when they were cast in the lost wax method so that the core of the horse was not necessarily from a modern technique.[1] The iron in the legs was probably used to support the wax legs of the original model—a further proof of lost wax.

The latest group of tests are not absolute in their verdict and are, in fact, disputed by Noble, who is now director of the Museum of the City of New York. Future investigations could lead to different conclusions, but for now the statue is considered an authentic work of the fifth century, B.C.

Comment

The bronze horse controversy illustrates the extent to which technology affects all aspects of our lives. Scientific analysis of art works of all types is not only of ever increasing importance to art historians for dating and classification, but is assuming an increasing role in authentication. On the other hand, art forgers are also using increasingly sophisticated technology in their work. Thus, art tends not to be recognized as art until it has been defined so by modern technological methods. Problems of authentication have multiplied greatly in recent years because of the enormous increase in the numbers of collectors of antiques and memorabilia. Increasing demands are being heard that the federal government somehow regulate and authenticate the sale of such collectors' items to protect the public from fraudulent products and to protect the investments of the owners of authentic rare items. But clearly the control of such items—including, for example, the campaign buttons and bumper stickers from the aborted vice-presidential bid of Thomas Eagleton in 1972—will require not only sophisticated analytical technologies, but also substantial amounts of time of regulatory, law enforcement and judicial officials.

REFERENCES

1. *Kansas City Star*, p. 10 (December 31, 1972).
2. "Museums," *Time*, **90**, 84 (December 15, 1967).
3. "Of a Different Color," *Newsweek*, **70**, 94 (December 18, 1967).
4. *Kansas City Times*, p. 2 (December 26, 1972).
5. "Crackpots and Forgeries in Art and Science," *Sci Digest*, **66**, 40 (November 1969).

Synthetic Turf and Football Injuries

Abstract

Since the introduction of a synthetic turf playing surface at Houston's Astrodome in 1966, over one hundred football fields (high school, college and professional) have been covered with a synthetic grass surface.

In 1971 a controversy concerning the safety of synthetic turfs broke out. Some studies showed a much higher injury rate for players on synthetic turf rather than natural grass. The controversy soon involved the National Football League Players Association, the Congress and the public. Further statistics are being compiled, and methods of reducing the number of injuries are sought while play continues.

Background

In the spring of 1965 the construction of Judge Roy Hofheinz's huge $35 million enclosed, air-conditioned sports palace, the Astrodome, was completed in Houston, Texas. One of the unique features of this "wonder of the world," as it was called at the time, was the use in the domed roof of windows made of a special plastic[1] which transmitted the correct wavelengths of light needed to grow grass on the enclosed ground.

The baseball season opened in the Astrodome to considerable excitement, not only because this was the first time major league games had ever been played indoors, but because, in addition, it soon became apparent that it was a terrible place to play baseball on a bright sunny afternoon; the small white ball could not be seen well against the mosaic ceiling of light windows and the dark grid-supporting beams.[2] A national television audience was treated to the spectacle of major leaguers missing routine fly balls by twenty

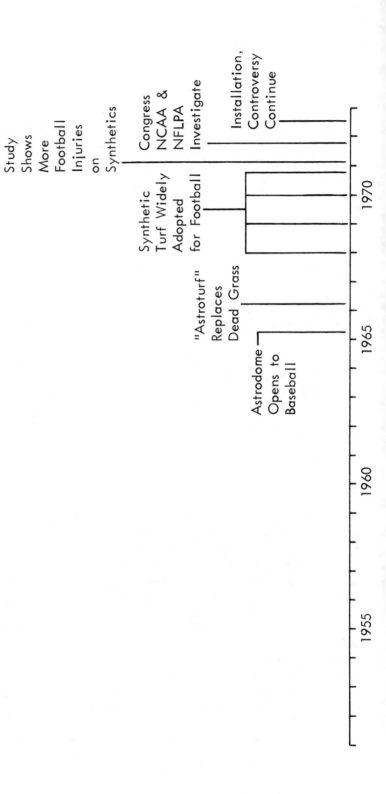

feet or even being hit on the head by the ball, to the consternation of baseball officialdom. Players and coaches denounced the conditions. In order to make the Astrodome acceptable for day baseball, all the windows of the dome were painted black (to provide a uniform background) before the end of April.[3]

The paint, however, prevented sunlight from reaching the grass, which died. Experimentation with new strains of grass were unsuccessful and periodic resodding was necessary to prevent the unthinkable appearance of a brown dirt baseball "pasture." Before the 1966 season, Major League Baseball gave permission to install an artificial "grass" surface on the infield of the Astrodome for test purposes and on July 19, the entire field was covered.* The synthetic grass selected was a nylon plastic product which the Monsanto Chemical Company had developed primarily for the bare inner-city playground market. Monsanto reportedly installed the synthetic turf for the cost of the material alone and promptly renamed it "AstroTurf." Although some baseball men, such as the widely quoted Leo Durocher,† immediately denounced the AstroTurf and many complained that the hard surface changed the style of play, synthetic turf became accepted in professional baseball. It was soon installed in new open-air stadiums, such as the one in St. Louis in mid-1966, although the surface temperatures on a summer day were found to rise above a hot 120°F.

Key Events and Roles

The synthetic turf had a considerable advantage over natural grass in reducing maintenance costs and in resisting damage in rainy weather; in multiple-purpose stadiums, the field could be readied for baseball the day after an otherwise muddy football game. In fact, synthetic turf appeared ideally suited for football. Under an energetic sales campaign by Monsanto and the two other leading artificial turf producers, 3M Company ("Tartan Turf") and American Biltrite ("Polyturf"), the synthetics were widely adopted for professional, college and even high school athletic fields during the late 1960's. By the fall of 1971, 113 synthetic football fields had been installed at a cost of about $250,000 each and were being used

---

*The quick approval to paint the ceiling and to install artificial turf was in contrast to baseball's long-standing rule against making any changes in the dimensions of a given park during the course of a playing season.

†Interestingly enough, he was named manager of the Houston Astros in late 1972.

by over 1 million athletes.[4] Projections were made[5] that the use of artificial turf and rubberlike surfaces in sports would total a $1.2 billion market during the 1970's.

Suspicions that synthetic turf increased the already large number of football injuries began to arise by 1970. In the spring of 1971, Dr. James Garrick, of the University of Washington's School of Medicine, announced the results of his one-year study of football injuries in 229 high school games, including 80 on AstroTurf and 149 on natural grass: nearly twice as many severe injuries (nineteen) occurred on the synthetic turf as on grass, and total injuries averaged 0.7 per game on the synthetic compared with 0.5 per game on grass. Total injuries were less on wet AstroTurf than dry, but still higher than on grass.

The statistical validity of Garrick's results were immediately challenged by Monsanto, which had 75 percent of the synthetic turf market, but the results were sufficiently impressive to John E. Moss, Chairman of the Subcommittee on Commerce and Finance, U.S. House of Representatives. He announced an investigation in September 1971[6] and requested performance and safety test data from the manufacturers. The National Collegiate Athletic Association (NCAA) also started a study of collegiate football injuries.[7]

Professional football also became involved. By the 1971 season, synthetic turf was in use in twelve of the National Football League stadiums, and the fourteen remaining were expected to switch within five years.[8] In September the Players' Association (NFLPA) formally demanded of the club owners a moratorium on artificial turf installation until the incidence of injuries could be investigated.[8] Monsanto disputed the charges that artificial turf had caused an alarming rise in player injuries as "irresponsible and without basis in fact," and claimed that "detailed injury data compiled every year by a number of colleges do not support the allegations made by the players union."[7] The club owners continued with plans to install synthetic turf for the 1972 season.

Disposition

The results of a poll (reported in January 1972) of the National Football League Players Association showed that 84 percent of the 756 players responding preferred natural grass to artificial turf; a total of 72 percent of the players in the league voted in the poll.[9] The results, however, appeared to be of little influence on the use of synthetic turf.

Instead, attempts were being made to reduce the number of injuries to an acceptable level by modifying the players' shoes or other equipment, and perhaps by modifying the qualities of the synthetic surface. For example, Dr. Richard Thompson, team physician for the Detroit Lions, stated in a recent seminar[10] that moisture and thicker padding can make synthetic turf fields less hazardous. He also pointed out that technological advances in shoe manufacturing will be an advantage for artificial turf. Although he was not opposed to artificial turf, he believed that more studies should be made of the injury factor attendant to it. As of September 1972, the controversy over synthetic turf was continuing. Most observers agreed that more study and additional data were needed to help settle the issue.

While producers of synthetic turf were noting the prospect of shrinking stadium markets in November 1971, they were enthusiastic[11] about other outlets for their products (as synthetic ground coverings on highway medians, for shopping centers and private homes) which were growing rapidly, in part no doubt because of the nationwide publicity afforded by football use.

In late 1973, the new Consumer Product Safety Commission denied the NFLPA's petition to regulate use of the synthetic turf.[12]

Comment

Under normal circumstances major league baseball would probably never have permitted the adoption of synthetic turf playing surfaces: it was introduced not as a technological means of improving player performance, but rather as a means of recovering from a technological blunder. Subsequently, it was permitted even where natural grass could have been used; because of aggressive marketing efforts it is rather widely used at present. It has changed the nature of the game to some extent, but has not been a source of unacceptable side effects.

The artificial turf was widely adopted in football, on the other hand, not only by the professionals and major universities, but even by many high schools. While it also changed the nature of play to some extent, football is less linked to tradition than baseball and the turf was generally regarded as a technological improvement. It was hailed as a big saver in maintenance cost, permitted more days of practice, and is aesthetically superior to a muddy field on color TV. As the evidence mounted that the turf increased the number and severity of football injuries, the professional players have protested its further use, but the turf manufacturers have denounced the evidence. The club owners and athletic directors have apparently concluded that the possible injury of a few more players is an acceptable risk in view of the financial benefits—at least until technology can produce a better product.

REFERENCES

1. "Plastics Score in Stadiums," *Chem W*, **96**, 87 (April 17, 1965).
2. "Astrodome Glare Problem Bothers Outfielders," *Bsns W*, 30 (April 17, 1965).
3. "Touch Up for a Plastic Sky," *Chem W*, **96**, (May 8, 1965).
4. *Kansas City Star*, p. 145 (September 12, 1971).
5. "Artificial Grass: Good and Bad News," *C&EN*, **49**, 17 (May 24, 1971).
6. "Synthetic Turf Draws Fire," *C&EN*, **49**, 27 (September 20, 1971).
7. "AstroTurf Injuries Disputed," *C&EN*, **49**, 13 (October 11, 1971).
8. *TV Guide*, 14 (October 13, 1971).
9. *Kansas City Star*, p. 10 (January 28, 1972).
10. *Kansas City Star*, p. 45 (March 5, 1972).
11. "A Very Rough Season for Synthetic Turf," *Bsns W*, 35 (November 6, 1971).
12. *Federal Register*, **38**(239), 34361 (December 13, 1973).

Disqualification of Dancer's Image

Abstract

Dancer's Image won the Kentucky Derby May 4, 1968, and was then disqualified when tests revealed that the horse had residues of an illegal painkiller, phenylbutazone, in his system. In December 1968, the $122,600 purse was awarded to the second place horse, but after appeals and a change in the law, Dancer's Image was again awarded the money and trophy in December 1970.

Background

Horse racing is a leading spectator sport and over $3 billion are wagered on it annually. The Kentucky Derby has one of the largest audiences, officially announced as 100,000 every year, not including national radio and television coverage. The race is so steeped in prestige and tradition that it was considered beyond "fixing" or foul play.

Phenylbutazone, a drug classified as an analgesic and effective in alleviating inflammation of the joints, was developed in Europe and introduced into the United States in 1951 for treating humans and sometimes race horses. Doctors prescribe it for human ailments such as arthritis and bursitis, and athletes have taken it to relieve aches, pains and stiffness. Star baseball pitchers Sandy Koufax and Whitey Ford both took the medication for pain and stiffness in their throwing arms.[1] Most states banned its use on horses *in a race* because the presence of the drug could affect performance, although it is neither a stimulant nor a depressant. The drug can, however, be legally used for the treatment of a horse in training as long as all traces of the drug are expelled from the animal's system before a race.[2] In 1962, Kentucky was one of the last states to ban the drug.[1]

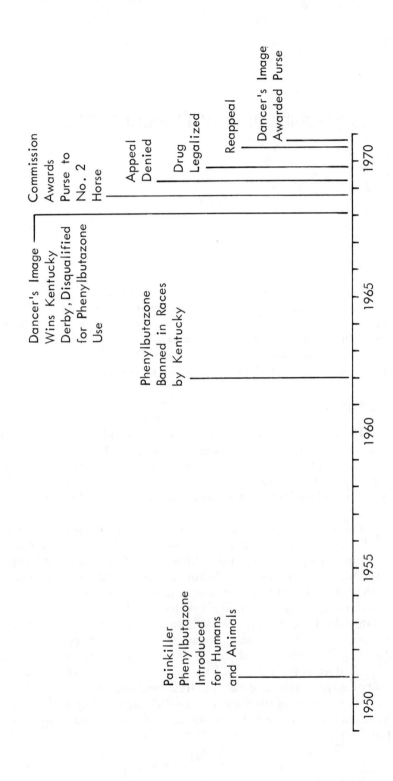

Key Events and Roles

When Dancer's Image arrived at Churchill Downs on April 25, 1968, he was put under the care of Dr. Alex Harthill by owner Peter Fuller. On April 28, it was apparent that the horse had sprained his ankle while being exercised the day before and Dr. Harthill then administered four grams of phenylbutazone, 152 hours before the Kentucky Derby.[3] Dr. Harthill and Lou Cavalaris, the horse's trainer, later testified that it was the only time the horse had received phenylbutazone.[3]

On May 4, 1968, Dancer's Image won the Kentucky Derby. Immediately following the award ceremony, the horse was taken to the detention barn where required urine and saliva specimens were collected. Samples were identified by numbered tags, and tags with the number and name were sealed in an envelope by the track stewards and locked up. At 7:30 that evening, tests of the urine samples showed a positive reaction and further testing revealed the drug present was phenylbutazone. The chemist in charge called Lewis Finley, Jr., commission steward and told him that they had a positive on Saturday's racing card. Finley decided they would treat it as any other day and told him to submit his report Monday morning when all three stewards would open the envelope. On Monday, May 6, at 11:00 a.m., they opened the envelope and the positive number belonged to Dancer's Image. The steward immediately sought Lou Cavalaris (because the trainer is held responsible at all times for the horse) who had returned home to Canada. They told him the news over the phone that evening and asked him to return to Louisville immediately.[3]

At about noon on Tuesday, the news became public with tremendous impact. The headlines which appeared in most papers were "Derby Winner Doped" or "Drugged" or variations on that theme. The first day after the announcement a great deal of confusion existed about the composition and effect of the drug, and over the status of the winner, the purse of $122,600 and the bets. The announcement, by mistake, was first called an official steward's ruling but later that was retracted pending a hearing which was scheduled for the next week. During the intervening period, the accusations and insinuations by the principals received wide media coverage.

Here is a summary of the accusations:[1-3]

- The tests were wrong—there was some mix-up of the samples

- The horse was "got at" because
  - Track security was lax
  - Kentuckians don't like out-of-state horses winning their Derby
  - Fuller had given an earlier purse to Mrs. Martin Luther King, Jr., and some Ku Klux Klan types wanted revenge
- The veterinarian did it
- The horse had an unusual metabolic system and retained the drug an extraordinarily long time
- The trainer and/or assistants did it knowingly and risked being caught

Disposition

A few weeks after the Derby, trainer Lou Cavalaris and his assistant Robert Barnard were given thirty-day suspensions.[4] Under those circumstances and considering the charges, the punishment was the equivalent of a slap on the wrists. The hearings dragged out through the year. Then, on December 23, 1968, the stewards ruled that Dancer's Image was the official recordbook winner, but that the first-place money and trophy would go to the second-place horse, Forward Pass. Additional fines were levied (on February 7, 1969) against Dr. Harthill and a trainer Douglas M. Davis, Jr., found guilty of "improper conduct" for adding aspirin to the horse's feed two days after the Derby.[5]

In 1969, the Kentucky State Legislature legalized phenylbutazone for use in race horses. Mr. Fuller continued to appeal the ruling and finally in December 1970, Dancer's Image was again declared the full winner and awarded the purse.[6]

Comment

The case of Dancer's Image illustrates one of the ways in which modern technology can impact on and nearly disrupt some of our older traditions. Because too little was known about it, phenylbutazone was a time bomb, once regulations were established that approved its use at some times and totally banned its use at others. After the violation of the regulations was discovered, efforts of the horse-racing industry and state regulatory agencies to manage the crisis did little to inspire the confidence of the public. While the use of drugs on race horses is presumably now under more realistic rules, the use of drugs such as pep pills and anabolic steroids by human athletes has proved even more difficult to regulate, and must yet be resolved by society.

REFERENCES

1. "Drug at the Derby," *Time*, **91**, 60 (May 17, 1968).
2. "Shocker in Kentucky," *Newsweek*, **71**, 64 (May 20, 1968).
3. "It Was a Bitter Pill," *Sports Illus*, **28**, 21 (May 20, 1968).
4. "The Dancer Plot Thickens," *Sports Illus*, **28**, 18 (May 27, 1968).
5. *Facts on File*, p. 516 (1969).
6. *Facts on File*, p. 983 (1970).

The AD-X2 Battery Additive Debate

Abstract

The case of battery additive AD-X2, a storage battery rejuvenator developed in 1947 by Pioneers, Inc. of Oakland, California, was essentially an effort to determine the worth of the product. Embroiled in the conflict were the FTC, National Bureau of Standards (NBS), Post Office Department, Senate Small Business Committee, Department of Commerce, Massachusetts Institute of Technology, National Better Business Bureau, and, of course, Pioneers, Inc. From 1948 to 1961, all the above organizations tried to determine whether or not AD-X2 was a fraud or a worthwhile product. On April 1, 1953, the issue came to the nation's attention when Dr. Allen V. Astin, Director of NBS, was publicly requested to resign by Mr. Sinclair Weeks, Secretary of Commerce, after each had taken opposite sides in the controversy. Dr. Astin was later reinstated after a loud outcry of indignation from the nation's scientific community. The value of AD-X2 was never clearly resolved, but before the case ended, numerous questions of scientific and political significance arose.

Background

In 1946 Jess M. Ritchie returned to Oakland, California after bossing a construction job in the Philippines. The job was plagued with trouble in the maintenance and use of storage batteries. He was determined to find a way to increase the useful life of storage batteries, which commonly needed replacement in two years or less.

With the advice and assistance of Dr. Merle Randall, a physical chemistry professor at the University of California, he mixed and

418

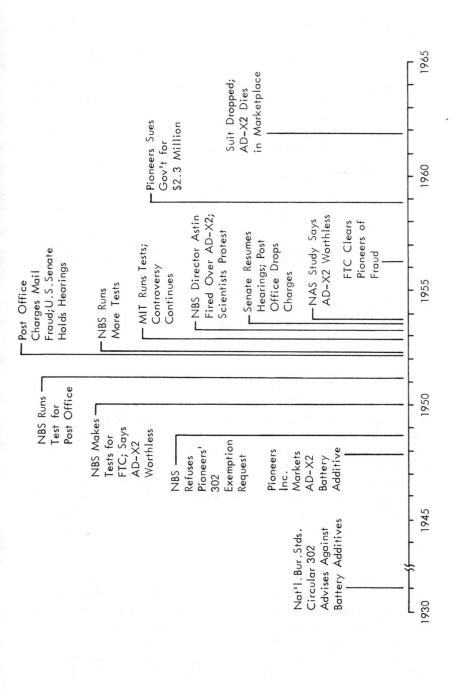

Nat'l. Bur. Stds. Circular 302 Advises Against Battery Additives

Pioneers Inc. Markets AD-X2 Battery Additive

NBS Refuses Pioneers' 302 Exemption Request

NBS Makes Tests for FTC; Says AD-X2 Worthless

NBS Runs Test for Post Office

Post Office Charges Mail Fraud; U.S. Senate Holds Hearings

NBS Runs More Tests

MIT Runs Tests; Controversy Continues

NBS Director Astin Fired Over AD-X2; Scientists Protest

Senate Resumes Hearings; Post Office Drops Charges

NAS Study Says AD-X2 Worthless

FTC Clears Pioneers of Fraud

Pioneers Sues Gov't for $2.3 Million

Suit Dropped; AD-X2 Dies in Marketplace

1930   1945   1950   1955   1960   1965

tested hundreds of compounds in an attempt to restore the lives of old batteries. In September 1947, he found a compound which seemed to work. A month later he incorporated as Pioneers, Inc., and began selling his product, AD-X2.[1]

Battery additive AD-X2 was a white powder which, when added to a lead storage battery, was advertised as capable of increasing the life of new batteries and of rejuvenating old batteries. Although AD-X2 was a new product, battery additives had existed for years. The National Better Business Bureau (NBBB) had claimed for years that battery additives were fraudulent, and the National Bureau of Standards (NBS) in 1931 had recommended (in NBS Circular 302) against the use of any battery additive and had named sodium sulfate and magnesium sulfate as worthless additives.[2]

Circular 302 and the NBBB's widely distributed pamphlets which claimed that battery additives were worthless, adversely affected Mr. Ritchie's business prospects, even though he backed AD-X2 with a money-back guarantee and numerous testimonials of satisfied users.[3] He therefore launched a vigorous campaign to prove that his product was a valuable one, and in doing so, began a controversy which lasted over a decade.

Key Roles and Events

Mr. Ritchie assembled numerous testimonials from industrial users and individual consumers of AD-X2 to support his argument that AD-X2 should be excepted from Circular 302 and other pamphlets which stated that battery additives were worthless. He presented his case to the Oakland Better Business Bureau and the Oakland Chamber of Commerce, who sided with Pioneers, Inc. in its struggle to gain acceptance for the battery additive. During 1948, Dr. Merle Randall and these two organizations wrote letters to the National Bureau of Standards on behalf of Pioneers, Inc., requesting that NBS give AD-X2 a fair test to evaluate its worth as a battery additive.[2]

The NBS did not do any commercial testing or identify tested products by name. It tested a product only when a test was requested by another government agency. As a consequence, the above appeals went unanswered in the laboratories of NBS.

During 1949, the NBBB pressed the NBS for a statement as to the worth of AD-X2, so that it could properly inform a confused public. In addition, the NBBB pamphlets were adversely affecting the business of Pioneers, Inc., and Mr. Ritchie's claim that his product was different from all others could have considerable legal ramifications.[2]

The issue was finally forced on the NBS in 1950 by the Federal Trade Commission (FTC), which had received complaints from both the NBBB and the Association of American Battery Manufacturers regarding the worth of AD-X2. In response, the NBS analyzed the compound and found it to be a mixture of sodium sulfate and magnesium sulfate with several trace elements. The AD-X2 was also tested on several batteries. NBS issued a report to the FTC in May 1950 that the additive was worthless, and reported its findings in NBS Circular 504, *Battery Additives*, issued in January 1951. The FTC requested that Pioneers, Inc. modify some of its advertising claims. Mr. Ritchie refused to do so. The case was then to remain suspended on the FTC docket, "pending further investigation" until 1954.[2]

In the meantime, the U.S. Post Office Department had received complaints during 1951 that Pioneers, Inc. was conducting an unlawful enterprise through the mails. It appealed to NBS for guidance on the usefulness of AD-X2, and whether fraud was involved. NBS then ran more extensive laboratory battery tests and again concluded that AD-X2 did not improve performance. In April 1952, the Post Office asked Mr. Ritchie to appear in Washington to answer charges involving unlawful mailing of his product.[2]

At this point, Mr. Ritchie decided to do considerably more. At a hearing before the Senate Select Committee on Small Business, Ritchie presented numerous testimonials of satisfied customers and claimed that his small business was being assailed by the large battery manufacturers. The Committee, which included Senators Hubert H. Humphrey and Richard Nixon, requested that NBS perform further tests to clarify the matter.

Extensive tests were run by the NBS in May and June of 1952 with Mr. Ritchie present. Again, about sixty batteries were tested after the adding of AD-X2. They showed no evidence of having been improved. However, the test procedure, which had been proposed by Mr. Ritchie, was altered slightly by the NBS scientists, and he again claimed that his additive was not given a fair trial. The NBS scientists, of course, claimed that their tests were fair and objective.[4]

In an effort to settle the matter, the Massachusetts Institute of Technology, one of the most prestigious scientific laboratories in the country, agreed to test the additive. It tested the product on numerous batteries and offered a report without comment in December 1952. The report stated that the batteries were definitely affected by the additive, and that in some cases, the properties of the battery were improved. No conclusions were made, however, as to whether or not the additive affected the performance of a battery as its manufacturers claimed. Mr. Ritchie immediately interpreted the MIT report as favorable to his case. The NBS countered that the MIT tests revealed nothing about the performance of the batteries under actual operating conditions.[2]

The AD-X2 controversy came to the whole nation's attention in March 1953 when Sinclair Weeks, the newly appointed Secretary of Commerce, asked Dr. Allen V. Astin, the Director of the NBS, to resign. Mr. Weeks felt that the testimonials of consumers of AD-X2 could not be discounted, and that the product should be given a fair chance in the marketplace. He charged the NBS had not been sufficiently objective in testing the battery additive AD-X2. In addition to asking Dr. Astin to resign, Secretary Weeks had Circular 504 withdrawn from sale,[5] and persuaded the Post Office Department to suspend the order banning AD-X2 from the mails (which they had imposed the previous month).

The scientific community was outraged at the firing of Dr. Astin in what they felt was a battle between science and politics. If such a policy were allowed to exist in the federal government, future scientific endeavors would be subject to finding "the right answers." News reports indicated that four hundred employees at NBS intended to resign if Dr. Astin were forced to leave. The Federation of American Scientists and numerous other scientific groups protested the political assault on NBS, which had been a nonpolitical scientific bureau for over fifty years.[6]

Secretary Weeks and Dr. Astin appeared before the Senate Small Business Committee to present their views. With pressure mounting from the scientific community, Secretary Weeks asked the advice of the Visting Committee of the NBS, a group of eminent scientists. On April 14, 1953, the Visting Committee recommended that Dr. Astin be retained until the AD-X2 issue had been studied, and on April 17, 1953, Secretary Weeks announced that Dr. Astin's resignation was temporarily suspended. To settle the issue, Secretary Weeks appealed to the National Academy of Sciences to investigate the technical aspects of the case, and asked Dr. Mervin J. Kelly, President of Bell Telephone Laboratories, to form a group to evaluate the general situation of NBS.[7]

The Senate Small Business Committee resumed its hearings in

June 1953, to gather further evidence in the AD-X2 case. Testimony was given by: Mr. Ritchie; Dr. Astin; Dr. Harold C. Weber, Professor of Chemical Engineering at MIT; four industrial technologists supporting AD-X2; a battery shop manager; and six technologists reporting on military tests and field experience with AD-X2 (only one reported unfavorable results). Data presented to the committee included the MIT tests, the NBS tests, three sets of military tests, and affidavits from satisfied users of AD-X2. The voluminous quantitative data presented from the various tests were unmanageable and inconclusive. The committee reached no firm conclusion of its own on the merits of AD-X2, nor on the ability of NBS to test battery additives. Instead, it decided to defer the issue to the findings of the Committee on Battery Additives of the National Academy of Sciences, which was investigating the technical aspects of the case at the request of Secretary Weeks.[2]

After reviewing the case with Dr. Astin and discussing the matter with Dr. Kelly, Secretary Weeks decided that Dr. Astin's handling of the NBS was commendable under the circumstances. On August 21, 1953, he announced that Dr. Astin was being retained as the director of NBS indefinitely.[8]

The Committee on Battery Additives of the National Academy of Sciences reported their findings on October 30. They concluded that NBS had done excellent work in battery testing, that the NBS personnel were objective in their findings, and that AD-X2 had no merit as a battery additive. That same month the committee, headed by Dr. Kelly, recommended that the Department of Commerce handle the policy and procedures involving the nontechnical nature of commercial product tests, and that NBS handle only policy on the technical content of a problem.[2]

Disposition

Since the Senate Small Business Committee could not reach a firm decision on the AD-X2 case, no regulatory legislation evolved from the episode. In August 1953, the Post Office Department cancelled its fraud order against Pioneers, Inc.[9] The FTC complaint against the advertising of AD-X2 became active in 1954, but dragged on until May 1956, when the commission cleared Pioneers, Inc., of all charges of falsely advertising its product.[10]

Pioneers, Inc., brought suit against the U.S. government for $2,369,000 in damages for declaring AD-X2 worthless. The suit was filed in a U.S. Court of Claims in September 1958.[11] After the Justice Department served notice that it would fight the case to the end with evidence gathered in three years of investigation, Pioneers, Inc. abandoned the claim in December 1961. The court dismissed the suit with prejudice, meaning that Pioneers, Inc. and Mr. Ritchie could not take it to the U.S. Court of Claims again.[12]

Although the legal hurdles to the marketing of AD-X2 had been surmounted, the product apparently could no longer develop a market because of the publicity, the increasing technical sophistication of the public, and perhaps most important of all, the increasing affluence of the public and availability of new batteries. AD-X2 does not appear to be marketed anywhere today.

Comment

The AD-X2 affair illustrates the difficulty that government regulatory agencies can encounter (and even help generate) in trying to exercise control over even a very simple technology. In this episode a resourceful and aggressive individual was able to present a case, supported primarily by unscientific testimonials, that tied up several important governmental agencies and made it appear that the two most prestigious scientific institutions in the country were reaching opposite conclusions. These results, coming soon after the technological advances of the 1940's and just as "big science" was starting to grow, decreased considerably the scientific community's confidence in the capabilities of government administration in scientific areas; it probably tempered the public's confidence in the scientists themselves.

The Senate committee's unsuccessful effort to come to grips with the technology has been—to various degrees—repeated in other instances. Conflicting technical testimony appears frequently to bewilder those without a grasp of the scientific approach; its impartial evaluation to be perhaps unusually difficult for those trained in an adversary approach—which the majority of our lawmakers use. In addition, the legislative and regulatory functions are characteristically (or at least ideally) concerned with trying to increase benefits and minimize risks for the public or to reach an acceptable balance. In the AD-X2 case, neither the benefit nor any hazard was ever clearly defined for the Senate and our institutions appeared to be poorly equipped to provide meaningful documentation for a course of action. Thus, no action resulted and the public was left to wonder what the fuss was all about.

REFERENCES

1. "New Life for Batteries," *Newsweek*, **36**, 58 (December 11, 1950).
2. "Technical Information for Congress," Report of the Legislative Reference Service, Library of Congress, to the House Committee on Science and Astronautics, U.S. Government Printing Office, 14–60 (April 25, 1969).
3. "NBS on the Spot," *Newsweek*, **40**, (December 22, 1952).
4. "Was There a Deal in the Astin Case?" *New Repub*, **128**, 9 (May 4, 1953).
5. "The Dynamite in AD-X2," *New Repub*, **128**, 5–6 (April 13, 1953).
6. "The Astin Uproar," *Fortune*, **47**, 108 (May 1953).
7. "Washington Notes," *Bul Atomic Sci*, 100–102 (February 1953).
8. *New York Times*, p. 1 (August 22, 1953).
9. *New York Times*, p. 22 (August 21, 1953).
10. "Science Frowns on AD-X2, But FTC Bows to Customer Approval," *Bsns W*, 36 (May 26, 1956).
11. *New York Times*, p. 27 (September 25, 1958).
12. "AD-X2: The Case of the Mysterious Battery Additive Comes to an End," *Science*, **134**, 2086 (December 29, 1961).

Chapter IV

Synopses of Other Cases of Technology and Social Shock

In this chapter fifty-five additional cases of public concern over technology are described briefly. These cases illustrate several points and examples of alarm that are not seen in the preceding case histories. They are presented in an order that emphasizes the human need served by the technology involved—rather than the threat that is posed—except that several cases that involve far-reaching environmental concerns are grouped together. The grouping of cases is shown below in the table of contents for this chapter. The cases are not individually referenced, but a compilation of important sources and suggested further reading is provided at the end of the chapter.

The Alaska Pipeline Controversy

In 1968 oil was discovered on Alaska's North Slope at Prudhoe Bay on the Arctic Ocean. Within a short time seven oil companies had bought $912 million worth of oil leases in the region and had formed a consortium to build a $1 billion pipeline 789 miles across the middle of Alaska to Valdez, a port on the Pacific Ocean. A year later, a forty-eight-inch pipe that would carry 2 million barrels of hot crude oil had been designed, and an application was made to the U.S. Department of the Interior to build it. But in late 1969 the National Environmental Policy Act was passed, one provision of which called for the filing of an environmental impact statement (EIS) on proposed projects that might seriously affect the environment.

Three environmentalist groups quickly challenged the trans-Alaska pipeline in the courts on the grounds that it could cause irreparable damage to the delicate permafrost environment, inhibit the migration of animals, and produce catastrophic oil leaks if it broke because of earthquakes or ground shifts. The environmentalists were soon joined in court by a group of Alaskan natives who claimed land along a twenty-mile segment of the route. The project was halted by a court injunction and Interior began preparation of an EIS. Completed in January 1971, the EIS concluded that environmental risks could be held to an acceptable minimum and that the project should proceed.

The EIS was immediately attacked from any quarters as being inadequate on the grounds that it had understated the hazards, overstated the advantages, and neglected to consider alternatives such as a trans-Canadian line. At the court's insistence, Interior renewed its EIS study and returned in March 1972 with a nine-volume report which it said represented $9 million of effort. Although the EIS stopped short of recommending approval of the specific permit application, it concluded that early completion of some kind of pipeline was an important national security objective. The public debate was continued with vigor, but in May, Secretary of Interior Rogers C. B. Morton announced his intention to approve the application for the pipeline. The design had been much improved during the 2-year litigation, and the cost was now said to be $2.5 billion.

The environmentalist groups contended, however, that the EIS was still inadequate, and continued the court fight over this cause célèbre. In February 1973, the U.S. Supreme Court ruled that Interior could not grant permission to construct the pipeline, not because of environmental aspects, but rather on the grounds that the proposed right-of-way was wider than the 50-ft maximum permitted on federal lands under the 1920 Minerals Leasing Act. An act of Congress would now be required!

Under the impetus of a gasoline shortage which suddenly intensified, the Senate in July 1973 passed a pipeline bill that would clear both right-of-way and environmental barriers, and in November under impetus of a partial Middle East oil embargo the law was enacted. Although construction of the pipeline is expected to require at least three years, talk of a second pipeline is already being heard.

The Four Corners Power Plant—Black Mesa Coal Issue

The growing rate of electric power consumption in California and other Southwestern states during the 1950's stimulated a search for new power sources. In the mid-1950's sizable coal deposits were discovered in the Black Mesa formation in northeast Arizona near the unique Four Corners junction of Arizona, New Mexico, Utah and Colorado. A consortium of twenty-three southwestern

utilities and public agencies soon began laying plans for building
six mammoth power plants—two each in New Mexico and Utah,
one each in Arizona and Nevada—all to be fueled by coal that
would be strip-mined from Black Mesa by the nation's largest coal
producer, Peabody Coal Company. First, however, approval had to
be obtained from the Department of the Interior and particularly its
Bureau of Indian Affairs: three of the proposed sites were on the
Navajo Reservation; the Black Mesa mines would be largely in the
Hopi Reservation; and the electric transmission lines and a proposed coal slurry line (to Nevada) would cross Indian lands. In addition, the mining, slurrying, and generating facilities would require
enormous amounts of water that would have to come from the Colorado River.

Interior soon approved the proposed project, and leased mineral rights on 65,000 acres of Black Mesa to Peabody. Construction
was quickly started on the giant complex, which was expected
eventually to furnish 13,000 megawatts of power to nine million
people—nearly half of them in southern California. Many of the
Indians regarded Black Mesa as sacred, and the strip mining and
power lines a desecration, but other Indians welcomed the employment opportunities; opposition to the project was weak.

The first of the six plants to come onstream was the Four Corners plant in Arizona which started operating its first three units in
late 1963 and units four and five in 1969 and 1970. With a capacity
of 2,208,500 kilowatts, it was the largest coal-burning facility in the
United States. It consumed nearly 25,000 tons of coal per day and
lost nearly 20 million gallons of cooling water per day—this in an
arid region where water was precious. The Black Mesa coal was
high in ash content and the Four Corners plant had only the simplest of pollution control devices; it immediately began pouring
200–250 tons per day of particulate pollutants into the air from its
thirty-story smokestack and added about 240 tons per day each of
corrosive sulfur and nitrogen oxides.

The plant's operators, Arizona Public Service and Southern
California Edison, were quickly surprised to find that not even the
remote Four Corners region could absorb this much pollution: a
pristine desert area that included one hundred-mile vistas was soon
clouded, and sharp increases in respiratory illnesses were soon reported at communities sixty miles away. Under intense public pressure the companies began in 1966 to make a series of agreements,
first with Interior and later with the Environmental Protection
Agency, to install increasingly costly pollution-control devices

(over $40 million were contracted in 1970 and 1971). Even so, Four Corners will continue to emit nearly thirty tons of particulates per day and SO_2 will not be reduced until 1977.

The Four Corners experience had caused great national concern over the Black Mesa project, and even more over plans for forty-two new power plants scattered throughout the West: mineral rights to 773,000 acres of public lands had been leased to coal companies and other private interests in 1969–1970. In 1971, Senate hearings were held publicly at Four Corners that generated two thousand pages of testimony. But the basic conflict remains between the electrical power demands of distant Southern California and the environmental and human needs of the immediate Southwest desert region.

The Great Northeast Power Failure

On November 9, 1965, an electrical power failure developed in the Northeastern United States just as millions of people were headed home from work at dark. Within minutes some 25–30 million people in parts of eight states and Ontario, Canada, were without electric lights or power. Emergency generators were started wherever available, but a massive blackout was to engulf most of the affected areas for five to fifteen hours. New York City was hit especially hard. Some 800,000 people were stuck in subways and elevators, most surface and air transportation was halted or badly jammed, and almost all normal activities were disrupted. Only the existence of transistor radios and emergency generators at broadcasting stations and telephone companies enabled the public to get information on the nature and scope of the blackout, that may have prevented a panic.

The blackout caused a sensation with the public and in the media: they demanded to know why it had happened. President Lyndon B. Johnson and two Congressional committees called on the Federal Power Commission (FPC) and the electric utilities (especially the giant Consolidated Edison Company) for an explanation. The ensuing probe determined that the failure was actually triggered by an event in Canada, as briefly as summarized below.

The electric power companies tend to have natural monopolies in each community, but as the local companies grew during the 1920's they found it advantageous to interconnect their individual power plants so that they could trade or import power at times when local demand exceeded generation capacity for some reason.

At government urging during and after World War II, the approximately 3,600 power systems became increasingly interlinked in groups, and in 1964 the FPC urged that all systems be linked in one giant national grid. By 1965, 97 percent of them were woven into five major grids, of which the blacked-out area was one. What had happened was that a transmission line in Canada that was carrying a near capacity load had unexpectedly tripped a telephone-sized circuit breaker—a safety device. The load was automatically switched to four other lines, but these were already at capacity also and their circuit breakers tripped. The result was that 1.7 million kilowatts of electricity was suddenly switched from Canadian lines to U.S. lines and activated a chain reaction of safety breakers and relays throughout the grid. Entire power plants were automatically shut down and many hours were required to reestablish the complex system.

The incident, amplified by closely following blackouts in the Southwest and in England, generated a great controversy over the desirability of the grids and spurred intensive efforts to improve their reliability.

The Storm King Project Controversy

In 1963 the Consolidated Edison Company proposed to build a $150 million power plant near Storm King Mountain (about forty miles north of New York City along the Hudson River) with a special arrangement to handle the problem of peak daytime power demand, which can be twice that of night. A reservoir would be built several hundred feet above the level of the river. During the night, when the plant had power to spare, the reservoir would be pumped full of water. During the day, water from the reservoir would be sent back to the river through supplemental hydroelectric generators to give a peak capacity of two megawatts power.

The project met stiff opposition from environmental groups (especially the Scenic Hudson Preservation Conference), because it would flood acres of scenic land and possibly damage breeding areas of fish, and from the City of New York, which feared that construction activities might damage the city's aqueduct. The Federal Power Commission (FPC) approved the project in 1965, but the U.S. Court of Appeals ordered FPC to review environmental considerations. In 1970 the FPC again approved the project and again the case was taken to the courts. On June 19, 1972, the U.S. Supreme Court declined to review the FPC decision. On March 14,

1973, an appeal in the New York State Courts was denied and plans were made to start construction of the project, the cost of which had tripled during the ten-year litigation. But in early 1974, the U.S. Court of Appeals—in response to litigation brought by the Hudson River Fisherman's Association—ordered FPC to hold new hearings to consider whether the plant should be allowed to operate during the striped bass spawning season.

The Coal Strip Mining Dilemma

The industrialization of America after the mid-1800's was fueled largely by coal, but a century of mining left an ugly legacy of human and environmental damage. Deep mines have taken dreadful tolls of the lives and health of miners by explosions and accidents, have collapsed after residential areas were built over them, and have polluted the air by long-smoldering fires. Even more objectionable to many people have been the surface mines, usually of the strip type.

In the rugged Appalachians, contour strip mining has left thousands of benched, pitted, and augered hillsides that yield acidic drainage wastes to pollute streams, mudslides to endanger homes, and scenes of desolation. On the Midwestern plains, area strip mining has converted thousands of acres of farmland into sterile, ridged and trenched devasation. During the 1900's the use of strip mining increased as the development of bigger and bigger machines made it more economical than deep mining. The coal companies took the coal and moved on.

By 1940, the environmental consequences of strip mining were obvious, but the states moved individually, slowly, and with much difficulty to control the practice and to require that the derelict lands be reclaimed to some extent. After the Donora Air Pollution Episode of 1948, less polluting fuels such as gas and oil were increasingly sought and the production of coal did not keep pace with the surging consumption of energy in the United States. But the increasing awareness of the strip mine problem generated great pressure by conservationists and environmentalists to ban the practice entirely: many charged that reclamation could never be complete and that the ecological damage done was irreparable.

In January 1969, as energy consumption increased at the unprecedented rate of 5 percent annually, the U.S. Department of the Interior issued the first national regulations requiring reclamation of strip mined lands. Coal producers objected to the increased costs and some production was further curtailed. The Interior regulations were not strongly enforced, however, according to a 1972 General Accounting Office study, and a bill was introduced in Congress to ban all strip mining in the nation. By 1973, however, the supply of gas and oil was becoming insufficient, while coal reserves sufficient for five hundred years remained untouched. The dilemma could no longer be ignored, and in mid-1974 the Congress began debate on a controversial bill to regulate strip mining on a national basis for the first time.

The Calvert Cliffs Decision and the Nuclear Power Plant Debate

The surging demand for electrical energy and the increasing awareness of both the environmental objections to coal-fired power plants and the limited availability of gas and oil has focused attention on the development of nuclear power. The Atomic Energy Commission (AEC), which was set up in 1947 to oversee nuclear weapons development, to promote peaceful uses of atomic energy, and to regulate all these activities in the public interest, was empowered in 1954 to interact with private interests in developing new uses. Enthusiastic efforts to build nuclear-fueled electric power plants were soon under way. The first commerical nuclear power plant in the U.S.* was completed at Shippingport, Pennsylvania in 1957, and by 1970, seventeen nuclear power plants were operating, 54 were under construction, and forty-seven others were being designed.

This use was not without opposition: Some companies in the coal, oil, and gas business opposed it on economic grounds,† and many members of the public and the scientific community were fearful of emissions or even a catastrophic loss of radioactive materials. But the AEC exercised complete regulatory control over the plants: in 1969 Minnesota found that it could not even set its own strict standards on radioactivity emissions. In 1970 two AEC employees publicly denounced the AEC's standards as a nationwide menace and three other states joined Minnesota in the court battle, although it did not appear promising (and was eventually rejected by the U.S. Supreme Court in April 1972). In late 1969, however,

*The world's first plant was completed in Britain in October 1956.
†Major companies in fossil fuel industries were quick to enter the nuclear field.

the National Environmental Policy Act (NEPA) was enacted, one provision of which required an environmental impact statement (EIS) on projects that might pose environmental danger, and another that established an Environmental Protection Agency (in December 1970) with regulatory power over emissions of pollutants (including radioactivity) to the environment.

Among the nuclear power plants under construction at the time was that of the Baltimore Gas and Electric Company near Chesapeake Bay at Calvert Cliffs; it was soon to become the focal point of the fight over nuclear power. An environmentalist group, the Calvert Cliffs Coordinating Committee, took the AEC to court on the charge that they had not filed an adequate Environmental Impact Statement. On July 23, 1971, the Federal Court of Appeals scathingly accused the AEC of making a mockery of NEPA. The AEC declined to appeal the Calvert Cliffs' decision, which affected over one hundred other plants as well, and set about preparing an EIS (in which such factors as thermal pollution of the Bay, disposal of spent radioactive fuels, and alternatives to the plant were considered) and tightening up its standards and procedures. In March 1972, the AEC decided to let construction at Calvert Cliffs and at ten other plants resume while it completed its study. But when the AEC presented its report a month later the EPA said it lacked sufficient data on environmental effects of thermal pollution and that further study was needed.

In the meantime the whole question of nuclear power plants had been taken up by AEC in January 1972 at public hearings in Washington that were to run for 105 days and to yield over 20,000 pages of testimony. The hearings stimulated the AEC to shift the EIS burden to the electric utilities (a move they denounced), but has not resolved the battle over the nuclear power plants.

The Automobile Recalls

The automobile is a complex technological achievement that is, however, subject to myriad kinds of mechanical and electrical failure. Over the years, owners often found that even new cars were defective because of improper assembly or design: the first "recall" of a new auto model is said to have been for a defective gas tank mounting in the 1916 Buick. But auto manufacturers issued few such recalls, preferring instead to repair individual autos on a case-by-case basis when they were presented by the complaining owners. As the assembly lines at Detroit and around the country poured forth increasing numbers (almost one per minute) of ever more complicated cars (that quickly became dated by annual styling redesigns), the complaints from owners grew louder.

In 1966, the National Traffic and Motor Safety Act was passed, one feature of which required that manufacturers notify new car buyers of defects that posed a safety hazard. In 1966 also, Ralph Nader's book, *Unsafe at Any Speed,* appeared. It roundly denounced the nation's largest auto manufacturer, General Motors Corporation, and its products.

The auto giant's blundering counterattack on Mr. Nader—which included hiring private detectives to snoop into his personal life—made him a martyr in the media and a folk hero in many quarters. The nation became keenly aware of the poor quality of its glorious chariots.

The result was a crescendo of reported defects and numerous campaigns by the manufacturers to recall and repair new cars. From 1966 to 1970, thirteen million vehicles—38 percent of the total production—were recalled for one reason or another by General Motors, Ford Motor Company, Chrysler Motors Corporation, American Motors Corporation, and several other manufacturers. Still the crest had not been reached: recalls began to involve mammoth numbers of vehicles, as GM announced in December 1971 the recall of 6.7 million Chevrolet cars and trucks because of motor mount troubles. In 1972, more vehicles were said to have been recalled than were built, as many were recalled two or three times to check for separate possible defects. And although defects were largely repaired at the expense of the manufacturer, this practice had not always been followed, and by late 1973 Congress was considering legislation to make the manufacturers liable.

Studded Snow Tire Question

In 1963 a new type of snow tire for automobiles was introduced in which tough tungsten carbide "studs" could be inserted to provide better traction, braking, and turning ability on snow and especially on ice. Most states already had laws that prohibited the use on highways of vehicles with wheels containing metal lugs (such as early farm tractors and construction equipment), but many quickly passed enabling legislation to permit the new studs. By the mid-1960's, the heavily advertised studs were quite popular, especially in the northern states and in other areas that had heavy snow-falls.

By 1970, however, it had become evident that the studs were not an unqualified blessing. Heavily traveled streets and highways of northern cities and states soon began to show increased wear. Minnesota estimated that $55 million would be needed there to repair stud damage to pavements, which included not only loss of paint markings and traction groves but actual rutting of the concrete so badly that water collected in the wheel paths. By April 1971, a congressional committee heard testimony from the Federal Highway Commission that the studs' damages were not compensated for by their benefits, but that it was powerless to ban them. A month later Minnesota refused to renew its enabling act and studded snow tires became illegal there, even on cars from out of state. Other northern states began to impose partial or complete bans and most southern states continued to refuse permission to use them. They are still permitted in many states, however, and an owner of studded tires risks breaking the law when traveling out of state.

Passive Restraints in Automobiles — The Air Bag

The 50,000 or more traffic deaths per year caused the government to require that seat belts be installed in all autos in 1967. Research has determined, however, that most people do not use their seat belts. Increasingly complicated interlock systems have been devised to make use of the belts necessary. The government now feels some type of passive restraint is needed, such as the "air bag" that inflates instantly upon a crash. The Department of Transportation (DOT) has specified that air bags must be standard equipment in 1976 cars. The decision has created a hectic race to perfect a reliable system and a great wave of public controversy.

The sensor devices and long-term effects of the system have not been tested adequately; much concern is expressed over accidental inflations and over the lack of protection against sideswipes, sudden stops, or rollovers. In addition, DOT advocates omitting seat belts in cars that are equipped with air bags, but the auto industry and safety groups disagree. Finally, the consumer feels that he will have to bear the additional cost of a safety feature of dubious value, over which he has no control, and which he may not even want. Much public resentment is heard that the passive restraints and the increasingly complicated interlock systems are further examples of Big Brother (the government) infringing on personal liberty.

Carbon Monoxide in the Corvair

In 1960, General Motors Corporation introduced a new compact sedan, the Corvair, in which an air-cooled engine was mounted in the rear. Sales of the Corvair were considered excellent: 250,000 in its first year and 303,000 in 1962. Even in the face of competition and numerous consumer complaints about its reliability, Corvair sales held at over 200,000 through 1965. But one of these early Corvair has been sold to a Mr. Ralph Nader. His unhappy experiences with it led to the book, *Unsafe at Any Speed* in 1966. Widespread doubts developed over the mechanical safety of the Corvair, and sales plummeted to only 2,328 units in 1969 when production was terminated.

On January 14, 1971, a startling new development in the Corvair story occurred. The National Highway Traffic Safety Administration (NHTSA) issued a special consumer protection bulletin warning an estimated 1.4 million Corvair owners that their cars might pose a health hazard. Tests had found that carbon monoxide and other fumes from the engine could enter the passenger compartment via the heater in 1961–1969 models (1960 models used a special gasoline-burning heater). The NHTSA explained that the Corvairs (many of which were in the hands of second and third owners by now) were not being recalled, but that owners who smelled fumes should keep a window open and seek repairs. The NHTSA did not say how one could detect carbon monoxide (which is odorless), what repairs would be needed, or how much repairs would cost. Nor did it note that General Motors had settled a 1962 suit involving carbon monoxide poisoning out of court for $125,000 (the settlement to include transfer of all the plaintiff's files on the case to GM).

By the fall of 1971, *Consumer Reports* and the Automobile Club of Missouri reported that over 10 percent of the Corvairs tested had serious or dangerous levels of carbon monoxide. These results prompted NHTSA and GM on Novemeber 22, 1971, to issue a warning to nearly 680,000 Corvair owners; while GM would inspect the Corvairs, the owners were apparently left to bear the $150–200 cost of repairs.

Safety and Suicide on Interstate Highways

The construction of the interstate highway system that began in 1956 called for thousands of new bridges as overpasses. The bridges are supported with large circular pillars, usually in the median strip and on both sides of the highway.Their proximity to the flow of traffic posed much danger of accident. Even before the interstate system was completed, fatalities involving the pillars became so numerous—many of which were suspected to be suicides—that a strong public protest developed.

Guardrails have since been placed around all the pillars to prevent vehicles from hitting them head-on. When the interstates were constructed, good safety practice had provided that guardrails be placed on curves and at places where a steep bank fell away from the road. Numerous fatalities were required, however, before traffic engineers recognized the danger of the pillars and action was taken to safeguard motorists.

Highway Deicers

The use of deicers on highways and city streets has caused much public concern. The amount of salt used has risen from 0.5 million tons in 1947 to 9 million tons in 1970. In earlier years, abrasive substances like cinders or sand were used more often than salt. These do not lower the melting point and thus do not melt the snow, but do aid traction. These materials are insoluble and leave a dirty street, although they are generally inert and are not a serious environmental problem. As the use of coal as a fuel decreased, cinders became less available, but the increased use of salt resulted primarily from its greater efficiency.

The use of massive amounts of salt has caused serious problems, however, including contamination of drinking water supplies and farm ponds, deterioration of highways, corrosion of vehicles, and destruction of vegetation along roadways. Some of the effects tend to be cumulative; the extent of the problem has only been recently recognized after many years of salt use. The salt itself is sometimes not the only problem, as substances such as chromium compounds and phosphates may be added for corrosion protection, and cyanides may be added to prevent caking of the salt.

Recommendations have been made by the EPA to improve methods of spreading salts, to use salts only when most needed, to decrease the use of additives, and to reduce hazards by several other means. Clear roads provide advantages in safety and uninterrupted commerce, but these must be balanced against environmental dangers.

The Skyjacking Nightmare

During the 1940's and 1950's incidents occurred in which commandeered airplanes were used in flights from the East European nations to the West to gain political asylum. But on July 24, 1961, the United States was shocked when an American plane was hijacked at gunpoint and taken with crew and passengers to Cuba, a country with which the United States did not have diplomatic relations. The hijackers were granted asylum in Cuba; the passengers were returned to the United States the next day and the plane somewhat later. But a precedent had been set and several other cases of "skyjacking" as it was to become known, occurred in the next few years—not only from the United States to Cuba, but throughout the world—as political and criminal fugitives fled their would-be captors, and several psychopaths acted for their own bizarre reasons. The first skyjackings caused instant alarm in the United States. The Federal Aviation Administration (FAA) quickly authorized the arming of flight crews, but the airlines opposed this idea. By September 6, 1961, a law was enacted making skyjacking a federal crime punishable by death, but the penalty was to have little deterrent effect. In 1964, the FAA ordered that all cockpit doors be locked during flights to prevent the skyjackers from reaching the pilot, but the rule had little effect when hostages were taken.

In 1968, the incidence of skyjackings escalated rapidly, to the point that even Cuba began backing international efforts to halt the problem: twenty planes were commandeered from the United States to Cuba that year and armed "sky marshals" were added to many flights in high-incidence areas. The following year the International Federation of Airline Pilots Association adopted a resolution to boycott countries that failed to punish skyjackers. And as some of the skyjackers began demanding ransom money, the airlines began to screen passengers by a composite "skyjacker profile" which had been developed on the basis of studies of previous skyjackers. Even so, a plane was skyjacked from the United States to Rome, Italy, in November 1969. In June 1970, a skyjacker over the Midwest demanded the astounding sum of $100 million ransom (he was shot by the FBI when the plane landed at Washington, D.C.). In September, the world was further shocked as three jets from different airlines were hijacked over Europe by Arab extremists, flown to the Jordanian desert and burned, after allowing the passengers to disembark.

The situation was now becoming a nightmare and efforts were increased to find a way to stop the abuse of this complicated transportation technology. President Nixon authorized 2,100 armed guards for flights of U.S. airlines (1,300 were added immediately), ground security guards stepped up their "profiling" of passengers and searching of suspects, and a crash research program was initiated to develop reliable metal detectors. Still the skyjackings and international conferences continued. The first passenger was killed in June 1971, and a plane with three sky marshals and an FBI agent was skyjacked to Cuba in October. (Most sky marshals were soon shifted to ground security.) On November 24, a lone skyjacker fascinated the nation by parachuting with $200,000 ransom from a Boeing 727 somewhere between Seattle and Reno. D. B. Cooper, as he was known, was never found and became the subject of a folksong, a brief sweatshirt fad, and a host of unsuccessful imitators. There were numerous charges that news coverage of the skyjackings were simply inspiring more of them. In February 1972, the FAA ordered the airlines to screen passengers more closely and to use the magnetometers. The American Civil Liberties Union complained that searches for other purposes (such as for narcotics violators) were being conducted under the pretext of searching for weapons: six thousand airport arrests had been made in a twenty-two-month period. of which fewer than 20 percent were related to possible hijacking.

Then came the final straw. On November 10, 1972, the thirty-first successful skyjacking of the year culminated in a wild, circuitous, twenty-nine-hour flight over several states and Canada before landing in Cuba. The skyjackers circled Washington, D.C., threatened to crash the plane into the nuclear reactor facility at Oak Ridge, Tennessee, obtained $2 million of the $10 million ransom they demanded, and were arrested in Cuba. But the episode had far-reaching results. Less than a month later, President Nixon ordered that everyone of the 170 million airline passengers per year walk through the magnetometers, have their carry-on luggage personally searched, and their checked luggage X-rayed at the nation's 531 commercial airports.

The personal-searching ruling drew numerous complaints from

the airport operators (who thought the government should pay for the cost of equipment and guards) and from thousands of passengers (including columnists and Senator Vance Hartke (D-Indiana)), who charged constitutional violations. A federal judge ruled that skyjacking constituted an emergency, however, and the program was fully implemented with resounding success; not a single skyjacking was made in the United States during 1973.

The SST Debate

The supersonic transport (SST), a proposed giant airplane, became controversial in 1970, a decade after it had been conceived and after $1 billion had been spent by the government in developing it to the state of a plywood mock-up. Intended primarily for commercial use, the SST would require a substantial further effort to develop a flying prototype. But ultimate production of the SST was expected to bolster the sagging aerospace industry and especially the depressed Seattle area. Backed by President Nixon, by government officials (who looked to the SST to provide 200,000 jobs in the United States and a better balance of payment from hundreds of foreign sales), and by the industry (which anticipated financial benefits), further funding of the SST seemed reasonably well assured.

But the rise of environmental concern in the late 1960's, coupled with wide disenchantments over military expenditures for the Vietnam War, brought a surprisingly swift turn of events. Almost overnight the SST became the subject of great debate as some economists doubted its economic benefits to the public and environmentalists deplored the intolerable engine noise which they claimed would result from a fleet of SST's. But the most telling blow was the highly technical argument that developed over the effects of the SST's on the upper atmosphere: critics contended that exhaust from the high-flying planes might destroy the ozone layer and permit increased ultraviolet radiation that would cause skin cancers and disrupt weather patterns.

One could not really tell if these proposed effects would happen without flying the SSTs. In December 1970, the Congress, after acrimonious debate, declined to fund the SST further until the absence of harmful effects could be shown. The debate over the SST continued, and although the National Academy of Sciences issued a report in late 1972 that indicated environmental risks were acceptable, 189 members of that prestigious organization petitioned that the report be repudiated. In early 1973, the NAS issued a second report calling for utmost concern over the effects of the SST. Construction of an American SST during the 1970's appears remote.

The Sea-Level Panama Canal

The Panama Canal is becoming obsolete. For both commercial and strategic reasons, a new sea-level canal would be desirable. An important problem in building such a canal would be the enormous cost. The use of nuclear explosives to excavate the canal has been proposed to reduce the cost, but this idea is not without objections. The extra "clean" thermonuclear charges are still to be perfected and their use in this application might violate the nuclear test-ban treaty. Strong objections to the sea-level canal come from marine biologists, who say that it would permit intermixing of species from the Pacific Ocean side with those from the Caribbean Sea and thus pose grave dangers to delicate ecosystems.

The Fiberglass Pole Controversy

Many athletic and sports activities intimately involve the use of a technological device (such as a rubber ball), but this use in pole vaulting was minimal: a hickory, ash, or bamboo pole was all that was needed. The bamboo pole gradually became the standard, and in 1942, Cornelius Wamerdam used it to establish the world vaulting record at 15 feet 7-¾ inches. In the 1952 Olympics, the decathlon champion, American Bob Mathias, did well with the new fiberglass pole, but these poles were tricky to use, and steel and aluminum poles came into vogue in the U.S.: in 1959, Don Bragg broke Wamerdam's record with the aluminum pole.

But fiberglass technology had been improving substantially. In 1971 Bragg's record was broken by George Davies with a fiberglass pole, and although the record was only inches greater than that set in 1942, Americans switched rapidly to fiberglass. Then in early 1962, John Uelses astounded the athletic world by easily surpassing one of sport's supposedly unbreakable barriers—the sixteen-foot pole vault—with a fiberglass pole. The International Amateur Athletic Federation (IAAF) debated whether or not to accept the new mark, and Bragg complained bitterly that fiberglass made the sport catapulting, not vaulting. Dramatic photos showed that the new pole could bend nearly 90 degrees before flinging its occupant skyward.

No consensus could be reached on banning fiberglass, however, and the record soon began to escalate rapidly—to over seventeen feet in 1963 and then to over eighteen feet. In the trials in Oregon for the 1972 Olympics, Bob Seagren used a newly designed, lightweight fiberglass pole to set the world record at 18 feet 5-¾ inches in July. But in August, the United States was shocked when the IAAF ruled that the new pole would not be allowed at the Olympics in Munich, although the new record would be accepted. A furor ensued and the IAAF reversed its decision in late August as the games drew near. But when East European countries protested, the IAAF again banned Seagren's pole. On September 3, Wolfgang Nordwig of East Germany took the Gold Medal with a vault of 18 feet ½ inches, while Seagren with an unfamiliar pole, failed to clear 17 feet 10-½ inches. It was the first time that an American had failed to win the event since the Olympics began in 1896. The new pole became acceptable after the Games and heights of nineteen feet were being anticipated.

Disqualification of the Soap Box Derby Winner

The winner of the 1973 Soap Box Derby was disqualified when it was found that his vehicle had an illegal magnetic device in its nose. The device, which enabled him to get a faster start, had been installed by the fourteen-year-old boy's uncle and guardian, whose own son had been the 1972 winner. The uncle had given the boys many helpful tips and contended that many of the cars were actually designed and built with advanced engineering know-how, not by the boys alone as the rules required. The Derby officials began considering termination of the annual Derby because of the unfavorable publicity received and the evidence that competition for the top prize had become excessive.

High-Rise Public Housing

During the late 1950's the federal government embarked on a program to provide low-cost housing for many of the nation's poor. The focus, of course, was on the older, inner city and ghetto areas; the housing would be a part of urban renewal programs in major cities across the land. Because space and funds were limited and the objective was to provide as much housing as possible (about 250 people per city block), the high-rise design was widely adopted by planners and architects. Luxury high-rise apartments were at that time becoming quite fashionable, and the planners assumed that tenants would be easily attracted to this kind of public housing. High-rise housing projects were soon under way in many cities, with the stimulus of $2.3 billion of funding under the Federal Better Communities Act.

One of the larger of these was the Pruitt-Igoe complex in St. Louis—a collection of eleven-story buildings intended to house over two thousand families. Pruitt-Igoe's counterpart in Kansas City was the 738-unit Wayne Miner project, dedicated in 1960. In other cities, equally large housing facilities went up, but even before many of these high-rise apartments were fully occupied it had become painfully apparent that "the projects," as they were called, left much to be desired as homes.

In the first place, the planners had failed to anticipate correctly the characteristics of the people who would be attracted to these buildings. Not only did they have more children than expected—the 556 families in Wayne Miner contained nearly two thousand in 1973—but a substantial percentage posed serious socioeconomic-moral issues. The children soon made obvious such deficiencies of the project as play areas, safety features, basic plumbing needs, and—especially in Pruitt-Igoe where the elevators stopped only on every other floor—basic transportation needs. One of the interrelated problems was the low incomes of the tenants. At Wayne Miner, for example, 75 percent of the tenants in 1973 were on either public welfare or social security. The average rent paid was only $23 per month, indicating an average income per family of less than $100 per month (the Brooke amendment to the Act prohibits charging more than 25 percent of income). The high-rises thus attracted a substantial number of people who had little or no steady employment because of poor health, little education, criminal records, or drug problems. Inevitably the high-rises were beset with vandalism and high crime rates; bad tenants tended to drive out the good. In most projects a substantial number of apartments always stood vacant, in need of repair; in Wayne Miner nearly 25 percent were empty by 1973, and in Pruitt-Igoe, vandals broke windows and even ripped out and sold bathroom and electrical fixtures before many apartments could be rented for the first time.

By the 1970's the federal government was prepared to concede that these "monuments to mediocrity" were untenable. In St. Louis, authorities attempted to make Pruitt-Igoe more livable, selected buildings were demolished with explosives to decrease congestion. The effort proved futile and the public housing project was abandoned in May 1974. In some other cities, funds were sought to convert the high-rises from housing for families to housing exclusively for senior citizens. For Wayne Miner, authorities estimated the cost of this conversion as $8 million for the physical plant and $2 million for relocation of present residents. The great experiment was coming to a close in many cities across the land, but some of the technological remnants were proving suitable for other social purposes.

Plastic Plumbing Pipe Controversy

Plastic plumbing pipes were introduced after World War II for a variety of purposes and began slowly to replace some of the uses of metal pipe. By 1958 the several plastics manufacturers held about $50 million of the $2 billion pipe business. Compared to metal pipes the plastics were inexpensive, light-weight, and corrosion resistant, offered low resistance to fluid flow, and could be installed in about half the time required for metal pipe. The use of the plastics were restricted, however, by plumbing codes in the states and in many large cities. Some of those codes had been written over fifty years before the advent of the plastic pipes; they could not be easily changed by advocates of the plastics because of the powerful lobbying by makers of metal pipe and the plumbers unions, and the uncertainty or apathy of local government officials. Critics contended the plastics were breakable and not resistant to heat or impact. The economic interests of the metal pipe manufacturers and the plumbers (who feared decreased employment) were evident, but numerous failures of misapplied or poor quality plastics could be cited.

The Department of Commerce had published standards for plastic pipe in the 1950's; the Federal Housing Administration and national code organizations approved certain plastics for drain-waste-vent systems in the early 1960's. Industrial use of the plastics increased rapidly during the 1960's as polyethylene, ABS, PVC, and other plastics were selected for specific applications. The battle over building codes for commercial and residential applications was taken up largely on a local basis.

By 1964, plastic pipe sales reached $100 million per year, and resistance intensified from the plumbers' unions, who saw losses not only in the building trades, but also in the lucrative home repair market. But by 1968, twelve states and 500 local communities in thirty-four states had revised building and plumbing codes to permit some uses of plastics. In 1972, the U.S. Department of Housing and Urban Development forced San Francisco (under threat of the loss of millions of dollars in federal grants) to revise its sixty-seven-year-old code to permit the plastic pipe. During 1972, total sales of the plastic pipe neared $550 million and 1.5 billion pounds. An estimated 218 million pounds of this went into drain-waste-vent systems, including 25 million pounds for mobile homes and 193 million pounds for new construction and repair of housing, as HUD worked to impose uniform plumbing standards across the nation.

Illegal Use of Copying Devices

Xerox copying machines and tape recorders are examples of new technologies which were primarily beneficial to man but which were readily susceptible to abuse. The new copying machine allowed businesses to reproduce various documents and became a time- and money-saving piece of equipment. The tape recorders allowed individuals a new selectivity in their musical tastes and the ability to record sounds that previously went unrecorded. However, elements of society also abused these devices. "Tape pirates" began making unauthorized reproductions of phonograph records and musical shows and then selling the tapes at a profit. Proprietary documents have been reproduced with copying machines for illegal use, and government classified documents have been copied and given to the news media for political purposes. The federal government is one of the biggest duplicators of scientific articles in the world.

The courts have ruled that the present copyright laws, which were passed before development of modern copying equipment, do not cover duplication of proprietary documents by these copying machines. Congress is currently working on new copyright laws to cover this area of technology.

Although the abuse of these innovations might have been anticipated, society and its institutions had not looked sufficiently ahead to be prepared for such abuse.

Concern Over Video Violence

Congressional committees and parent groups are exploring the effects on children of televised violence, which has become a big business. The possibility of adverse effects has been a matter of

serious concern and controversy among many parents. In 1972 the
U.S. Surgeon General's report included several significant findings.
Programs on three major networks averaged about eight episodes of
violence an hour during prime time. Despite increasing concern
about such violence, the amount had increased from 1967 to 1969,
particularly in the programs most frequently seen by children.
Much TV violence is committed against strangers or casual
acquaintances—suggesting the possibility to children that they can
become unwitting victims—and is typically depicted as a way to
reach a personal goal (for example, for private gain or power or to
perform a "duty"), rather than because of broad moral or social
issues.

The Surgeon General's study stated that TV entertainment may
be essentially a national mythology in which violence has the special function of explicating the values and norms held by our society with regard to power and influence. According to the report,
violence on TV is seen not as a mirror image of our society, but as an
idealized representation—social ideology in the garb of fantasy.

The oft-heard contention that children (or any viewers for that
matter) are not influenced by the violence that they see on TV is
easily refuted by consideration of the demonstrated vast influence
of TV advertising: society simply has not been able to measure the
influence of the televised violence.

Weather Modification

Cloud-seeding techniques, the basis of most weather
modification attempts, have been employed both scientifically and
commercially for about twenty-five years, with little or no governmental control. The small particles (nuclei) used in seeding may
influence the amount and type of precipitation, the electrical
phenomena in the clouds, and wind movements, thus providing a
potential for controlling rainfall, hail, lightning, and even hur-
ricanes.

Conflicting public opinion has often arisen in areas where
weather modification attempts have been made. Residents in
Ouray, Colorado, for example, objected strongly to an experimental
seeding program in the San Juan Mountain area; they feared that
additional snowfall might increase the number or severity of av-
alanches. Another common concern is that successful seeding in
one region might cause decreased precipitation in downwind areas.
Experiments in Florida in 1970 created a serious public relations

problem; vegetable growers, conservationists and cattlemen became so alarmed when local floods occurred a short time before proposed cloud-seeding tests that part of the program had to be postponed.

In mid-1972, rumors spread that the United States military had used rainmaking as a war tactic in Vietnam, and this possible use generated much animosity against the technology. The animosity turned to fear in many quarters when evidence suggested that cloud seeding had triggered a devastating flood at Rapid City, South Dakota, on June 9, 1972, that took over 200 lives and caused over $100 million in property damage. The National Oceanic and Atmospheric Administration ruled in late 1972 that all private cloud-seeding firms must report their activities and moved to require similar reports on the $100 million of weather modification research being done by federal agencies.

Deep Well Injection and Denver Earthquakes

Beginning in 1962, the Army began to dispose of some of its nerve gases and nerve-gas byproducts at its Rocky Mountain Arsenal near Denver by the method of deep well injection. By this method, liquid wastes are pumped into underground cavities in areas where they cannot get into underground water supplies. The quantity of wastes that the Army planned to dispose of was quite large—160 million gallons over the first four years.

About a month after the injection started, Denver had its first earthquake of the century. In the next five years, there were 1,500 tremors. All originated within a six-mile zone around the injection well. Several groups made studies of the earthquake problem; most concluded that the well was responsible. By June 1969, the Army had been forced to stop using the deep well. The big question then was what to do with the wastes already injected: some thought that the Army should be compelled to pump it out again, but others were afraid this reverse could cause a severe earthquake.

Phosphate Detergents

Washing detergents that combined phosphate builders with organic surfactants (many of them having as much as 50 percent phosphate by weight), were introduced in 1947 and were an enormous success. But by the late 1960's strong pressures were being brought by environmentalists upon manufacturers and politicians to halt the use of phosphates in detergents. They felt phosphates were the primary factor in eutrophication of lakes and streams, and that detergents were a major contributor of the phosphates.

Detergent manufacturers explored two approaches to reducing phosphates in their products. One was the use of NTA, but it was voluntarily removed from products in December 1970 because of feared environmental health effects. The second was to substitute carbonates and metasilicates for phosphates, but early in 1971 officials became aware that these products were strongly alkaline and therefore could burn the eyes, nose and throat if they were inhaled or ingested. Throughout the summer of 1971, testing showed the nonphosphate detergents would need more strict labeling because of their inherent danger to humans. In the meantime, utter confusion reigned in many local communities as phosphate detergent bans were issued, some of which were soon rescinded. The government apparently agreed that phosphates were environmental hazards, but it recommended in September 1971 that phosphate detergents were the best alternative available.

Fertilizer Runoff Hazard

During the 1960's the increasing alarm over water pollution focused attention on the agricultural "runoff" problem, that is, pollution contributed to America's streams and lakes by rainfall-carried sediments from tilled fields, wastes from huge livestock feedlots, pesticides, and the phosphate and nitrogen fertilizers. The nitrate types of fertilizers were of particular interest. Consumption of nitrate-contaminated farm water supplies had produced numerous cases of "blue babies," a serious condition in which formation of methemoglobin in the blood causes a depletion of oxygen. In addition, fears existed that nitrates could be reduced to nitrites in humans which, in turn, could form carcinogenic nitrosamines.

In the summer of 1971, the state of Illinois began hearings on the extent of nitrate fertilizer runoff, the hazard it posed, and what regulations, if any, should be placed on the use of fertilizers. Nationwide news coverage was given to the nitrate problem, and particularly to the claims of well-known environmentalist Dr. Barry Commoner. He reported that his measurements of nitrogen isotope ratios had shown that nitrate fertilizers were a greater contributor to nitrate levels in surface waters than were natural sources, such as soil organic matter. Dr. Commoner's analytical methods have been questioned, however, and the use of the nitrates and the controversy continue.

The Leaded Gasoline Problem

Tetraethyl lead has been used in gasoline since 1923 as an antiknock agent, to raise the "octane rating" of the fuel and thus permit higher compression in the engine before ignition and higher efficiency of combustion. The organolead compounds (tetramethyl lead is also used) decompose in the engine and pass out of the engine with the exhaust.* A "premium" gasoline may contain three or four grams of lead compound per gallon and even "regular" gasoline may contain 0.5 to one gram of lead compound per gallon. The use of lead compounds in gasoline increased rapidly after about 1940, and reached 524 million lb in 1968.

Lead contamination of the air is now a matter of nationwide concern. Although industry spokesmen and a controversial NAS report claim that present levels are not a problem, many authorities consider atmospheric lead a serious health hazard. Starting about 1970, efforts have been made to reduce the amount of lead used in gasoline. Cars built since 1970 have had lower compression engines that permit the use of regular rather than premium gasolines. One company introduced a lead-free gasoline with a considerable advertising campaign, and although U.S. government cars now use it, it is more expensive and consumers have not switched to it in large numbers. Some states and localities have considered banning the use of leaded gasoline.

The EPA has proposed that lead-free gasoline be made available at all service stations by 1974 and that all lead be removed from gasoline by 1975, a move which opponents claim would cost consumers five to seven cents a gallon because of increased refinery costs. Another reason given for the need to remove lead from gasoline is that it contaminates the proposed catalytic converters to which the auto industry has committed itself as a control device for other auto emissions, such as hydrocarbons, nitrogen oxides, and carbon monoxide. A requirement that lead-free gasoline be used in all 1975 cars poses numerous questions: will the oil companies be able to switch over enough refineries to make it? Will the lower

---

*Ethylene dibromide, tricresylphosphate, and other additives may be added to the gasoline to prevent the buildup of lead deposit in the engine. These additives can cause corrosive acids when burned.

efficiency require too much additional gasoline in a time of shortage? And will (as suggested by experimental data) the removal of lead from gasoline cause the emission of greater amounts of aromatic hydrocarbon pollutants? These complex problems must yet be resolved.

The Strontium-90 Alarm

Strontium-90 is the most potent fallout product of atmospheric nuclear bomb tests. During the 1956 election campaign the hazards of Sr-90 were frequently proclaimed to the public. In addition to its immediate radioactivity hazard, it replaces calcium in the bones, then decays and leaves the bones without the proper amount of calcium.

By 1959, most scientists were convinced that milk was the principal source of Sr-90 for humans. Then studies conducted in 1960 by Consumers Union, showed that Sr-90 is not selective and that some other foods were also significantly contaminated. To obtain an insight into the total Sr-90 intake of the American population, CU extended its sampling from milk to all the foods in a typical American diet.

In terms of strontium units, CU found the level of Sr-90 in the diet of young Americans to be significantly higher than had been indicated by previous tests made on milk alone. This information was a cause of concern to the public, raised widespread questioning of the adequacy of government monitoring, and identified an urgent need for systematic monitoring of Sr-90 in the total diet. The Sr-90 problem was effectively eliminated, however, after enactment of the atmospheric nuclear test ban treaty of 1963.

Uranium Mill Tailings in Grand Junction, Colorado

About 83 million tons of uranium mill tailings (sand-like material that is the residue from uranium refining) had piled up in nine Western states by 1969. These tailings, a waste byproduct of the atomic bomb manufacturing process, are radioactive and produce gamma and radon radiation. In the Grand Junction area of Colorado, about 200,000 tons of tailings were given away from one uranium mill between 1952 and 1966, and were used around homes, businesses, churches, and schools for construction, landfill, and even in concrete. The discovery that the rates of cancer incidence, birth defects, and cleft palate were higher in Grand Junction than in the rest of Colorado spurred an intense debate over the effects, if any, of the radiation and over whether or not the material should be removed. Costs of removing the tailings range from $45 to $15,000 for homeowners, depending on the individual situation. In January 1972, University of Colorado was awarded a contract by EPA to study effects on residents of the tailings contained in foundations of their homes. Chromosome defects and other health signs will be studied.

The Great Mississippi Fish Kill

In November 1963, a massive fish kill occurred on the lower Mississippi River. Several smaller kills had occurred during the autumns of 1960–1962, but these had usually been blamed on natural sicknesses. The size of the 1963 kill spurred intensive investigations and, subsequently, a substantial amount of controversy and confusion. Eventually, the U.S. Department of Health, Education, and Welfare, the U.S. Department of Agriculture, the U.S. Department of Interior, the U.S. Senate Government Operations Committee, and various regulatory, scientific, industrial and agricultural authorities in Louisiana, Mississippi, Arkansas and Tennessee became not only involved but often in conflict.

The cause of the fish kill could not be easily determined and the investigation headed by HEW's Public Health Service continued through the winter as samples of water, fish, and sediments were analyzed. The PHS team was lead by Dr. Donald Mount, a fisheries toxicologist whose doctoral research had been on the toxicity of the insecticide endrin, a chlorinated hydrocarbon that came into substantial agricultural use in 1958. By early March 1964, Mount and the PHS had concluded that minute amounts of endrin in the water had caused the fish kill. The PHS called in representatives of the only two American producers of this product to discuss the problem quietly. On March 19, PHS officials declared publicly that pesticides from farmland use had apparently killed the fish, and disclosed privately to reporters that endrin was the most likely cause, although dieldrin and two other still-unidentified chlorinated hydrocarbons were also found in the fish. Furthermore, both endrin and dieldrin were present in minute amounts in the treated drinking water of New Orleans.

The announcement set off a tremendous storm of public concern. Senator Abraham Ribicoff immediately pointed out that this was not a case of some accidental pesticide spill or overuse, but simply a result of "business as usual;" he concluded that this episode proved the need for the tough new pesticides legislation that he was pushing, and announced that his subcommittee would open hearings on April 7. USDA officials publicly doubted that agricultural use of endrin was the cause of the kill; although it is used on sugar cane and cotton in the area, it is used there before planting, that is, in the spring. The USDA announced it would open its own hearings on pesticides and on April 9, and USDA Secretary Orville L. Freeman went to testify before Ribicoff's committee to dispute the PHS conclusions. Secretary of Interior Stewart Udall also voiced concern over pesticide use.

The Ribicoff subcommittee hearings began with Assistant Secretary of HEW James Quigley testifying that the fish kill demonstrated that USDA should take steps to control the use of pesticides on farmland. By this time, however, PHS officials were indicating privately that they thought the endrin had come from an industrial source rather than from farmland runoff, because they had detected two chlorinated hydrocarbons ("x" and "y") believed to be intermediates in endrin production. In fact, they were already sampling the waters in and around Velsicol's plant at Memphis, Tennessee. On April 23, Ribicoff revealed publicly that PHS investigators had found heavy concentrations of endrin in sewer sludges near Velsicol Chemical Corporation's plant, and implied strongly that this was the cause of the fish kill.

At this turn of events, officials of the USDA and the rest of the pesticide manufacturing industry breathed a sigh of relief; the accusing finger swung away from agricultural pesticides in general—and toward a case of sloppy housekeeping that could be remedied. Velsicol officials, who at the USDA hearings had disputed the PHS conclusions that endrin use was to blame, were now forced to defend their own company at a PHS "Conference on the Pollution of Interstate Waters, Lower Mississippi" called by Secretary of HEW Anthony Celebrezze. At the conference, held in New Orleans, May 5–6, PHS officials pushed hard to convict endrin and Velsicol's plant. Numerous technical criticisms of PHS's data and conclusions were offered by spokesmen from Velsicol and the other manufacturer, Shell Chemical Company, and by several other scientists, but these were largely ignored by the lawyerly PHS official in charge. A conference conclusion was that industrial wastes and drainage in and near Memphis were sources of endrin in the Mississippi. Responsibility to remedy the matter fell on the Tennessee Pollution Control Board.

Critics had already claimed several weaknesses in the PHS case: the symptomology of the dying fish had been different from that resulting from endrin poisoning; the PHS had concentrated its analysis largely on catfish, which numbered some 175,000 in the kill, but had not checked the salt water menhaden, which numbered nearly 5 million dead; and the PHS had first said their analytical data implicated dieldrin as a factor in the kill, but dieldrin was not manufactured at Memphis (although it apparently was in previous years). In addition, how could the endrin kill fish four hundred miles downriver, and not at Memphis itself? On June 1, the USDA announced it had found no reasons to restrict the use of endrin or dieldrin. By the time Ribicoff called additional hearings of the subcommittee in June 1964, the PHS officials had further modified their position: they now contended only that endrin had been responsible for no more than one-third of the catfish killed. Velsicol sealed certain of its sewers and improved other waste disposal procedures. Efforts were made to reduce water contamination by other pesticide users along the river drainage system.

The fall of 1964 passed without a recurrence of the previous near-annual fish kills. To some, this demonstrated that Velsicol's endrin discharges had indeed been the cause of the kill. Subsequently, however, massive kills were observed along the Arkansas River and other rivers that were traced to such contaminants as organic wastes from livestock feedlots. Thus, the cause of the 1964 kill may have been from a combination of factors.

Polychlorinated Biphenyls

Polychlorinated biphenyls (PCB's) are used in industry in many ways, such as industrial heat transfer and dielectric fluids; fire retardants; softeners in plastics, paints and rubbers; and as additives in printing inks and in carbonless duplication papers. They are odorless, colorless, syrupy mixtures of related chemicals that are quite stable. Several PCB fluids have been manufactured by the Monsanto Company—it made an estimated 5 billion pounds between 1930 and 1970—and they were also made in Japan and Europe. Although not extremely toxic, PCB's can cause acne, impaired vision, abdominal pains, and even death after long-term exposure.

In the late 1960's PCB's were discovered as environmental contaminants in fish and birds. In 1968, one thousand Japanese became ill from "Yusho" disease, which was determined in 1969 to be caused by eating food that was cooked in rice oil contaminated with PCB's. In December 1970, 146,000 chickens in New York State were found to be contaminated with PCB's and were destroyed. In the summer of 1971, PCB's again made the headlines when they were discovered in turkeys, chickens, and eggs in excess of the 5-parts per million allowable concentration, including products sold in Washington, D.C. By this time, PCB's had been found throughout the environment and were suspected in some quarters of being a greater hazard than DDT. Under this impetus the government and the manufacturer soon agreed to restrict PCB use to closed systems, for example, electrical transformers, where it does not escape to the atmosphere and can be disposed of by incineration after use.

2,4,5-T Controversy

Developed in World War II as an anticrop agent, 2,4,5-T was first marketed as a domestic herbicide in 1948. The first indication of its teratogenicity came to light in June 1966 as a result of a general screening of pesticides done by the Bionetics Research Laboratories. Follow-up experiments were delayed two years, and in September 1968 the first report from Bionetics was delivered to the National Cancer Institute. Early in 1969, copies of the report were sent to the FDA, the Department of Agriculture, and the Defense Department. The 2,4,5-T was being used extensively in Vietnam for defoliation at this time, but none of the officials took steps to protect the people of Vietnam or the United States from the herbicide. The President's science adviser, Lee DuBridge, was informed of the report in October 1969 and made an official announcement of the report October 29. On April 15, 1970, the government suspended use of 2,4,5-T in Vietnam, and in May the herbicide was restricted for domestic use on cropland and around homes.

Investigations showed that some batches of 2,4,5-T caused blindness, anencephaly, cystic kidneys, and death in mice and rats, but that the major toxic effects were caused by chlorodioxin impurities rather than by the herbicide, itself. A committee of the National Academy of Sciences recommended restoration of 2,4,5-T to normal use in May 1971. EPA said further investigations were needed, and litigation continues.

The Gypsy Moth Dilemma

The gypsy moth is not native to America; it was accidentally released in Connecticut many years ago and developed into a serious threat to Eastern forests. The use of DDT during the 1950's and early 1960's was fairly effective in controlling the pest (and even eliminating it in Michigan), but use of DDT was halted by environmentalists. Some of these claim that no pesticides are required: parasites will develop to control the population of the moths. By the summer of 1970 about 260,000 acres of trees were defoliated by the gypsy moth caterpillar in New England and Northeastern states. In 1971, over 2 million acres were being devastated. Tourism, industries, and homeowners were adversely affected.

Debate centers upon methods of controling the moths without damaging the environment. New Jersey has attempted to develop parasites to control the growth of the gypsy moth. Ten species of parasites have been imported from Europe and Japan, and tests using parasites in 1971 had favorable results. Commercial interests are also studying methods and products to control the insects, such as a microbiological pesticide. No solution has yet been devised, however, and a nonpersistent pesticide, carbaryl, is being widely used in an attempt to confine the pest.

Snowmobiles and the Environment

Snowmobiles were originally built to provide overland transporation to remote areas for hunters and people needing such transportation. The concept caught on in the United States and became fashionable as a recreational pastime. Introduced in 1959 (when three hundred were sold), the number of snowmobiles in the United States today is almost 2 million.

The recreational models are high-speed vehicles of questionable stability and safety. The owners operate them on the streets in some areas, which is dangerous. They are noisy. Many people enter remote areas where few went in the past, upsetting the terrain and wildlife. Some operate them on frozen lakes, and several drownings have occurred. Many communities are now considering laws and regulations to control the operation of snowmobiles, which are now virtually uncontrolled.

Minibike Safety

The minibike has recently become a popular American vehicle, and with its growth in numbers have come several problems. Minibikes are particularly popular among young people, many of them not old enough to have a legal driver's license. Many bikes are not constructed to conform to the vehicle safety standards required for operation on public streets. The bikes are noisy when operated. Law enforcement agencies are now beginning to control the use of minibikes. Most minibikes are restricted to private property. Those that operate on the streets must be licensed, and operated by a licensed driver. In many communities a minibike must pass a safety inspection under standards set for motorcycles.

Plastic Bags and Child Suffocation

In 1957, dry cleaners first used thin plastic bags to protect cleaned clothes on their trip home; most dry cleaners soon adopted them. By June 1959, over fifty children had been suffocated by used bags that had either been reused in a hazardous manner (as a soil-proof mattress or pillow cover) or left around the home where children could get hold of them. Most of the victims were babies less than a year old. Although these were only a fraction of the approximately 1,200 children below four years of age that were suffocated that year, the public was greatly shocked.

The Public Health Service reported a total of sixty-three deaths from plastic bags in 1959, in spite of an extensive warning campaign that included attaching leaflets to all the bags. The manufacturers of the bags then tried to reformulate the plastic so that it would not build up the high static electrical charge that caused it to stick to the skin. Legislation was also passed that requied warnings on the bags and holes in the bags to permit a child to breathe if a bag got caught over a child's head; the holes also discouraged its use as a mattress cover. In spite of the government's and manufacturers' efforts, suffocations by the plastic bags still occur occasionally.

The Discarded Refrigerator Hazard

During the 1950's, accidents repeatedly occurred where children climbed into discarded refrigerators, became locked in, and died from suffocation. The number of incidents prompted the passage of a law in 1956 requiring that all new refrigerators be equipped with a type of latch that could be opened from the inside. A campaign was also launched to warn the public about the problem and to suggest that the doors be removed or holes drilled in the refrigerators before discarding them. Still the accidents continued and reached 160 in 1960. The rate has apparently declined as most of the older refrigerators were finally replaced.

Corrosive Liquid Drain Cleaners

The problem of stopped-up drains is familiar to most households. It has for decades inspired homeowners to pour lye or soda ash in their sinks periodically to ward off stoppages, and has supported thousands of sewer service businesses. While lye is extremely corrosive and soda ash highly caustic to skin or delicate tissues, these materials are solids and were seldom ingested in harmful amounts by children. But during the 1960's, new, highly concentrated (26 percent lye) liquid drain cleaners were introduced on the consumer market under a heavy barrage of advertising; many tragic reports soon appeared of children who had ingested and suffered burns and permanent damage to the esophagus from these dangerous products. In 1969 alone, eighty-nine children swallowed Clorox Company's *Liquid-Plumr* (thirty-four went to the hospital) and twenty-seven children swallowed *Industrial Strength Drano*. In 1970, a total of 169 cases of liquid drain cleaner ingestion were reported that resulted in fifty-one hospitalizations, many of which required several operations to repair the damage.*

In late 1970, the Secretary of Health, Education, and Welfare moved to limit the strength of liquid drain cleaners to a maximum of ten percent lye (or potassium hydroxide). Consumers Union, in a widely quoted article in the September 1971 issue of *Consumer Reports*, insisted that these products were still dangerous and decried the Secretary's timidity in not banning them from the consumer market under provisions of the Hazardous Substances Act of 1962. In 1973, child-proof caps were required on such products.

Petroleum-Base Furniture Polishes

Petroleum-base liquid furniture polishes can be lethal if swallowed in large doses, and even a small amount in the lungs can cause pneumonia. Manufacturers often masked the normal odor of

---

*One company settled suits for damage to six children for $1.8 million.

these polishes with fruit-flavored scents, and packaged them in beverage-like containers. Some brands were marketed in easy-to-open containers which toddlers often mistook for soda pop or juice bottles. In 1968, 782 cases of ingestion of petroleum-based polishes were received by FDA; 178 victims had to be hospitalized. Acceptable and less dangerous substitutes for these distillates are said to be available and have been recommended, and packaging changes will soon be required.

Glue and Aerosol Propellant Sniffing Hazard

In 1962, a new fad called glue sniffing developed among youngsters. The volatile solvents in glue, particularly airplane glue, are capable of causing a form of exhilaration or "high," but can also cause illness and even death. As public concern increased, several cities or states passed laws restricting the sale of glue. Some manufacturers then began adding oil of mustard, which causes coughing and sneezing, to the glue. The Adhesive and Sealant Council, however, claimed that this is an ineffective and potentially dangerous practice. The use of warning labels and in some cases less toxic solvents has been the primary solution to the technological abuse.

In 1967, another fad of sniffing aerosol cans began. Some of the deaths were due to the inhalation of the fluorocarbon propellant, while others were due to freezing of the larynx from the quick-freeze aerosol sprays. By 1971, one hundred deaths from aerosol inhalation had been recorded. In 1972, the FDA announced a plan to require a health hazard labeling on aerosol containers.

Leaded Paints

Lead was widely used in paints as a drying agent, a corrosion inhibitor, and a pigment, until 1955, when manufacturers voluntarily reduced the lead content for household paints to about one percent. It is estimated that 400,000 children suffer from lead poisoning, with annual costs of two hundred deaths and $200 million in lost earnings and medical care for victims. The major factor in child poisonings is the ingestion of flaking paint in older houses. Lead paint chips have a sweet taste and are attractive to children. In addition, lead poisoning is difficult to diagnose unless it is specifically tested for.

In March 1972, the FDA ordered reduction of lead in paint from one percent to 0.5 percent by the end of 1972, and to 0.06

percent by the end of 1973. This is largely a result of an increase in the number of child lead-poisoning cases recognized as such. The problem will remain with us a long time because of all the old buildings.

The Microwave Oven Safety Question

Microwaves are a relatively low energy form of electromagnetic radiation that have wave lengths between those of infrared rays and the longer radio waves. The first major application of microwaves was in radar, which was used during World War II for military purposes and subsequently adopted widely for auto and air traffic control, and for weather studies. Microwaves were also soon applied in communications. Although microwaves (unlike X-rays) are a nonionizing radiation, they can be absorbed by organic molecules or materials and cause an increase in temperature. While microwaves at low intensity levels pose little danger from brief exposure, at high levels they are capable of injuring biological tissues by their ability to penetrate and cause rapid heating throughout the exposed area. Early microwave devices were of relatively low power, however, and because of the warming sensation that exposure produced, serious injuries were apparently few.

By about 1950, microwaves had been applied for heating purposes, in diathermy machines, and for cooking foods in microwave ovens: at sufficiently high intensity levels they can heat many precooked foods in less than a minute and cook a small roast in half an hour. Microwave ovens became commercially available and were used for many purposes in industry (such as for quickly drying glue) and in certain establishments for the quick preparation of foods (for example, in hospitals). By the late 1960's, the price of the microwave ovens had decreased to the point that they were becoming more widely adopted in commercial establishments and even in the home.

Studies had shown that overexposure to microwaves posed special hazards to sensitive organs such as the eyes and the testes; in 1967, microwaves were added to the controversy over other radiation hazards. The Radiation Control for Health and Safety Act of 1968 empowered the Secretary of Health, Education, and Welfare to control microwave products (along with all other electronic products that gave off radiation). The controversy over microwave ovens continued to increase, however, as new cases of accidental injuries occurred (such as eye cataracts), and the Public Health

Service reported that radar technicians appeared to produce increased numbers of mongoloid babies. In 1970 a government–industry survey concluded that of the 200,000 microwave ovens then in use, about 10,000 of them might be leaking unlawful levels of microwave energy. In 1971, the Food and Drug Administration required the manufacturers to correct deficiencies on 11,000 ovens and in October of that year imposed strict new design standards. The debate continued into 1973, however, as Consumers Union reported that new ovens which had met the government specifications could, in everyday use, soon deteriorate to an unsafe condition, and as Congress prepared to review the effect of the 1968 Act.

The Acupuncture Debate

Medical authorities have historically been much concerned over questions of orthodoxy in those allowed to practice medicine and in the methods they used. In the United States, for example, the development of osteopathy, after 1874, and the less sophisticated chiropractic, after 1895, were regarded with great suspicion by orthodox medicine, since both of these practices placed emphasis on physical manipulations of the body (bones, muscles or nerves) rather than on drugs and surgery. Acupuncture, the insertion of very thin needles at specific points of the body, had apparently been used as a therapeutic treatment in China for centuries by small numbers of unorthodox medical practitioners. With the closing of parts and eventually all of China to occidental medicine under the Maoist regime after the late 1930's, the acupuncturists rose in prominence. By the late 1950's, acupuncture had been integrated into many Chinese therapy practices, and had even been employed as an anesthetic for dental work, tonsillectomies and other surgery. Acupuncture practioners also existed in other Asian countries, Europe, and even in the Chinatowns of New York and San Francisco. But acupuncture was little known in North American medical circles, and the word was not even listed in standard American desk dictionaries and encyclopedias in 1970.

In early 1971, the long frozen relations between China and the United States began to thaw, and Americans began to visit there for the first time in nearly a quarter of a century—first a ping-pong team accompanied by a few newsmen, and then two biology professors who visited research facilities. The biologists' reports that

Chinese doctors were performing major surgery on conscious patients who were anesthetized only by a few needles gained little serious attention in American medical circles. But the announcement in July that President Nixon would visit China early the next year generated great interest in everything Chinese. In the ensuing months, several American doctors visited China and other countries to view acupuncturists at work.

Acupuncture exploded into virtually an American fad after Nixon's visit in February 1972, and the revelation that James Reston, influential columnist and vice president of the *New York Times*, had undergone acupuncture for post-appendectomy discomfort while in China. Stories soon circulated of dozens of acupuncture experimenters in the U.S. and even of the marketing of acupuncture kits by entrepreneurs. Doctors were beseiged with calls from patients whose illnesses seemed hopeless. In July 1972, the National Institutes of Health announced that it would support research on the subject. By August 1972, acupuncture made the cover of *Newsweek* and a three-day meeting of the Acupuncture Society of America was held in Kansas City.

The medical profession was nevertheless greatly concerned about the new craze. American doctors were not trained to practice the new technique and Chinese practitioners in the United States were generally not licensed physicians. In November, a New York practitioner was arrested for performing acupuncture without a medical license. Throughout 1973, medical societies and many state medical boards pondered the question of what to do about acupuncture.

Aspirin—The Most Widely Used Drug

Aspirin (acetylsalicylic acid) has been part of the medical pharmacopoeia since 1899, and is the most widely used medicine in human history. Current American consumption of aspirin, its combinations, and related salicylates amounts to an estimated 20 billion tablets a year. Very little is known about how its beneficial effects are achieved, and its mode of action in the body is only now being slowly discovered.

Aspirin is not without adverse effects. It is known to slow blood-clotting ability in many people, and in a few cases aspirin can cause an erosion in the intestinal lining which can lead to intestinal bleeding. Dr. Vernon M. Smith, Professor of Medicine at the University of Maryland, in a recent study of one hundred persons treated for severe intestinal bleeding found that ninety-four had taken aspirin or a preparation containing aspirin within twenty-four hours of the start of the bleeding. Aspirin in large amounts is known, of course, to be toxic and its widespread and casual use has made it a leading cause of accidental child poisonings. Yet, when the Poison Prevention Packaging Act came into effect, many aspirin manufacturers chose, under a provision intended to help arthritics (who would ordinarily purchase a large-size package), to continue marketing the popular one hundred-tablet bottle without the child-proof cap.

The Heart Transplant Fad

In December 1967, the first semisuccessful heart transplant was performed in South Africa by Dr. Christian Barnard. The patient, Louis Washkansky, lived eighteen days—not long enough to be considered a great success, but better than the few hours obtained in previous operations. Washkansky died of pneumonia, when the medication used to prevent his body from rejecting the transplant caused it to lose its ability to fight disease.

Over fifteen other heart transplants were made within six months of the original transplant, and over ninety within the first year. Heart transplants were suddenly being made all over the world; they had become a virtual fad in the medical world. In addition, they stimulated more transplants of other organs. Although most people were excited by the news, these operations did raise some ethical, legal, and moral questions. One concerned the moment of death of the donor, as rumors circulated that one or two donors may still have been alive when their hearts were removed by the enthusiastic doctors. Professional concern also developed that the transplants would slow down the development of artificial hearts, which appeared to be of greater long-range importance.

Dr. Michael De Bakey, one of America's most famous heart surgeons, stopped making transplants after only two of twenty-one recipients in his practice survived more than two years. However, one American has now lived over five years with a transplant and some doctors are continuing with transplants. The fad, however, has run its course and the use of a transplant now seems dependent on the merits of each individual case, the doctor's discretion, and the availability of a donor.

The Methadone Maintenance Program

Methadone is a synthetic drug discovered in Germany in 1944 and developed to replace wartime shortages of morphine and other opiate painkillers. In sufficient doses, it can produce a "high" somewhat like that of heroin and is an addictive drug, although withdrawal from it causes less severe symptoms than from morphine or heroin. Methadone can be given orally to reduce the agonies of heroin withdrawal without producing a high; it was approved for this purpose and adopted during the late 1950's as a treatment. As with opiates, the body can develop a tolerance to high daily doses of methadone, and this tolerance carries over to the opiates.

In 1963, Dr. Vincent P. Dole and Dr. Marie E. Nyswander began federally approved studies in New York in which confirmed heroin addicts were given (orally) increasing amounts of methadone—up to a total of 80–120 milligrams per day, the so-called "blocking dose." When the addict was on a high-methadone regime, he no longer craved heroin and it would have produced little sensation anyway. The New York program, which used methadone under an investigational new drug (IND) application, was closely supervised and recorded. By 1970, some two thousand patients had been treated and the researchers concluded that the "methadone maintenance" was successful in eighty-two percent of the cases, that is, the addicts were off of heroin (although still on methadone) and living a life free of crime or other antisocial activity, and in some cases were leading socially productive lives in school or on a job.

By this time, however, word of the treatment method had spread. Over sixty IND-approved programs using methadone had begun operation around the country. It appeared that tens of thousand of addicts would soon be receiving the drug, if government agencies would just ease their restrictive rules. And here organizational and philosophical problems were involved. The National Institute of Mental Health had been designated to lead federal research on narcotics addiction and rehabilitation, while the Bureau of Narcotics and Dangerous Drugs (BNDD) in the Justice Department, and the Food and Drug Administration, in HEW, set the terms for the methadone maintenance programs. In addition, President Nixon had appointed a Special Action Office for Drug Abuse Prevention to coordinate matters in the drug area.

The FDA could not approve wholesale, long-term use of a drug for which the safety had not yet been satisfactorily proved. BNDD was much concerned about the narcotics control problem if the maintenance programs spread to thousands of physicians. In addition, many people complained that the use of methadone was merely substitution of one addiction for another. But under pressure, BNDD, FDA and NIH announced new liberalized regulations on methadone use in April 1971. Methadone manufacturers soon announced the development of methadone tablets that would be difficult to dissolve and inject by drug abusers, and that could be dispensed to patients in a full week's dose; this would reduce the high clinic operating costs incurred with daily patient visits ($1500 or more per patient annually). The use of methadone now increased rapidly. A program was launched in New York State to treat 20,000 addicts.

But by October 1971 a new phenomenon had been reported. Small children were being poisoned in the homes of addicts by methadone. In Detroit alone, forty cases including one death had been reported since a public clinic had opened there in March 1970. Reports of teenager deaths from methadone soon followed, and by the end of 1971, ninety-eight methadone overdose deaths had been recorded.

During 1972, the methadone dilemma created a wave of news headlines. Dozens of stories of abuse of the maintenance programs circulated and a black-market trade developed. Unscrupulous doctors conducted maintenance clinics for thousands of patients (some of which were apparently never heroin addicts). By April, FDA said it would have to stiffen the regulations for the over four hundred legal maintenance programs and their estimated 50,000 patients. By August, the BNDD was seeking to arrest the operator of the largest private program in New York, and fifty-seven methadone overdose deaths had been recorded in that state alone.

In December 1972, when 585 clinics were operating, FDA, BNDD and the President's Special Action Office announced new, more strict regulations on the methadone maintenance programs.

XYY Genotype and Antisocial Predisposition?

Scientists have discovered that some human males have an abnormal XYY genotype (that is, they have an extra male chromosome in each cell of their bodies). Publicity given in 1965 to early studies which suggested that such individuals had a higher delinquency rate gave rise to a popular stereotype of the XYY individual as physically aggressive and violent. False newspaper reports that Richard Speck (the convicted mass murderer of eight Chicago nurses) was an XYY created much concern among the public. The scientific findings raise fundamental legal, social and medical questions: can XYY's be held responsible for their actions; is preventive incarceration or surgery to be considered; should results of cytogenetic study be used in deciding to abort an XYY fetus; and should the doctor tell the individual or his family of an XYY diagnosis when he appears otherwise normal. The need to face such questions will depend on future studies to determine the risk factors involved.

The Surgeon General's Report on Smoking

In 1964 the Surgeon General of the United States issued a report linking lung cancer with cigarette smoking. The commission that made the study did no research of its own, but studied other papers on the subject. The evidence is mainly a concomitant relationship: no specific ingredient of smoke was named as carcinogenic.

Various regulations on cigarette smoking soon began to appear. Cigarette manufacturers were required to put a notice on packages: "Warning: The Surgeon General Has Determined That Cigarette Smoking Is Dangerous to Your Health." Cigarette advertising was also banned from television.

Many smokers simply ignored the warning. Some felt that the government does not regard the danger as too serious when it is subsidizing tobacco growers and spending millions of dollars on tobacco-growing research. However, the per capita consumption of cigarettes declined slightly in the last ten years, from 217 packs to 205 packs per year, although total consumption increased because of population gains.

Many people, including some nonsmokers, feel that cigarette smoking is a personal habit, that the government should not try to control it, and that smokers should be given the facts but be allowed to make their own decisions. Nevertheless a strong move is now under way to ban smoking in all public places where it could be objectionable to nonsmokers.

Questions of Life—And Death

The Merriam-Webster unabridged dictionary devotes over ten column inches to a score of definitions of the word "life"; perhaps no other word in the English language is at the center of so much controversy in our technological society. We are concerned not only with how much human life Planet Earth should support and what its quality should be, but also with questions of who should live, when are they actually "alive," and how are they to be treated. The specter of overpopulation has immense social, economic, political, psychological, and moral implications; modern technology has contributed not only to lengthening life, but also to methods of preventing new life. The "quality of life" has recently become of much interest—at least in the affluent countries where people can afford to worry about that kind of thing—but just how one measures it remains controversial. In many ways modern technologies seem to have added to the quality of life, but in some ways they clearly do not; people keep trying "to get away from it all."

Profound legal, medical, and ethical questions exist over when should (or does) life exist, and what norms of conduct toward that life should be practiced by those with power over its fate. The taking of human life has been practiced formally for criminal justice purposes in most countries of the world—although this practice has been stopped in some—and less formally for political purposes, eugenics, or a humanitarian rationale. In the United States, the use of capital punishment has been severely limited recently by the Supreme Court and is still being debated in the individual states. A more subtle dilemma has been posed by recent shifts in thought over the use of prison inmates (or others who may not be able to give their free, informed consent) for experimental medical research. The 1972 revelation that the U.S. Public Health Service had continued an observational study of 430 syphilitic black men long after effective penicillin treatment was available (the Tuskegee Study) has added greatly to pressures to control and restrict "human guinea pig" experiments of all types.

Of even greater controversy has been the question of how long and to what extent efforts should be made to support the life of those who may be terminally ill or suffering from uncurable diseases. Technological innovations in medicine and in medical equipment have made it possible to keep alive many persons who would otherwise die. Indeed, as the biomedical machines have become evermore sophisticated, and transplants of hearts and other organs have become widespread, several states tried to establish legal meanings for the word "death." But the American Medical Association's 244-member policymaking body refused in late 1973 to set a definition of death; it merely left that decision to the clinical judgment of the doctor in each case. While many people—from doctors to theologians—seem to agree that "heroic" and possible economically ruinous measures need not be employed in a hopeless case, little agreement exists on how extensively mercy killing or euthansia should be practiced. Despite wide agreement that "death with dignity" is a desirable goal, the question of "who should pull the plug?" (that is, make the decision to stop life-sustaining treatment) for the helpless patient continues to haunt medical moralists.

Equally great ethical and moral delimmas are centering on questions involving human conception and birth. As we have already seen, controversy has surrounded the technologies of artificial insemination and contraception. Similar controversies have raged over some state and federal laws authorizing sterilization at government expense or recommendation of persons with certain serious mental or even socioeconomic problems. Particularly troublesome has been the question of when does the unborn fetus receive the protection of our criminal and civil laws—when is it a human life under the law? A recent U.S. Supreme Court ruling that the states could not prohibit abortions during the first six months of pregnancy has stimulated nationwide protests of citizens participation "right-to-life" groups and efforts by many states to bring their laws on the subject into conformity with the Court decision while rejecting "abortion on demand."

The host of problems associated with research on either aborted or intended-to-be aborted fetuses, and on the question of whether efforts should be made to save the life of an aborted baby are the focal points of much current concern. Serious ethical questions also surround the use of the amniocentisis technique to determine the genetic qualities of the fetus: should it be used routinely, for example, to search for such tragic genetically-caused diseases as Tay-Sachs disease or sickle cell anemia (which occur primarily among Jews and Blacks respectively) so that abortions could be performed? And should it be used merely to predetermine the sex of the fetus, so that the parents can have a child of the sex they choose?

Further controversies are rising rapidly over such new biomedical technologies as test-tube fertilization of a human egg with sperm; implantation of a fertilized egg in a "surrogate mother"; and genetic manipulation—changing the gene structure of a developing human. Each of these biomedical technologies poses grave challenges to ethical and moral standards and to our societal institutions. A serious future challenge may be posed by "cloning," a word still incompletely defined even in medical dictionaries. In the present context, it refers to the possible asexual generation of a new human individual from a single cell of an existing person. (Such reproduction occurs naturally among some of the lowest species and has been produced experimentally on small amphibian species.) The two persons would have exactly the same genetic makeup and would presumably be virtually indistinguishable (at comparable ages) by physical or chemical criteria—although their different ages and experiences in life may lead to different mental makeups and interests. Through cloning, literally hundred of duplicates of the original might be produced. Undoubtedly enormous controversies will arise over whether this should be allowed, and who would be permitted to practice it, on whom, and for what purpose. Because our system of laws is based largely on individual responsibility, we would clearly need entirely new appraoches simply to identify people for purposes of criminal justice, recording educational attainment, ownership of property, inheritance, and the like, if cloning of humans is successfully developed and extensively practiced. And Webster's will need still other definitions of life—and death.

The Coal Tar Dyes Problem

Coloring materials have been used by man since the beginnings of civilization to adorn his body, his home, his clothing, and even his food. In 1771, the first synthetic dye was made, and during the later half of the nineteenth century an increasing number of dyes based on coal tar chemicals were commercialized. By World War II, hundreds of millions of pounds of synthetic coal tar dyes were consumed annually in the United States. Under the Pure Food and Drug Law of 1906 "harmless" coal tar dyes were to be permitted in foods and drugs, and nineteen colors ("FD&C colors") were eventually approved. Dyes that had been found to be toxic when ingested were banned. Under the federal Food, Drug, and Cosmetic Act of 1938, coal tar dyes used in cosmetics were also brought under regulation and over one hundred "D&C colors" were approved.

In 1950, many children became ill after eating Halloween candy that contained an unusually high concentration of an approved orange color. The event triggered a reexamination of the toxicities of all the FD&C colors and some of the D&C colors. By 1954, serious concern had developed over the use of three coal tar dyes in foods and in cosmetics. But when the FDA in 1955 tried to ban the use of these colors (including Orange No. 32 used by citrus producers), the matter was taken to the courts and the Congress. In 1956, Congress passed a bill to rescind a USDA ban on Orange No. 32 to color oranges artificially, and the FDA agreed to allow its use for three more years while tests continued. During 1957 and 1958, the proposed ban on the three coal tar dyes continued to be a matter vexing the FDA and the courts (even the Supreme Court). But during the early 1950's evidence had been developing that certain coal tar chemiclas (such as the flavorings, safrole and coumarin) were carcinogenic. The passage of the 1958 Food Additive Amendment, with its Delaney Clause on carcinogens, raised new concern about coal tar dyes.

The FDA announced in April 1959, that its studies had found seventeen D&C colors unsafe and proposed to ban their use. The proposal, which involved thirteen colors used in lipsticks, created a furor in the $80 million cosmetics industry. The Senate attempted to respond by passing a bill that would have permitted the use of "safe" amounts of coal tar dyes, but in October FDA announced it had limited the thirteen colors to "external" uses (excluding lipstick), banned the other four D&C colors (which were little used) and banned seven of the FD&C colors.

The food and cosmetics industries protested vigorously and the matter went into public Congressional hearings during the first half of 1960. But in July 1960, a bill was finally passed that banned unsafe color additives in foods and cosmetics. It shifted the burden of proof of safety to the manufacturers and permitted FDA to establish curbs. Since 1960, a succession of colors made news as new evidence of hazard has been reported or a ban has been proposed, protested or enacted.

Fortified Breakfast Cereal Issue

On July 23, 1970, Robert B. Choate, Jr., a former government consultant on hunger (but trained as a civil engineer), testified before a Senate subcommittee that forty of the sixty leading brands of dry breakfast cereals were so low in nutritional value that they were "empty calories" that would "fatten, but do little to prevent malnutrition." Choate presented a chart that ranked the sixty brands on the basis of nine nutrients present—a chart that showed many of the highly-advertized, popular brands far down on the list—and displayed a series of TV commercials to illustrate his contention that the cereal manufacturers were pushing sales of the lower-quality products via children. Choate's testimony immediately drew the wrath of the food industry (which already had large numbers of critics), touched off a debate that splashed on front pages across the nation, and boosted sales of high-rated (but previously slow-moving) brands.

A few days later, the cereal company executives and industry nutritionists testified. They challenged Choate's qualifications, disputed his claims and contended that a breakfast of cereal and milk was as nutritious as one of bacon and eggs. Critics quickly countered that the milk alone would be largely responsible for the value in a breakfast of many of the best-selling cereals, but other nutritionists noted that much research would still be needed before reliable nutritional standards could be set. The cereal industry failed, however, to refute the charges that its $87.5 million worth of TV advertising annually* was misleading the children. Following the hearings, the Federal Trade Commission began to take a more active interest in food nutrition claims in advertising, and the Food and Drug Administration imposed new standards on nutrients allowed in foods, including cereals.

Irradiation Sterilization of Food

Since 1960 the Atomic Energy Commission has had an active research and development program in food irradiation, that is, exposure of food to ionizing gamma or beta rays to sterilize or pasteurize food products. Examples include killing the trichinae organisms in pork or insect infestations in grain, and sterilizing a variety of foods sufficiently to extend their life substantially without refrigeration.

The chemical reactions which occur in food as a result of the ionization process are exceedingly complex, but FDA experts state that there is no introduction of radioactivity into the food. Questions exist, however, as to whether irradiated food might cause cystogenetic or genetic damage in animals or humans. Although research is being conducted by the AEC to answer this question,

*Cereal sales were only $700 million/year.

resolution of the issue to the satisfaction of everyone is doubtful. To prove a complete absence of genetic damage would require raising many generations of animals with many animals in each generation, a long-term program. The problem of appraising the safety of irradiation sterilization is thus complex and of wide scope; a great deal of work remains to be done to establish the safety of this process. Meanwhile, public concern exists with regard to the safety of this method of sterilization.

Sodium Nitrite in Meat

Sodium nitrite is used as a food additive in such meats as frankfurters, bologna, spiced ham, Vienna sausage and smoked salmon. It serves two functions: it maintains the red color of meat and acts as a preservative by preventing the growth of *Clostridium botulinum* spores, which can produce lethal botulinus toxin. Concern over the use of nitrite in food arose when experiments in animals revealed that if large amounts were fed in combination with secondary amines, the two react to form nitrosamines, which are carcinogenic.

The FDA contended that fears about nitrites were greatly exaggerated and that no reduction in the allowable levels was necessary. FDA noted that no one had yet established whether sodium nitrite is or is not carcinogenic, claimed that additional data must be developed, and said that they are working to develop this data. Critics, on the other hand, contended that the wide assortment of meats that are heavily laced with nitrites were a hazard, and claimed that the FDA was already withholding important test results in violation of the Freedom of Information Act. In the face of a court suit brought by the Environmental Defense Fund, the FDA revised it policies and announced that it would reveal previously guarded test results on many food additives, including sodium nitrite. In 1972, a report prepared by the House Committee on Governmental Operations was highly critical of FDA's handling of the nitrite problem, while a report from the National Academy of Sciences concluded that the nitrite levels allowed in foods by FDA appeared to be safe. Experimental data to resolve the issue is accumulating rapidly.

Food Vending Machines

Vending machines have become an $8 billion a year business, $400 million of which consists of prepared foods such as sandwiches, pastries, salads, and hot meals. Rising consumer complaints during the 1960's concerning the sanitary conditions of vending machines prompted some health officials to see them as potential sources of food poisoning. Improper food preparation, improperly planned transportation of the food, and unclean or defective vending machines could cause problems.

The vending machine business has outgrown local and state health departments, and federal laws and inspections may be required. Included might be laws to regulate distribution of the machines and spot-checking the food in them. The most serious problems arise from "fly-by-night" operators who invest money from newspaper ads and don't even license the machines in many cases. Most of these people are average citizens seeking part-time income or starting a new business who are not familiar with food regulations. Another serious problem is created by firms that rebuild old machines. Many of the rebuilt models, and there are many of them in operation today, are not built to specifications of the FDA. Vending machines still represent an inadequately controlled industry in the consumer market.

The Polywater Controversy

In 1968, Dr. Boris V. Derjaguin, a noted physical chemist in the Soviet Union, discovered a substance condensed in freshly drawn glass capillary tubes. Upon examination of the substance, he concluded it was a previously undiscovered "anomalous" form of water.

A controversy immediately developed over whether it really was a new compound or merely a contaminated form of water; dozens of research projects were soon initiated to study the new phenomenon. Scientific papers soon appeared describing the difficulties and best methods of making it, and theoreticians proposed various possible structures, such as a polymeric form. The controversy became quite intense; in general it was kept within the scientific community, but several articles on it appeared in the popular press. At one point it was suggested that if the scientists' "polywater" were loosed to the environment, it might serve as a condensing point, and turn all water into polywater, thus destroying all life.

Despite repeated efforts, only tiny amounts of the mysterious substance could ever be made and evidence increased that it was merely a viscous water solution of substances from the glass. In

1973, Dr. Derjaguin, one of the staunchest defenders of its being a new form of water, concluded that polywater had never existed.

Project Mohole

Early in 1957 a plan was conceived to explore through the earth's crust to the heavy underlying layer by drilling. Drilling on land was not feasible, but drilling through the ocean floor appeared to be practical since the earth's crust is thin at the ocean's bottom. The National Science Foundation, the sponsor of the project, felt that useful scientific data could be obtained by examining the earth's core.

In 1961, a National Academy of Sciences team made preliminary studies, funded by NSF, of deep-water drilling from a floating platform. Their success, coupled with the fact that Russia was attempting the same thing, setting up a "race," led to further exploration. As time went by it became obvious that the goal was going to be most difficult to achieve. A much larger organization, considerable further development in the state of the drilling art, and investment in a large floating platform capable of sustaining life and stability over a 2½-year drilling campaign would be required. After a sizable investment in time and drilling technology, Congress and the NSF abandoned the project in the summer of 1966. What started out to be a scientific conquest had resulted in a dubious distinction of failure by a prestigious governmental science body.

Suggested Additional Readings

The Alaska Pipeline Controversy
Science, March 19, 1971, pp. 1130–1132
Science, March 9, 1973, pp. 977–981
Science, July 27, 1973, p. 326

The Four Corners Power Plant–Black Mesa Coal Issue
Saturday Review, June 3, 1972, pp. 29–34

The Great Northeast Power Failure
Business Week, Nov. 13, 1965, pp. 41–46
Time, Nov. 19, 1965, pp. 36–43

The Storm King Project Controversy
Science, Oct. 1, 1971, pp. 44–46

The Coal Strip Mining Dilemma
Appalachia, Feb.–March 1972, pp. 1–21

The Calvert Cliffs' Decision and the Nuclear Power Plant Debate
Science, Sept. 1, 1972, pp. 771–775
Science, Sept. 8, 1972, pp. 867–871
Science, Sept. 15, 1972, pp. 970–975

The Automobile Recalls
Newsweek, March 10, 1969, pp. 79–80
U.S. News and World Report, March 17, 1969, p. 14
Time, Jan. 26, 1970, p. 58

Studded Snow Tire Question
Popular Mechanics, Nov. 1971, p. 36

Passive Restraints in Automobiles —The Air Bag
Chemical Week, March 22, 1972, p. 13

Carbon Monoxide in the Corvair
Consumer Reports, Sept. 1971, pp. 572–574

Highway Deicers
Water Pollution Control Research Series: Environmental Impact of Highway Deicing, Environmental Protection' Agency, printed by the Government Printing Office

The Skyjacking Nightmare
U.S. News and World Report, Sept. 21, 1970, pp. 17–19
U.S. News and World Report, Dec. 28, 1970, pp. 15–16

The SST Debate
Science, Aug. 6, 1971, pp. 517–522
Science, Oct. 6, 1971, pp. 386–388

The Sea-Level Panama Canal
Business Week, Feb. 15, 1964, pp. 45–46
Science, Jan. 29, 1971, pp. 355–356
Bio. Science, Vol. 19, No. 1, pp. 44–47

The Fiberglass Pole Controversy
Sports Illustrated, Feb. 26, 1962, pp. 10–13
Sports Illustrated, Feb. 25, 1963, pp. 39–42

Plastic Plumbing Pipe Controversy
Chemical Week, Aug. 9, 1958, pp. 77–80
Chemical Week, Sept. 21, 1963, pp. 89–96

Concern Over Video Violence
Consumer Reports, Feb. 1970, pp. 109–111
Science, Feb. 11, 1972, pp. 608–611

Cloud Seeding and Weather Modification
Science, Sept. 14, 1973, pp. 1043–1045

Science, June 29, 1973, pp. 1347–1350
Science, June 16, 1972, pp. 1191–1202
Deep Well Injection and Denver Earthquakes
Environmental Science and Technology, Feb. 1972, pp. 120–122
Phosphate Detergents
Chemical Week, April 28, 1971, pp. 10–12
Science, Oct. 20, 1967, pp. 351–355
Consumer Reports, Sept. 1970, pp. 528–531
Chemical Engineering, June 1, 1970, pp. 70–72
The Leaded Gasoline Problem
Consumer Reports, March 1971, pp. 156–160
EPA Report, by A. P. Altschuller, Feb. 1972, "The Effect of Reduced Use of Lead in Gasoline on Vehicle Emissions and Photochemical Reactivity."
The Great Mississippi Fish Kill
Science, April 3, 1964, pp. 35–37
The New Republic, May 2, 1964, pp. 4–5
Science, Dec. 24, 1965, pp. 1732–1733
Science, June 3, 1966, pp. 1388–1389
Chemical Week, July 25, 1964, pp. 19–26
Polychorinated Biphenyls (PCB's)
Science, Sept. 29, 1972, pp. 1191–1192
Science, May 5, 1972, pp. 533–535
2,4,5-T Controversy
Science, May 1, 1970, pp. 544–554
Science, Dec. 24, 1971, pp. 1358–1359
The Gypsy Moth Dilemma
Science, July 7, 1972, pp. 19–27
Chemical Week, Feb. 2, 1972, p. 35
Snowmobiles and the Environment
Consumer Reports, Jan. 1973, pp. 45–49
Plastic Bags and Child Suffocation
Newsweek, June 1, 1959, p. 82
Life, June 8, 1959, pp. 117–118
The Discarded Refrigerator Hazard
McCalls, Nov. 1961, pp. 114–115
Good Housekeeping, March 1964, p. 160
Corrosive Liquid Drain Cleaners and Petroleum-Based Furniture Polishes

Consumer Reports, Sept. 1971, pp. 529–531
Chemical and Engineering News, Aug. 23, 1971, pp. 12–16

Glue and Aerosol Propellant Sniffing Hazard
Science, Nov. 20, 1970, pp. 866–868
Consumer Reports, Jan. 1963, p. 40

Leaded Paints
Environmental Science and Technology, Jan. 1972, pp. 30–35
Science, Nov. 19, 1971, pp. 800–802
Science, April 30, 1971, pp. 466–468

The Microwave Oven Safety Question
FDA Papers, May 1972, pp. 14–30

The Acupuncture Debate
Newsweek, Aug. 14, 1972, pp. 48–52
Science, Aug. 18, 1972, pp. 592–594

The Heart Transplant Fad
Newsweek, Dec. 15, 1967, pp. 64–72
Business Week, Jan. 6, 1968, pp. 98–100
Reader's Digest, Oct. 1968, pp. 275–279, 284–306

The Methadone Maintenance Program
Science, May 8, 1970, pp. 684–686
Science, May 26, 1972, pp. 881–884
Science, Feb. 23, 1973, pp. 772–775

The Surgeon General's Report on Smoking
American Scientists, March–April 1971, pp. 246–252
Consumer Reports, Feb. 1968, pp. 97–103

Questions of Life—And Death
Medicine, Technology, Ethics, Fortress Press, 1970, p. 199
Science, Feb. 12, 1971, p. 540
Science, Aug. 20, 1971, p. 694
Science, Nov. 19, 1971, p. 779
Science, May 25, 1973, p. 840
Chemical and Engineering News, July 31, 1972, p. 14
Science, March 3, 1972, p. 949
Research on Human Subjects, Russell Sage Foundation, 1973, p. 264
Scientific American, Dec. 1970, p. 45
Science, Jan. 7, 1972, p. 40

The Coal Tar Dyes Problem
Consumer Reports, Feb. 1960, pp. 96–98

Fortified Breakfast Cereal Issue
U.S. News and World Report, Aug. 17, 1970, p. 45
Business Week, Oct. 3, 1970, p. 59
Sodium Nitrite in Meat
Science, July 7, 1972, pp. 15–18
Polywater Controversy
Chemical and Engineering News, June 29, 1970, pp. 7–8
Chemical and Engineering News, Nov. 9, 1970, pp. 7–8
Science, April 16, 1971, pp. 231–242
Scientific American, Nov. 1970, pp. 52–64
Chemical and Engineering News, July 16, 1973, pp. 13–14

Chapter V

Analysis and Reflections

Science and technology have become woven inextricably into the fabric of modern society, but—as many observers have noted—not without some losses as well as gains in the quality of our lives. Indeed as the preceding cases illustrate dramatically, technologies affect today almost every phase of our lives. Modern technologies have influences on the individual from conception and birth to death and burial—whether awake or asleep, at work or at play—and have direct effects on the community, on society's institutions, and even on the global environment. And as we have seen, these technologies are frequently controversial; their impacts often adverse. On the other hand, new technologies are almost certain to develop and grow, because of man's inherent inventiveness and because science and technology have contributed substantially to the improvement of human comfort and the economy. How then does society go about channeling its technological growth to obtain additional benefits, without having—in today's jargon—the accompanying "disbenefits"? In this chapter we examine and compare the preceding cases of social shock over technology to see how they originated, what and who they involved, what was done about the problems, and what they have in common. In addition, we want to examine briefly the effects that such cases of public alarm and of technology in general have on society's institutions. The chapter is organized as shown by the table of contents below.

Academic/Professional Precursor of the Technology
Important Foreign Technological Inputs
The Scale of the Technology

Technology Assessment and Social Shock
   TA to Avoid Technological Threats
   TA to Reduce Social Shock

Reflections on the Impacts of Technology on Society's Institutions
   Legislative Branches of Government
   Executive Branches of Government
   Judicial Branches of Government
   National Defense Systems
   Law Enforcement Systems
   Agriculture and Food Production Systems
   Industry and Commerce
   Labor Unions
   Financial Systems
   Communication Systems and the News Media
   Health Care Delivery Systems
   Educational Systems
   Cultural/Entertainment/Recreational Systems
   Religious Organizations
   A Parting Note

   Let's look at and compare the study cases to see what threads of commonality they have and what differences existed among them. In particular, we will want to examine the technology that was involved in each case (where it came from and why and how it was used), the kind of threat it posed, why this threat had developed, the impacts of the discovery of this threat on the public and on society's institutions, what was done about it, and how the whole episode might have been handled differently. In making these analyses, we are well aware that the forty-five case histories, or even the one hundred total cases, do not provide as large a sample as one would like for statistical purposes, and that the criteria and method for selection of cases (described in Appendix A) have introduced a degree of bias. In addition, as the reader will note, some aspects of the analyses have required substantial measures of subjectivity on the part of the project team (details of the analysis methods are given in Appendix A). Despite these limitations, we hope that these analyses are both interesting and stimulating to the reader.

A. The Technological Origins

A technology does not develop and exist of its own accord; a scientific discovery is applied by a segment of society to serve some human cause. Let's look at the technologies involved in our case histories and see why they were developed, what they were, where they came from, and how big or important they were.

1. *The kinds of technology implicated:* Biomedical-related technologies were involved in thirteen of the forty-five study cases while food and beverages, consumer products, manufacturing, and military technologies were involved in seven cases each, as shown in Table VII. Because of the selection criteria, the observed wide range of implicated technologies was expected. With a more extensive or a different list of cases (even the fifty-five cases synopsized in Chapter IV), somewhat different results would be expected. These categories, however, would probably be heavily represented in any study of social shock over technology.

2. *Human needs that the technology served:* The food, medical care, and national defense areas predominate as the major needs served in the forty-five case histories, as shown in Table VIII. Industrial/commercial needs, waste disposal, energy requirements, communications, and recreation/entertainment were each involved at least four times. Note that some duplication of needs occurs (for example, detergent technology serves both a clothing need and a waste disposal need).

A survey of the study cases suggests that when a societal need or activity exists, a variety of technologies will be utilized or proposed to fulfill it. Each of these technologies has, however, a potential for having effects other than that intended. As we will see, the cases of concern often arise when there is a conflict of values—a technology is introduced to serve one of society's needs, but later poses a threat to another of society's segments.

3. *Academic/professional precursor of the technology:* The academic or professional sciences and disciplines that can be regarded as the forerunners of the technologies that were implicated in the forty-five study cases include fifty-five citations in four major categories as follows: Physical Sciences (twenty-two cases), Engineering Sciences (sixteen cases), Health Professions (ten cases) and Business Administration (seven cases). The most-cited specific discipline by far was chemistry with medicine, physics, and military science being cited frequently. The complete results are shown in

TABLE VII

TECHNOLOGIES IMPLICATED IN THE STUDY CASES

No.	Technology	Number of Cases
	AGRICULTURE-RELATED	
1.	Animal and Plant Breeding	(1)
2.	Cultivation, Fertilization, Feeding, Land Management Techniques	(1)
3.	Farm Equipment	(0)
4.	Pest Control	(2)
	FOODS AND BEVERAGES	
5.	Processing and Manufacturing	(4)
6.	Packaging and Distribution	(1)
7.	Preparation and Serving	(1)
8.	Marketing and Franchising	(1)
	BIOMEDICAL-RELATED	
9.	Experimental Research	(2)
10.	Medicines and Drugs Production	(2)
11.	Medical Equipment Manufacture	(1)
12.	Personal Health Care Delivery	(3)
13.	Public Health and Safety	(2)
14.	Population Control	(3)
	CONSUMER PRODUCT-RELATED	
15.	Appliances Manufacture	(1)
16.	Clothing	(1)
17.	Cosmetics	(1)
18.	Detergents and Soaps	(3)
19.	Furniture	(0)
20.	Lawn and Garden Equipment	(0)
21.	Service and Repair of Consumer Products	(0)
37.	Street and Highway Maintenance	(0)
38.	Traffic Engineering	(0)
39.	Railroad Operations	(1)
40.	Shipping	(2)
	COMMUNICATIONS	
41.	Telephone/Telegraph Transmission	(0)
42.	Radio/Television	(1)
43.	Recorders/Replayers	(0)
44.	Printing/Reproduction	(0)
45.	Information Storage/Retrieval	(0)
	CIVIL ENGINEERING PROJECTS	
46.	Dams and Reservoirs	(1)
47.	Draining and Channeling, Harbor Improvement	(1)
48.	Surveying and Mapping	(0)
49.	Landscaping	(0)
	ENERGY-RELATED	
50.	Coal Production and Utilization	(0)
51.	Gas and Oil Production, Transportation, Refining and Use	(2)
52.	Hydro-energy	(0)
53.	Nuclear Energy	(1)
54.	Solar, Geothermal, Wind and Tidal Energy	(0)
55.	Electricity Production and Transmission	(0)
56.	Energy Conversion and Storage	(1)
57.	Ultrasonic Microwave Technologies	(0)
58.	Explosives and Propellants	(0)

TABLE VIII

THE NEED OR WANT THAT THE TECHNOLOGY SERVED OR
AIDED

No.	Human Need or Want	Number of Citations
1.	Water Supply	1
2.	Food Supply and Agriculture	9
3.	Sleep	1
4.	Reproduction, Human	2
5.	Medical Care, Public Health	7
6.	Care of Elderly, Infirm, Underpriviledged	0
7.	Clothing	4
8.	Housing and Buildings	1
9.	Communication	4
10.	Education	0
11.	Transportation	0
12.	Security and National Defense	8
13.	Labor Supply	0
14.	Money Supply and Availability	0
15.	Energy Supply	4
16.	Waste Disposal	5
17.	Industry and Commerce, NEC	6
18.	Control over Forces of Nature (floods, etc.)	1
19.	Recreation/Entertainment/Cultural Values	4
20.	Religion	0
21.	Government	0
22.	Other	0
		57

Table IX. In many of the cases the technology had two or more precursors or related disciplines. Some subjectivity was involved in making choices.

In addition to sciences or disciplines being precursors to the technology, they were often involved in the later development of the case, as will be discussed subsequently. In all, approximately one-half of the forty-five cases ultimately involved chemistry or medicine (or more precisely, chemicals and medicinals). This high proportion probably occurred* because the development of the chemical and pharmaceutical industries came about largely in this century; they were built upon the basic scientific discoveries of the

*The preponderance of backgrounds among project members in chemistry and chemical engineering may have contributed.

nineteenth century, but became sufficiently large that their adverse effects caused significant alarm only after about 1945. In addition, all material things are made up of "chemicals," and their compositions have become generally well known during this century. In contrast, mechanical engineering was involved only tangentially in a few of the forty-five cases. This low proportion probably occurred because the introduction of new mechanical devices was in its heyday during the nineteenth and early twentieth centuries; improvements in occupational safety have helped reduce the terrible tolls of accidents of earlier times: occasional accidents are now more or less accepted, and generate little media interest. In the opposite direction, however, are such new disciplines as genetics, biomedical engineering, and nuclear engineering, which have been moving from the basic-science level to the applied-technology level during the last quarter century. Many of the adverse effects of such new technologies may still be unsuspected or of uncertain size. Many academic disciplines are not intimately involved in the development of technology, of course, but all become involved in using technology or are affected by it, and members of some disciplines may become involved in the abuse of a technology which their discipline did not develop.

4. *Important foreign technological inputs:* In at least fourteen of the cases (31 percent) important inputs from foreign sources clearly occurred, and an additional five cases (11 percent) had more remote foreign inputs. In a few cases a foreign input may not have been recognized, while in others it was a tenuous connection or had long preceded the problem (for example, the x-ray cases; asbestos).

5. *The scale of the technology:* The scale of the technology, in terms of both its "size" (number of units, millions of pounds of production, dollars of services, or percent of the maximum market, etc.), and the numbers of people involved in its use (either as producers or consumers), is an important factor in its potential to cause social shock. Each of the study cases was rated from 1 to 5 (where 5 is very large) in terms of these two scales. The number of cases rated at each relative scale is shown below.

Factor	Relative Scale	Number of Cases				
		1	2	3	4	5
Size of Technology		5	14	16	7	3
Number of People Involved		17	8	11	7	2

TABLE IX

ACADEMIC AND PROFESSIONAL SCIENCES AND DISCIPLINES; FORERUNNERS OF THE TECHNOLOGY

No.		Number of Cases
	AGRICULTURAL AND RELATED SCIENCES	
1.	Agronomy	(2)
2.	Animal Husbandry	(1)
3.	Farm Management	
4.	Fish and Wildlife Management	
5.	Forestry	
6.	Horticulture	
7.	Soil Science	
8.	Veterinary Medicine	(1)
	BIOLOGICAL SCIENCES	
9.	Anatomy	
10.	Biochemistry	
11.	Biology	(1)
12.	Biophysics	
13.	Botany	
14.	Ecology	
15.	Entomology	
16.	Genetics	(1)
17.	Microbiology (including Bacteriology, Mycology, Parasitology, Virology, etc.)	
18.	Pathology	
19.	Pharmacology	
20.	Physiology	(1)
21.	Plant Pathology	
22.	Plant Physiology	
23.	Zoology	
	BUSINESS AND ADMINISTRATION	
24.	Accounting	
25.	Advertising, Public Relations	
26.	Clerical Work, Office Work	(3)
27.	Finance (Banking, Capital Management)	

No.		Number of Cases
	BUSINESS AND ADMINISTRATION (*Continued*)	
28.	Industrial Relations, Personnel Work	
29.	Management	
30.	Marketing, Market Research	(2)
31.	Production, etc.	(2)
32.	Public Administration	
33.	Purchasing	
34.	Sales (Retail or Wholesale, Real Estate, Insurance, etc.)	
35.	Secretarial Science	
	ENGINEERING SCIENCES	
36.	Aeronautical	
37.	Automotive	(1)
38.	Biomedical	(1)
39.	Civil (including Agricultural, Architectural Sanitary)	
40.	Chemical (including Ceramic)	(3)
41.	Electrical	(3)
42.	Industrial	
43.	Mechanical (including Marine, Welding, Textile)	
44.	Metallurgical	(2)
45.	Mining (Geological, Geophysical, Petroleum)	(3)
46.	Nuclear Engineering	(3)
47.	Traffic Engineering	
	EDUCATION	
48.	Elementary	
49.	Secondary	
	English	
	Foreign Languages	
	History, Social Studies	
	Mathematics	
	Natural Sciences	

TABLE IX (Concluded)

No.		Number of Cases
	EDUCATION (*Continued*)	
50.	Specialized Teaching Fields	
	Agricultural Education	
	Art Education	
	Business Education	
	Counseling and Guidance	
	Education Administration and Supervision	
	Educational Psychology	
	Education of Exceptional Children (Speech)	
	Home Economics Education	
	Industrial Arts Education (Nonvocational)	
	Music Education	(1)
	Physical Education, Health, Recreation	
	Trade and Industrial Education (Vocational)	
	HEALTH PROFESSIONS	
51.	Dentistry	(1)
52.	Medical Technology	(1)
53.	Medicine	(8)
54.	Nutrition	
55.	Occupational Therapy	
56.	Optometry	
57.	Pharmacy	
58.	Physical Therapy	
	HUMANITIES	
59.	Classical Languages and Literatures	
60.	English, Creative Writing	
61.	Fine and Applied Arts	(1)
	(Art, Music, Speech, Drama)	
62.	History	
63.	Modern Foreign Languages and Literature	
64.	Philosophy	
	PHYSICAL SCIENCES	
65.	Astronomy	
66.	Chemistry	(14)

No.		Number of Cases
	PHYSICAL SCIENCES (*Continued*)	
67.	Geography	
68.	Geology	(2)
69.	Metallurgy	
70.	Meteorology	
71.	Oceanography	
72.	Physics (including Nuclear Physics)	(6)
	PSYCHOLOGY	
73.	Clinical Psychology	
74.	Counseling and Guidance	
75.	Educational Psychology	
76.	Experimental and General Psychology	
77.	Industrial and Personnel Psychology	
78.	Social Psychology	
	SOCIAL SCIENCES	
79.	Anthropology, Archeology	(1)
80.	Area and Regional Studies	
81.	Economics	
82.	Political Science, Government, International Relations	
83.	Public Administration	
84.	Social Work, Group Work	
85.	Sociology	
	OTHER SCIENCES AND DISCIPLINES	
86.	Architecture	(3)
87.	City Planning	
88.	Home Economics	
89.	Journalism, Radio-Television, Communications	
90.	Law	
91.	Library Science, Archival Science, Information Storage and Retrieval	
92.	Mathematics and Statistics	
93.	Military Science	
94.	Theology, Religion	
95.	Sciences and Disciplines, NEC	(6)

The relative size of the technologies has a roughly Gaussian distribution, but a larger than normal number of cases involved relatively small numbers of the total populace. The cases involving the largest numbers of people, however, also involved the largest technologies. Ratings of individual cases are in the appendix.

B. The Threat Develops

The social shock in each of these cases began when a proposed or existing technology was perceived by someone to threaten an adverse effect. We want to examine the manner in which such a threat is recognized.

1. *Early warnings missed:* In nearly 40 percent of the forty-five study cases, an early warning signal that may have reduced the public alarm over the technology was apparently missed. In the case of the oral contraceptive safety question, for example, the manufacturing companies and the medical profession had sufficient information to justify including a warning statement to the user long before the pressure of the hearings forced them to do so. In numerous cases involving environmental pollutants (asbestos, taconite, foaming detergent, mercury), information that was not utilized was available to the industrial companies involved. Government regulatory and military agencies have also missed warning signals, for example, the diethylstilbestrol case and the nerve gas disposal problem. In short, we frequently have early warnings, but don't notice or don't recognize them; in fact, we have a propensity for overlooking early warnings.

2. *A problem technology allowed to grow:* In nearly half of the cases it can be said that the technology continued to grow after evidence of a problem had been observed. These results clearly suggest that we pay too little attention to early warning signals.

3. *Technology used irresponsibly:* A survey of the forty-five study cases suggests that in over half of them, the technology had been employed, in part, with less than adequate responsibility by its users. (Examples include use of the unapproved herbicide in the cranberries case; the rush to use fluoridation before testing was complete; several cases involving environmental pollution; occasional abuse of the chemical mace; and the overuse of the medical X-rays). In some cases the users were operating properly under governmental regulations (cyclamates; diethylstilbestrol; the tuna industry; DDT) which were later revised.In several other cases, a technology was clearly being abused (skyjacking; disqualification

of the soap box derby winner; illegal uses of copying devices; the X-ray shoe-fitting machines; and some of the uses of MSG). In only a few of the one hundred cases can the precautions in adopting the technology be considered exemplary (for example, the NTA detergent case; irradiation sterilization of food). This last result was predictable because of the selection criteria and does not imply that our use of all technologies is so irresponsible. In fact, the opposite could be true for a host of technologies that have never produced widespread public alarm.

4. *New technological information defines the threat:* In nearly two-thirds of the cases, new information of a technological nature (including statistical analysis) played an important or central role in the discovery that a threat existed. In many instances, the nature of the technology involved in the discovery of the threat had little if any relation to the technology that produced it, for example, analytical chemistry and marine and avian biology were used to establish the build-up of DDT in ocean-based food chains that were quite distant from the terrestrial point of DDT's use. The public may thus be presented with news stories that describe not only a complex technological problem, but also an unfamiliar analytical technique that is defining the problem for society.

C. Factors that Influence the Impact of the Threat

The public impact of a report of a new technologically derived threat depends on several factors, including the historical setting, the kind of threat, and other circumstances existing at the time. These factors may either increase or decrease the social shock of the case. Nontechnological factors (such as the coincident occurrence of an unusual event or a problem of national or international significance) may, of course, affect the media's coverage of the case or the public's recognition of it, but here we wish to focus on technology-related factors.

1. *Preceding related alarms:* A survey of the forty-five study cases indicates (despite some difficulty in definition of terms) that in at least half of them, a previous alarm over a related technology had been raised. A succession of cases has arisen over the hazards of the several forms of radiation we use, the military hardware, and the chemicals in food and medicine. The existence of a previously related case apparently has a great sensitizing effect on the news reporters and the public. The previous publicity over the Huckleby family that was poisoned by eating mercury-treated seed grain

doubtlessly increased the concern over the mercury discharges by industry, and contributed to the overconcern about trace amounts of mercury in tuna fish. Similarly, the *Torrey Canyon* had already familiarized the press and public with the ecological problems of an oil spill before the Santa Barbara leak, and both experiences contributed much to the fears of the proposed Alaska Pipeline.

Even apparently unrelated cases of alarm over technology probably have influenced the effects of a given case, because these cases, all taken together, might be expected to lead to a general suspicion of technology and a loss of confidence in those who produce it, use it, or regulate it. Not only the number of cases, but the duration that they remain in the arena of public controversy must be considered. The period of time that each of the forty-five study cases was of public concern is shown in Figure 5; not only do some of the cases remain controversial for extended periods of time, but some are revived after once appearing settled. Just what the public feels about such a recurring case is unknown, but one suspects they must wonder what the fuss is about and why public officials cannot take care of it. Not only do similar previous cases have an influence on a given case's impact, but a considerable mania can build up over a topic such as the environment.*

2. *The kinds of threat perceived:* In over half of the forty-five study cases, an adverse effect of a technology was presented to the public—if not always perceived so—as a threat to someone's personal health and comfort. In nearly as many of the cases, an economic threat was involved, either to the public at large, or to users of a technology which was about to be limited by governmental action. In seventeen of the cases a threat to the physical environment existed. The remaining four categories used in the analysis were threatened in a total of fourteen instances, as shown in Table X.

Additional categories could, of course, have been used, and because of the criteria used in selecting the cases, examples of all of these threats were expected. The degree of shock probably depends greatly on the nature of the threat, but the relationship undoubtedly vaires with different segments of the public and is beyond the scope of the present study.

3. *The number of people threatened:* A very large number of people (over 1 million) appear to have been threatened directly in

---

*Figure 5 gives the impression that the frequency of these cases of social shock over technology is increasing with time; this may well be true, but it must be remembered that the forty-five study cases were rather arbitrarily selected.

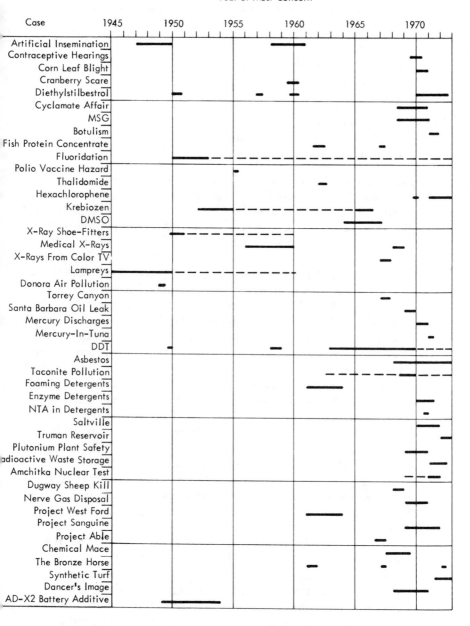

Year of Most Concern

Case	1945	1950	1955	1960	1965	1970
Artificial Insemination						
Contraceptive Hearings						
Corn Leaf Blight						
Cranberry Scare						
Diethylstilbestrol						
Cyclamate Affair						
MSG						
Botulism						
Fish Protein Concentrate						
Fluoridation						
Polio Vaccine Hazard						
Thalidomide						
Hexachlorophene						
Krebiozen						
DMSO						
X-Ray Shoe-Fitters						
Medical X-Rays						
X-Rays From Color TV						
Lampreys						
Donora Air Pollution						
Torrey Canyon						
Santa Barbara Oil Leak						
Mercury Discharges						
Mercury-In-Tuna						
DDT						
Asbestos						
Taconite Pollution						
Foaming Detergents						
Enzyme Detergents						
NTA in Detergents						
Saltville						
Truman Reservoir						
Plutonium Plant Safety						
Radioactive Waste Storage						
Amchitka Nuclear Test						
Dugway Sheep Kill						
Nerve Gas Disposal						
Project West Ford						
Project Sanguine						
Project Able						
Chemical Mace						
The Bronze Horse						
Synthetic Turf						
Dancer's Image						
AD-X2 Battery Additive						

Figure   5. Time of Occurrence of the Cases of Social Shock

TABLE X

THE KINDS OF THREATS POSED IN STUDY CASES

No.	Value Threatened	Number of Cases
1.	Personal Health and Safety	25
2.	Physical Environment	17
3.	Economic Security	21
4.	Social Systems	4
5.	Legal/Political System	5
6.	Ethical or Religious Values	1
7.	Cultural Values	4

nine of the study cases and indirectly in seven other cases. A moderate number of people (10,000 to 1,000,000) were directly threatened in fourteen cases, and only a small number (less than 10,000) in sixteen cases. In five cases, few if any people were directly threatened, although many persons may have felt indirectly threatened.

4. *Special characteristics of the threatened group:* The threats in most of the cases were unequally perceived by various segments of the public. In only about thirteen cases can the threat be seen as applying equally to all segments of the public. The threat appeared to be greater for the young in four cases, for women in three cases, for certain occupations in five cases, for a particular race in one case, for a special region in twelve cases, and other special segments in seven cases. The threats in most of these cases are probably more fully perceived by persons of above average education. In a few cases such as the Amchitka and SST debates, serious concern over an adverse effect of the technology was almost limited to an intellectual elite. Similarly our list of cases contains very few examples—the High-Rise Public Housing is the best—of issues involving predominantly those of lower socioeconomic status. In fact, a preponderance of the cases are of most concern to a middle-upper-class public—perhaps best exemplified at that time by the color TV recall.

5. *Avoiding the threat:* In at least 75 percent of the forty-five study cases, the concerned public probably felt that it could not avoid the threat that was presented—either because of its pervasive nature or because control resided in a distant bureaucracy. The degree to which this inability to avoid the threat increases social shock has not been determined, but intuitively a significant correlation would be expected.

6. *The regulatory climate:* The actions of governmental regulatory agencies before and during an episode of public alarm over technology should have a significant effect on the degree of social shock in a case. Two questions related to the locus of responsibility within government would appear to be influential on the public's attitude: had some government agency been "in charge" of the technology from which the threat arose? and is someone identifiable really "in charge" of the problem now that we know about it? In most cases, and especially in those of recent origin, one can say almost without exception that someone was nominally "in charge" but in fact, of course, the degree of government control may have been lax or at least perceived to be so by the public. In addition, suspicions may have existed that the regulating agency had a conflict of interest or did not necessarily represent the public interest. Hence an analysis of these questions has an inherent measure of subjectivity, and the questions may even require opposite answers at different stages in the development of the problem, that is, someone may have been put in charge of the problem after a time. With these limitations in mind and trying to look at the cases through the eyes of the public, we must conclude that in about 66 percent of the forty-five study cases, an agency of government had nominally been in charge of the technology, but in 72 percent of the cases no one really seemed to be on top of the problem at the time it arose.

In nearly half of the cases, either a conflict of interest existed within a government agency or a conflict of opinion existed between government agencies. The former is best illustrated by cases where the agency had responsibility not only to regulate the technology, but also to use it or to help a societal element that could use it, for example, AEC regulation of nuclear energy and the USDA's regulation of DDT.

In several of the cases, the regulatory agency was already under pressure to act in a given problem area when a specific problem in that area arose. Thus the pressure which began to build in about 1950 over food additives affected many subsequent regulatory decisions. Similarly, the concern over environmental pollutants, and especially over phosphates which were a component of laundry detergents, led to unusually quick action on the use of NTA as a detergent ingredient.

In a surprisingly large number of cases (more than 50 percent), a previous law, regulation or ruling was either a partial cause of the controversy or complicated the problem.

7. *The age of the technology:* The length of time that the technology has been in use before an adverse effect was discovered varied widely among the study cases. In general, however, the longer the technology had been extensively used, the greater the social shock that resulted. The degree of shock is probably related in part to the opportunity that the technology has had to grow large and become integrated into daily life or with other complex technologies, and in part to psychological factors—a familiar technology has "betrayed" us.

In twenty of the forty-five cases the technology can be said to have been in use for ten to twenty-four years before the adverse effect became of public concern. In another sixteen cases, the technology had been in use twenty-five years or more and in five cases one hundred years or more. In only eight cases was the technology less than ten years old, and in half of these an episode of concern arose during the first two years.

8. *The image of the party "at fault":* A real "villain" in each case of social shock is not always readily identified and, in many cases, there may not really be one; an unexpected adverse effect was suddenly identified with little if any previous warning to the public. But in most cases, some group or someone in the public or private sector gets his or her name in the news stories as being at least connected with the cause of the problem, if not actually presented as the "heavy" of the story. The forty-five study cases were screened against a short list of "societal elements" to determine the element that utilized or regulated the technology and was most likely to have been perceived by the public as being primarily responsible for the alarm. The number of cases involving each element is shown in Table XI. While the element categories are oversimplified and the argument could be made for additional assignments of responsibility, it is already clear that very often the cases arise from two or more societal elements (in nearly half of the forty-five). It seems reasonable enough that a characteristic of such episodes is that a division of responsibility or interest has existed: when no one is looking at the total picture of what is happening, the technology may very well "go astray."

The social shock that the case generates will probably depend on the attitudes of various segments of the public toward that societal element (or elements) that are regarded as being primarily the cause of the controversy,—on the "image" of the party "at fault." During the 1950's and 1960's substantial negative changes occurred in the attitudes of much of the American public toward such societal elements as the military, industry, the farmer, and

TABLE XI

SOCIETAL ELEMENTS PERCEIVED AS RESPONSIBLE FOR ORIGIN OF STUDY CASES

No.	Element	Number of Citations
1	Agriculture and Agri-Business	4
2	Food and Beverage Industry	4
	Medical Establishment	
3	Medical Profession	6
4	Pharmaceutical Industry	3
5	Public Health Practice	2
	Industry	
6	Heavy Industry	8
7	Light Industry	4
8	Commerce	7
	Government Agencies	
9	Military and National Defense	8
10	Regulatory and Law Enforcement	14
11	Other Governmental Agencies	2
12	Commercial Communications Industry	1
13	Education Establishment	1
14	Recreation, Entertainment and Cultural Industries	3
15	Religious Organizations	1
16	Basic Scientific Research Establishment	9
17	Other Institutions, NEC	0

probably the medical doctor. In some measure, these changes resulted from alarm over some of the technologies that these elements were using, and in turn, the alarm over one technology of a given agency tended to bring quick suspicion on other technologies of that agency or group.

D. The Technical Information Available

Sufficient reliable information about the technology and its impacts are important not only to the decision makers (who may be either in the government or the private sector) but also to the public.

1. *To the decision makers:* A survey of the forty-five study

cases suggests that at the time each became of public concern, the decision makers' information could be classified as sufficient in only nine cases, timely in only eleven cases, and undisputed in only twelve cases. In the majority (thirty-two) of the cases, however, the technical information was apparently critical in making the decisions. One must realize of course, that in a typical case, several decision makers are involved at different points in time with different information available, so that these data are rather subjective. It is of interest, however, that in seventeen cases where the information was important to the decision, and in ten cases where it was not judged critical, the information was neither timely, sufficient nor undisputed. In several cases, strong polarizations developed among the technological experts. These polarizarions sometimes reflected existing disputes between the conflicting parties at interest or among governmental figures, but at other times reflected the diverse technical backgrounds of the experts. In a few cases these strong polarizations may have speeded up the acquisition of necessary new information, but in most cases, it is probably safe to say, they often added unnecessary elements of confusion for the decision makers and the public, without appreciably helping in the resolution of the problem.

A particularly troubling area is that of technical analytical methods. Inadequate methods have been a problem in a host of cases, while the development of new, more sensitive methods has led to a sudden recognition of a problem in others. Examples involve chemicals (DDT, NO_2, etc.), biological methods (teratogenicity, carcinogenicity, how even to determine death), physical methods (X-rays from color TV), and environmental problems (effect of phosphates, space orbital mirrors). Because of the widespread but nonrandom exposure of the populace to unsuspected technological hazards in our lives, such as X-rays and chemicals (DES, HCP, cyclamates), it is becoming increasingly difficult even to find a true control population for epidemiological studies. A poor understanding of analytical limitations is often a source of concern to the public, to its legislative and regulatory bodies, and occasionally to scientists.

2. *To the public—media interest:* Much if not most of the public's technical information on a case is received through the news media. Hence, the ability of the news media to collect, organize, and present this information accurately and with sufficient detail and coherence that the layman can understand it is an important factor in public participation. The degree of technical informa-

tion that is transferred to the public on each case is not necessarily proportional to the degree of news coverage, but the latter is nevertheless of importance in the transfer. We have used the degree of news coverage as one direct measure of public concern in the study cases. A "media index" was calculated for each case, based on a combination of number and length of stories on the subjects in selected standard printed media sources.* In Table XII the forty-five study cases are ranked in the order of decreasing media index (that is, public concern) and "concern rating" categories have been assigned by letter designations. In Table XIII the public concern ratings of a dozen other interesting cases that were synopsized in Chapter IV are listed. In addition, certain characteristics of the coverage (pictures, cartoons, editorials, etc.) are tabulated in Appendix A.

Not surprisingly, the long-running controversies over DDT and water fluoridation top the list, whereas the regional controversies over the taconite pollution, the plight of Saltville, and the Truman Reservoir are at the bottom in Table XII.

The well-justified, health-related alarms over the Salk Vaccine and Thalidomide are near the top, as might be expected, but the high indexes of the scares over cyclamates and cranberries reflect the impact of the national recalls rather than the true health hazards. The exceptionally high rating of the Santa Barbara oil leak, compared to other environmental problems, reflects in large measure the location of the spill (a tourist community near Los Angeles) and the unusual suitability of the event to pictorial presentation. Similarly, the Northeast Power Failure has a very high concern rating, not only because it affected 25 to 30 million people, but also because it directly affected the Eastern-based news establishment; the *New York Times*, with its 101 stories on the five-to fifteen-hour failure and its aftermath, shows elements of provincialism, as do other city newspapers. The ongoing skyjacking issue has an even higher rating than DDT, one that reflects not only the physical and economic dangers posed, but also the statures of the businesses and people (especially executives, jet-set tourists and reporters) being affected. It may also suggest that the public is still fascinated with,

*The media index is calculated as the sum of the entries in the *Readers' Guide*, *Business Periodicals Index* and *New York Times Index*, plus 10 percent of the total column inches in the nine selected "standard periodicals" (see Appendix A). Because of indexing inadequacies, some cases may be underrepresented. In addition, the *BPI* was not published before 1958; therefore, the totals for older cases (AD-X2, Donora, Krebiozen, Salk Vaccine, and most of Lampreys) have a low bias.

TABLE XII

RANKING OF PUBLIC CONCERN OVER STUDY CASES

Rank		Media Index[a]	Concern Rating[b]
1	The Rise and Fall of DDT	1236	A
2	The Fluoridation Controversy	968	A
3	Salk Polio Vaccine Hazard Episode	326	B
4	The Santa Barbara Oil Leak	297	B
5	The Thalidomide Tragedy	261	B
6	Amchitka Underground Nuclear Test	247	B
7	The Cyclamate Affair	230	B
8	Krebiozen—Cancer Cure?	226	B
9	The Great Cranberry Scare	196	C
10	Human Artificial Insemination	192	C
11	Abuse of Medical and Dental X-Rays	173	C
12	The Torrey Canyon Disaster	166	C
13	The Diethylstilbestrol Ban	152	C
14	Project West Ford—Orbital Belt of Needles	144	C
15	Southern Corn Leaf Blight—A Genetic Engineering Problem	142	C
16	The AD-X2 Battery Additive Debate	133	C
17	Nerve Gas Disposal Controversy	118	C
18	Botulism and Bon Vivant	115	C
19	Foaming Detergents	112	C
20	Oral Contraceptive Safety Hearings	100	C
21	Mercury Discharges by Industry	95	D
22	MSG and the Chinese Restaurant Syndrome	92	D
23	Introduction of the Lampreys	89	D
24	The Mercury-In-Tuna Scare	87	D
25	The Fish Protein Concentrate Issue	79	D
26	Asbestos Health Hazard	72	D
27	The Chemical Mace	67	D
28	The Dugway Sheep Kill Incident	56	D
29	Disqualification of Dancer's Image	51	D
30	Enzyme Detergents	48	E
31	Hexa, Hexa, Hexachlorophene	48	E
32	NTA Detergents	44	E
33	The Donora Air Pollution Episode	37	E
34	X-Radiation From Color TV	28	E
35	X-Ray Shoe-Fitting Machine	28	E
36	DMSO—Suppressed Wonder Drug?	28	E
37	Synthetic Turf and Football Injuries	26	E
38	Storage of Radioactive Wastes in Kansas Salt Mines	23	E
39	Project Sanguine—Giant Underground Transmitter	21	E
40	Plutonium Plant Safety at Rocky Flats	10	E
41	The Bronze Horse—A Technological Definition of Art?	10	E
42	Taconite Pollution of Lake Superior[c]	9	F
43	Project Able—The Space Orbital Mirror	8	F
44	Saltville—An Ecological Bankruptcy?	1	F
45	Truman Reservoir Controversy	1	F

[a]See text for method of calculation.

[b]A = Index of > 800; B = Index of 200–799; C = Index of 100–199; D = Index of 50–99; E = Index of 10–49; F = Index of 0–9.

[c]As of mid 1972 only, and does not include asbestos-related news.

TABLE XIII

PUBLIC CONCERN RATING OF OTHER SELECTED CASES

Case Title	Media Index	Concern Rating
The Skyjacking Nightmare	1,370	A
Heart Transplant Fad	915	A
Northeast Power Failure	315	B
Phosphate Detergent Issue	305	B
Coal Tar Dye Problem	117	C
Great Mississippi Fish Kill	104	C
Polychlorinated Biphenyls	68	D
Glue and Aerosol Propellant Sniffing	55	D
Fortified Breakfast Cereal Issue	54	D
Sea Level Panama Canal	47[a]	E[a]
Discarded Refrigerator Hazard	31[a]	E[a]
Studded Snow Tire Question	13[a]	E[a]

[a]These are minimum values because portions of the literature search was incomplete.

if not in awe of, this newest method of mass transportation. The extremely high media interest in the heart-transplant fad, on the other hand, partially reflects fascination that this symbol of emotion and soul (so long considered inviolate) was so amenable to medical technology. In addition, of course, much serious concern was generated over the ethics and morals of the technology as it was being practiced.

The very high rating of the Amchitka Test Case resulted from the unusual degree of involvement of the congressional and judicial branches of government in Washington, as well as from the low level of confidence which many people apparently had in the Atomic Energy Commission. The AD-X2 Case is rated very high because of the extreme conflicts that it generated between governmental agencies, and is almost unique in that overtones of health, environment, and sociolegal questions were absent; the only significant question was the technological-economic one of efficiency. In contrast, in the case of the fifteen-year athletic controversy over the fiberglass pole for vaulting, not even the technological efficiency was questioned; essentially the only objection was the aesthetic one—old records were made obsolete by the new technology—and this fact makes the case almost unique among those studied here (but hardly unique in real life). The very high rating of the Krebiozen case reflects not only the unusual degree of involvement of legislative, regulatory and judicial bodies, but also the statures of the central figures, mankind's fervent hope for a cancer cure, and the inability of either side to prove its case over a fifteen-year period.

In reviewing these cases, one concludes that the association of a specific chemical, medicinal, or piece of equipment with a problem permits ready identification and focus by the media. This focus, in turn, leads not only to public awareness of the problem, but also to improved indexing of the case so that nearly all stories about it can be retrieved from media indexes. In particular, the media focus best on chemicals, medicinals, or projects with short "headline-size" names, or even better, three-letter abbreviations.* Thus, DDT, DES, NTA, the Pill, and Project Able are easily identified in news headlines and can be repeated often in the text.

Cases involving chemicals or medicinals, foods, or other consumer products often involve an immediate national (or even global) impact, and therefore receive much attention by the media. In contrast, cases such as the siting of a nuclear power plant or radioactive waste disposal facility may be of regional concern; cases such as studded snow tire damage or plastic plumbing pipe acceptance may be resolved on a state-by-state or city-by-city basis; and cases such as the invasions of privacy by computer data banks, electronic bugs, snooperscopes, photocopying devices, etc., tend to be diffuse and more difficult to focus upon. In such cases, the news coverage may never become intense or of a national character.

In about half of the study cases, an individual can be identified in the news stories as providing key information or taking a key action or making a key announcement. In a few cases, the technical person who discovers an adverse effect releases information directly to the news media (as in the mercury discharge discovery) or is widely quoted after presenting his results at a hearing or in a technical publication. In other cases, a government official makes a dramatic announcement (as in the cranberries and cyclamate cases), while in others a citizen or consumer advocate group successfully publicizes the discovery of the technical person. In a few cases, a dramatic story by a news reporter is a very important factor in the case, for example, the thalidomide case. On the whole, the effects of the "instant expert" spokesmen and the "elite opinion makers" in causing social shock are not clear, but they do appear to be

*Thus, the far-sighted company might prefer to choose a name of intermediate length of its products, as a hedge against unexpected adverse side effects.

influential in bringing change. More important perhaps is the observation that clamor in the scientific world oftens precedes public concern over a problem in technology.

The popular news sources that were drawn upon in the preparation of the study cases varied substantially in the amount and detail of technical information they presented. Certain news magazines appeared to show consistently better ability than others to report factual data accurately and to explain it evenhandedly, according to our preliminary evaluation. We were not surprised to find that some news magazines presented certain stories in a way that reflected the bias that one might anticipate in their readerships. In addition, several examples were noted in which the story appeared to reflect a bias (or even a strong polarization) on debatable social implications of a technology (such as birth control). Some individual writers have adopted such dubious techniques as the "composite man" to whom all the worst "or best" things seem to happen, to help present their biased message to the public. Others frequently resort to the unnamed "authoritative sources" bit, and others to photographic displays that emphasize the reporter's point.

On the whole, the news media have not done a particularly creditable job of reporting scientific and technological events to the public over the last thirty years. They tend to overdo the bizarre or the scare aspects at the beginning of a case and seldom follows through to summarize adequately the resolution of an issue. They tend to over-focus on the catchy phrase (at congressional hearings) and the spectacular event (the moon voyages) while neglecting the general trends that may be changing our lives. Some needed fundamental changes appear to be occurring slowly in the media's attitudes on reporting scientific and technological news. The printed media's coverage appears to be far more extensive and possibly more responsible today than it was thirty years ago. The roles of the media and the public opinion polls in assessing our technology are almost sure to grow.

E. Governmental Units Involved

Several elements of government often become involved after a technologically related threat is identified. The nature of these elements and the manner and extent of their involvement have a bearing on the degree of social shock that develops. In this section we look at the major government sectors that became involved in

TABLE XIV

MAJOR FEDERAL REGULATORY AND ADVISORY AGENCIES

A. CABINET LEVEL AGENCIES

DEPARTMENT OF AGRICULTURE

International Affairs and Commodity Programs
Agricultural Stabilization and Conservation Service
Commodity Credit Corporation
Federal Crop Insurance Corporation
Marketing and Consumer Services
Agricultural Marketing Services
Animal and Plant Health Inspection Service
Commodity Exchange Authority
Food and Nutrition Service
Packing and Stockyards Administration
Rural Development and Conservation
Farmer Cooperative Service
Farmers Home Administration
Forest Services
Rural Electrification Administration
Rural Telephone Bank
Soil Conservation Service
Science and Education
Agricultural Research Service
Cooperative State Research Service
Extension Service

DEPARTMENT OF COMMERCE

Bureau of Census
Domestic and International Business Administration
Maritime Administration
National Bureau of Standards
National Oceanic and Atmospheric Administration
National Weather Bureau
Office of Product Standards
U.S. Patent Office

DEPARTMENT OF DEFENSE

Army Corps of Engineers

DEPARTMENT OF JUSTICE

Antitrust Division
Bureau of Narcotics and Dangerous Drugs
Civil Division
Civil Rights Division
Criminal Division
Immigration and Naturalization Service
Land and Natural Resources Division
Law Enforcement Assistance Administration
Tax Division

DEPARTMENT OF LABOR

Employees Compensation Appeals Board
Employment Standards
Manpower Administration
Occupational Safety and Health Administration
Office of Equal Employment Opportunity
Office of Labor Management Relations

DEPARTMENT OF STATE

DEPARTMENT OF TRANSPORTATION

Federal Aviation Administration
Federal Highway Administration
Federal Railroad Administration
National Highway Traffic Safety Administration
National Transportation Safety Board
Saint Lawrence Seaway Development Corporation
Urban Mass Transportation Administration
United States Coast Guard

DEPARTMENT OF TREASURY

Bureau of Alcohol, Tobacco and Firearms
Bureau of Customs
Internal Revenue Service
Office of Domestic Gold and Silver Operations
Office of Tariff and Trade Affairs
U.S. Secret Service

505

DEPARTMENT OF HEALTH, EDUCATION, AND WELFARE

Office of Civil Rights
Office of Education
  National Institute of Education
Public Health Service
  Alcohol, Drug Abuse and Mental Health Administration
  Food and Drug Administration
  National Institutes of Health
Social and Rehabilitation Service
Social Security Administration

DEPARTMENT OF HOUSING AND URBAN DE-VELOPMENT

Community Planning and Management Office
Federal Insurance Administration
Government National Mortgage Association
Housing Management Office
Housing Production and Mortgage Credit Office
  Federal Housing Administration
  Office of Interstate Land Sales
Technical and Credit Standards

DEPARTMENT OF THE INTERIOR

Bureau of Indian Affairs
Bureau of Land Management
Bureau of Mines
Bureau of Outdoor Recreation
Bureau of Reclamation
Fish and Wildlife Service
Geological Survey
Mining Enforcement and Safety Administration
National Park Service
Office of Energy Conservation
Office of Energy Data and Analysis
Office of Land Use and Water Planning
Office of Oil and Gas
Office of Territorial Affairs

B. INDEPENDENT AGENCIES

Arms Control and Disarmament Agency
Atomic Energy Commission
Civil Aeronautics Board
Civil Service Commission
Commission on Civil Rights
Consumer Product Safety Commission
Cost of Living Council
Environmental Protection Agency
Equal Employment Opportunity Commission
Farm Credit Administration
Federal Communications Commission
Federal Deposit Insurance Corporation
Federal Energy Administration
Federal Home Loan Bank Board
Federal Maritime Commission
Federal Mediation and Conciliation Service
Federal Power Commission
Federal Reserve System
Federal Trade Commission
General Accounting Office
General Services Administration
Indian Claims Commission
Interstate Commerce Commission
National Credit Union Administration
National Labor Relations Board
National Mediation Board
Occupational Safety and Health Review Commission
Securities and Exchange Commission
Selective Service System
Small Business Administration
Subversive Activities Control Board
Tariff Commission
Tennessee Valley Authority
United States Postal Service
Veterans Administration

C. EXECUTIVE AGENCIES

Council on Environmental Quality
National Security Council
Office of Management and Budget

the study cases (excluding, if present, the promoters of the offending technology).

1. *State and local government institutions:* State and local agencies and governmental bodies became involved in over half of the cases. This involvement ranged from regulatory decisions by city public health agencies to lengthy hearings in the state legislatures.

2. *Federal regulatory agencies:* In nearly 90 percent of the cases, one or more of the federal regulatory agencies (Table XIV) became involved. The Department of Health, Education and Welfare ultimately became involved in over half of the cases. The Department of Agriculture, the Atomic Energy Commission, and the Department of Defense became involved in some cases both as a regulator of the technology and as its user or promoter. In several cases the attention of the President of the United States was personally directed to the question of how to deal with the technological problem.

3. *Concern in the Congress:* The Congress of the United States became concerned at one point or another in two-thirds of the study cases. This concern ranged from speeches and hearings on the subject to the enactment of remedial legislation.

4. *The courts:* Half of the study cases contained questions that resulted in court litigation of some sort. The courts' involvement ranged from requests for temporary restraining orders against a proposed technology to the settlement of mammoth damage suits over complex technological relationships.

5. *International relations:* In at least eleven of the forty-five cases, the threat posed by the technology transcended national boundaries and was a potential strain on international relations. These cases range from the court actions involving parties in different countries, to movement of pollutants across national borders and to the production of an earth-orbital band of copper needles.

F. Private Sector Parties Involved

In addition to companies and individuals that were using the technology or were directly threatened by it, societal organizations in the private sector became involved in many of these cases.

1. *Citizens' participation groups:* In over half of the study cases some kind of public interest group played a role. These groups ranged from small, locally oriented groups (such as the Save Lake Superior Association) to much larger conservation groups

(such as the National Wildlife Federation), and to the rather professional "consumer advocate" spokesmen. In relatively few of the cases, however, did a large proportion of the individuals who were directly threatened participate in the solution of the problem.

2. *Academic and professional association:* In over 40 percent of the study cases, professional or academic associations or societies became involved in the proceedings. These groups ranged from the powerful American Medical Association, which voiced its stand on numerous health-related issues, to the less politically powerful International Astronomers Union. In addition, the prestigious National Academy of Sciences was brought in on many of the cases to conduct a study of the problem.

3. *Other groups:* Labor unions became concerned in several of the cases that involved occupational exposure to alleged hazards (such as artificial football turf) or potential closure of an existing plant (such as the Taconite Plant). Farmer associations and industry trade associations became concerned in several cases involving agricultural or consumer products, for example, diethylstilbestrol and the color TV recall, respectively. Organized religions' expressed concerns in the forty-five study cases and the fifty-five synopsized cases have been largely limited to those that involved ethical aspects of human reproduction, experimentation or death.

G. Impacts and Dispositions of the Cases

The social shock that occurred in most of our study cases depended on several interrelated factors. The nature of the threat posed and the media presentation of what the problem was about were important, but the real impacts of the case, how the immediate problem was handled, and what was done about the underlying causes were also significant. The ultimate impacts often depended on how a case was resolved, but the disposition, in turn, depended at times on the public concern that already existed over the immediate threat. Thus public opinion—which as we have seen may depend on previous cases of alarm, the images of various parties, and the like,—may be influential in determining the impacts and outcome of the case. On the other hand, the degree of public concern over a reported adverse technological impact and the response of the decision makers to this concern appears to be only indirectly related to the severity of the effect already produced. An examination of the impacts and dispositions of the cases is of interest.

1. *Influence of public opinion:* In virtually none of the

forty-five study cases does it appear that initial public concern and/or media attention led directly to a removal of the threat. In contrast, the initial concern very often led, after a time, to a hearing by a government agency (or something equivalent to one) and sufficient publicity resulted from this platform of expression, that change was initiated. In some cases, the change was agreed to by decision makers in the private sector without official regulatory or legislative action (removal of MSG from baby food; reduced use of enzymes in detergents). Similarly, public opinion played significant roles in ultimate decisions to terminate certain planned projects of governmental agencies (the Amchitka tests; Project West Ford). In a large number of cases, however, either the immediate decisions were little affected by public opinion (the Nerve Gas Disposal Controversy; the Synthetic Turf Debate; the Fish Protein Concentrate Issue; and the automobile interlock and passive restraint systems*), or the ultimate decisions were evolved from protracted legislative or judiciary processes (the AD-X2 Battery Additive Debate; the Diethylstilbestrol ban; several cases involving environmental questions, such as the taconite discharge, Truman Reservoir, Alaska Pipeline, and strip mining episodes; and several cases involving legal or ethical aspects of human life, such as artificial insemination, abortion, and fetal research). Some elements of the public may have felt that they had played a role in these delayed decisions, but most may have become confused over why "the fuss" was still going on.

2. *Governmental action:* One or more of the agencies or branches of the federal, state and local governments became involved in virtually all of the one hundred cases.† In about half of the forty-five study cases, one could say that a government agency took a definitive action, (although this action may have been after the fact, as in the case of the ban on ocean disposal of war gases), but in fewer than ten of these cases could one describe the government action as prompt, and in two of these ten (cranberry scare and mercury-in-tuna scare) the very speed and timing of the government action added unnecessarily to the social shock. New legisla-

---

*Public opinion may eventually force a reversal by government agencies in this last case.

†Exceptions are: the bronze horse (where a movement to involve government in authentication of art and memorabilia has arisen); the fiberglass pole and soap box derby cases (the Consumer Product Safety Commission did rule on synthetic turf); and the polywater controversy (the government did contract for research in this case).

tion and regulations have at times added greatly to the potential for social shock for example, the zero tolerance concept on pesticide residues and the zero discharge concept in water pollution; a zero threshold level (occupational exposure) may be set for certain industrial chemicals.

Definitive remedial action (or at least what was assumed to have been remedial action) by government can be said to have been long-delayed in many cases (Diethylstilbestrol; Fluoridation; Krebiozen; air pollution at Donora; strip mining; the cigarette ban; and medical X-ray regulation). In general, a delay in initiating remedial action (or in determining if it is needed) is relatively short (one to five years) when it results from a need for more technical information (cyclamates, studded snow tires; microwave ovens), but much longer when it results from a conflict of powerful interests (the cigarette ban, the diethylstilbestrol ban; the breakfast cereal issue, the nuclear power plant safety issue). New legislation has been passed on the state or national levels in areas of (or related to) nearly 30 percent of the forty-five cases, and new regulations were made in nearly 70 percent of the cases.

3. *Adverse health effects:* While an adverse health effect was a threat in over half of the cases, in almost none of these cases is the actual number of deaths and illnesses precisely documented. The best documentation is probably available on the botulism episode (where one death is certain, although the number of illnesses is less certain) and the thalidomide tragedy (where the number of deformed living children was fairly well counted, although the number of stillbirths or other complications is less certain). In the hexachlorophene case, the deaths of thirty-five French babies from the contaminated talcum powder seems to be well established, but the numbers of other deaths and injuries from the once-approved uses of hexachlorophene will probably never be determined. In the Salk and Sabin polio vaccine cases, the numbers of vaccinated children who later developed polio was probably determined, but some of these may have already been infected before vaccination. In the MSG case, no deaths are known, but the discomforts of the "Chinese Restaurant Syndrome" were very serious to those that experienced it.

The degree of public concern is probably not directly related to the number of deaths caused in a case. The point is well illustrated by the much greater amounts of research money that have been spent on cancer, which accounts for 17 percent of the total deaths (Table XV), than for cardiovascular diseases (which account for over

TABLE XV

DEATH RATES FOR MAJOR CAUSES IN THE UNITED STATES IN 1969[a]

	% of Deaths Subtotal	% of Deaths Total
Major Cardiovascular Diseases		52.8
Diseases of the heart	38.6	
Active rheumatic fever and chronic rheumatic heart disease	0.74	
Hypertensive heart disease with or without renal disease	0.70	
Ischemic heart disease	35.2	
Chronic disease of endocarbium and other myocardial insufficiency	0.28	
All other forms of heart disease	1.7	
Hypertension	0.41	
Cerebrovascular diseases	10.8	
Arteriosclerosis	1.67	
Other diseases of arteries, arterioles and capillaries	1.32	
Malignant Neoplasms, including neoplasms of lymphatic and hematopoietic tissue		17.3
Accidents		5.8
Motor vehicle accidents	2.74	
All other accidents	3.04	
Influenza and Pneumonia		2.93
Pneumonia	2.86	
Influenza	0.07	

	% of Deaths Total
Cirrhosis of the Liver	1.67
Suicide	1.20
Homicide	0.92
Congenital Anomalies	0.81
Other Infective and Parasitic Diseases	0.45
Nephritis and Nephrosis	0.44
Peptic Ulcer	0.42
Infections of the Kidney	0.39
Hernia and Intestinal Obstructions	0.38
Tuberculosis—All Forms	0.23
Cholelithiasis, Cholecystitis, and Cholangitis	0.18
Enteritis and Other Diarrheal Disease	0.12
Hyperplasia of Prostate	0.09

TABLE XV (*Continued*)

Certain Causes of Mortality in Early Infancy		2.06	Meningitis	0.07
Diabetes Mellitus		1.96	Acute Bronchitis and Bronchiolitis	0.05
Symptoms and Ill-Defined Conditions		1.76	Syphilis and its Sequelae	0.026
Bronchitis, Emphysema, and Asthma		1.56	All Other Causes	6.38
Emphysema	1.16			
Chronic and Unqualified Bronchitis	0.28			
Asthma	0.11			

[a]Source: Adapted from *Statistical Abstract of the United States, 1973*, United States Department of Commerce.

50 percent of all deaths). The latter even strikes a high proportion of individuals at the peak of their productive years, while the former strikes a high proportion of the elderly. The potential of cancer to produce disfigurement and prolonged pain surely are major reasons for this anomaly in expenditure of resources.*

Similarly, deaths from frequently occurring types of accidents (Table XVI) do not cause concern in proportion to their numbers. We are rather inured to annual tolls of 18,000 deaths from accidental falls, or 50,000 deaths from motor vehicle accidents. Catastrophic accidents (involving large numbers of deaths per event) (Table XVII), on the other hand, tend to generate much comment in the news media, but their impact on the public tends to be transient. The real concern that such an event causes is difficult to evaluate; it may be affected strongly by factors other than the number of deaths or injuries.

A large proportion of accidental injuries and deaths occur around the home. A survey of the major residential hazards (Table XVIII) is of interest in that the bicycle heads the list, yet has until recently created relatively little concern—just what one might forecast for a relatively old technology. On the other hand, household chemicals are of very great current concern, not only because the accidents they cause can be very serious (even disfiguring), but because they often happen to small children; in addition, this is the era of concern over chemical technology.

The degree of controversy is often greater precisely where the extent of adverse health effects is most difficult to prove, for example, the fluoridation case; the several cases involving suspected carcinogens, etc.; and cases involving possible radiation hazards. In contrast, a large percentage of the population at risk has tended to discount or ignore the substantial evidence that cigarette smoking is hazardous. Thus the uncertainty of the health risk is an important factor in adding to the degree of controversy over a case and thus possibly to the degree of social shock.

4. *Adverse economic effects:* Of the twenty-one cases in which an economic threat was posed, actual losses were sustained by individual members of the public in relatively few cases (certain farmers and investors in the corn blight case, certain farmers in the DDT case, perhaps certain doctors or technicians in the medical X-ray case, and some residents of Saltville and Santa Barbara).

*Cardiovascular disease research received increased funding in 1973. The public relations efforts of cancer researchers and even the short name, "cancer," may also have contributed to the proportion of funds obtained.

TABLE XVI

DEATH AND DEATH RATES FROM ACCIDENTS
IN THE UNITED STATES IN 1969[a]

Type of Accident	Deaths	Rate[b]
All accidents	116,385	57.6
Railway accidents	884	0.4
Motor-vehicle accidents	55,791	27.6
Traffic	54,636	27.1
Nontraffic	1,155	0.6
Other road-vehicle accidents	236	0.1
Water-transport accidents	1,743	0.9
Air and space transport accidents	1,778	0.9
Accidental poisoning by—		
Solid and liquid substances	2,967	1.5
Gases and vapors	1,549	0.8
Accidental falls	17,827	8.8
Fall from one level to another	4,952	2.5
Fall on the same level Unspecified falls	12,875	6.4
Blow from falling object	1,271	0.6
Accidents cause by—		
Machinery	(NA)[c]	(NA)[c]
Electric current	1,148	0.6
Fire and flames	7,163	3.5
Hot substances, etc.	288	0.1
Firearms	2,309	1.1
Inhalation and ingestion of objects	3,712	1.8
Accidental drowning	6,181	3.1
Excessive heat and insolation	271	0.1
Complications due to medical procedures	2,572	1.3
All other accidents	8,695	4.3

[a] Source: Adapted from *Statistical Abstract of the United States, 1973*, United States Department of Commerce.

[b] Per 100,000 resident population (estimated as of July 1).

[c] NA = Not available; deaths and death rate from machinery are approximately 2,000 and 1.1, respectively.

TABLE XVII

CATASTROPHIC ACCIDENTS AND DEATHS, 1941–1970[a]

Type of Accident	1941–1950		1951–1960		1961–1970	
	Accidents	Deaths	Accidents	Deaths	Accidents	Deaths
All types[b]	1,050	13,213	1,483	13,790	1,338	12,328
Motor vehicle	292	1,985	666	4,037	561	3,553
Bus	59	539	32	264	31	313
Collision with railroad train	12	139	5	35	5	90
Motor vehicle other than bus	233	1,446	634	3,773	530	3,240
Collision with railroad train	69	431	94	574	55	346
Air transportation	88	1,371	112	2,133	173	2,775
Water transportation	89	974	82	719	45	416
Railroad[c]	45	861	24	368	6	52
Fire and explosion	330	4,529	419	3,093	417	3,062
Dwellings, apartments	166	1,034	311	1,911	323	1,970
Hotels, boarding houses, rooming houses	36	515	24	175	25	241
Houses for aged, convalescent, hospitals, etc.	16	281	17	254	12	200
Other	112	2,699	67	753	57	651
Tornados, floods, hurricanes, etc.	111	2,292	98	2,682	80	1,919
Mines and quarries	53	870	23	346	15	298
All other	42	331	59	412	41	253

[a] Source: Adapted from *Statistical Abstracts of the United States, 1973*, United States Department of Commerce.

[b] Catastrophic accidents defined as those in which five or more persons were killed; data for Alaska and Hawaii and military aviation accidents are excluded.

[c] Collisions of railroad trains with motor vehicles are classified as motor vehicle accidents.

Some losses in these and in certain other cases may have been reimbursed by private or governmental funds.

In a larger number of cases a greater economic damage was incurred by industrial or commercial organizations, either by expenses incurred in recalling a product (automobiles or diet foods), by losing a market (NTA, swordfish), by paying claims (Thalidomide), by installing corrective measures (control of asbestos emissions; airport passenger searching equipment) or in a few cases by closing down a plant site or company (in the Saltville and botulism cases).

In a few cases, a government agency paid injured parties directly for damages that resulted from its actions (the nerve gas sheep kill; sonic booms). In a few other cases a government agency paid an indemnity to certain parties who were injured because of action of a regulatory agency and were judged to be "innocent by-standers" in the affair (some cranberry growers). Firm rules do not seem to eixst, however, on when an indemnity should be paid; each individual case seems to require congressional action with uncertain results. Thus, when some 9 million chickens in Mississippi became contaminated with the pesticide dieldrin, through a contaminated commercial feed, and had to be destroyed in early 1974, a bill to indemnify the growers passed the Senate, but failed in the House. In several other cases, the taxpayer could see that a large government expenditure had been made with little, if anything, good to show for it (Project Sanguine; the SST; Project Mohole). Damages were reportedly reimbursed monetarily in at least six of all the forty-five cases. Fines were levied in only two cases.

5. *Other impacts:* Sociological, psychological, and political effects are more difficult to quantify than are economic or health effects, but clearly numerous examples of such effects occurred in our cases. The recent widespread loss of confidence in societal institutions by many elements of the public may be, in large measure, a result of the manner in which our institutions have used and abused its technology, and in the inadequate ways in which they have responded to the kind of technological scares that we have been looking at here. In over one-fourth of the forty-five study cases, it can be said that the threat was not as great as originally described by "opponents" of the technology, but in over half of the cases, the threat was probably greater than first admitted by the proponents of the technology. The dispositions of some cases probably appeared to be much more orderly, and may well have been

TABLE XVIII

THE TOP 25 HAZARDS AROUND THE HOME[a]

Product of Activity	Injuries Treated Yearly[b]	Pattern of Injuries
1. Bicycles	372,000	Fractures and lacerations from mechanical failure, foot caught in chain or spoke, loss of control.
2. Stairs, ramps, landings	356,000	Fractures, other injuries from falls.
3. Nails, tacks, screws	275,000	Puncture wounds, lacerations by stepping on, striking or swallowing.
4. Football	230,000	Broken bones, muscle and joint injuries, broken teeth from inadequate protective equipment, poor condition of playing surface.
5. Baseball	191,000	Cuts, bruises, broken bones from being struck by bat or ball.
6. Basketball	188,000	Bruises and fractures, primarily from falls and player contact.
7. Glass doors, windows, tub and shower enclosures	178,000	Punctures and lacerations from breaking of glass by falls, slips, or failing to see closed door.
8. Cutlery	172,000	Cuts from carelessness or improper use.
9. Doors (excluding glass)	153,000	Minor injuries from hitting door, or being caught by closing garage door.
10. Tables (excluding glass)	137,000	Lacerations, other injuries from contact with sharp corners.
11. Swings, slides, etc.	112,000	Concussions, fractures, bruises from falls, equipment failures, sharp edges, protruding bolts.
12. Beds	100,000	Fractures, burns, other injuries from falling from or against beds, or from burning fabric.

517

TABLE XVIII (*Continued*)

13. Chairs (excluding upholstered)	68,000	Broken bones, bruises, punctures from tipping and falling.
14. Storage furniture	68,000	Broken bones, cuts, bruises from sharp edges and corners or from tipping over.
15. Lawn mowers	58,000	Amputations, broken bones, other injuries from contacting blade, ejected objects, malfunctions.
16. Baths, showers	41,000	Bruises, cuts, fractures, scalds, electric shocks from slipping, falling, hot-water burns.
17. Cleaning agents	35,000	Chemical burns, poisonings from ingestion, splashing, inhalation.
18. Swimming pools	32,000	Drownings, injuries from falls.
19. Liquid fuels	25,000	Burns, explosion injuries from vapor ignition, improper use or storage.
20. Cooking ranges, ovens	25,000	Burns, blisters, bruises and scalds from contact with hot surface, fabric ignition, explosions.
21. Heaters, stoves	22,000	Burns from contact with hot surface, fabric ignition, explosion, carbon-monoxide poisoning.
22. Paints, solvents	14,000	Poisonings and burns from ingestion or vapor ignition.
23. Household chemicals	11,000	Chemical burns from ingestion, flames, inhalation.
24. Coins, paper money	9,000	Choking from swallowing.
25. Clothing	6,000	Burns from ignited fabric.

aSource: The U.S. Consumer Product Safety Commission, see: *U.S. News & World Report*, p. 69 (October 22, 1973).
b"Injuries treated" are those cared for in hospital emergency rooms, based on reports from 119 hospitals across the nation.

much more emotionally satisfying to the public, than the disposition of other cases.

In several cases, a frenzied atmosphere developed which tended to force governmental institutions into hasty actions and to cause an unnecessary degree of concern in the public. In a few cases, federal jobs were apparently lost because of the way the case was handled (Salk vaccine case, cyclamates).

A most interesting feature of some individual cases has been the way in which they have caused an increase in concern over technology on a larger scale. This increase appears to be of two kinds: "horizontal" spreading to related technologies and "vertical" growth to related users. Thus, concern over radioactive fallout spread to concern over burial of radioactive wastes in old salt mines. Similarly the concern over occupational hazards of asbestos spread to concern over asbestos in ambient air and in taconite tailings. In contrast, concern over the drug, thalidomide, increased vertically to become concern over the practices of the entire pharmaceutical industry and much of the medical profession. Concern over DDT increased not only horizontally to other pesticides, but also vertically to generate suspicion of many practices of the farm community. In short, the recognition of a specific problem tends to identify not only similar new problems, but others that are only slightly related. In addition, the development of new uses for an existing technology may lead to the discovery of an existing but unrecognized hazard (as in the hexachlorophene case).

H. Technology Assessment and Social Shock

In Chapter I we noted that "technology assessment"—a systematic study of the potential impacts on society of a new technology, and the ways that society might best influence the course of events—has been proposed as a new approach in avoiding unpleasant side effects of our technologies. The cases we have been studying all involve side effects of technology that were perceived by some segments of the public as unpleasant; they beg the question, "Could TA have helped?" In attempting to answer this question, one must differentiate between the threat that was posed and the social shock that developed when this threat was discovered.

1. *TA to avoid technological threats:* A survey of the forty-five study cases suggests that a good technology assessment could probably have foreseen the future threat in at least 40 percent of the cases (for example, the Santa Barbara Oil Leak; the taconite case; nerve gas disposal; the synthetic turf; Project Sanguine; a partial

TA was effective in the NTA case). On the other hand, a TA would likely have been of little help in about 15 percent of the cases (the cases on botulism; thalidomide; Krebiozen; x-ray from color TV; the bronze horse). In the remaining 40 percent of the cases, a TA might have identified a future problem correctly if the group doing the TA had asked just the right question (and then answered it correctly), but the likelihood of this having happened appears to be remote (we assume here a time period when the TA might logically have been done; today we might well be sure to raise such a question). Examples of such cases are: the corn leaf blight; the diethylstilbestrol ban; the fluoridation controversy; the *Torrey Canyon;* the DDT case; the lampreys; and the Truman Reservoir). A most striking fact of the analysis is that in those thirty-seven cases where TA might or would likely have been effective, the TA could well have been done many years before the threat appeared, for example, ten years or more in twenty-three of the cases and probably one hundred years in the case of artificial insemination.

Assuming that a TA is of a perceptive quality, its results still have to be used by someone, if it is to be beneficial to society. In the thirty-seven cases where TA might have been useful in avoiding a subsequent threat, the group that probably could have best used the TA results was as often in the private sector as in the governmental sector (in several cases groups in both sectors could be major uses). Of all of society's institutions, perhaps our judicial system is most limited in ability to take quick advantage of the results of TA's; the structure yields trial cases generally involving narrowly defined charges that are judged largely on the basis of precedent in the law.

The time that the TA is performed may be critically important in identifying future threats; a TA done very early in the development of the technology may miss important features, while one done after some technologies are undertaken may be too late to reverse an undesired event (the lampreys) or certain long-delayed threats (persistent global pollutants and carcinogens). Very obviously even a "good" TA may need frequent updating as a new technology grows or as new information becomes available. We need several decision points in the development of a technology that indicate when additional research on potential hazards should be done, for example, when the scale of production reaches certain orders of magnitude (units sold or pounds manufactured*).

*The government and private sector parties often do not have such numbers available from industry today.

In summary, then, one can say that the failure to do TA's is very likely to bring unpleasant surprises; doing TA's may avoid many of these, but the observed strange cause and effect links in some of our cases indicate strongly that it will be most difficult to even guess at some of the future adverse effects of a new technology.

2. *TA to reduce social shock:* Although a threat from a given technology has already been identified, a survey suggests that a TA may still be useful in reducing the social shock in over 60 percent of the forty-five study cases. The situation here, however, is markedly different from using TA to avoid technological threats, in that the time available to do the TA is short—rarely more than five years before the storm of public controversy and sometimes less than six months—and the primary users of the results of the TA appear overwhelmingly to be governmental authorities. In many instances of this kind, the discovery of the threat leads to a "management-of-crisis" atmosphere within the responsible segment of government. Such an atmosphere may lead to hasty decisions, or even to unsound decisions that may have to be later reversed on the basis of more complete information—information that even a brief, systematic TA might have indicated as critically important. In practice, of course, these cases frequently embroil some kind of special study committee, presidential advisory panel, or the like, and their efforts may in many ways resemble a TA; ordinarily, however, the committees and panels are not only under severe time limitations, but are directed toward solving problems rather than toward taking a societal overview.

I. Reflections on Impacts of Technology on Society's Institutions

The cases studied herein reveal serious effects of technology on society's institutions and systems. Some thoughts on these follow.

1. *Legislative branches of government:* Legislative bodies at the local, state, and national levels spend vast amounts of their time today on matters arising from technology, its development and control. The water fluoridation issue, for example, has involved the governments of city after city in controversy, while the regulation of human artificial insemination has been resolved on a state-by-state basis. The Congress in Washington has held hearing after hearing on complex technologies ranging from additives for foods to additives for old batteries, from pumping oil beneath the sea to orbiting giant mirrors around our planet, from the legal use of electronics (to

televise pap and violence into our living rooms) to its illegal use (to record the sights and sounds of private citizens and documents).

The legislators are obviously having a very hard time of it. While they must attempt to grasp these complex, technologically derived problems and draw a balance that is best for society (or at least satisfies their constituents), they must usually act either upon incomplete information, or on a surfeit of conflicting testimony developed during a hearing (and recorded in a manner that usually is no easier to retrieve for study than material in the Old Testament). And for the most part our legislators are lawyers, trained in the adversary approach, with little formal training in the sciences or engineering, and modest exposure to scientific methods and language. All too often the people's representatives in government drop key, but controversial, technical points during their hearings with the comment, "We'll let you experts argue that outside."

Legislative regulatory action tends to lag well behind the development of technology and is, perhaps more often than not, reactive in nature. In general, the technology is developed in the academic-industrial-agricultural-military complex, and regulatory government becomes involved only after it presents a problem, although at times legislative action is taken to encourage the adoption of a technology. The more closely tied the technology is to the basic needs of society or the more clearly it threatens the individual, the more its users will be regulated. Unfortunately, the vision of those in government, like those in the private sector, is too often narrow or of very limited range. The legislative branch appears to think in terms of about a decade; the executive-branch agencies, the President's four-year term of office; business, perhaps even less; and agriculture, not much more than a growing season. When a crisis erupts, the legislative-regulatory agencies may act very hastily and with minimal foresight.

Technology, on the other hand, tends to exert its more profound influences over longer periods of time. A half-century is not an uncommon length of time before a significant new technology becomes diffused into the fabric of life. Hence, it is not at all surprising that legislative-regulatory actions are taken that do not fully address the problems. They may even generate new problems for the future. The present energy crisis, for example, was produced in large part by past governmental actions* that produced artificially

---

*In particular: the imposition of low natural gas prices by governmental regulation at the well-head, the granting of generous tax allowances for "oil depletion," and certain production costs.

low fuel prices and hence the development of extremely large and often wasteful energy-use practices.

But how can the legislative branches of government get better technical information and make better decisions? Consideration of these needs recently led the Congress to establish an Office of Technology Assessment to assist it in developing better informatin on long-range impacts of selected technologies. Even so, probably not all technology-related problems can be studied systematically. And the Congress will probably retain its capacity for moving with great unison (but little forethought) on some easy decisions (such as the 1973 law that bans local television blackouts of professional football games that are already sold out), while ignoring far too long more pressing matters of either a technological nature (such as the developing energy crisis or the pending controversies over biomedical research applications) or of a nontechnological nature (such as reform of campaign financing practices and seniority practices of the Congress itself).

2. *The executive branches of government:* The growth over nearly two hundred years of the executive branches of government (particularly at the Federal level) reflects the growth of our technology: as new technologies appear, new regulatory agencies are generated to control them, whether the technology involves air transportation, nuclear energy, agricultural pesticides, or public water supplies. Unlike the legislative and judicial branches of government, the executive branch at the federal level has tended to collect in its regulatory agencies and special advisory offices or panels, the technological and scientific talents that were needed to understand the complex technical problems.

On the other hand, as the agencies have grown over the years, they have all too often tended to be dominated by—or at times even subservient to—the technology-based industries that they were established to help regulate; special advisory panels have often been charged with conflicts of interest among its members. The basic problem is that *people* are involved: the agencies or panels must often draw their experts from the subject technological field (be it industry, medicine, education, agriculture, or whatever); many of the agency experts go to (or return to) the industry they regulated when they leave government service. Completely unbiased "instant experts" are rare.

But communication, between the technical experts in and out of government and the political administrators within the executive branches, has grown difficult as the government has grown. Our mayors, governors, and the President clearly need the inputs of the technical and scientific community, but how best to accomplish this? How especially at the federal level, where agencies have often grown large and unwieldy? And how, if important individuals in government are at times more concerned with technological stunts such as putting a man on the moon than with the deteriorating human condition, or with surreptitious bugging and taping schemes than with an impending energy crisis?

3. *Judicial branches of government:* The conflicts over technologically derived problems often end up in our courts. Here the judges must decide whether a case involving some new question is to be heard, must rule on how various technical aspects of the cases can be presented, and must render judgments, verdicts, or sentences, based on the complex and often conflicting technical evidence. The juries, whose members are probably more often picked because of their lack of technical knowledge than because of it, must also evaluate this conflicting complex technical evidence and reach a verdict.

Just how many of the cases of our overloaded courts stems from technological impacts is uncertain, but the number is surely large, and is probably increasing as the sophistication of our technology increases. Indeed, many issues which are at one point apparently resolved by the higher courts later reappear in more complex forms as the technology changes.

An example is the recent decision on the abortion issue by the U.S. Supreme Court. In it, the Court said that the states did not have authority on the subject until the fetus was six months old, that is, when it could be viable outside the mother. Thus, the Court did not rule on the basis of legal, philosophical, or ethical grounds. What the Court unwittingly did was to tie the time period to the *technology* for keeping the fetus alive. Inevitably, our technology for doing this will improve—perhaps to three months or even to zero months (true test-tube babies)—and the Court will be faced with the question again.*

4. *National defense systems:* War and intense social conflict throughout history have been leading factors in the development and adaption of technology, and in the present century especially,

---

*As experimentation toward test tube babies proceeds, it seems likely that live aborted fetuses will be used; inevitably one will survive and grow. Thus, we will have the problem that "something" which was once ruled by the highest court in the land to be "not a person" is in fact a person. The courts will then have to decide, "What will be the legal rights and responsibilities of all those involved?"

in the growth of scientific knowledge. In turn, such new technological developments as the stirrup, the cross-bow, gunpowder, poison gas, and the atomic bomb, have in successive ages produced military advantages for the powers that introduced them—advantages which, more often than not, have been turned to aggressive rather than merely defensive purposes. Today, the military forces of even the less advanced nations are based on complex weapons, communications, and transportation systems that require lengthy training periods for those in military service. In the more advanced countries exceedingly complex systems have been developed, based on the capabilities to deliver nuclear attacks at intercontinental distances or from global orbit. The publicity and glamour that have so often attended technological warfare have, almost as much as the fear of conquest, produced strong support for those who favored expenditure of much of the nations' resources on national defense.

But the development of these modern systems, the training of the millions of technicians needed to operate them, and the maintenance of them in a readily available status are not without grave risk. As the public has repeatedly suffered or feared accidental exposures to radioactive atmospheric fallout, escaping nerve gas, exploding carloads of rocket propellant, giant radio transmitters, and SST's during the past two decades, its infatuation with the military turned noticeably to distrust. And the distrust has extended to the very top—the Commander-in-Chief, the President. In fact, in the presidential election of 1964 a major contention of the Democrats' successful campaign to defeat Republican presidential candidate Barry Goldwater was that they did not trust him to have his "finger on the button" of technological warfare. Elaborate and presumably "fail safe" procedures have been established to prevent accidental or unauthorized use of our more devastating technological systems, but public unease remains. It is doubtful that the military will ever regain the degree of public confidence it once held as long as such vast power rests in the hands of one individual in a time of international tensions.

5. *Law enforcement systems:* The establishment of the first civilian police force in England in the 1820's proved to be very popular with the public, and systems based on the English system were gradually adopted throughout much of the Western world. The policeman often held wide affection as well as respect in earlier times, but the situation has changed greatly in the present cen-

tury. One suspects that much of this change has occurred because of technology.

While it is true that the criminal element widely adopted the technological developments of warfare,* it is the degree to which the police departments themselves have adopted technology that appears to cause much public resentment.

For not only have the police adopted increasingly effective weapons to fight the hard criminal element, but they have had much difficulty in moderating their technologies for other purposes, for example, crowd control. Not only do the police car and helicopter help spot criminal activity, but they generate many citizen complaints by racing noisily through our streets† and hovering over our homes, spotlights ablaze, and they help generate millions of traffic tickets every year that are widely disseminated among the citizenry. But possibly the worst effect of all is the isolation of the policeman from the public that the police car produced; too often the only communication a citizen really hears from the policeman is the alarming wail of his siren and obnoxious crackle of his radio.

The law enforcement departments and investigative bureaus across the land must perform under many handicaps; they need the help and cooperation of every citizen. But at the same time, these departments represent a sizable market for companies that sell technological devices that range from chemical mace and cannons to computers and data banks; all these devices have the potential for causing public concern. Clearly, these agencies, like many other sectors of our nation, need to do a better job of technology assessment.

6. *Agriculture and food production systems:* Agriculture is one of man's oldest technologies, and its introduction about eight thousand years ago changed not only his diet, but stimulated vast changes in his way of life and in his social structure. For the domestication of the cereal grasses and root crops permitted an efficiency in obtaining food that had never before existed; it led to permanent settlements and a specialization in labor. This domestication directly produced a great increase in the amount of starch in the diet of primitive man, who had been primarily a carnivorous hunter. He was able to accommodate to this change, however, because he had retained the ability of his primate ancestors to subsist, if necessary,

*Particularly after World War I, when submachine guns and even armored army vehicles were employed by gangs.
†Streets were designed with civilian transportation in mind.

on fruits, nuts, and other vegetable foods, and because he soon
domesticated animals to yield the meat and skins that he no longer
obtained by hunting.

The pace of change in agriculture was relatively slow until the
Industrial Revolution, but has increased rapidly in the last 150
years. These latter-day changes in agriculture have again made vast
changes in our diets and in our social structure. For the intensive
use of machinery, plant and animal hybridization, fertilization, and
pest control methods, have led to prodigious productions of food
and fiber by a decreasing labor force and an increasingly corporate
system of management. And increasingly, the marketing, packaging, preparation, and even production of food have become industrialized.

The technological aspect is well illustrated by the use of food
sweeteners. As recently as 1750, sugar was a luxury few people
could afford regularly. The development of sugar plantations made
sucrose sugar more available, but even by 1850 the average person
is said to have consumed only about twenty pounds per year in
Britain. With the further adoption of heating, centrifuging and
vacuum-drying technologies, sugar became readily available, and
consumption increased to an average one hundred pounds per year
by 1950—a tremendous dietary change over a two-century period,
the adverse effects of which medical science has only started to
measure.

But even as the use of sugar skyrocketed, industrially produced
synthetic sweetners such as saccharine and the cyclamates were
added heavily to many diets. The increasing use of industrially
produced artificial flavors, colors, thickeners, extenders, and the
like (even entirely synthetic foods such as the chemical colas), and
the development of synthetic fibers, have blurred the distinction
between agriculture and industry. This distinction may be gradually eliminated as the family farm disappears and all food and fiber
production is carried on by integrated industries.

But as the controversies over fertilizer and feedlot runoff, pesticides in the environment, DES and MSG, and genetic susceptibility indicate, far better technology assessments are needed than
have been done heretofore by those who have stimulated agricultural technology and overseen its adoption.

7. *Industry and commerce.* Modern industry and commerce
have grown symbiotically with technology, and industrial and
commerical leaders believe, with considerable justification, that
they know how to put technology to work for them. But as the

technologies have become increasingly complex and the scales of their utilization have grown ever larger, it has become almost impossible for any one individual in a company to really know what is going on, what the company is doing with its technology.

Increasingly, communications within a company or industry appear to segregate into horizontal layers, and the individuals in top management tend to become isolated from the technological base. They are thus greatly surprised when the public becomes concerned over the oil spills, mercury leaks, foaming rivers, and unsafe or shabby products that their companies produce. They are, it would seem, often poorly aware of even the large technologically related trends in the world, or else callous of the public interest. How else, for example, does one account for the construction of a huge lake-polluting iron ore facility at a time when environmental concern is increasing rapidly? And how could the oil companies continue their heavy advertising of gasoline right up to the time that shortages existed,* or how could they fail to convince the auto producers that an incipient energy crisis dictated smaller engines? And why would Detroit continue to make autos with speedometers that register up to 120 mph even as speed limits are being reduced to fifty five miles per hour across the country?

In cases where public alarm over industrially related technology has arisen, the record of big business has been generally dismal. The accused firm as often as not appears to expend more effort on smothering the charges with a public-relations program than on getting the facts and making them available to the public. But since the news media and the public expect this, the company's denials and announcements tend to be discounted. Business, in general, tends to unite behind the accused firm—whether it is innocent or not—against those bringing the charges, rather than demand that it make an accounting and end the suspicions that it may have cast on its fellow companies. One would think that occasionally it would be in a company's interest, as well as in the public interest, to point out the adverse effects of another company's technology, but this is almost invariably left to government, consumer advocates, environmentalists, academic researchers, or the news reporter.

A troubling factor about modern industry is the tremendous scale of its efforts in moving materials about the face of the earth. When one considers that about sixteen tons of concrete is poured

---

*When the public finally became aware of the energy situation, the oil companies launched a mammoth advertising campaign to exonerate themselves.

yearly for every man, woman and child in the United States, it is easy to see that even the supply of sand can be locally exhausted.* When persistent pesticides can be made in quantities of several hundred million pounds per year, one must conclude that not even the oceans have an infinite capacity for their dilution.

An alarming feature of modern industry and commerce is the tendency of smaller companies and corporations to be merged into ever larger corporations and conglomerates; they become so technologically complex as to be nearly unmanageable. We see railroads, airlines, an auto manufacturer, "drugstore" chains, and fast-food franchises suddenly beset by severe financial distress; some must then be subsidized or even operated by the government in the public interest. A further alarming feature is the growing technological interdependence of businesses of all types on each other. Companies that can manufacture something and market it to the public without substantially utilizing other technology-based businesses are growing fewer. As this interconnection continues, one cannot help but wonder if some day the system will collapse, as one system did during the great Northeast power failure.

The Industrial Revolution came at a time when little thought was given to potential adverse effects of new technologies. Even in modern industrial societies, too little technology assessment has been done. As the "postindustrial society" is entered, where information resources are the keys to development, the need for better technology assessment is greater than ever.

8. *Labor unions:* The divisions of labor have historically been structured along the lines of technological developments; since the development of the guilds and the craft unions in medieval times the structure has been rather formal. Even the more recent industrial unions are closely aligned to the overall technologies upon which their respective industries are based.

The labor unions are repeatedly affected by technology as older trades and industries are displaced by those based on newer technological developments. The unions' keen sensitivities to some of the potential immediate effects of the introduction of newer technologies have often led to conflict, expressed by thousands of jurisdictional strikes. Occasionally, all-out efforts have been made to prevent the use of new products, such as the plastic plumbing pipe, which can be installed more quickly than metal

---

*In San Antonio, Texas, for example, "synthetic sand" is made by crushing rocks.

pipe, and can even be installed by the homeowner with few of the plumber's special tools.

The union's long-range vision, on the other hand, has been very poor; older technology-based industries, such as the railroads, are often weighed down by union "featherbedding" practices.* The union thus helps strangle the very industry that provides its members with jobs.

The union leadership for the most part is still tuned to the battle with management. Labor needs help in recognizing that technology is a factor to which it should be giving more thoughtful, long-term attention, a factor which both labor and management should consider jointly for their mutual benefit.

9. *Financial systems:* The financial systems of the advanced countries have become greatly dependent on modern technologies, and particularly, during the past two decades, on the computer. Punched-card billings, plastic credit card payments, magnetically coded checks, and the existence of giant data banks of personal information, have suddenly become features of our daily lives. The technology has grown so rapidly and so large that few persons can see its full implications. Complaints are heard on many sides of the errors inherent in such complex new technologies.

Because these technologies directly involve money, and human avarice is what it is, they are extremely susceptible to abuse. Already we have seen a successful public demand for liability limits on lost or stolen credit cards, but even this may not save the owner from a harrowing experience: an abused card may still damage his computer-owned credit rating to a frightening degree.

While new technologies have been introduced in the financial world from Wall Street to the neighborhood store, assessments are lacking on what these will mean in terms of traditional norms, for example, of buying, savings and indebtedness habits, or on other aspects of life. The effects may be surprising.

10. *Communications systems and the news media:* Modern communications systems are obviously built on some of man's most ingenious technologies. Whereas a century or so ago a California rancher might not hear who won a presidential election until a month afterward, today he may learn who won even before he has voted, thanks to modern communications, statistical analysis, and the wonders of the computer. One is besieged by information today

*That is, to require an employer to maintain a workman after his work has been displaced by technological developments, or to limit production to a level below that readily obtainable with new technology.

as never before in history. The effects of all this communication on
our lives are largely unknown, but theories have been proposed
that link overcommunications of bad news to such serious effects as
an increase in the suicide rate.

Although the printing press has been undoubtedly one of the
greatest technological inventions, the societies of the world have
even yet enormous difficulty in adapting to its presence. In most
countries of the world attempts are made to control what is printed.
Even in a superpower such as the Soviet Union, renowned authors
smuggle critical manuscripts to the outside for publication at the
risk of imprisonment or death. In the United States, the courts of
the land are engaged in never-ending debate over what constitutes
unacceptable pornography.

But the main interest here is on the electronic and printed
media that bring the daily and weekly news, how these media are
handling technologically related news, and how the method of presenting this news affects the public. The present study concludes
that much room for improvement exists; while the news media have
carried thousands of technologically related stories over the past
few decades, they have had great difficulty in presenting them
evenhandedly, factually, and credibly.

The difficulty in presenting these stories evenhandedly arises
in part from the very nature of journalism. While reporters may seek
several views on a story, they will emphasize the unusual, the
bizarre, to make it more interesting, more salable. For television
network reporters, whose salaries depend in part on the number of
minutes their stories command on air time, a direct financial incen-
tive to glamorize the issue is involved.

A factor that may affect both the evenhandedness and the factuality of many technologically related stories is the relatively small
amount of science training that most journalism students have received. In recent years, however, increasing numbers of science
writers appear to be working for the major newspapers, news services, and networks, and a National Association of Science Writers
now exists. Periodicals such as *Time, Newsweek,* and the *Saturday
Review* have carried regular "Science" columns. Recently, a few
daily newspapers such as the *Kansas City Star* have introduced a
weekly "Science Page" —a situation rather like the planned introduction of the "Sports Page" by Hearst fifty years ago, and not
altogether unlike the introduction of public libraries one hundred
years ago. Whether this partial segregation of science news is for
the best remains to be seen, but stories presented therein are usu-

ally more carefully researched and factual than news stories that attempt to cover rapidly breaking events. The factuality of a story is often limited by the reporter's sources; data may be incorrect, or unavailable. A particular problem for the news media in getting information is its own poor reputation for factuality among the scientific community: many scientists avoid reporters for fear that what they say will be quoted incorrectly, taken out of context, or greatly exaggerated. What should, or can, the news media do to counteract this image?

Another problem is that data may be only partially revealed by an "expert" or spokesman who, in fact, may have a vested interest in the subject. This condition may also affect the story's credibility. For, while the press may cover a story and give it an identity by repetition, the public may simply think that it's biased, untruthful, or irrelevant. The credibility problem is really not that of the media so much as it is of the spokesman for the government agency or private sector who is trying to get information to the public via the media. Thus, when a crisis erupts, the news reporters may listen with justified skepticism to a government spokesman who is in essence competing for news space with every other spokesman on the subject. The public may get a flood of conflicting information which it cannot evaluate. What seems to be needed is an independent scientific organization that can rapidly accumulate and evenhandedly organize whatever facts are known about an erupting crisis, and could present the results to the news media under conditions that would inspire confidence. But how can such an organization be developed and who could operate it without being charged with bias or with controlling the news? Efforts have been made, but improvements are obviously needed.

Finally, one must conclude from the present study that the recording of the ways in which sociotechnological information is transferred in our culture is poor. While original scientific work is published in technical journals in retrievable form, access to the popular press coverage of technologically related stories is limited, and access to that of the electronic media is almost impossible. We must do a better job of indexing if we want to do really adequate technosocial research. But who should set up the repositories and computer retrieval systems? And who will assure that adverse stories are not intentionally left out of the system—as could in fact be happening today?

Many problems are clearly retarding the communication of technological information to the public, but the public needs this

information to make its inputs through the democratic processes,
and much thought and effort on the part of many segments of society will be required to achieve success.

11. *Health care delivery systems:* Modern medical practice is
based on an extremely complex mixture of technologies that include huge pharmaceutical companies, sophisticated instrument
and equipment manufacturing firms, diverse hospital facilities, and
large biological and medical research and training institutions, all
supervised by overlapping layers of city, state, and federal regulatory agencies, and under the coordination of rather tightly knit professional associations. As diagnoses and therapeutic health care
have become more technologically complex, the professional has
found it impossible to keep abreast of everything and increasingly
necessary to specialize, as indeed have the hospitals and other
health care delivery institutions. Complaints are heard on many
sides that as the medical technology becomes more complex, people find the distance between the patient and doctor lengthening,
the hospitals bewildering, and the costs of health care financially
unbearable.

While medical technology has led to an unprecedentedly high
level of health care for the public, episodes of alarm over the discovery of its unexpected or potential adverse effects occur with
disturbing frequency. Although the public has more confidence in
our medical system than in most of our societal institutions, the
level of confidence is still quite modest; it reflects the public's
suspicions that medical ethics and foresight have not been able to
keep pace with the technology, and that better assessments of these
technologies are in order.

12. *Educational systems:* The growth of technology in the present century has been so fast that the educational system has been
hard pressed to keep up. While the schools have been quick to
adopt new technological aids ranging from tape recorders and electronic calculators to closed-circuit television and computers, they
have had far more difficulty in deciding what new technological
developments to incorporate into their curricula and how to do it.
Even on the elementary level, textbooks are superseded with diz-
zying speed. At the university level, the classical structured departments respond poorly to the need for interdisciplinary cooperation in training the student to recognize and moderate the social
implications of technology. What is taught in the classroom tends to
lag behind technological growth; if we are to make the best use of

our technology, our educators must assist our students in learning how better to assess its potential impacts.

13. *Cultural/entertainment/recreational systems:* Technology's impacts on our entertainment and cultural systems are myriad —some quite obvious and others so subtle that they have gone largely unrecognized. Technological developments have been widely adapted to existing and even new cultural, amusement, and recreational purposes—some to good effect (acrylic paints, stereo recordings, fiberglass boats), and others with mixed effects (synthetic football turf, the turbine engine racer, the snowmobile). Our technologies have the capability of greatly changing our cultural/entertainment/recreational practices and values (the technological definition of art, the music "synthesizer," the fiberglass vaulting pole), and of generating entirely new forms of public entertainment. They have also the potential for being much abused (deafening electronic amplifiers, magnet-equipped soapbox derby racers, athlete-stimulating drugs). Heretofore these technologies have been largely adopted with little thought to their effects on the aesthetic qualities of our lives; better assessments will be demanded in the future if the public resists the ever-quickening pace of change.

14. *Religious organizations:* The history of man's religions is filled with conflicts over the control of his developing science and technology. Through the ages denouncements have been made of such technologies as the crossbow and napalm, coffee and tobacco, the farm tractor and televised violence, medical treatments and abortion. Although the concern has largely arisen over the effect of the technology on the temporal practices of the church membership, numerous conflicts have developed over the degree to which technology should be introduced into worship practices. Technology can even be abused to strike at the basic practices of a faith, as in the recent incident in Italy where tape recorders were brought surreptitiously into the confessional.

The religious leaders' foresight in identifying the coming impacts of technology has been relatively weak, even by comparison with the poor records of many other societal institutions. More often than not, religious leaders have been reactive to new technologies after the fact, and then in considerable disunity. If technology assessment is to produce results that are satisfactory to a large proportion of the public, it must consider the ethical norms of that public, but inputs regarding new technologies must be received more quickly from the religious leaders.

15. *A parting note:* As we have seen, the unstructured type of technology assessment usually follows the introduction and development of a technology and the subsequent discovery of an adverse effect.

Technologies that are growing rapidly (such as color TV, diet beverages) or changing extensively in size or nature (such as oil transport or drilling, mining operations) tend to develop alarm-producing side effects rather rapidly.

In contrast, older technologies, such as farming, the clock, or the railroads, tend to become acclimated to societal and environmental limitations and are less prone to develop alarms. On the other hand, these older technologies have become so integrated into our way of life that they can serve as the basis for enormous controversies. Thus, as we have frequently seen also, a given case of alarm may not arise from a single technology. Each technological advance is built on earlier and interconnecting technological developments in a kind of "technological tower," as shown for a small segment in Figure 6. Sizable portions of the tower may rest on a small and fragile bases. Sometimes we do not recognize or we forget this base until its existence is dramatically made known, for example, when our susceptible corn is suddenly stricken with blight or a portion of our oil imports is suddenly shut off. If we do not recognize and allow for these fragile bases in the technological tower, we may unexpectedly find that large portions of it can collapse like a house of cards when an appropriate stress appears.

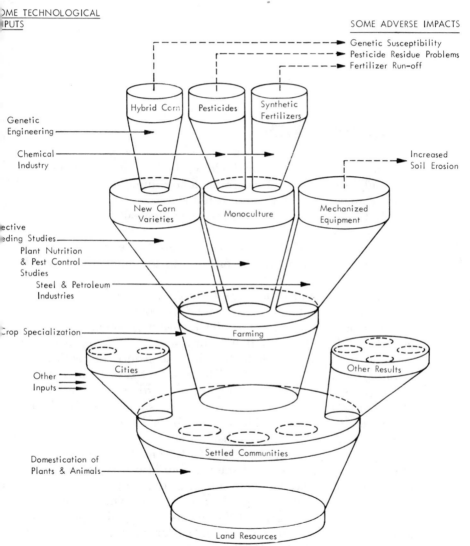

Figure 6. A Segment of the "Technological Tower"

APPENDIX A

Methodology

In this appendix, the methodology that was used in conducting the study is described. It is arranged according to the five tasks of the research approach that are listed below:

1. Identify areas and cases of technologically related concern or alarm.
2. Develop selection criteria and select cases for study.
3. Develop standardized procedures for conducting case history studies.
4. Develop individual case histories.
5. Analyze collective cases and draw conclusions.

The processes and criteria for identifying potential cases, for classifying them for operational purposes, for selecting particular cases for study, and for conducting the case history studies (Tasks 1, 2, and 3) are described in detail. The results of Task 4 form the basis of Chapters III and IV, and only a brief further comment on it is made here. The results of Task 5 have been summarized in Chapter V, but details are provided here. Some brief definitions and ground rules as used in the study are listed first.

Technology. Technology is broadly considered to include not only all varieties of applied science, but also basic research, the results of which may soon form the basis of a proposed technological development, device, product, or system. One definition of technology is "the totality of the means employed to provide objects necessary for human sustenance and comfort" (*Webster's Seventh New Collegiate Dictionary*).

Episode. A concentrated period of public concern over a specific technology whose perceived adverse cultural impacts stimulated action by society's institutions. Of particular interest are episodes in which alarming new information, either technological in nature or related to an already adopted technology, was given rapid dissemination and widespread visibility to the public.

Representative. The group of cases studied should reflect diverse technological and cultural origins and impacts, and include problems that involved a wide range of scientific and technological disciplines and societal institutions.

Recent. Approximately since World War II.

Public. Every case will have a different "public." Some controversial issues have a truly national public. Some episodes get nationwide publicity, but the public may be regional, monolithic, or relatively small in number. In a few cases, the public that has its values directly threatened may be very tiny, but a larger public may become interested because of news coverage.

Concern or Alarm. A marked interest or apprehension by segments of the public, usually over a perceived threat to their personal values by the technology. The concern does not have to be of panic proportions, but it must have elicited interest by the news media and one or more of society's institutions.

Society. An enduring and cooperating group of people whose members have developed organized patterns of relationships with one another; a cultural group having common traditions, institutions, and collective activities and interests.

Institution. A significant practice, relationship, or organization in a society or culture.

Culture. The characteristic features and behaviors in a stage of advancement in a civilization or society.

Case History. A description of the significant events and conflicts in the development of the episode of public concern, its impacts on our society, individuals and institutions, and the manner in which society came to grips with the problem that the technology presented.

Task 1. Identify areas and cases of technologically related concern or alarm. Considerable attention was given to the selection of cases. In particular, a wide selection of cases was sought which would be representative of the many diverse technologically related problems that arise in our society. At the beginning of the program a goal was set to prepare a sufficiently large number of case histories to permit inclusion of examples of all the major types of problems that might be identified. Two preliminary (Task 1) steps were required before the selection of representative cases could be made (Task 2). These steps involved (a) the development of a logical classification system for the case so that a fair representation could be assured, and (b) the development of a large list of potential cases from which actual study cases would be selected.

Numerous cases of public alarm over technology were, of course, in mind at the beginning of the study, but in order to systemize the study, the identification of cases was conducted concurrently with the development of an acceptable classification system. The development of several trial classification systems was based in part upon consideration of cases already identified and, in turn, considerations of various classification systems directed attention to areas in which new potential cases were identified.

a. *Identification of potential cases:* An attempt was made to identify a large number of cases in which technologically related information had received coverage by the popular press and mass communications media and had concerned or alarmed segments of the public. Potential cases were initially listed as suggested by project members, the project advisory committee, and other individuals. The suggestions were usually based on personal recall and experience, or were noted during scans of popular and technical publications. Considerable flexibility was allowed in accepting cases for this list (see criteria below). Additional cases were added, as identified during the course of the program.

Issues were added to the preliminary list of cases if they involved some sort of technologically related aspect, had received some coverage in the news media, and were apt to have caused concern in a segment of the public. A liberal definition was used for "technologically related," but technology was generally limited to the medical and physical sciences and areas involving material things.* The "concern" or "alarm" did not have to be of nationwide panic proportions to qualify. The criterion used was that the issue did cause either direct concern with some segment of the public (or its governmental agencies), or that it was of such a nature and had received sufficient news coverage that it may have caused concern regarding our closely held values or institutions (for example, it posed a threat to certain freedoms or cast doubts on the safety of medical practice). A further criterion was that the case had to be recent—within the last twenty-five years, approximately.

Because of the emphasis on public alarm, a large percentage of

---

*For example, debate over genetic effects on IQ measurements was not included.

the cases was expected to involve issues concerned with some sort of direct physical threat to man (such as those involving hazardous consumer products, medicines, or environmental contaminants). Hence consideration was given early in the study to individual and societal needs which, if threatened, would produce various degrees of alarm. A systematic outline of needs was developed (see Appendix B) based on about seventy-five identified potential cases, which was then used as a guide to search for additional cases in other areas.

Well over two hundred potential cases were identified that appeared to have reasonably distinct histories, although in a few instances the case title covered a series of closely related episodes, or a respresentative example of a compound case was used as the focus. One suspects that by more diligent search and the use of liberal definitions of "concern," a list of one thousand cases might be developed. Each of the cases was given a tentative descriptive title and short summaries were prepared for most of them.

b. *Classification of the cases:* In order to study and discuss the many technologically related cases that have caused public concern or alarm, a grouping of cases according to systematic criteria was desired. Because these cases have many causes, effects, and facets, however, they are susceptible to grouping according to several approaches. Several rationales were considered for grouping the cases: by the type of product or service involved; by the industry or sponsoring government agency which directly caused the problem to arise; by the scientific or technological discipline involved; by the regulatory governmental unit involved; by the human need or activity which gave rise to the case; by the type of threat perceived in the alarm; and by chronological considerations. However, a superior grouping system, that is, one which would accept all the diverse cases in a logical, orderly and self-consistent manner, was difficult to identify.

Inevitably, one appeared to encounter problems of definition, the need to generate meaningful subcategories, and the overlap of numerous cases into more than one category or subcategory. Thus, one may attempt to group the cases according to a broad criterion such as "causes." One then must decide if he means the social institutions involved (such as industry, government, universities, medical institutions, and the like), the basic human needs which gave rise to the case, the technologies involved, the scientific disciplines involved, or possibly some other subgroup. Alternatively, one might group the cases under broad criteria, such as the kind of threat perceived by the public; the impacts which occurred; the type of industry, product, service, or project involved; the type of controversy involved; or the government or regulatory agencies which get involved. One must then differentiate between threats, impacts, industries, and other factors.

None of these systems, after a survey of the facts of the cases under study, appeared to be completely suitable. For example, if one considered the five cases involving detergents, one noted that all the detergents could be called household products, served a basic human need of sanitation, could be considered as being indirectly "caused" by industry (or perhaps the user), and have a chemical technology aspect. But, the *reasons* why they were objectionable and were producing concern or alarm were different in each case:

- Early foaming detergents—Primarily objectionable on aesthetic grounds.
- Phosphate detergents—Vague and uncertain environmental effect—that is, eutrophication.
- Enzyme detergents—Direct but generally mild toxic (allergic) effect on some users.
- NTA detergents—Fear of an indirect toxic effect: potentiation of heavy metals that are already present naturally in the environment.
- Carbonate/silicate detergents—Direct caustic effect on user or children.

In addition, the government agencies involved differed for the different detergent cases.

Similarly, the many cases involving poisonous substances in food and drink products have quite diverse causes. Some involved direct additives which were previously government-approved for the purpose; others involved materials approved for use on animals, growing crops, or packaging materials, or in the food preparation and serving areas; still others involved accidental contamination of foodstuffs by man-made pollutants in the environment or by biological organisms during large-scale preparation and handling.

Conversely, a number of other cases caused alarm for different reasons (for example, some related to personal health concerns; some to concern for the global environment), but had a factor in common in that they arose from some planned or existing government-financed project. Still others had a commonality in that they were essentially spectacular accidents.

A working classification system was therefore adopted, as

TABLE A-I

WORKING CLASSIFICATION SYSTEM FOR CASES

1. Food and Drink Additives
2. Food Contaminants
   a. From pretreatment of animals, plants or packaging materials
   b. By indirect or environmental sources
   c. By biological organisms during preparation and handling
   d. By preparative methods
3. Food Efficacy, Wholesomeness and Nutrition
4. Drugs, Medicines, and Medical Treatments
   a. Unexpected side effects
   b. Problems of quality control and "supervised" accidents
   c. Efficacy
   d. Abuse
   e. Ethical, legal, and moral aspects
5. Cosmetic Products and Cosmetic Treatments
6. Household Products
7. Other Hazardous or Substandard Consumer Products
8. Shelter (Clothing and Housing)
9. Environmental Pollutants or Impacts
   a. Long-term pollutants
   b. Big accidents
   c. Problems and effects of pollution-control movement
10. Agriculture and Land Management
11. Energy and Power
12. Transportation
13. Communications
14. Law Enforcement
15. Recreation, Entertainment, and Cultural Values
16. Governmental "Grand Projects"
17. Miscellaneous

shown in Table A-I. This system was based in large measure on
the concepts of threats to the basic human needs and activities
discussed in Appendix B. The system also noted, however, features
which might affect the degree of alarm (whether an undesirable
food constituent was an accidental contaminant or an intentional
and previously approved additive), and considered certain areas of
commonality (household detergents and big government projects)
for operational convenience.

   Task 2. Develop selection criteria and select cases for study.
An attempt was made to define the procedures used in selecting
cases for study, but the development of rigorous, systematic criteria

was precluded for several reasons. In the first place, because of the problem survey and evaluation nature of this program, a collection of cases was desired which would be "representative" of (a) the different problems besetting our society, and (b) a number of different viewpoints (such as the basic human needs and activities involved, technological origins, governmental or industrial segments involved, regional versus national interests, and nature of attention by news media). Beyond this general guideline of diversity, a number of other selection criteria were used, including certain general areas of exclusion and certain positive criteria. Finally, the selection was limited by time and budget constraints; each case must be "manageable" within the total number selected.

Criteria for the selection of issues for case history studies fall into two broad categories, as follow:

  a. *Reasons for exclusion of potential cases from present study:*
  • They involved man-made problems of primarily an accidental nature, but which have occurred rather frequently (flash floods resulting from breaking dams or levees; railroad, bus, or airplane crashes; the annual 50,000-plus auto-accident fatalities; and large industrial fires).
  • They involved problems which were primarily of an occupational accident nature, particularly those which have occurred many times (such as mine disasters), or those which involved a small number of workers (such as those who applied radium paints to make luminous watch dials during the 1920's).
  • They involved problems where a large proportion of the information may have been unavailable because of governmental security regulations (such as rocket tests that might emit poisonous, persistent pollutants such as beryllium).
  • They involved questions which were primarily of a policy nature rather than of a direct technological origin (such as no-fault auto insurance).

In general also, cases have been excluded if they were essentially a one-time occurrence arising from a mistake by an individual. Exceptions were allowed, however, if the case had unique features or if its occurrence reflected on the integrity of a governmental or social unit, or a larger practice (for example, the Dugway sheep kill incident).

  b. *Reasons for choosing cases for study:* An important consideration in the selection of cases was that the composite list would represent a cross section of the many different kinds of origins, impacts and public sectors affected, and the range of media interest shown. A number of criteria were applied, but in general, these centered on two factors, "public concern" and "technology." Thus, primary consideration was given to cases which were basically "people-oriented" in their effects since such cases would be likely to cause concern to the public and to be sufficiently reported by the news media that some sort of content analysis could be made of news stories. Cases which had involved a sizable national or local public or those involving a conflict of interest were considered likely to have received coverage.

An attempt was also made to select cases which would be representative of: the various human needs or activities which give rise to the issues; the technological origins of the problems; the governmental or industrial organizations which may have contributed to the problem; the scientific disciplines involved; and the governmental regulatory agencies involved.

A further important criterion was that a selected case should be essentially "closed" rather than ongoing, so that the full impact could be better judged. However, experience soon showed that many cases that were believed to be closed were merely dormant, and could reappear in the news again as further research, legislation, or new events occurred.

The application of the preceding criteria in the selection of the cases could not be made rigorous, however, for several reasons. Most important, perhaps, is that significant aspects of most cases were unknown to the project team until the literature search was underway. Second, the criterion of diversity in the composite selection probably overrode certain other criteria. The selection of some cases was desirable even before all the criteria could be developed and agreed upon, because of timing and staffing requirements and the need to develop standardized procedures for actually conducting the case studies. Finally, the selection of a limited number of cases from a list of over two hundred potential cases (or from a list of possibly even one thousand cases if the preliminary search had been exhaustive) is a bit like selecting the ten best-dressed women (or the ten worst-dressed men) in the country—personal opinion inevitably enters into the selection process.

Task 3. Develop standardized procedures for conducting case history studies. The study of individual cases of public concern over technology required information of two types: (a) specific details about each case, how it developed technically, who it affected, how it was handled, etc.; and (b) information about the public concern that developed over each case.

In order to make the collection of this information both

efficient and sufficiently thorough to permit subsequent analysis, an effort was made at the beginning of the project to develop standardized procedures. As the subsequent discussion will show, the printed news media ultimately became the basis for collecting both types of information, but the rationale for this choice may be of interest to the reader.

   a. *Sources of information:* While the present program focused on the transfer of technological information that would alarm the public, a brief survey was made of the literature on the transfer of technological news to the public or between members of various segments of the public and of news in general. The media showed little interest in "science news" before World War II according to early studies, and even in 1955 less than 5 percent of the space in daily newspapers was devoted to this category.[1] Coverage of science news increased greatly after Russia launched its Sputnik in 1957 and doubled in many newspapers—although still lagging well behind, "sports," "women's news" and most other categories. Only those science articles that dealt with humans attracted more than 20 percent of newspaper readers* according to one study conducted for the National Association of Science Writers.[1] This and other studies[2,3] during the late 1950's and early 1960's found that the public's primary source of news about developments in science was the printed media (35–40 percent from newspapers, 20–30 percent from magazines); the electronic media were a secondary source (25–35 percent from television,† 5–10 percent from radio). Persons with more education generally relied on magazines more and TV less than the average, and also had more accurate recall. The effectiveness of communication may, of course, be influenced by the greater previous knowledge of the recipient of the information and also by his attitude toward the communicator. Statements attributed to "prestige" sources were found to be more credible in one study[4] than those attributed to less well-known sources. But with the passage of time, the recipient tended to forget which statements came from "untrustworthy" sources,‡ and could often

---

*Articles that attracted a high (more than 50 percent) readership when presented under a "drab" headline aroused little more interest when the headline was in "journalese" (even when a threat was implied).

   †About 48 percent of health-related information, however, was received via TV.

   ‡Limited studies have indicated that a certain scientist is regarded as a credible and respected source by some elements of the public, for example, the agricultural scientist by the farmer.[5]

recall previously uncreditable information better than he could the "truth."

Many investigators[6,7] have concluded that the mass media and interpersonal communication play rival but complimentary roles in a two-step flow of information to the public, although some have contended that the news media by themselves have only a limited effect on public attitudes.[8] "Opinion leaders" are widely held to be influential in the interpersonal communications route, and several studies have found that they are important in the transfer of a technology within a group of technologists. Thus, when a new technological development is announced (for example, a new drug to doctors[9] or hybrid seed corn to farmers[10]) it is first tried by a small percentage of its potential users, then adopted rapidly by most of the users (who learn about it through conversations with the innovators), and then slowly by the other users who may have little contact with their peers. Similar patterns have been observed when some new products have been introduced on the consumer market where advertising plays a role.

Interpersonal communications in the diffusion of dramatic news have been studied by many investigators with results that varied considerably with the nature of the event and with time; they play a greater role in the case of extremely shocking events (such as the assassination of a national figure) but perhaps not as great a role now as formerly. On the other hand, as we have seen above, interpersonal communications have a significant role in the formation of attitudes. Few, if any studies, however, have been made of the roles of interpersonal communications or opinion leaders in the transfer and evaluation of unpleasant news over technology, and these factors did not appear amenable to analysis on the present program.

Since the primary routes by which the public receives technological information appears to be the formal news media, we evaluated the roles of each of these in informing the public, reflecting or generating concern, and as a tool for measurement. A significant problem, as we will see, was the multiplicity of news sources that are in use, but the major problem was that most of these sources are insufficiently indexed to permit rapid and complete retrieval of information. On the other hand, the similarity of training received by journalism students, whether they become employed by periodicals, newspapers, or television, lends a degree of uniformity across media lines. A brief discussion of each of these media and of public opinion polls follows.

Television and radio: Although the major networks maintain news reporters and newscasters, the Associated Press is probably the main supplier of news for television and radio. Most local stations, in fact, generate very few news stories of national interest. On the other hand, the major network and locally produced newscasts are a primary source of news and often of opinions for millions of people. Because the television and radio newscasts are extremely short compared to a newspaper, the number of stories carried is limited. Thus, in a typical evening newscast, fifteen to twenty stories are used from the hundreds or thousands of potential stories available and these few must be selected to try to balance newsworthiness with viewer appeal, program format (typically: big stories—little stories—leav'em laughing) and possibly other considerations (even advertisers).

The major deterrent to historical study of news presentation via TV or radio is the unavailability of indexes, compounded by the general unavailability of a permanent record of what was said, when, and how. Transcripts and videotapes of newscasts are usually very difficult to obtain for study, even when a specific date is of interest. Fortunately, studies have shown that TV time devoted to a story correlates fairly well with its news space.

A number of network and local specials have dealt with technologically related issues. These specials did not appear to be sufficient in number or to be sufficiently available for analysis to be of statistical value in the present program, but information on specials has been noted in several cases.

Newspapers: About 1,700 daily newspapers are published in the United States. The *Standard Periodicals Directory* lists about 325 which are considered to be "major" newspapers. Those newspapers of largest circulation tend to represent the major population centers, but a paper's circulation is strongly affected by the absence of other major papers in the area—the need to achieve fast distribution and close reader identification tends to give most newspapers a strong local orientation. The circulations of the largest daily newspapers are listed in Table A-II.

The *New York News*, with over 2 million circulation, is the largest, while the *Wall Street Journal*, at over 1 million, is atypical in that it has a nationwide, but rather select and very influential readership. The *New York Times* is also somewhat atypical in that it has a more nationwide circulation than most city papers and is a standard reference source in schools and libraries across the country.

Not listed in Table A-II, but well known for its analysis and commentary on the news, is the *Christian Science Monitor* with a nationwide circulation of 205,200. Other newspapers which may be important in forming opinions by a segment of the public are certain tabloids, for example, the *National Inquirer*, which is distributed in uncertain circulation via supermarkets and similar outlets, but probably has the largest circulation of any weekly newspaper. It has primarily a lower-middle-class readership, but combines a surprisingly large amount of news about science and technology (albeit often the more alarming news) with much gossip about prominent individuals and a touch of mysticism.

In the present program, the usefulness of newspapers for comparison of public alarm of the many different cases was limited by three considerations. First of all, the total number of papers is large and each tends to give strong emphasis to locally important issues. Thus, while local impacts and alarm are very important factors, the scope of the present program did not permit a systematic analysis of so many papers. Second, most of the newspapers do not have their own news services for nonlocal stories; most rely on the news wire services of Associated Press or United Press, and on the news services of a few of the very large newspapers such as the *New York Times* and the *Chicago Tribune*. A third and very important consideration which limited the usefulness of most newspapers in the present study was the lack of suitable indexes. Apparently, the *New York Times* is the only newspaper which has a widely available *Index* which covers the over-twenty-five-year period of interest. Typically, an entry in the *New York Times Index* is a short abstract of the article. Articles considered of major interest are now (since about 1952) indicated by use of boldface type and cartoons and editorials are also noted. This *Index* was started in 1851 and is a literature standard.

Few, if any, other papers publish any index, although the *Wall Street Journal* has had an index available since 1958 and the *Washington Post* started one recently. A common practice appears to be that stories are clipped and filed by subject in a morgue for future reference, but these are unavailable to the public. Microfilm of back issues are usually available for major newspapers, but search of these is prohibitively time-consuming for all the cases in the present study, even if the date could be pretty well pinpointed.

The news wire services fill a special role in the communications field and require further comment. The large wire services regularly discuss (over a conference "hot-line" between various

TABLE A-II

CIRCULATION OF MAJOR DAILY NEWSPAPERS[a]

Newspaper	Circulation
1. New York Daily News	2,112,200
2. Wall Street Journal	1,038,100
3. Los Angeles Times	861,400
4. New York Times	840,500
5. Chicago Tribune	805,900
6. Los Angeles Herald-Examiner	731,500
7. Detroit News	700,300
8. Philadelphia Bulletin	671,500
9. New York Post	628,100
10. Detroit Free Press	590,500
11. Chicago Sun-Times	552,200
12. Philadelphia Inquirer	516,700
13. San Francisco Chronicle	493,000
14. Washington Post	467,500
15. Chicago News	462,900
16. Boston Record American	455,900
17. Chicago American	446,900
18. Garden City Newsday (Long Island, New York)	421,400
19. Cleveland Press	389,700
20. Cleveland Plain Dealer	388,300
21. Milwaukee Journal	368,500
22. St. Louis Post-Dispatch	366,900
23. Miami Herald	364,000
24. Long Island Press	350,900
25. Pittsburgh Press	344,600
26. Kansas City Times (morning edition of the Star)	334,700
27. St. Louis Globe-Democrat	327,600
28. Washington Star	309,000
29. Buffalo News	284,600
30. Minneapolis Star	282,200
31. Houston Chronicle	280,500
32. Newark News	278,200
33. Houston Post	274,300
34. Rock Island Argus	262,800
35. Atlanta Journal	252,700
36. Denver Post	252,000

[a]Source: *Standard Periodicals Directory*, 1970

offices around the country) what events or people are potential subjects of stories for the day or week and should be covered, and by whom (that is, a value system is inherent in selection). Stories are then classified according to presumed interest and are transmitted, in the AP for example, on the "A" wire, "B" wire, or "budget" wire. (News wires may be international, national, regional, or local.) A further important feature is that while the local newspaper receives the "story" over the wire, it may be received in a cryptical skeletal, outline form which is then fleshed out by the local writer, possibly shortened in line with the editor's interests and space or other considerations, and run under a locally supplied headline. Thus, a given wire story may receive different presentations or emphasis and have considerably different impacts as it appears in various newspapers around the country. This difference would make evaluation of impact difficult.

In addition to the wire services, most newspapers also use material supplied by syndicated columnists and social commentary cartoonists. These vary widely, of course, among papers, and standardized measurements of their effects are impossible. Yet some of the columnists may reach an audience far greater than the circulation of any newspaper and perhaps greater than any TV commentator. Some of the "Dear Abby" or "Dear Ann" columns are said to have circulations of over 50 million.

Indexes are generally unavailable on the wire service stories, syndicated cartoons, etc. Local papers, for example, apparently do not file stories or cartoons which are received, but not used soon. No central index for cartoons appears to exist. Many of the better cartoons are frequently picked up by the periodicals, however. At times, cartoonists and columnists also publish books which are selected collections of their works. Information on the number of active social cartoonists and the size of the audience they reach does not appear to be readily available.

Periodicals: The *Standard Periodicals Directory* currently lists about 53,000 periodicals, the criteria being that publication is on a regular basis with a minimum of one issue in two years or less. Of these periodicals, 325 are major daily newspapers and most of the rest can be described as magazines. The *Standard Periodicals Directory* also classifies the periodicals according to area of interest into 186 categories. A more specialized compilation of information on periodicals is *Bacon's Publicity Checker*, which lists about 3,900 periodicals (with an emphasis on advertising). *Bacon's* utilizes a system of about ninety-nine categories for classification of periodicals.

The major index to magazines is the *Readers' Guide to Periodical Literature* (started in 1890), which covered 159 major publications in 1972. Of these, twenty-eight periodicals had circulations of over 1 million in 1970 and another forty-five had circulations of over 100,000 as shown in Table A-III. The subject areas covered by the publications listed in *Readers' Guide* are quite diversified as indicated in the fifty-seven classifications represented in Table A-IV. Hence, these publications serve a wide cross section of the reading public (one suspects that these readers are above average as local opinion leaders). The index to the *Readers' Guide* should provide one valid tool for measurement and comparison of public concern with the various cases. Typically, an entry in the *Readers' Guide* gives the article's title, a word of explanation if necessary, the author if one was listed, and a notation if the article is illustrated. Editorials are not always indicated, unfortunately.

A second index to the magazines, the *Business Periodicals Index* (started in 1958), covers 162 publications (1972). None of these, however, had circulations of 1 million and only eleven had circulations of over 100,000, but a total of seventy-two had circulations of 14,000 or more as shown in Table A-V. Of these, however, five of the eight largest (and eight of the 72) are also included in the *Readers' Guide*. These 162 publications serve a wide range of interests as shown in Table A-VI, with coverage particularly on the economic aspects. The readership of these publications is undoubtedly influential in decision-making and opinion-forming processes in this country. Hence, the *BPI* should be a legitimate criterion of interest of an important segment of the public.

One should note that the widely distributed (but somewhat regionally oriented) *TV Guide* is not included in *Readers' Guide*, but may at times be influential in forming opinions about certain technological issues of national interest.

Books: Over 25,000 new books are now being published annually in the United States, and nearly half as many new editions of older books are also issued each year. Among the twenty three classifications of the Dewey Decimal System, by far the most books are published in the area of "Sociology and Economics," as shown in Table A-VII. Publications in "Science" rank fourth (behind "Fiction" and "Literature"), but substantial numbers of technology-related books are tabulated under the classifications of "Medicine," "Technology," and "Agriculture."

The degree to which books might contribute to public awareness or concern over technology is difficult to estimate. A large and

TABLE A-III

CIRCULATION OF MAJOR PERIODICALS LISTED IN READERS' GUIDE TO PERIODICAL LITERATURE, 1972

Rank	Periodical	Circulation[a]	Rank	Periodical	Circulation[a]
1	Reader's Digest	17,500,000	38	Senior Scholastic (Teacher Ed.)	540,000
2	McCall's	8,500,000	39	Motor Trend	525,000
3	Life	7,540,000	40	World Week	520,000
4	Better Homes and Gardens	7,250,000	41	Saturday Review	510,000
5	Ladies Home Journal	6,965,000	42	Fortune	505,000
6	National Geographic Magazine	6,000,000	43	The New Yorker	475,000
7	Good Housekeeping	5,685,000	44	Vogue	445,000
8	Redbook	4,000,000	45	Harper's Bazaar	435,000
9	American Home	3,696,900	46	Scientific American	419,000
10	Time	3,300,000	47	Home Garden and Flower Grower	415,000
11	Farm Journal	3,000,000	48	Popular Photography	404,000
12	Parent's Magazine and Better Family Living	2,150,000	49	Popular Electronics	400,000
13	Newsweek	2,128,000	50	Organic Gardening and Farming	400,000
14	Sports Illustrated	1,700,000	51	National Wildlife	340,000
15	Field and Stream	1,650,000	52	The Atlantic	325,000
16	U.S. News and World Report	1,600,000	53	American Heritage	300,000
17	Popular Science Monthly	1,600,000	54	Harper's Magazine	291,000
18	Today's Education	1,500,000	55	Travel	280,000
19	The New York Times Magazine	1,495,000	56	Modern Photography	280,000
20	Popular Mechanics	1,493,000	57	Flying	268,000
21	Consumers Report	1,325,000	58	Ramparts	220,000
22	Seventeen	1,321,000	59	Christianity Today	201,700
23	Successful Farming	1,300,000	60	House and Garden Incorporating Living for Young Homemakers	200,000
24	Changing Times	1,250,000	61	Natural History	192,000
25	Mechanix Illustrated	1,200,000	62	Dun's	175,000
26	Holiday	1,110,000	63	Scholastic Teacher Jr/Sr High Teacher Ed.	170,000
27	Ebony	1,012,000	64	Radio-Electronics	150,000
28	Outdoor Life	1,000,000	65	Sports Car Graphic	150,000
29	House Beautiful	980,000	66	Science Digest	149,000
30	Esquire	956,000	67	Science	140,000
31	Hot Rod	844,000	68	The New Republic	123,000
32	Sunset	839,000	69	Rod and Custom	115,000
33	Nation's Business	815,000	70	Horticulture	115,000
34	Today's Health	701,000	71	Motor Boating and Sailing	110,000
35	Mademoiselle	650,000	72	Yachting	110,000
36	Business Week	578,600	73	Consumer Bulletin	100,000
37	Forbes	550,000			

[a]Source: *Standard Periodical Directory*, 1970.

TABLE A-IV

CLASSIFICATION OF PERIODICALS IN READERS' GUIDE TO PERIODICAL LITERATURE ACCORDING TO STANDARD PERIODICALS DIRECTORY, 1970

Classification	No. of Periodicals	Classification	No. of Periodicals
Agriculture	2	Library	4
Antiques and Art Goods	1	Management	1
Architecture	2	Meteorology	1
Art and Sculpture	5	Motion Pictures	1
Astronomy	1	Music Trade	3
Automotive	4	Natural History	1
Aviation and Aerospace	3	Negro Interest	1
Boating	1	News	4
Business and Industry	4	Oceanography	1
Chamber of Commerce	1	Outdoors	2
Chemicals and Chemical Engineering	1	Parks and Recreation	1
Children	2	Photography	2
Conservation	5	Physics	1
Craft and Hobbies	5	Political and Literary Reviews	6
Dance	1	Political Science and World Affairs	9
Educational	8	Pollution Control	1
Electronics and Electrical Engineering	1	Psychology and Psychiatry	1
Forestry	1	Public Management and Planning	1
Gardening	3	Publishing	1
General Interest	12	Radio-TV	1
Geography	2	Religious	3
Geriatrics	2	Science	5
Health	1	Sociology	1
History	4	Sound Engineering and Reproduction	1
Home	5	Sports and Sporting Goods	2
Home Economics	2	Theatre	1
Journalism	2	Travel	1
Labor	1	Women and Fashions	7
		Youth	2

Note: Not all of the periodicals in the *Readers' Guide to Periodical Literature* were listed in the *Standard Periodicals Directory*, 1970.

CIRCULATION OF MAJOR PERIODICALS LISTED IN BUSINESS PERIODICALS INDEX, 1972

Rank	Periodical	Circulation[a]
1	Nation's Business	815,000[b]
2	Business Week	578,600[b]
3	Forbes	550,000[b]
4	Fortune	504,900[b]
5	First National City Bank, Monthly Economic Letter	360,000
6	Barron's	250,000[c]
7	Business Management	200,000
8	Dun's	175,000[b]
9	AFL–CIO American Federationist	140,000
10	Journal of Accountancy	120,000
11	House and Home	106,300
12	Aviation Week and Space Technology	97,400[b]
13	Burrough's Clearing House	97,000
14	Engineering News-Record	97,000
15	The Office	95,000
16	Harvard Business Review	92,000[b]
17	Marketing/Communications	85,500
18	Industrial Research	80,000
19	Iron Age	78,500
20	Progressive Grocer	74,900
21	Financial World	70,000
22	American Druggist	66,000
23	Fleet Owner	63,400
24	Datamation	63,000
25	Graphic Arts Monthly and the Printing Industry	62,600[c]
26	Advertising Age	62,000
27	Personnel	61,000
28	Management Accounting	60,000
29	Management Review	60,000
30	Chemical Week	59,900
31	Electronic News	58,000
32	Food Processing	55,000
33	Supervisory Management	51,000
34	Administrative Management	50,500
35	Oil and Gas Journal	48,000
36	Quick Frozen Foods	42,500
37	Banking: Journal of the American Bankers Association	41,800
38	Best's Review (Life/Health Edition)	41,000[c]
39	Business Automation	40,000
40	Merchandising Week	40,000
41	Purchasing	40,000
42	Sales Management, the Marketing Magazine	38,000
43	Credit and Financial Management	36,000
44	Automotive Industries	35,000
45	Chain Store Age (Combination)	34,900
46	Broadcasting	30,000
47	World Oil	27,700
48	Bankers Monthly	27,300[c]
49	Textile World	25,800
50	Air Conditioning, Heating, and Refrigeration News	25,700
51	Editor and Publisher	25,300
52	Airline Management and Marketing	25,200
53	Industrial Distribution	25,000
54	The National Underwriter (Fine Edition)	25,000[c]
55	Industrial Marketing	24,000[c]
56	Publishers Weekly	23,200[b]
57	Best's Review (Property/Liability Edition)	23,000[c]
58	National Petroleum News	21,600
59	The Accounting Review	20,000
60	American Economic Review	20,000
61	Journal of Marketing	20,000
62	Michigan Business Review	20,000
63	Financial Analysts Journal	19,000
64	Appraisal Journal	18,000
65	The New York Certified Public Accountant	18,000
66	Forest Industries	16,800
67	Inland Printer/American Lithographer	16,500
68	Industrial Development and Manufacturers Record	16,000
69	Trusts and Estates	15,500[c]
70	Journal of Taxation	15,000
71	Railway Age	14,500
72	Oil, Paint, and Drug Reporter	14,000

[a]Source: *Standard Periodical Directory*, 1970, except as noted by [c].
[b]Periodical also covered by *Readers' Guide to Periodical Literature*.
[c]Source: *Bacon's Publicity Checker*, 1970.

TABLE A-VI

CLASSIFICATION OF PERIODICALS IN BUSINESS PERIODICALS INDEX
ACCORDING TO STANDARD PERIODICALS DIRECTORY, 1970

Classification	No. of Periodicals	Classification	No. of Periodicals
Accounting	5	International Trade	1
Advertising and Marketing	8	Labor	5
Aeronautics and Astronautics	2	Library	13
Appliances	1	Management	1
Automation and Computers	5	Manufacturing	1
Automotive	1	Mathematics	1
Banking and Finance	12	Metals and Metalworking	1
Book Trade	1	Newspaper Industry	1
Building and Construction	3	Office Management and Equipment	2
Business and Industry	13	Packaging	2
Chamber of Commerce	1	Paper	3
Chemicals and Chemical Engineering	3	Petroleum and Natural Gas	2
Department Stores	2	Printing	1
Drugs and Pharmaceuticals	2	Public Utilities	2
Economics	8	Purchasing	1
Electronics and Electrical Engineering	1	Radio and TV	1
Food	2	Railroads	2
Forestry	6	Real Estate	1
Government Publication	1	Science	1
Grocery	2	Selling	3
Heating, Plumbing, A-C, and Refrigeration	4	Taxes	1
Industrial Relations and Personnel	6	Textiles	2
Insurance		Traffic and Transportation	1
		Vending Machines	

Note: Not all of the periodicals in the *Business Periodicals Index* were listed in the *Standard Periodicals Directory*, 1970.

TABLE A-VII

RANKING OF NUMBERS OF BOOKS PUBLISHED BY CLASSIFICATION[a]

Classifications	1971 New Books	1971 New Editions	1971 Totals	1972 New Books	1972 New Editions	1972 Totals	Total 1971–1972
Sociology, Economics	4,268	1,827	6,095	4,688	1,727	6,415	12,510
Fiction	2,066	1,364	3,430	2,109	1,151	3,260	6,690
Literature	1,383	1,603	2,986	1,398	1,127	2,525	5,511
Science	2,225	472	2,697	2,143	443	2,586	5,283
Juveniles	1,991	232	2,223	2,126	400	2,526	4,749
Biography	853	944	1,797	1,086	900	1,986	3,783
History	949	1,029	1,978	906	723	1,629	3,607
Medicine	1,252	403	1,655	1,404	435	1,839	3,494
Religion	1,140	427	1,567	1,233	472	1,705	3,272
Travel	950	659	1,609	972	519	1,491	3,100
Poetry, Drama	932	562	1,494	883	601	1,484	2,978
Technology	1,057	252	1,309	1,184	241	1,425	2,734
Art	932	314	1,246	1,097	373	1,470	2,716
Education	1,020	230	1,250	1,041	251	1,292	2,542
Philosophy, Psychology	947	407	1,354	829	335	1,164	2,518
General Works	715	297	1,012	802	246	1,048	2,060
Sports, Recreation	645	245	890	686	255	941	1,831
Business	550	150	700	529	155	684	1,384
Law	415	246	661	418	298	716	1,377
Home Economics	381	96	477	479	117	596	1,073
Language	400	136	536	354	125	479	1,015
Music	214	188	402	215	187	402	804
Agriculture	241	83	324	286	104	390	714

[a]Adapted from: Publishers Weekly, p. 48 (February 15, 1973).
Notes: Biographies include those placed in other classes by the Library of Congress; 1972 figures are probably subject to upward revision.

TABLE A-VIII

PRELIMINARY COMPARISON OF INDEXES FOR SELECTED CASES

(Numbers of Entries)

	Readers' Guide Index	Business Periodicals Index 1958 to Date	New York Times Index	Wall Street Journal Index 1959 to Date
Southern Corn Leaf Blight—A Genetic Engineering Catastrophe	34	5	55	31
Krebiozen—Cancer Cure?	8	NA	21	NA
DMSO—Suppressed Wonder Drug?	6	0	2	0
The Donora Air Pollution Episode	6	NA	17	NA
The Mercury-In-Tuna Scare[a]	35	14	72	17
Foaming Detergents	32	101	100	41
Nitrilotriacetic Acid (NTA) Detergents	7	7	17	1
The Dugway Sheep Kill Incident	9	1	17	1
Project West Ford—Orbital Belt of Needles	49	5	34	0
Project Sanguine—Giant Underground Transmitter	6	4	5	0
Project Able—The Space Orbital Mirror	4	0	0	0
The AD-X2 Battery Additive Debate	43	NA	69	NA
Northeast Power Failure	25	55	101	0
Sea Level Panama Canal	19	incomplete	14	0

[a] Primarily the Mercury-In-Tuna Scare, but related articles included here.

TABLE A-IX

SELECTED STANDARD PERIODICALS

Periodical	Circulation[a]	Classification[a]	Comment
Reader's Digest	17,500,000	General Interest	Wide, middle-class appeal
Life[c]	7,540,000	General interest	Wide readership
Time	3,300,000	News	Middle- and upper-class readership
Farm Journal	3,000,000	Agriculture[b]	Primarily rural readership
U.S. News and World Report	1,600,000	News	Readership largely in upper-to-middle education and economic brackets
Consumer Reports	1,325,000	Home economics[b]	Influential, middle-class readership
Today's Health	701,000	Health[b]	Fairly broad readership
Business Week	578,000	Business and industry	Select, influential readership
Science	140,000	Science	Select, influential readership

[a] According to the *Standard Periodicals Directory*.
[b] *Bacon's Publicity Checker* classifies *Farm Journal* as general farm, *Consumer Reports* and *Today's Health* as general, and *Business Week* as business and commercial.
[c] *Life*, unfortunately, ceased regular publication with the issue of September 29, 1972, and has appeared since only as occasional special issues.

probably even a major portion of the technology-related books are textbooks or reference works for a technical audience. In recent years, however, increasing numbers of books (many are paperbacks; some are hardcover) seem to be directed toward the lay adult market that have as their subjects aspects of technology, its impacts, or its control. In fact, books have been published about several of the cases studied herein—a few even written by individuals in the center of the controversy. In general, however, books are useful for getting specific information on cases, but not as comparative measures of public concern.

Technical literature: A vast technical literature is available to and used by the scientific and technological community that is generally unfamiliar (and of little interest) to the general public—all who are not interested in the particular field. This technical literature includes not only books and encyclopedic compendia, but over 15,000 serial publications. Access to particular subject matter in this literature is generally much better than to that in the popular literature, for example, via *Biological Abstracts, Chemical Abstracts, Physics Abstracts, Electrical Engineering Abstracts, Nuclear Science Abstracts, Applied Science and Technology Index, Biological and Agricultural Index, Index Medicus, Psychology Abstracts, Social Sciences and Humanities Index, Air Pollution Abstracts,* etc. These sources are useful for collecting technical information on specific cases, but not for comparative information on public concern.

Public opinion polls: Public opinion polls are taken on a multitude of topics ranging from national affairs to local issues and consumer preferences. Among the better-known polling organizations are Louis Harris and Associates (New York City); the Roper Organization (New York City); the Gallop Poll (Princeton, New Jersey); the Opinion Research Corporation (Princeton, New Jersey); the Office of Public Opinion Research (Princeton, New Jersey); the National Opinion Research Center, University of Chicago (Chicago, Illinois); the Survey Research Center, University of Michigan (Ann Arbor, Michigan); the Minneapolis Poll (Minneapolis, Minnesota); the Iowa Poll (Des Moines, Iowa) and the California Poll (San Francisco, California). The results of many of these polls are published in magazines and syndicated newspaper columns. The results of early polls were summarized in Hadley Cantrils' *Public Opinion, 1935–1946* (Princeton University Press), and the results of polls on various topics of interest are summarized regularly by Hazel Erskine in *Public Opinion Quarterly.* In addition, the Roper Public Opinion Research Center (Williamstown, Massachusetts) has compiled a data bank of over ten thousand poll results from around the world.*

A survey of the *Public Opinion Quarterly* and contacts with the Roper Public Opinion Research Center revealed a substantial number of surveys taken at various times on topics related to case histories being prepared on the present program. Particularly well represented were issues such as cigarette smoking, fluoridation, pesticides, radioactive fallout, oral contraceptives, and, more recently, pollution. An evaluation of the breadth of the polls and of the way questions were often asked on a given subject, however, did not indicate that public opinion polls were well suited to our purposes on the present program, and an exhaustive search was not undertaken.

b. *Information collecting:* Because of the emphasis on "public" concern or alarm, the news media were used both as a primary information source and as a means of measuring public concern. An assumption was made that news media coverage of a story is roughly proportional to public interest and, in the cases under study here, to public concern. The printed news media were the focus, since information retrieval procedures are so poorly developed for the electronic media. The periodicals were selected as the principal media for study and measurement in the public sector, but were supplemented as described below:

The *Readers' Guide to Periodical Literature* (*RG*) and the *New York Times Index* (*NYT Index*) are the only indexes that cover the approximately thirty-year period of interest. The *Business Periodicals Index* (*BPI*) and the *Wall Street Journal Index* (*WSJ Index*) have been available since 1958.

A preliminary comparison of the *WSJ Index* was reasonably well duplicated by the combination of the *BPI* and the *NYT Index* as shown in Table A-VIII. Therefore, only the *RG*, *BPI* and *NYT Indexes* were utilized for systematic searches on every case. In addition, nine periodicals were selected as "standard periodicals" to be used on every case. These periodicals were selected on the basis of several criteria, such as their total circulations and the amount of news and commentary that they carry; they represent a cross section of areas of interests served and types of readerships. These periodicals and some of their characteristics are summarized in Table A-IX.

*L.L. Lederman used the data bank to prepare "An Annotated Bibliography on Public Opinion of Science and Technology," Battelle Memorial Institute, 1965.

Collection procedures: A systematic search was made through the *RG, BPI* and *NYT Index* for each selected case. Special note is made here that the natures of many of the cases and the inconsistencies of the indexes complicated the search. For example, a given case might be indexed under a number of different headings, even in the same index (especially a case that ran on for several years). In other cases the entries were not specific enough (for example, DDT in Coho Salmon). In a few cases, almost no entries could be found (for example, Carbon Monoxide in the Corvair*, Studded Snow Tires, Silicone Injection Safety), and the case was deemphasized unless an alternate information source was available. The literature search was conducted by three project members who worked closely together and used a standardized checklist.

A reading file was prepared for each case study. It contained lists of entries (headlines and references) on the case from the *RG*, a list of references and comments (or entire abstracts) on entries in the *NYT Index*, and all pertinent articles from the nine "standard periodicals." These were supplemented by other articles of popular or technical interest (either identified through the search, or in some cases through a search of other abstracts or indexes); entries on many cases from *Facts on File;* clippings from local newspapers; published hearings before the Congress; and copies or excerpts from pertinent books, reports, and pamphlets.

Measurements of concern: The nine "standard periodicals," the *RG*, the *BPI* and *NYT Indexes* were used to measure media interest in each case. As has been shown in Table VII, the following data were determined:

- Nine standard periodicals. Total column inches (including pictures); total number of pictures; total number of cartoons; total number of editorials; total number of cover stories.
- *Readers' Guide.* Total number of entries; total number of illustrated articles.
- *Business Periodicals Index.* Total number of entries.
- *New York Times Index.* Total number of entries; the number of entries in boldface type; the total column inches of abstracts; total number of editorials; total number of cartoons.

In summary then, systematic procedures were established for collecting the factual and technical details required to construct the history of the case as it involved the public, and for obtaining objec-

*On this case even some known articles in periodicals normally covered by the *RG* failed to turn up in the index search.

tive measures of public concern so that the cases could be compared with each other. The printed news media were the focus, because information retrieval procedures are so poorly developed in the electronic media. Even in the printed media, indexes were barely adequate for our program. *Readers' Guide to Periodical Literature*, the *New York Times Index*, *Business Periodicals Index*, and *Facts on File* were the principal sources of information on popular news stories.

Task 4. Develop individual case histories. The file of information on each case was reviewed. This file was then supplemented in some of the specific cases by pertinent published articles, books, hearings before the Congress, reports, and other sources that were identified, by personal contacts with individuals, companies, associations, or other unpublished information sources, and, in a few cases, by partial searches of *Biological Abstracts*, *Chemical Abstracts*, or the *Wall Street Journal Index*. A project member then prepared a draft case history. A goal was set to include as much objective, factual information as possible, to supplement this with limited amounts of subjective evaluation as necessary, and to document the case thoroughly so that the total results will serve as a useful data bank of information for interested researchers. In order to make the results as amenable as possible to subsequent analysis and study, the many diverse cases were presented in a consistent form: in addition to the title and abstract, each case contained a time line, narrative account (background, key events and roles, and disposition), a comment and references.

The time line is a visual depiction of the significant events in the case. The narrative account of each case is, the authors hope, an objective description of the who, what, when, and where of the episode. The narratives are essentially chronological, and include:

Background—Technological, economic, legislative or regulatory, social, psychological (previous similar cases, etc.), political, early warnings of trouble.

Key events and roles—Organization, agencies, people, public statements, decisions, legislations, news stories, technological features (accidents or incidents, dates that the products were found objectionable, etc.).

Disposition—Outcome, impacts, consequences for principal parties involved; legislation enacted; products banned; projects stopped.

Comment—A statement of the authors' views of what the case was really all about (the why of the case), how it was handled by the various parties concerned, what its overall impacts on our society have been, and what future events may occur as a result of the case.

The references are documentation of critical events and news stories in the case, particularly those involving technical details and important statements or decisions. A list of abbreviations for literature citations is appended. The draft case histories were reviewed by other project members and, in some cases, by members of the advisory committee or other reviewers; they were then revised by the project leader to incorporate clarifications and suggestions or to emphasize points of interest and reviewed editorially.

Task 5. Analyze collective cases and draw conclusions. An analysis of the collective cases was made to compare several aspects, such as statistical coverage by the news media, to determine their general characteristics, and to examine the implications of technology for our society and its institutions. This analysis is anticipated to be the forerunner of further subsequent analysis by a host of users, including various governmental, industrial, and academic scientists and agencies, the news media, consumer advocate groups, and, of course, those interested in doing better Technology Assessments.

The analysis and conclusions sections of the report, Chapter V, was prepared primarily by the project leader, with inputs and comments from many individuals. Of particular value were the comments of the twenty-four persons who reviewed a preliminary draft of the manuscript at the request of the National Science Foundation.

The analysis of the forty-five case studies discussed in Chapter V are based in part on the preliminary analyses presented here. Additional information on the tabulated data* is given below.

- A broad list of technologies was developed as shown in Table VII, and was screened against the cases. The one or two technologies judged to be most implicated are indicated in Table A-X by the order numbers used in Table VII.

- A list of human needs and wants was developed as shown in Table VIII and was screened against the cases. The one or two needs that were judged to be most served by the technology involved are indicated in the appropriate column in Table A-X by the code numbering system used in Table VIII.

- A list of academic and professional sciences and disciplines

---

*A dash in the tables indicates either that the question could not be answered with the information in hand or was not applicable to the particular case. A question mark denotes a substantial uncertainty in the answer.

was developed as shown in Table IX, and was screened against the cases with the results in Table A-X. The one or two sciences/disciplines judged to be most related in each case are indicated using the numbering system in Table IX.

• Each study case was screened to determine if an output from outside the United States was important in the development of the technology. The results are shown in Table A-X.

• The "scale" of the technology implicated in each case was considered in terms of its "size" and the "numbers of people involved" in its use. The size factor was a subjective composite of pertinent factors such as the number of units produced, millions of pounds of production, cost of services or equipment, percentage of market, tons of earth moved, etc. The people factor was a subjective composite estimate of the number of users of the technology as producers and as consumers. Each factor was rated from 1 to 5, where 5 is very large. The results are shown in Table A-X.

• The case histories were reviewed to determine if an early warning signal that was missed could be identified. The results are indicated by "yes" or "no" in Table A-X.

• The case histories were reviewed to determine if a technology that was perceived to be potentially troublesome had been allowed to grow. The results are indicated in Table A-X.

• The case histories were reviewed and a subjective evaluation made of whether the technology had been used responsibly, with the results shown in Table A-X.

• The case histories were reviewed to determine if new information of a technological origin or nature helped define the threat. The results are shown in Table A-X.

• The natures of the alarms and the times of occurrence of these alarms were compared for the study cases. A judgment was then made for each case as to whether it had been preceded by a closely related or similar alarm, with the results shown in Table A-XI. The time periods that the cases were most in the limelight were then evaluated with the results shown in Figure 5.

• The seven values listed in Table X were screened against each study case to determine if they were threatened. The results are shown in Table A-XI (the number system is the same as in Table X).

• The number of people threatened directly and indirectly in each case were estimated with the results shown in Table A-XI. A "L" indicates more than 10^6 people, "M" indicates 10^4–10^6, and a "S" indicates fewer than 10^4 people.

• Special characteristics of the threatened group in each case

TABLE A-X

TECHNOLOGICAL ORIGINS OF THE STUDY CASES

Case	The Technology Implicated	Implicated	The Need Served	By the Technology	Professional/Academic Precursors		Foreign Technology Input	Scale of Use — Relative Size	Scale of Use — People Involved	An Early Warning Missed	Problem Allowed To Grow	Use Was Very Responsible	New Technological information
1. Human Artificial Insemination	14	—	4	—	36	53	—	2	1	No	No	—	No
2. Oral Contraceptive Safety Hearings	14	—	4	—	20	53	—	3	3	Yes	Yes	No	Yes
3. Southern Corn Leaf Blight	1	—	2	—	16	—	No	5	4	Yes	Yes	No	No
4. The Great Cranberry Scare	4	14	2	—	1	66	No	2	3	Yes	Yes	Yes	Yes
5. The Diethylstilbestrol Ban	2	—	2	—	2	66	No	5	5	No	Yes	No	Yes
6. The Cyclamate Affair	5	—	2	—	25	66	No	3	4	No	No	No	Yes
7. MSG and the Chinese Restaurant Syndrome	5	7	2	—	88	66	Yes	2	2	No	No	No	Yes
8. Botulism and Bon Vivant	6	8	2	—	31	88	No	4	4	No	No	—	No
9. The Fish Protein Concentrate Issue	5	—	2	5	30	88	No	1	1	No	No	No	No
10. The Fluoridation Controversy	13	—	1	—	51	—	No	3	3	No	No	No	Yes
11. Salk Polio Vaccine Hazard Episode	10	—	5	—	53	—	No	4	4	No	No	No	Yes
12. The Thalidomide Tragedy	10	17	3	—	52	—	Yes	3	2	Yes	Yes	No	Yes
13. Hexa, Hexa, Hexachlorophene	12	—	5	—	53	66	Yes	3	4	No	No	No	Yes
14. Krebiozen—Cancer Cure?	9	—	5	—	53	—	Yes	1	1	No	Yes	No	No
15. DMSO—Suppressed Wonder Drug?	9	—	5	—	53	66	Yes	2	3	No	Yes	No	Yes
16. X-Ray Shoe-Fitting Machine	16	—	7	—	72	—	—	2	5	Yes	No	No	Yes
17. Abuse of Medical and Dental X-Rays	11	12	5	19	53	72	—	5	3	No	Yes	No?	Yes
18. X-Radiation From Color TV	15	42	9	—	31	72	No	4	2	No	No	No?	Yes
19. Introduction of the Lampreys	47	—	17	—	39	—	No	2	1	No	Yes	No?	No
20. The Donora Air Pollution Episode	25	27	17	—	40	44	No	4	3	No	Yes	No	Yes
21. The Torrey Canyon Disaster	51	40	15	—	45	—	Yes	3	2	Yes	No	—	No
22. The Santa Barbara Oil Leak	51	40	17	—	45	68	No	4	1	Yes	Yes	No	No
23. Mercury Discharges by Industry	27	—	2	—	40	66	Yes	3	1	No	Yes	No	Yes
24. The Mercury-In-Tuna Scare	5	—	2	5	1	—	No	3	3	Yes	No	Yes	Yes
25. The Rise and Fall of DDT	4	13	8	—	39	66	Yes	4	4	No	Yes	—	Yes
26. Asbestos Health Threat	26	—	17	—	44	—	—	3	2	Yes	Yes	—	Yes
27. Taconite Pollution of Lake Superior	24	—	17	—	45	45	No	2	1	Yes	Yes	No	Yes
28. Foaming Detergents	18	—	7	16	30	66	Nb	4	3	Yes	Yes	No	No

No.	Title	(1)	(2)	(3)	(4)	(5)	(6)	(7)	(8)	(9)	(10)	(11)	(12)	(13)
29.	Enzyme Detergents	18	—	7	16	11	25	No	2	2	No	Yes	No	Yes
30.	Nitrilotriacetic Acid (NTA) in Detergents	18	—	7	16	66	—	No	2	1	No	No	Yes	Yes
31.	Saltville—An Ecological Bankruptcy?	27	59	17	—	40	—	No	3	—	Yes	Yes	No	Yes
32.	Truman Reservoir Controversy	46	—	18	15	39	—	No	3	1	No	No	—	Yes
33.	Plutonium Plant Safety at Rock Flats	66	—	12	16	46	68	Yes	3	—	No	Yes	No	Yes
34.	Storage of Radioactive Wastes in Kansas	64	53	15	—	46	—	No	3	1	No	No	No	Yes
35.	Amchitka Underground Nuclear Test	66	—	12	16	46	66	Yes	3	2	No	No	—	No
36.	The Dugway Sheep Kill Incident	67	—	12	12	93	66	Yes	2	1	Yes	Yes	No	No
37.	The Nerve Gas Disposal Controversy	67	39	12	12	93	72	No	3	3	No	Yes	No	No
38.	Project West Ford	73	—	9	12	93	72	No	2	1	Yes	Yes	No	Yes
39.	Project Sanguine	73	—	9	—	93	72	Yes	3	—	No	No	—	No
40.	Project Able	73	—	9	—	93	66	Yes	2	2	Yes	No	No	Yes
41.	The Chemical Mace	73	—	12	—	61	79	No	2	1	No	Yes	No	No
42.	The Bronze Horse	25	—	19	—	50	66	Yes	1	3	No	No	—	Yes
43.	Synthetic Turf and Football Injuries	22	—	19	8	53	53	No	2	1	No	Yes	No	Yes
44.	Disqualification of Dancer's Image	12	—	19	—	53	53	Yes	1	1	Yes	No	No?	Yes
45.	The AD-X2 Battery Additive Debate	36	56	17	—	25	37	No	1	1	No	No	No	Yes

See text for explanations and column numbers.

TABLE A-XI

FACTORS INFLUENCING THE IMPACTS OF THE THREATS

Case	Related Case Preceded	Value Threatened	People Threatened Directly	People Threatened Indirectly	Special Group Threatened	The Threat Could Be Avoided	A Gov't Agency Had Been "In Charge" of the Problem Technology	Someone Was Really "In Charge" of the Problem	A Conflict of Interest or of Opinion Existed in Gov't Agencies	Previous Law or Regulation Was a Complicating Factor	Age of Technology	The Party Perceived as the Cause
1. Human Artificial Insemination	No	6	S	—	Ot	Yes	No	No	No	Yes	>100	3,10,15
2. Oral Contraceptive Safety Hearings	Yes	1	L	—	S	No	Yes	Yes	Yes	Yes	15	3,4,11,12
3. Southern Corn Leaf Blight	No	3	M	L	Oc	No	Yes	No	Yes	No	15	1,8,10,16
4. The Great Cranberry Scare	No	1	M	—	Ot	Yes	Yes	Yes	Yes	Yes	10	1,10
5. The Diethylstilbestrol Ban	Yes	1	L	—	,S	No	No	No	Yes	Yes	20	1,10
6. The Cyclamate Affair	Yes	1	L	—	—	Yes	Yes	No	No	Yes	30	2,10
7. MSG and the Chinese Restaurant Syndrome	No	1	M	—	Ot,A	No	Yes	No	No	No	60	2
8. Botulism and Bon Vivant	No	7	L	—	—	No	Yes	Yes	No	Yes	>100	1
9. The Fish Protein Concentrate Issue	No	1	S	—	—	Yes	No	Yes	Yes	Yes	10	2,5,10
10. The Fluoridation Controversy	No	1	L	—	A	No	Yes	No	Yes	No	5	5
11. Salk Polio Vaccine Hazard Episode	No	1	S	—	S	No	Yes	Yes	Yes	No	2	4,10
12. The Thalidomide Tragedy	No	1	M	—	—A,S	No	Yes	Yes	No	No	12	4
13. Hexa, Hexa, Hexachlorophene	Yes	1	—	—		No	No	No	No	No	30	3
14. Krebiozen—Cancer Cure?	Yes	1	L	S	—	Yes	Yes	No	No	Yes	8	3,16
15. DMSO—Suppressed Wonder Drug?	No	1	L	S	A	No	No	Yes	No	Yes	5	3,10,16
16. X-Ray Shoe-Fitting Machine	Yes	1	L	—	—	Yes	No	No	Yes?	Yes	20	8
17. Abuse of Medical and Dental X-Rays	Yes	1	S	—	SS,A	No	Yes	No	No	No	35	3,7
18. X-Radiation from Color TV	Yes	1	S	—	Oc	No	Yes	No	No	No	15	7
19. Introduction of the Lampreys	No	3	—	L	Re	No	No	No	Yes	No	>100	7
20. The Donora Air Pollution Episode	Yes	1	M	L	Re	No	No	No	No	No	30	6
21. The Torrey Canyon Disaster	No	2	M	L	Re	No	No	No	No	Yes	20	6
22. The Santa Barbara Oil Leak	Yes	2	—	L	—	No	Yes	No	Yes	Yes	70	6
23. Mercury Discharges by Industry	Yes	1	S	—	Ot	No	No	No	No	No	70	6
24. The Mercury-In-Tuna Scare	Yes	1	M	L	-,Oc	No	Yes	Yes	No	Yes	>100	10,16
25. The Rise and Fall of DDT	No	2	M - L	—		No	Yes	No	Yes	Yes	25	1,6,8,10,16

#	Title													Code numbers
26.	Asbestos Health Threat	Yes	1	3	L	—	Oc	No	No	No	Yes	Yes	100	6
27.	Taconite Pollution of Lake Superior	Yes	2	3	M	M	Re	No	Yes	No	Yes	Yes	12	6
28.	Foaming Detergents	No	2	3	M	M	—	No	No	No	No	No	10	7,8
29.	Enzyme Detergents	Yes	1	—	M	L	Ot	No	No	Yes	No	Yes	3	8
30.	Nitrilotriacetic Acid (NTA) in Detergents	Yes	2	4	S	—	—	No	Yes	Yes	Yes	No	1	8,16
31.	Saltville—An Ecological Bankruptcy?	No	3	—	M	M	Re	No	Yes	No	No	Yes	75	6,10
32.	Truman Reservoir Controversy	Yes	3	2	S	M	Re	Yes	Yes	No	Yes	Yes	15	11,16
33.	Plutonium Plant Safety at Rock Flats	Yes	1	2	S	M	Re	No	Yes	No	Yes	Yes	20	9
34.	Storage of Radioactive Wastes in Kansas	Yes	2	—	S	S	Re	No	Yes	No	Yes	No	15	9
35.	Amchitka Underground Nuclear Test	Yes	2	—	S	S	Re	No	Yes	No	Yes	Yes	20	9
36.	The Dugway Sheep Kill Incident	No	1	5	S	S	Re	No	Yes	No	Yes	No	25	9
37.	The Nerve Gas Disposal Controversy	Yes	1	2	M	M	Re	Yes	Yes	No	Yes	Yes	20	9,16
38.	Project West Ford	Yes	2	—	—	S	Oc	No	Yes	No	Yes	No	2	9
39.	Project Sanguine	Yes	2	—	S	M	Re	Yes	Yes	No	Yes	No	10	9
40.	Project Able	Yes	2	—	—	S	Oc	No	Yes	No	Yes	Yes	1	10
41.	The Chemical Mace	No	1	5	M	M	Ra,SS	No	No	No	No	Yes	50	14
42.	The Bronze Horse	Yes	7	—	—	S	Ot	No	No	No	No	No	—	
43.	Synthetic Turf and Football Injuries	No	1	7	M	M	Oc	No	No	No	Yes	No	15	13,14
44.	Disqualification of Dancer's Image	No	3	7	S	M	Ot	Yes	No	Yes	Yes	Yes	15	10,14
45.	The AD-X2 Battery Additive Debate	Yes	3	5	S	—	—	No	Yes	No	Yes	Yes	20	8,10,16

See text for explanations and code numbers.

were identified if possible with the results shown in Table A-XI. Categories where a special risk was felt by part of the people are coded as follows: "A," age; "S," sex; "Ra," race; "Re," regional populace; "Oc," occupation; "Ss," socioeconomic status; and "Ot," other.

• The concerned public in each case was evaluated and a subjective judgment was made as to whether the group was likely to have felt that it could avoid the threatening technology. The results are shown in Table A-XI.

• The regulatory climate that existed at the time the threat came to the public's attention was evaluated in each case. In each case four questions were asked: Had a government agency really been "in charge" of the technology before the crisis? Was someone really "in charge" of the problem at the time of the alarm? Did a conflict of interest or opinion exist between government agencies? Had a previous law or regulation contributed to the controversy or complicate the problem? The results are shown in Table A-XI.

• The age of the technology involved at the time that the public alarm developed in each case was determined. The results shown in Table A-XI are the ages in years.

• The societal elements listed in Table XI were screened against the study cases to determine which element utilized or regulated the problem technology or was otherwise most likely to have been perceived by the public as being primarily responsible for the alarm. The results are shown in Table A-XI (the numbering system is the same as in Table XI).

• The technical information that became available to the decision makers in government or in the private sector when alarm developed in each case was subjectively evaluated as to whether (a) it was sufficient to understand the problem; (b) it was available when needed; and (c) it was accepted as factual by the parties at interest. A judgment was then made as to whether the information was critical to the decision that was to be made. The results are shown in Table A-XII.

• The study cases were surveyed to determine if a key spokesman or dramatization had brought the threat or issue to the public's attention. The subjective judgments are shown in Table A-XII.

• The results of the literature search on the fourty-five study cases and on twelve other cases are shown in Tables A-XIII and A-XIV, respectively. The news media index was calculated for each case as the sum of the number of entries in the *Readers' Guide*, the

THE TECHNICAL INFORMATION THAT WAS AVAILABLE
WHEN THE CASE ERUPTED

Case	The Decision Maker's Information				The Public's Information
	Was Sufficient	Was in Time	Was Undisputed	Critical to Decision	Key Spokesman or Dramatization
1. Human Artificial Insemination	Yes	Yes	Yes	No	No
2. Oral Contraceptive Safety Hearings	No	No	No	Yes	Yes
3. Southern Corn Leaf Blight	No	No	No	Yes	No
4. The Great Cranberry Scare	No	No	No	Yes	Yes
5. The Diethylstilbestrol Ban	No	No	No	Yes	Yes
6. The Cyclamate Affair	No	No	No	Yes	Yes
7. MSG and the Chinese Restaurant Syndrome	No	No	No	Yes	Yes
8. Botulism and Bon Vivant	Yes	Yes	Yes	Yes	No
9. The Fish Protein Concentrate Issue	Yes	Yes	Yes	No	No
10. The Fluoridation Controversy	No	No	No	No	No
11. Salk Polio Vaccine Hazard Episode	No	No	Yes	Yes	Yes
12. The Thalidomide Tragedy	No	No	Yes	Yes	Yes
13. Hexa, Hexa, Hexachlorophene	No	No	No	Yes	Yes
14. Krebiozen—Cancer Cure?	No	No	No	No	Yes
15. DMSO—Suppressed Wonder Drug?	No	No	No	Yes	Yes
16. X-Ray Shoe-Fitting Machine	No	No	No	Yes	No
17. Abuse of Medical and Dental X-Rays	No	No	No	Yes	Yes
18. X-Radiation from Color TV	No	No	No	Yes	Yes
19. Introduction of the Lampreys	No	No	Yes	Yes	No
20. The Donora Air Pollution Episode	No	No	No	No	No
21. The *Torrey Canyon* Disaster	No	No	No	No	No
22. The Santa Barbara Oil Leak	No	No	No	No	No
23. Mercury Discharges by Industry	No	No	Yes	Yes	Yes
24. The Mercury-In-Tuna Scare	No	No	No	No	Yes
25. The Rise and Fall of DDT	No	No	No	Yes	Yes
26. Asbestos Health Threat	No	No	No	Yes	Yes
27. Taconite Pollution of Lake Superior	No	Yes	Yes	No	No
28. Foaming Detergents	No	Yes	Yes	Yes	No
29. Enzyme Detergents	No	No	Yes	Yes	No
30. Nitrilotriacetic Acid (NTA) in Detergents	No	No	No	Yes	No
31. Saltville—An Ecological Bankruptcy?	Yes	Yes	Yes	Yes	No
32. Truman Reservoir Controversy	No	No	No	Yes	Yes
33. Plutonium Plant Safety at Rock Flats	No	No	No	Yes	Yes
34. Storage of Radioactive Wastes in Kansas	No	No	No	Yes	Yes
35. Amchitka Underground Nuclear Test	Yes	Yes	No	Yes	Yes
36. The Dugway Sheep Kill Incident	Yes	Yes	No	Yes	No
37. The Nerve Gas Disposal Controversy	Yes	Yes	Yes	Yes	No
38. Project West Ford	No	No	No	No	No
39. Project Sanguine	No	No	No	Yes	No
40. Project Able	Yes	Yes	No	Yes	Yes
41. The Chemical Mace	Yes	Yes	No	Yes	Yes
42. The Bronze Horse	No	No	No	No	Yes
43. Synthetic Turf and Football Injuries	No	No	No	No	No
44. Disqualification of Dancer's Image	No	No	No	Yes	No
45. The AD-X2 Battery Additive Debate	No	No	No	No	No

See text for explanations

TABLE A-XIII

NEWS MEDIA COVERAGE OF 45 STUDY CASES

Case Title	Selected "Standard Periodicals"					Readers' Guide		Business Periodicals Index	New York Times Index					
	Column Inches	Pictures	Cartoons	Editorials	Cover Stories	Entries	Illustrated Articles	Entries	Entries	Entries in Boldface	Column Inches, Abstract	Editorials	Cartoons	
Human Artificial Insemination	435.5	4	0	0	1	74	20	1	73	4	24	0	0	
Oral Contraceptive Safety Hearings	518[a]	8	1	0	0	35	10	3	10	4	24.5	0	0	
Southern Corn Leaf Blight—A Genetic Engineering Problem	482.7	16	1	0	0	14	18	5	55	3	24	1	0	
The Great Cranberry Scare	400.5	20	0	3	1	17	6	83	56	4	14	1	3	
The Diethylstilbestrol Ban	551	11	1	0	0	50	12	11	36	5	16.5	0	0	
The Cyclamate Affair	725	20	0	0	0	59	24	49	50	5	20.5	0	0	
MSG and the Chinese Restaurant Syndrome	499	18	0	0	0	16	8	16	10	2	4.2	0	1	
Botulism and Bon Vivant	473.7	12	0	0	0	12	8	5	51	6	29.2	0	0	
The Fish Protein Concentrate Issue	116.7	1	0	0	0	14	5	19	34	0	10.2	0	0	
The Fluoridation Controversy	1,662	30	0	5	0	310	64	24	468	16	118.7	20	1	
Salk Polio Vaccine Hazard Episode	1,278	38	0	0	0	41	14	NA	157	33	117.5	7	2	
The Thalidomide Tragedy	1,081	22	0	2	1	47	16	15	91	6	24.2	2	0	
Hexa, Hexa, Hexachlorophene	76.2	0	0	1	0	7	1	21	12	1	11.5	2	0	
Krebiozen—Cancer Cure?	1,118	8	0	0	0	54	11	10	43	1	9.7	0	0	
DMSO—Suppressed Wonder Drug?	185.2	4	0	0	0	6	5	0	13	0	4.3	2	0	
X-Ray Shoe-Fitting Machine	62.5	4	1	0	0	9	3	1	12	1	4	0	0	
Abuse of Medical and Dental X-Rays	646.5	15	0	0	0	47	20	4	57	1	19.7	1	0	
X-Radiation From Color TV	136.7	3	0	0	0	21	11	41	28	0	10	2	0	
introduction of the Lampreys	313.5	9	0	0	0	35	20	1[b]	22	2	4	0	0	
The Donora Air Pollution Episode	129.5	9	0	0	1	6	2	NA	18	6	8	2	1	
The Torrey Canyon Disaster	798.5	36	0	1	0	19	11	7	60	10	18	2	0	
The Santa Barbara Oil Leak	893.5	48	0	0	0	35	24	50	118	2	58.5	8	0	
Mercury Discharges by Industry	222	4	0	0	0	25	12	9	39	4	23	1	0	
The Mercury-In-Tuna Scare	304.5	2	0	0	0	16	5	7	34	4	19	1	0	
The Rise and Fall of DDT	3,905	22	0	1	0	254	103	94	497	14	143.8	3	19	

Asbestos Health Hazard	5	150.5	0	0	12	5	20	25	1	28	0	0
Taconite Pollution of Lake Superior[c]	0	20	0	0	5	2	2	0	0	0	0	0
Foaming Detergents	8	224	1	0	37	18	29	24	0	7.2	1	0
Enzyme Detergents	5	105	0	0	13	11	22	13	2	13	0	0
Nitrilotriacetic Acid (NTA) Detergents	0	111.5	0	0	8	3	7	17	2	9.5	0	0
Saltville—An Ecological Bankruptcy?	0	0	0	0	1	1	0	0	0	0	0	0
Truman Reservoir Controversy	0	0	0	0	0	0	0	0	1	0	0	0
Plutonium Plant Safety at Rocky Flats	1	210.5	0	0	0	7	1	0	0	2.7	0	0
Storage of Radioactive Wastes in Kansas Salt Mines	17	48	0	0	11	24	10	6	0	58.5	13	1
Amchitka Underground Nuclear Test	6	666	0	0	41	6	1	129	10	4.2	1	1
The Dugway Sheep Kill Incident	6	285	0	0	9	10	10	17	2	19.2	2	0
The Nerve Gas Disposal Controversy	3	296	0	0	22	12	5	56	7	12.2	4	0
Project West Ford—Orbital Belt of Needles	1	549.2	1	1	49	3	4	35	2	2.2	0	1
Project Sanguine—Giant Underground Transmitter	3	60	0	0	6	2	0	5	0	0	0	0
Project Able—The Space Orbital Mirror	2	48	0	0	3	3	3	0	0	38.0	0	0
The Chemical Mace	0	37.7	0	0	9	2	0	51	9	4	0	0
The Bronze Horse—A Technological Definition of Art?	1	19	0	0	2	7	7	6	1	7	0	0
Synthetic Turf and Football Injuries	0	27	0	0	9		1	7	0		0	0
Disqualification of Dancer's Image		10	0	0	8	8		41	5	9.2		0
The AD-X2 Battery Additive Debate	2	212.7	2	0	43	9	NA	69	6	20.2	4	1

[a] 1968–1970 only.
[b] Largely NA.
[c] To mid 1972 only.

TABLE A-XIV
NEWS MEDIA COVERAGE OF OTHER SELECTED CASES

Case Title	Selected "Standard Periodicals"					Readers' Guide		Business Periodicals Index	New York Times Index				
	Column Inches	Pictures	Cartoons	Editorials	Cover Stories	Entries	Illustrated Articles	Entries	Entries	Entries in Boldface	Column Inches, Abstract	Editorials	Cartoons
The Skyjacking Nightmare	1,276	108	10	1	2	190	88	115	937	33	321	38	8
Heart Transplant Fad	2,883	143	1	0	2	113	58	9	505	14	127	12	1
Northeast Power Failure	1,337.5	46	0	1	3	25	17	55	101	11	55	7	—
Phosphate Detergent Issue	716.5	19	1	1	0	32	16	101	100	9	57	5	2
Coal Tar Dye Problem	63.5	3	0	0	0	10	1	55	46	2	11	0	0
Great Mississippi Fish Kill	247	3	0	0	0	21	5	17	41	1	15	3	0
Polychlorinated Biphenyls	263.5	4	0	0	0	13	7	13	16	0	10	0	1
Glue and Aerosol Propellant Sniffing	139	1	0	0	0	8	3	2	31	0	8	0	0
Fortified Breakfast Cereal Issue	110.2	6	1	0	0	13	6	12	18	3	8	0	0
Sea Level Panama Canal	198.5	7	0	0	0	19	0	Inc.	8	—	—	—	—
Discarded Refrigerator Hazard	0	0	0	0	0	3	0	Inc.	28	2	5	0	0
Studded Snow Tire Question	Inc.	—	—	—	—	6	1	6	Inc.	—	—	—	—

TABLE A-XV

GOVERNMENT UNITS AND PRIVATE SECTOR PARTIES THAT BECAME INVOLVED IN STUDY CASES

Case	State and Local Government Agencies	Federal Regulatory Agencies	The Congress	The Courts	Other Nations	Citizens' Groups	Academic/Professional Associations	Organized Labor	Farm of Industry Associations	Organized Religion
1. Human Artificial Insemination	Yes	—	—	Yes	Yes	—	Yes	—	—	Yes
2. Oral Contraceptive Safety Hearings	—	Yes	Yes	Yes	—	Yes	Yes	—	—	Yes
3. Southern Corn Leaf Blight	—	Yes	—	—	—	—	—	—	Yes	—
4. The Great Cranberry Scare	—	Yes	Yes	—	—	—	—	—	Yes	—
5. The Diethylstilbestrol Ban	—	Yes	Yes	Yes	Yes	Yes	—	—	Yes	—
6. The Cyclamate Affair	—	Yes	Yes	Yes	—	—	—	—	—	—
7. MSG and the Chinese Restaurant Syndrome	—	Yes	Yes	Yes	—	Yes	Yes	—	—	—
8. Botulism and Bon Vivant	—	Yes	—	—	—	—	—	—	Yes	—
9. The Fish Protein Concentrate Issue	—	Yes	Yes	—	Yes	—	—	—	—	—
10. The Fluoridation Controversy	Yes	Yes	Yes	Yes	—	Yes	Yes	—	—	—
11. Salk Polio Vaccine Hazard Episode	Yes	Yes	Yes	Yes	—	Yes	Yes	—	—	—
12. The Thalidomide Tragedy	—	Yes	Yes	Yes	—	—	—	—	—	—
13. Hexa, Hexa, Hexachlorophene	—	Yes	—	—	—	—	Yes	—	—	—
14. Krebiozen—Cancer Cure?	Yes	Yes	Yes	Yes	—	—	Yes	—	—	—
15. DMSO—Suppressed Wonder Drug?	—	Yes	Yes	—	—	Yes	Yes	—	—	—
16. X-Ray Shoe-Fitting Machine	Yes	Yes	Yes	—	—	Yes	Yes	—	Yes	—
17. Abuse of Medical and Dental X-Rays	Yes	Yes	Yes	—	—	Yes	Yes	—	Yes	—
18. X-Radiation from Color TV	Yes	Yes	Yes	—	—	—	—	—	Yes	—
19. Introduction of the Lampreys	Yes	Yes	—	—	Yes	—	—	—	—	—
20. The Donora Air Pollution Episode	Yes	Yes	Yes	Yes	—	Yes	—	—	—	—
21. The *Torrey Canyon* Disaster	—	—	Yes	Yes	Yes	—	—	—	—	—
22. The Santa Barbara Oil Leak	Yes	Yes	Yes	Yes	—	Yes	Yes	—	—	—
23. Mercury Discharges by Industry	Yes	Yes	Yes	Yes	Yes	Yes	—	—	—	—
24. The Mercury-In-Tuna Scare	Yes	Yes	—	—	Yes	Yes	Yes	—	Yes	—
25. The Rise and Fall of DDT	Yes	Yes	Yes	Yes	Yes	Yes	Yes	—	—	—
26. Asbestos Health Threat	Yes	Yes	—	Yes	—	Yes	—	—	Yes	—
27. Taconite Pollution of Lake Superior	Yes	Yes	Yes	Yes	Yes	Yes	—	—	Yes	—
28. Foaming Detergents	Yes	Yes	Yes	—	—	—	—	—	—	—
29. Enzyme Detergents	—	Yes	—	—	—	Yes	Yes	—	—	—
30. Nitrilotriacetic Acid (NYA) in Detergents	—	Yes	Yes	—	—	—	—	—	—	—
31. Saltville—An Ecological Bankruptcy?	Yes	Yes	—	—	—	—	—	—	Yes	—
32. Truman Reservoir Controversy	Yes	Yes	Yes	Yes	—	Yes	—	—	—	—
33. Plutonium Plant Safety at Rock Flats	Yes	Yes	Yes	—	—	Yes	Yes	—	—	—
34. Storage of Radioactive Wastes in Kansas	Yes	Yes	Yes	Yes	—	Yes	Yes	—	—	—
35. Amchitka Underground Nuclear Test	Yes	Yes	—	Yes	Yes	Yes	—	—	—	—
36. The Dugway Sheep Kill Incident	—	Yes	Yes	—	—	—	—	—	—	—
37. The Nerve Gas Disposal Controversy	Yes	Yes	Yes	Yes	—	Yes	—	—	—	—
38. Project West Ford	—	Yes	—	—	Yes	Yes	Yes	—	—	—
39. Project Sanguine	Yes	Yes	Yes	Yes	—	Yes	—	—	—	—
40. Project Able	—	—	Yes	—	—	Yes	—	—	—	—
41. The Chémical Mace	Yes	Yes	Yes	—	—	Yes	—	—	—	—
42. The Bronze Horse	—	—	—	—	—	—	—	—	—	—
43. Synthetic Turf and Football Injuries	—	Yes	Yes	—	—	—	—	—	Yes	—
44. Disqualification of Dancer's Image	Yes	—	—	Yes	—	—	—	—	—	—
45. The AD-X2 Battery Additive Debate	Yes	Yes	Yes	Yes	—	Yes	—	—	—	—

See text for explanations and code numbers.

574

TABLE A-XVI

IMPACTS AND DISPOSITIONS

Case	Threat Was Overemphasized by "Opponents" of the Technology	Threat Was Overly Minimized by "Proponents" of the Technology	A Definitive Government Action Was Taken	Remedial Action Was Prompt	New Legislation Passed In Problem Area	New Regulations Made In Problem Area	Damage Claims Paid	Fines Assessed	Indemnities Paid
1. Human Artificial Insemination	—	Yes	—	—	Yes	Yes	No	No	No
2. Oral Contraceptive Safety Hearings	Yes	Yes	Yes	—	No	Yes	Yes	No	No
3. Southern Corn Leaf Blight	No	Yes	—	—	No	?	No	No	No
4. The Great Cranberry Scare	Yes	—	Yes	Yes	No	Yes	No	No	Yes
5. The Diethylstilbestrol Ban	Yes	Yes	Yes	—	Yes?	Yes	?	No	?
6. The Cyclamate Affair	Yes	Yes	Yes	—	No?	Yes	No	No	?
7. MSG and the Chinese Restaurant Syndrome	—	Yes	—	—	No	Yes	No	No	No
8. Botulism and Bon Vivant	—	No	Yes	Yes	No	?	Yes?	No?	No
9. The Fish Protein Concentrate Issue	—	No	Yes	Yes	Yes	Yes	No	No	No
10. The Fluoridation Controversy	Yes	Yes	—	—	Yes	Yes	No?	No	No
11. Salk Polio Vaccine Hazard Episode	No	—	Yes	Yes	No	Yes	Yes	No	No
12. The Thalidomide Tragedy	No	No	Yes	Yes	Yes	Yes	Yes	No	No
13. Hexa, Hexa, Hexachlorophene	No	Yes	Yes	—	No	Yes	?	No	No
14. Krebiozen—Cancer Cure?	No	Yes	Yes	—	No	Yes	No	No	No
15. DMSO—Suppressed Wonder Drug?	Yes	No	Yes	Yes	No	Yes	No	No	No
16. X-Ray Shoe-Fitting Machine	No	Yes	—	—	Yes	Yes	No	?	No
17. Abuse of Medical and Dental X-Rays	Yes?	Yes	—	—	No	Yes	No	No	No
18. X-Radiation from Color TV	Yes?	Yes	Yes	—	Yes	Yes	No	No	No
19. Introduction of the Lampreys	—	—	Yes	—	?	Yes	No	No	No
20. The Donora Air Pollution Episode	No	Yes	—	—	Yes	Yes	No	No	No
21. The Torrey Canyon Disaster	No	No	—	—	Yes	Yes	Yes	No?	No
22. The Santa Barbara Oil Leak	Yes?	Yes	—	—	Yes	Yes	Yes	Yes	No
23. Mercury Discharges by Industry	No	Yes	Yes	Yes	Yes	Yes	No?	?	No
24. The Mercury-In-Tuna Scare	Yes	No	Yes	Yes	No	Yes	No	No	No?
25. The Rise and Fall of DDT	—	—	Yes	—	Yes	Yes	No	No	No
26. Asbestos Health Threat	No	Yes	Yes	—	No	Yes	?	No	No
27. Taconite Pollution of Lake Superior	No	Yes	—	—	No	Yes	No	No	No
28. Foaming Detergents	No	—	Yes	—	Yes	Yes	No	No	No
29. Enzyme Detergents	Yes	Yes	Yes	—	No	Yes	No	No	No
30. Nitrilotriacetic Acid (NTA) in Detergents	No	No	Yes	Yes	No	No	No	No	No
31. Saltville—An Ecological Bankruptcy?	No	No?	Yes	—	No	No	No	No	No
32. Truman Reservoir Controversy	Yes	Yes	Yes	—	No	No	No	No	No
33. Plutonium Plant Safety at Rock Flats	No?	Yes	—	—	No	No	No	No	No
34. Storage of Radioactive Wastes in Kansas	No?	Yes	—	—	No	No	No	No	No
35. Amchitka Underground Nuclear Test	Yes	No?	—	—	No	Yes	Yes	No	No
36. The Dugway Sheep Kill Incident	No	Yes	—	—	No	Yes	Yes	No	No
37. The Nerve Gas Disposal Controversy	Yes	Yes?	—	—	Yes	Yes	No	No	No
38. Project West Ford	No?	Yes?	—	—	No	No	No	No	No
39. Project Sanguine	No?	Yes	—	—	No	No	No	No	No
40. Project Able	No	Yes?	—	—	No	No	No	No	No
41. The Chemical Mace	Yes	Yes	—	—	?	Yes	?	No	No
42. The Bronze Horse	—	No	—	—	No	No	No	No	No
43. Synthetic Turf and Football Injuries	No	Yes	—	—	No	No	?	No	No
44. Disqualification of Dancer's Image	Yes?	—	—	—	Yes	Yes	No	Yes	No
45. The AD-X2 Battery Additive Debate	No	Yes	—	—	No	No	No	No	No

See text for explanations

WOULD A TECHNOLOGY ASSESSMENT HAVE HELPED?

Case	TA Could Have Helped Avoid the Threat	Number of Years TA Could Have Preceded the Threat	Users of TA Results To Avoid Threat	TA Could Have Helped Reduce the Social Shock	Number of Years TA Should Have Preceded the Alarm	Users of TA Results To Reduce Shock
1. Human Artificial Insemination	Yes?	100	Many	Yes	10	C
2. Oral Contraceptive Safety Hearings	Yes	2	I	Yes	5	C
3. Southern Corn Leaf Blight	Yes?	20	AS	Yes	(0.3)	U
4. The Great Cranberry Scare	Yes	2	A,U	Yes	(0.1)	H
5. The Diethylstilbestrol Ban	Yes?	15	USDA	—	—	—
6. The Cyclamate Affair	Yes	10	F	Yes	1	H
7. MSG and the Chinese Restaurant Syndrome	Yes?	10	I	Yes	1	H
8. Botulism and Bon Vivant	?	1–40	I	—?	(0.1)	F
9. The Fish Protein Concentrate Issue	—	—	—	Yes	2	G
10. The Fluoridation Controversy	Yes?	5	P	Yes?	2	G
11. Salk Polio Vaccine Hazard Episode	Yes?	2	I&F	—	—	—
12. The Thalidomide Tragedy	—	—	—	—	—	—
13. Hexa, Hexa, Hexachlorophene	Yes?	—	I	Yes	2	G
14. Krebiozen—Cancer Cure?	—?	—	G	Yes	3	Many
15. DMSO—Suppressed Wonder Drug?	Yes	5	MS	Yes	1	F
16. X-Ray Shoe-Fitting Machine	Yes	30	I	Yes	5	G
17. Abuse of Medical and Dental X-Rays	Yes	30	M	Yes	10	G
18. X-Radiation from Color TV	—	—	—	Yes?	(0.1)	I
19. Introduction of the Lampreys	Yes?	5	G	—	—	—
20. The Donora Air Pollution Episode	Yes	30	I	Yes?	5	G
21. The *Torrey Canyon* Disaster	Yes?	10	I	—	—	—
22. The Santa Barbara Oil Leak	Yes	15	I	—	—	—
23. Mercury Discharges by Industry	Yes?	20	I	Yes	0.5	G
24. The Mercury-In-Tuna Scare	—	—	—	Yes	(0.2)	F
25. The Rise and Fall of DDT	Yes?	25	I,G	Yes	5	Many
26. Asbestos Health Threat	Yes?	50	I	Yes	5	G
27. Taconite Pollution of Lake Superior	Yes	20	I	Yes	5	G
28. Foaming Detergents	Yes?	15	I	Yes	5	G
29. Enzyme Detergents	Yes?	5	I	Yes	1	G
30. Nitrilotriacetic Acid (NTA) in Detergents	Yes	2	I,G	Yes	1	G
31. Saltville—An Ecological Bankruptcy?	Yes?	50	I	Yes	2	G,I,R
32. Truman Reservoir Controversy	Yes?	20	G	Yes	2	G,R
33. Plutonium Plant Safety at Rock Flats	Yes	15	G	Yes	(0.5)	G,I,R
34. Storage of Radioactive Wastes in Kansas	Yes	15	G	Yes	1	G,R
35. Amchitka Underground Nuclear Test	Yes	5	G	Yes	2	Many
36. The Dugway Sheep Kill Incident	—	—	—	Yes?	(0.1)	G
37. The Nerve Gas Disposal Controversy	Yes	20	G	Yes	2	G
38. Project West Ford	Yes	3	G	Yes	2	G
39. Project Sanguine	Yes	10	G	Yes	1	G,R
40. Project Able	Yes	1	G	Yes	1	—
41. The Chemical Mace	Yes	2	L	Yes	1	G
42. The Bronze Horse	—	—	—	Yes?	1	Many
43. Synthetic Turf and Football Injuries	Yes	3	I	Yes	1	G?
44. Disqualification of Dancer's Image	Yes?	6	MS	Yes?	(0.1)	Many
45. The AD-X2 Battery Additive Debate	Yes?	15	I	Yes	3	G

Abbreviations: A—Agriculture; C—Congress; F—Food and Drug Administration; G—Government, not otherwise defined; H—U.S. Department of Health, Education and Welfare; I—Industry; L—Law Enforcement; M—Medical Community; P—Public Health Service; R—Regional; S—Science and Technology; and U—U.S. Department of Agriculture.

Business Periodicals Index, and the *New York Times Index,* plus 10 percent of the total column inches in the nine selected "standard periodicals": *Reader's Digest, Life, Time, Farm Journal, U.S. News and World Report, Consumer Reports, Today's Health, Business Week* and *Science.*

• The study cases were reviewed to determine whether local, state, or federal agencies, the Congress, the courts or other countries became involved in them. The results are shown in Table A-XV.

• The study cases were reviewed to determine whether citizen participation groups (including "consumer advocates"), academic or professional associations, industry or trade associations, or organized religious groups became involved in them. The results are shown in Table A-XV.

• The study cases were surveyed to determine whether the threats turned out to be as real as stated, whether definitive, remedial actions were taken promptly, whether legislative or regulatory actions were taken, and whether damages, fines or indemnities were paid. In some cases our information was incomplete. The results are shown in Table XVI.

• The study cases were reviewed to determine if a technology assessment might have helped avoid the threat or the public shock, and, if so, how many years before could it have been done and who should have used the results of the TA. The results are given in Table XVII.

REFERENCES

1. Withey, S. B., "Public Opinion About Science and Scientists," *Public Opinion Quarterly,* **23,** 382 (1959).

2. Swinehart, J. W., and J. M. McLeod, "News About Science: Channels Audiences and Effects," *Public Opinion Quarterly,* **24,** 583 (1960).

3. Wade, S., and W. Schramm, "The Mass Media as Sources of Public Affairs, Science and Health Knowledge," *Public Opinion Quarterly,* **33,** 197 (1969).

4. Hovland, C. I., and W. Weiss, "The Influence of Source Credibility on Communication Effectiveness," *Public Opinion Quarterly,* **15,** 635 (1952).

5. Beal, G., and E. M. Rodgers, "The Scientist as a Referent in the Communications of New Technology," *Public Opinion Quarterly,* **22,** 554 (1958).

6. Lazarsfeld, P. F., B. Berelson, and H. Gaudet, *The Peoples Choice,* Columbia University Press, New York (1948).

7. Katz, E., "The Two-Step Flow of Communications," *Public Opinion Quarterly*, **21**, 62 (1957).

8. Bauer, R. A., "The Obstinate Audience," *American Psychologist*, **19**, 319 (1964).

9. Coleman, J., E. Katz, and H. Menzel, *Medical Innovation*, Bobbs-Merrill Company, Chicago (1966).

10. Griliches, Z., "Hybrid Corn: An Exploration in Economics of Technological Change," *Econometrica*, **26**, 501 (1957).

Appendix B

Technological Causes of Public Concern

The potential for public concern or alarm for any given case depends on numerous factors, but probably the most important one is the nature of the impact on the public involved. Therefore, a survey and evaluation of the areas where technology may have the kinds of impacts on man or on his society which can produce alarm was of interest. An assumption could be made that the potential for public concern or alarm will be greatest where the technology involves impacts on basic human needs and activities. One might assume further, that the degree of concern or alarm would be roughly proportional to the importance of the basic individual or societal need or activity, although a systematic ranking of these values may be nearly impossible. The system presented in this Appendix was first developed after about seventy-five potential cases had been identified; it was used as a guide to search for new cases and was later extended after the case histories were completed.

The categorization of basic human needs and activities has long been of interest. Abraham Maslow,* for example, developed a short list in which man's needs were ranked in order of decreasing necessity as follows:

- Physiological needs (hunger, thirst, sex, energy expenditure, rest).
- Safety needs (security, avoidance of fright, shelter).

*Maslow, A., as quoted in Lawrence S. Wrightsman's *Social Psychology in the Seventies*, Brooks-Cole Publishing Company, Monterey, California, p. 77, 1972.

- Belongingness needs (love, affection, intimacy).
- Esteem needs (praise, self-approval, self-esteem, status).
- Self-actualization needs (improvement, growth or development in one's job, upward mobility in the community, etc.).

In addition to Maslow, Henry Murry and Raymond Catell have evaluated basic human needs. Numerous other psychologists have done similar work but not entitled them human needs.* Human concerns, as revealed by public opinion polls, have also been reported.†

Reflection on cases of technologically related alarm indicates that most of them arise from physiological and safety needs, but that the above list is not nearly expansive enough for our purposes. In the first place, two levels of basic needs should be recognized: those of the individual and those of the society. Second, differentiation should be made between those needs which are basic to any society and those activities which have become "essential" to contemporary society. On the other hand, threats to activities in the latter category can also produce public alarm, so that some flexibility is needed in developing a listing of most important needs and activities. The listing in Table B-I appears to include many, if not most, basic needs and activities. Each of these categories is briefly discussed below.

1. *Basic needs of the individual:* If one ranks needs of the individual according to the length of time he could live without the need being filled, then at the top of the list is the internal needs of the maintenance of nerve and brain electrical activity and the blood circulatory system, and the externally dependent needs of the maintenance of suitable temperature and pressure. The circulatory system is, of course, inextricably tied to the basic need for a suitable oxygen content in the air we breathe. All the other physiological systems are important and can be causes of concern to some individuals. When something affects one of the body's systems, it may very well affect others also, of course, but because people tend *to think* of the major effect—the news media may focus on only one effect—the approach was to consider needs as separately as possible.

a. *Nervous system and brain:* Man's body cannot long tolerate serious imbalances in the electrical activity of his nervous system.

*Maddi, S., *Personality Theory: A Comparative Analysis,* Dorsey Press, Homewood, Illinois, 1968.
†Cantril, H., *The Pattern of Human Concerns,* Rutgers University Press, New Brunswick, New Jersey, 1965.

Electrocution, in which the large superimposed electric current rapidly and completely overwhelms the body's system, has been widely used in exacting capital punishment. The nerve gas war agents similarly act within minutes. Thus, unfavorable publicity about insecticides of the organophosphate (such as parathion) type which rapidly affect the nervous system, understandably generates strong emotions in the public. Similarly, toxic materials, such as mercury or hexachlorophene which are said to attack the brain, have the potential for causing great alarm. Drugs such as LSD, amphetamines, marihuana, and alcohol, all affect the brain and are therefore of special concern to many people. Diseases such as the encephalitis types which attack the nervous system are of similar special concern. Most recently, the revelation that vice-presidential candidate Thomas Eagleton had undergone electroshock treatment appeared to be the key consideration in the request by the news media and eventually by presidential candidate George McGovern for the withdrawal of his candidacy.

b. *Circulatory and respiratory systems:* the importance of the blood circulatory system needs no elaboration, and attack on this system is surely one of the oldest methods of killing. Hence, any technologically related cases involving detrimental effects to the blood or to the blood purification system (lungs and kidneys) are the subject of immediate concern to most people; for example, leukemias from radiation, or lung cancer from cigarettes or asbestos.

c. *Temperature, light, and radiant energy:* The human body cannot survive an internal temperature of about 110°F for more than seconds or minutes, or an external temperature of about 140°F for an hour or so. (Certain workers for NASA have been insulated to withstand temporarily, air temperatures up to 3000°F.) Burning at the stake or boiling in oil were long used as forms of execution and the fireball of the atomic bomb can kill large numbers of people by instantly vaporizing them. (Ancient myths and legends warn of the dangers of flying too close to the sun.) At the other extreme, humans have survived body temperatures temporarily as low as about 60°F and the limits of reviving people from deep-freeze is of current medical interest.

The need for individual temperature maintenance is, of course, as old as man, and threats to this need are not necessarily of technological origin. On the other hand, technological threats to the maintenance of global temperatures have become of recent concern: a feared "greenhouse effect" from increased conversion of

TABLE B-I

BASIC NEEDS OF THE INDIVIDUAL AND SOCIETY

1. *Basic Needs of the Individual*
    a. Nervous system and brain
    b. Circulatory and respiratory systems
    c. Temperature, light and radiant energy regulation
    d. Pressure regulation
    e. Oxygen content of the air
    f. Water
    g. Food
    h. Sleep and the diurnal rhythm
    i. Clothing and shelter
    j. Other basic needs of the individual

2. *Basic Needs and Activities of Society*
    a. Sexual activity, reproduction, and family organization
    b. Communication
    c. Medicine, drugs, and medical treatment
    d. Education of the young
    e. Cultural values
    f. Care of the elderly
    g. Housing
    h. Geographic mobility and transportation
    i. Political and governmental organization
    j. Labor services and industrial and agricultural production
    k. Monetary and financial systems
    l. Energy and power needs
    m. Waste disposal
    n. Regulation of weights and measures
    o. Measurement of time and recording of history
    p. Regulation of right- versus left-handedness
    q. Religion
    r. Other basic needs and activities of society

fossil fuels to carbon dioxide, with subsequent reduction of infrared radiation away from the earth; or a feared "ice age" effect from increased reflectivity of sunlight from particulate pollution of the atmosphere.

These fears of disturbing the levels of existing radiation were also at the heart of the SST controversy; the stratospheric emissions of nitrogen oxides would destroy the ozone layer which filters out much of the sunlight's ultraviolet radiation. Similarly, the experience of man with X-radiation, nuclear radiation, and microwave radiation, have been sufficient that news stories of cases related to these topics have great inherent scare potential. The analogies of the laser to the "ray guns" of science fiction are frequently pointed out in the popular press. The total level of visible light is also very important, because intense light is painful to the eye, while too little may make man's activities hazardous.

d. *Pressure:* The human body can survive a complete loss of air pressure—a vacuum—only for times on the order of seconds or at most a few minutes. The space program has, of course, demonstrated that man can operate very well for limited periods of time at pressures of about one-third of an atmosphere. Threats to the need for pressure maintenance have not been a cause for alarm to the general public, although with the advent of upper-altitude jet plane service and the space program, most people are now conscious of the need. Man can also work at pressures substantially above atmospheric, as has been demonstrated by underwater and underground projects. Although air pressure regulation has not been used as a formal method of execution, the shock-wave effects of bombs can cause the death of man and destruction of his buildings. The pressures of rocks and clubs are, of course, among the oldest of man's methods of killing.

e. *Oxygen content of air:* Man can survive only several minutes without breathing in oxygen (permanent damage occurs if the brain's oxygen supply is depleted for more than about two minutes). Hence any events which pose a threat to man's air receive immediate concern. On the individual level, technological threats of this type are usually encountered in events which cause suffocation (for example, the hazard for children of discarded refrigerators or plastic garment bags, or which cause the air to be unsafe to breath (for example, contamination with carbon monoxide or any other poisonous pollutants). The use of the gas chamber for executions depends, of course, on the reaction of hydrogen cyanide with hemoglobin to prevent oxygen uptake.

On the global scale, fear is frequently expressed that man's increasing combustion of fossil fuels will consume the atmosphere's approximately 20 percent concentration of oxygen faster than it can be replaced by the photosynthesis of a decreasing number of green plants on the surface of the earth. Similarly, fear is frequently expressed that man will somehow melt the polar ice caps and drown (suffocate) the world.

f. *Water:* Man can survive but a few days without water, and the need to maintain an adequate water supply is deeply ingrained. Hence, any technology which poses a threat to his water supply or its quality evokes an intense reaction. In general, most such threats have come from the side effects of technology (for example, water pollution by mercury discharges, pesticides or fertilizer runoff), the feared side effects (could a test tube of the reported "polywater" catalyze the polymerization of the entire ocean?), or the increased demands on limited water supplies that technology can make (such as agricultural irrigation and industrial use on surface or subterranean water sources). Other threats have come from a technological stimulation of too much water, which brings catastrophe (disastrous floods produced by a collapsing dam or by cloud seeding, as apparently occurred at Rapid City, South Dakota, on June 9, 1972). Serious concern has also been generated when news stories speculated that saboteurs might somehow add a dangerous drug or pathogen to the water supply of a large city, such as Chicago. But probably the greatest single controversy of this type arose over man's intentional addition of trace amounts of fluoride to water for medicinal (dental) purposes.

g. *Food:* In the complete absence of food, man's survival time is a matter of days, but with minimal amounts and quality of food, he may exist for many years in reduced states of health. Malnutrition has been the lot of large segments of mankind, and anything that indicates that our food supply has been made unwholesome or nonnutritious is sure to receive much coverage in the media and reaction from the public. Thus, many of the cases arise because some additive or contaminant which is present in a certain food is pronounced hazardous to health. The popularity of "health food" stores is one result of the repetition of such cases. Similarly, a statement that fortified breakfast cereals are really not as nutritious as claimed receives instant media attention.

h. *Sleep and the diurnal rhythm:* Man can endure only a few days without sleep before disorientation sets in. He requires a daily period of a few to several hours of sleep to maintain normal efficiency and health. Biologically, man's sleeping and waking periods are roughly synchronized with the night/day cycle—the diurnal rhythm. Factors which disturb or prevent sleep (the noise from an airport or a supersonic flight) or factors which interfere with the daily rhythm (the "jet fatigue" of the transcontinental flight, or the proposed space-orbital mirror to provide night light) cause ready complaint and concern. One of the cases that caused the most public alarm involved the sleeping pill, thalidomide.

i. *Clothing and shelter:* In most places on the earth, man requires shelter from the elements (sun, cold, wind, and rain) and from noxious species such as insects and dangerous animals. Clothing and housing have become greatly dependent on technology in most societies, and extremely so in the United States. Clothing needs have stimulated the growth of vast fiber technologies (ranging from natural wool and cotton, to complex chemical synthetics, and to paper throw-aways), and, since early times, intensive technological efforts to develop colored dyes for the clothing. Individual clothing has been the source of several instances of public alarm when hazardous materials were involved (for example, the flammable celluloid collars and the flammable children's night clothing issue); perhaps more often the concern arises in a societal context (clothing as a status symbol; clothing codes in school). Housing is even more of a societal concern, and will be considered in paragraph 2g.

j. *Other basic needs of the individual:* A number of other physical and physiological needs which are important to the individual can be identified, but these appear to be less basic or less definable than those preceding.

Certainly, general health is a basic concern to the individual, but the range of what constitutes good health is broad. Cases involving threats of deadly poisons or diseases can, of course, produce serious alarm. A consideration of the causes of human death in the United States is of interest. Table XV in Chapter V lists in decreasing order the major causes of death. The data on cancer-caused deaths (malignant neoplasms) are of particular interest in that although cancer causes approximately 17 percent of the deaths, cancer research has been receiving over 40 percent of the federal medical research funds. Hence, cases involving carcinogens are of particular interest because, while the threat is not immediate (and may even be statistical), the absence of a general cure and the inexorable course of the disease can lead to great concern. This concern is no doubt increased by the wide visibility via news media announcements, which give cancer a high identification with the public. Medical aspects are further discussed under societal needs.

A suitable sound level, neither too high nor too low, is necessary for man's health. Unsuitable occupational sound levels have been of much concern in our industrialized society. More recently, technological amplification of "rock" music has presented new hearing hazards to performers and perhaps even to the audience.

A reasonable amount of energy expenditure (work, exercise, etc.) is believed necessary to good health, although people have survived in a comatose state for a considerable period of time. While walking and jogging are used by many urban dwellers as a replacement for physical work, technological "exercise machines" and "vibrators" are used by hundreds of thousands, and while these do not generally cause public concern, some of them have been found to be useless and a waste of money.

Psychological needs are, of course, extremely important. They may vary substantially between cultures and individuals, however, and are quite complex, but many of them are affected by technology as threats to one's security. A need for a minimum "psychological space," in order to avoid severe emotional feelings of crowding may produce concern, as, for example, when architectural or technological circumstances cause this space to be violated at least temporarily. Similarly, most individuals desire strongly a certain amount of privacy; invasions of privacy can cause serious cases of alarm, at least in the United States. Some scientists have suggested that man, like other animals, has a genetically fixed need to live in or "own" a familiar geographical territory. Others have concluded than any environment that is too greatly different from the one we evolved in is apt to be stressful to humans—that we need to be able to feel elements of nature about us in order to maintain emotional balance. A recognition of and adaption to one's sexuality is an important physio-psychological aspect of every individual, but the technological aspects will be considered subsequently in a societal context.

Gravity, which is largely taken for granted, has been shown by the space flights to be expendable for several days, but man's tolerance limits are unknown. Certainly the antigravity device of fascination to the science fiction writer will be a cause of alarm to the public if it comes into existence.

2. *Basic needs and activities of the society:* As a social animal, man tends to organize his society into systems of structured groups and activities that reflect needs which are either basic to the existence of the society or are related to the quality of life in the society. Most improtant among these needs is the reproduction of the species, but beyond this, any attempt to rank the needs in order of importance becomes arbitrary in terms of present-day society; many basic needs are overfilled and many activities have become so common and widespread that they appear to fill basic needs.

a. *Sexual activity, reproduction and family organization:* Aging and death occur for all living things, and reproduction is obviously the most important need of any species. Nature has evolved complex varieties of ways and means of begetting and caring for the young of the species, and has intricate checks and balances designed to maintain the uniformity of the species. Hence, any events or practices which portend a threat to man's reproduction or care of children cause immediate and serious alarm. Technologically related cases involving materials which are mutagenic (cause genetic damage) or teratogenic (cause congenital deformities) receive wide coverage by the news media and attention by the public—the announcement that LSD may cause chromosome breakage apparently caused much more concern to its users than other stated hazards, and the thalidomide case is almost classic. Similarly, cases involving hazards to children receive strong identification, for example, the statement that DDT was present in mother's milk probably had a greater influence in causing its downfall than much of the other information. The discussion of the dangers of televised violence is nearly always in terms of its effects on children, although it surely affects adults also.

The maintenance of some kind of norms of sexual behavior has probably been of concern to every society of man which has existed, because of its unique importance in reproduction and social organization. Any attempts to change or negate these norms in a society inherently meet resistance, although changes obviously occur with time. Any technologically related cases which involve sexual activity are almost automatically considered newsworthy and of interest to the public.

b. *Communication:* The communication of thought from one individual to others of the species is a requirement in the formation of social groups. In man, the development of language with spoken and written words was a vast advance over the signal type of communication. While communication may have been a one-to-one process initially, evolution has been in the direction of one-to-the group and now even group-to-group. The development of the printing press made information and ideas available to a general public for the first time, and was of considerable concern to many people; an assessment of the effects of that technology on society at that time may well have concluded that it would be detrimental! The technological communications techniques have continued to be of concern to an increasing number of people. The concern has ranged from that of the individual, industry and government with right-to-privacy and security aspects, through that of jurists, behavioral scientists, parents, and others too numerous to mention with various practices of man's communications media, to the study of such questions as whether overcommunication affects man's suicide rate or other mental health problems. In the United States today, the practice of freedom of speech and the press may be greater than has even existed in a major country before, and a communications technology that is capable of hearing a man whisper on the moon is more available to more people than ever before. Hence, potential threats to many different people from many different sources will undoubtedly be perceived and argued before man's society can accommodate to his communications technology.

c. *Medicine, drugs, and medical treatment:* Man is inherently susceptible to disease and injury, and almost every society has individuals who are recognized as having special knowledge in medical areas and who generally have a special status in the community. A major problem with any medical system seems to be how to determine who practices medicine and what procedures, medicines, and drugs can be used. Medical quackery is probably as old as medical practice, and even in the United States today stories recur periodically of the exposure of a "doctor" who had been practicing for a considerable time without benefit of license, degree, or orthodox training (in some cases, the imposter may have worked with or even supervised bona fide orthodox doctors). Similarly, the problems of determining the effectiveness and safety of medical (or cosmetic) procedures and medicines are often the cause of uncertainty to the public and even to orthodox doctors and governmental regulatory agencies (for example, krebiozen, ineffective prescription drugs, thalidomide, hexachlorophene, heart transplants, accupuncture). Hence, any technologically related cases which raise questions of faith and trust in a medical matter will receive wide interest by the media and the public.

d. *Education of the young:* Almost every society has developed organized procedures for teaching its youth its social and cultural values, as well as its technologies. In modern industrial societies, however, far more technology is available than can possibly be taught to any given individual within a formal educational structure in the limited time that is available. Hence public concern arises regarding how much science and technology should be included at the expense of the society- or person-oriented courses. Claims are heard on many sides that our scientists, engineers and medical authorities are too narrowly trained, and others are heard that our social engineers, politicians, lawyers, executives, and journalists understand little of the technology all about them. Technology has impinged not only on the selection of what is taught, but also on how, when, and where the teaching is to be done, and by whom. Controversies arise over whether such topics as the technology of contraception should be included (and, if so, when); to what extent technologically based teaching devices (such as closed circuit TV and taping of lectures) should be utilized; how many of the available tax dollars should be devoted to new physical plants incorporating the latest technological developments; and whether the technological capabilities of modern transportation should be used to achieve racial balance in the schools at the expense of the children's time and in the face of much public opposition.

e. *Cultural values:* Every society develops cultural values that are important to the quality of life of its citizens, and which it attempts to pass on to each succeeding generation. These values include the accumulated body of facts, beliefs, and traditions of the group and also a set of activities of entertainment or emotional value to the physical and mental health of the individual and the group. These latter values have included from ancient times, art, music, literature, the stage, games, and athletic contests. In modern times, the motion picture, television, and professional sports industries, and the recreation and leisure time industries have become enormous and burgeoning economic enterprises.

Cases of public alarm from technological causes have developed in numerous areas involving the arts and letters (counterfeiting of ancient artwork; deafness induced by amplified rock music; unauthorized taping of performances and photocopying of printed matter), entertainment and athletics (decomposition of old movie films; football injuries from synthetic turf), and recreation (hazardous toys and games for children; loss of fishing and hunting areas from stream channelization; erosion of shorelines by high-speed boats). In some instances, the alarm arises because the introduction of new technology makes obsolete traditional performance standards (introduction of the fiberglas pole in vaulting or new drugs which boost athletic endurance of man or horse).

f. *Care of the elderly:* In societies from primitive to modern times, the elders have often been revered for their accumulated knowledge; the degree of care that it gives to its elderly and infirm members might in fact, be used as one measure of a society's civilization. Yet, nothing in an individual's life is probably as difficult as growing old gracefully, and possibly never has it been as difficult as it is today. For not only can modern technology provide comfort and medical care, it can keep the very old and terminally ill alive long after natural processes would have ended life. Unfortunately, these extra years are often accompanied by constant anxiety over the sufficiency of financial resources to provide for this extended life. For the elderly (and even for the not-so-elderly), the rapid technological pace has made obsolete much of their formal educations, and has filled their existence with new and dimly comprehended devices, many of which are producing vast changes in the life about them and thus causing concern.

g. *Housing:* The construction and regulation of housing for its citizens becomes a vital interest of organized societies, and particularly for urbanized communities. Not only must sufficient housing be available for most if not all of the people, but the materials and styles of construction must often meet certain norms. The geographical location of the individual units and organization of the collective units are usually regulated by the society. Modern housing is highly dependent on technological advances, and these advances have led indirectly to a host of problems that have concerned the public. Serious alarm has developed recently, for example, over the high incidence of lead poisoning in children who eat the peeling lead paints used many years ago in our older houses. Prolonged controversies can exist over aspects of building codes (for example, electrical wiring) or over methods and materials of construction (for example, sprayed asbestos fireproofing).

h. *Geographic mobility and transportation:* Even in the primitive hunter-gatherer societies, man moved from place to place on the earth in search of food. As societies grew more complex, man developed networks of transportation over land and sea to aid in commerce, communication, and so forth, and the methods of transportation became increasingly complex; today man is able to fly across continents in a few hours or orbit the earth in about ninety minutes. Almost everyone in the United States must personally interface daily with the complex technology of vehicles and roadways and the extensive regulations which apply to their use. A number of the technologically related cases of public alarm have involved safety and security of transportation (carbon monoxide in the Corvair and airplane hijacking), and the transportation networks (hazards of interstate highways and destructive effects of studded snow tires).

i. *Political and governmental organization:* Living in a social group predisposes toward the development of governmental organization of some kind which inevitably becomes more complex, it seems, the larger the group. Thus, various problems, such as who will determine the group rules or laws (political organizations), what rules will they adopt (legislative government), who will determine if the rules are being disobeyed (judicial and regulatory agencies) and, inevitably, who will enforce the rules of conduct (law enforcement ranges from the sergeant at arms to the military), are common to almost every society. These problems are addressed in different ways in different societies and at times have been pretty much handled by the religious organizations. The potential threats posed by technology to the social structures which relate to these problems may vary considerably with culture and time and are beyond a lengthy discussion in this work. However, technological developments related to detection and enforcement have probably given the greatest cause for public alarm (for example, wiretapping, alcohol breath analyzers, the chemical mace).On the other hand, the development of regulations and legislation regarding technological developments have long been the source of great controversies and much confusion.

j. *Labor services and industrial and agricultural production:* In every society a tendency apparently exists toward individual specialization in performing the services and producing the goods which the society uses. From ancient times through the Middle Ages, the crafts and arts developed increasing complexity, capability and organization, and in modern times enormous amounts of agricultural, consumer, industrial, and other products are produced by man's technology and labor. Quite frequently, the methods used to produce these products have side effects which result in public alarm (for example, runoff from agricultural fertilizers, pesticides, and animal wastes; smoke from steel plants; and tailings disposal from uranium mining). Not infrequently, the products themselves are the cause of alarm because of some inherently objectionable characteristic (for example, foods or cosmetics contaminated with hazardous substances; automobiles recalled for safety reasons; environmental persistence of polychlorinated biphenyls; and high flammability of celluloid), or because they are abused (for example, glue and aerosol propellant sniffing; aspirin poisonings; electronic eavesdropping; minibikes on streets; and criminal use of weapons such as submachine guns, grenades and silencers).

The labor used in producing these products or in providing services is less prone to be involved in technologically related alarms, although the entire field of replacing man by the machine and the computer (that is, automation) is of serious concern to the public. Jobs are essential to the quality of life.

k. *Monetary and financial systems:* All but the most primitive of societies develop a monetary system and, eventually, sets of rules and regulations involving financial matters (most certain to include a taxation system). Gold and silver have been known from ancient times and used as the base of monetary systems, but many other materials have also been used, and paper forms of money are most widely used today for circulation and in effecting financial transactions. Technologically related cases of public alarm involving money matters have ranged from the suspension of the use of silver in coins to the liability for misused credit cards (plastic money), and to the ruin of financial reputations by computer errors in credit ratings.

1. *Energy and power needs:* Man's strength, speed, and endurance have probably never been sufficient to satisfy him, and from earliest times he has invented tools, domesticated animals, and employed the energy of the wind, water, fire, and sun to accomplish what he could not do alone and to improve the quality of his life. With the advent of the steam engine, the electric battery, generator and motor, the coal-, gas- and oil-burning processes and devices, the internal combustion engine, and more recently, the jet engine, rocket engine, and nuclear reactor, man's daily life has become increasingly dependent on the utilization of huge amounts of external energy. The energy and power needs are now so great that all available sources are being strained and an "energy crisis" has developed. Several technologically related cases of public alarm have recently involved specific energy problems, including the great New York power failure, the recurrent brownouts, and the numerous controversies over siting, safety, and thermal pollution of nuclear power plants.

m. *Waste disposal:* The need to dispose of wastes is a basic problem which arose as soon as mankind adopted permanent locations for living; many caves inhabited in prehistoric times have been found to have an almost characteristic ridge in front, a result of accumulations of bones and litter. Many ancient settlements show evidence of a layer-upon-layer buildup of refuse (even in cities such as Paris early cobblestone streets may be found several feet below present structures). A significant concern in civilized societies has been the disposal of their dead; elaborate customs and sophisticated technologies have been developed for this purpose, ranging from the funeral pyre to elaborate crematorial procedures, from simple burial to complex mummification and embalming techniques, and to cemeteries hundreds of acres in size. With the advent of large cities, the industrial revolution, and our throwaway society, the problems of waste disposal have become immense.

Technological advances are often adopted as solutions to waste disposal problems, but in turn these often generated new problems, for example, household drain cleaners have had tragic consequences for many children; household detergents have added to environmental phosphate levels; huge solid waste disposal sites have

become, at times, dens of vermin and disease; municipal trash in-
cinerators may blacken the skies; and industrial and agricultural
wastes have polluted streams. All of these results have caused public alarm.

n. *Regulation of weights and measures:* Society's need for the
adoption of standards for the measurement of distances and weights
probably became obvious as soon as primitive groups began to settle in permanent communities and to barter with neighboring
groups. As technology became more complex, the need for standards grew rapidly, and required the attention of the highest levels
of authority—from kings and the National Bureau of Standards to
international agreements. The adoption of different sets of standards by different peoples has been a source of difficulty over the
ages; fears that the conversion of a highly technological society
from one system (English) to another (metric) will be accompanied
by great public concern, and confusion has long retarded international standardization.

o. *Measurements of time and recording of history:* Man's efforts to measure and record time have existed for so long and have
been so intensive that their existence and success can be used as a
measure of civilization itself. Equally important is the recording of
the events of time—the histories of civilizations, the earth, and the
universe. Varied and astonishingly complex technologies were
used to measure the passage of time by early societies. The development and adoption of clocks and calendars (of months and
years) has involved scholars and kings, popes and prophets. Increasingly intricate technologies have been developed by modern
man, until he is now able to measure events that occur in one
millionth of a second or less and those that occurred eons ago. And
he can record events on the micro and cosmic levels perhaps even
more accurately than he can record the events of human history as
they are occurring. Modern society's symbiosis with clock and
calendar leads, however, to high levels of anxiety. Much public
confusion is generated recurrently in the United States over the
switches of "daylight" time and of movable dates, such as holidays.
Much chagrin develops over the efforts of some men, organizations,
and governments to change the recording of history.

p. *Regulation of right- versus left-handedness:* Substantial
degrees of asymmetry exist in nature from the molecular to the
cosmic level; man's own asymmetry has assumed increasing importance as his technology has advanced. In industrial countries, standards have been adopted for the regulation of handedness at many

levels—screws normally have right-handed threads; clock hands move "clockwise." The standards are usually adopted to favor the right-handed majority at the expense of the left-handed minority. Considerable public confusion could occur when the handedness of an established custom is changed, because the original decision has shaped many subsequent decisions. Thus, when driving on the left side of the road is adopted, automobiles, roadways, and even streetside architecture may be designed on this basis. If one later decides to change from the left side of the road to the right side, as has been done in Sweden and Japan and is being considered in the United Kingdom, much effort is required.

    q. *Religion:* Almost every society apparently has a god figure and persons recognized as religious authorities, although in some cases these are both combined in the structure of the state or its titular head. For vast numbers of people, their valuations of the qualities of their lives are more dependent on their religious well-being than on their possession of material goods, social position or cultural skills, and for many people religion is more important than family and friends or even personal health and security. Hence, any technological developments that impinge on strongly held religious beliefs or practices are sure to raise concern in that part of the public that is affected.

    r. *Other basic needs and activities of the society:* A number of other needs and activities that exist in most societies might be identified, but they are probably less basic than those preceding. For example, a confidence in themselves and in what they are doing (their goals) appears to be important for at least certain influential segments of any society. Technological developments may cause concern in such areas, but are not so apt to be of the type which lead to technology assessments, either structured or unstructured. It may also be true, as Alvin Toffler has proposed in *Future Shock,* that society cannot stand change at too fast a pace. Certainly when a technological society impinges on a more primitive one, some of the effects on the latter tend to include a breakdown of kinship strengths, a disruption of religious values, and a degree of time disorientation, because of the faster pace of the new life. Perhaps similar effects may threaten us as the degree of technology in our own society increases rapidly. Prudence dictates that our society should find out.

Appendix C

Technology Assessments

There is little doubt that many people are concerned today about our science and technology, where they are leading us, and who, if anyone, is planning ahead. The case histories presented herein have illustrated some of the unexpected and undesirable impacts that science and technology are making, and the difficulties in handling the problems which arise from our technology. From experiences of these kinds has grown the concept of "technology assessment." In this Appendix, we present a brief summary of the background of the technology assessment movement, pertinent literature on the subject, and how one might undertake doing a "TA" on a specific subject.

In 1967 Congressman Emilio Daddario (D—Connecticut) introduced a bill to establish a Technology Assessment Board. The idea of the Board was that it would help the Congress in evaluating the increasingly complex legislation involving technology with which the Congress was faced, and also to balance the technological expertise already available to the executive branch. The technology assessment concept was widely discussed in a series of reports,[1-6] papers,[7-12] and at least three books.[13-15] This discussion led to the establishment of the new Office of Technology Assessment in late 1972, and to an International Society for Technology Assessment which publishes a journal, *Technology Assessment*, quarterly and sponsors conferences in many countries of the world.

Technology assessment has been most succinctly described as "the technological information input to the political decision-making process."[16] A technology assessment is a systematic, purposeful, and iterative search for information on the significant secondary (or higher order) consequences on man, on society and its institutions, and on the physical environment of the development of a technological device or system. The technology assessment would logically be made in time so that the results—a formal organization of what is known and unknown about the potential impacts—can be used in the evaluation of public policy options, in the development of wise decisions in the legislative and regulatory area, in the expenditure of research funds by agencies of government, and (ideally) in the important decision-making levels in the private sector. The thrust of the technology assessment movement, then, is on the study of how technology exerts influence outside itself, in the social, legal, political, economic, and academic spheres. In short, a "TA" is a societal impact statement.

The major types and subtypes of consequences or impacts which might result from the introduction of a new technology can be illustrated with the example of the introduction of television.

1. *First order consequence:*
   - *The intended effect;* for example, rapid pictorial communication for information and entertainment, coupled in many countries with the profit motive of the industry.
2. *Second order consequences:*
   - *Recognized and acceptable impacts;* for example, increased competition for radio and movies.
   - *Unanticipated indirect impacts;* for example, sports are tailored to the desires of the camera and sponsors.
   - *Accidental or statistical impacts;* for example, a national scare erupts over x-ray emissions from color TV sets.
   - *Abuse of the technology;* for example, invasion of privacy.
3. *Higher order consequences:*
   - *Chains of cause- and effect-relations;* for example, national networks reach millions of viewers at the same time; entertainment and news programs may have vast influence on social, political behavior of the public, national controversies erupt over TV violence, unidentified news sources.
   - *Interactions with other technologies;* for example, popular "TV dinners," based on refrigeration advances, spurs development of convenience food industry.

Despite the wide interest in the technology assessment concept, very few attempts were made before 1972 to conduct wide-scope technology assessments.[17-20] More recently, the National Science Foundation has initiated a series of technology assessments on issues of national interest. A study of off-shore oil and gas production was just completed[21] and TA's are underway of solar energy, geothermal energy, alternative strategies and methods for conserving energy, biological substitutes for chemical pesticides, integrated hog farming, conversion from the English to the metric system in the U.S., the checkless-cashless society, and alternative work schedules. The new Office of Technology Assessment has initiated other TA's. And none too soon. Because, as the present study has amply shown, the failure to do TA's is almost sure to produce unpleasant surprises.

A Methodology of Technology Assessment

A variety of techniques were used in part or in whole in the performance of partial technology assessments, including elements of technological forecasting, cost-benefit and benefit-risk analyses, systems analysis, operations research, decision or relevance tree approaches, and matrix methods including cross-impact[22-24] and trimatrix.[15] Forecasting techniques range from historical analogy trend extrapolation to the Delphi technique,[25] simulation-optimization,[26] scenario preparation for policy action effects,[27] and again, matrix methods. Still other approaches to technology assessment have involved simulation gaming, model-building, and public hearings. A useful guideline is that *the subject for the technology assessment tends to structure the study.* For a rich subject, a systems approach is needed to insure comprehensive, iterative coverage in a reasonable period of time by a diversified technology assessment project team.

A detailed description of how to perform a technology assessment is beyond the scope of the present work and, as already indicated, a considerable literature is available. But for the curious reader, we outline a generalized procedure for conducting a technology assessment by the systems analysis approach below and in Figure C-1.

1. *Define the technical boundaries* of the technology to be assessed. Is it a simple device or procedure, several related technological advances, or a complex, interlocking network of technological capabilities? Collect data base.

2. *Analyze the driving forces* which would bring the technology into existence, adoption, expansion, or restriction. Are they from a government program, commercial interests, health needs, environmental requirements, a consumer product, or some other source?

Survey applicable laws, regulations, policies and societal trends. Make a preliminary "walk-thru" of the whole TA to identify factors and parties at interest.

3. *Define the state-of-the-art* of the technology. Establish the geographical distribution of the technology. Establish who will be the owners of the technology, who will be its beneficiaries, and who, if anyone, will be directly hurt by the technology. Prepare state-of-art review.

4. *Make technological forecasts* for two or three levels, for example, the next ten, twenty-five, and possibly one hundred years. Estimate and evaluate the limiting technical factors pertinent to the technology. Prepare interim report for review.

5. *Identify the areas* where the technology will have an impact outside itself. Screen the technology (developed as forecast or at two or three levels of development) against a standard list of potential impact areas as shown in Table C-I. Make general state-of-society assumptions.

6. *Make preliminary impact analyses.* For each of the areas identified in Step 5, make an analysis of sufficient depth to rank the impact as higher or lower priority and to indicate what second order implications the impacts might portend.

7. *Make more detailed analyses* of significant impact areas. Bring in expert advice and consultants at this point. Analyze the tangible and intangible benefits and disbenefits and make benefit/cost analyses. Identify public policy options and make preliminary evaluations. Identify those impacts which are likely to develop the most public concern, because this factor affects the available policy options.

8. *Recycle to Steps 4–7* to identify more fully secondary or higher order consequences of the technology. Feed information from each impact analysis group to other groups. *Identify problem areas, critical points of public concern, and political and institutional realities.* Refine forecasts and conclusions.

9. *Identify and evaluate* public- and private-sector action alternatives and options to be applied to problem areas. Bring in political, legislative, regulatory experts at this point. *Recycle to Steps 4 and 7* to evaluate the effect of policy decisions on the technology forecasts and impacts. Here, we must warn the reader, may be the most difficult of all tasks in the technology assessment, because as Forrester has pointed out,[28] the human mind is not adapted to the way social systems behave. Not only are social systems more complex than technological systems, but our proposed

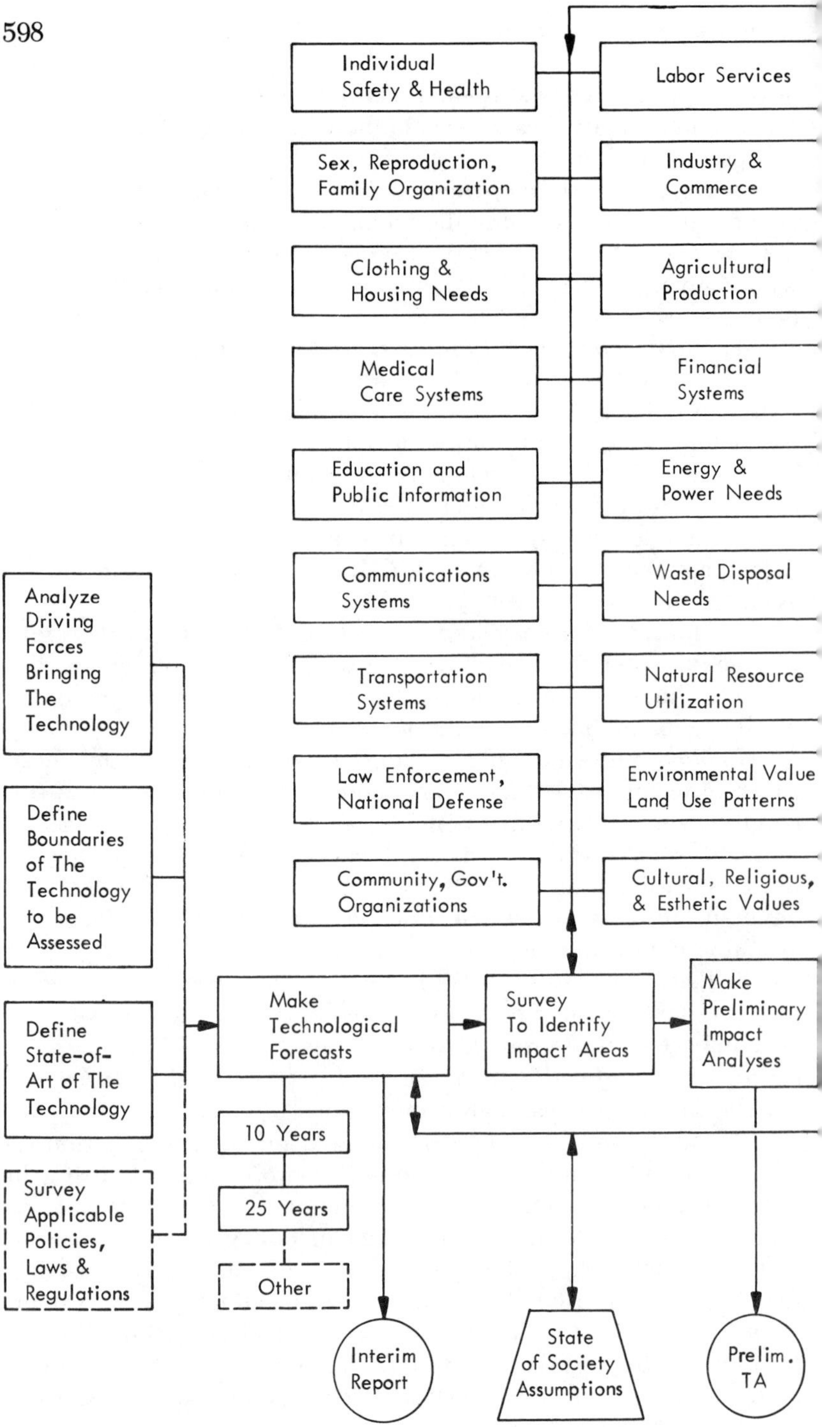

Figure C-1. Generalized Flow Chart for Technology Assessment

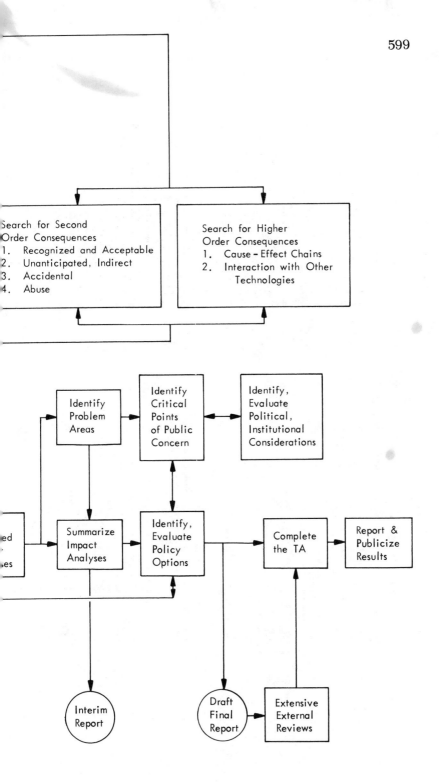

TABLE C–1

POTENTIAL ADVERSE IMPACTS OR THREATS THAT MAY BE POSED BY A TECHNOLOGY

A. To the Individual's Safety and Health
1. Mechanical Injury Hazard
2. Loss of Air Pressure, Oxygen Content, or Atmospheric Purity
3. Loss or Contamination of Water Supply
4. Reduction or Contamination of Food Supply
5. Loss of Sleep, Diurnal Rhythm, or Control of Light Intensity
6. Loss of Control over Temperature
7. Loss of Control over Noise Level
8. Biochemical Hazards
   a. Affects nervous system or brain, or alters the mind
   b. Affects circulatory or respiratory systems
   c. Carcinogenic effects
   d. Other poisonous or corrosive effects
9. Electrical Hazards
10. Radiation Hazards
11. Psychological Hazards
    a. Affects sense of security
    b. Affects sense of belonging to social unit
    c. Invades privacy
    d. Causes fear or regulation or appetites for food, sex, etc.

B. To Society and Its Institutions
1. Hazard to Sexual Activity, Reproduction, Family Organization
   a. Causes sterilization, impotency, frigidity, etc.
   b. Mutagenic (genetic) or teratogenic (causes deformities) effects
   c. Affects parental responsibility
2. Disbenefits in Medical Care
   a. Decreases availability of medical services
   b. Decreases quality of medical services
   c. Decreases confidence in the medical establishment
   d. Increases contagious diseases
3. Disbenefits in Communications
   a. Retards information transfer and retrieval
   b. Abuses or circumvents normal communication channels
   c. Decreases public confidence in what is communicated
   d. Provides too much communication
4. Disbenefits in Educational Institutions
   a. Disjoints population from school locations
   b. Disrupts processes
   c. Creates conflict over teaching practices
5. Conflict over Cultural and Entertainment Values
   a. Introduces new practices or changes traditions
   b. Casts suspicion on basis of values

6. Disbenefits in Housing
   a. Decreases quality
   b. Limits availability
7. Disbenefits in Transportation
   a. Decreases effectiveness
   b. Limits availability
   c. Increases risks
8. Hazards to Political and Governmental Systems
   a. Disrupts normal processes and functions
   b. Requires entirely new legislative or regulatory actions
   c. Decreases confidence of the public in the process
9. Threats to Labor Services
   a. Displaces workers, increases unemployment
   b. Changes required technical skills
   c. Produces unsatisfying work environment
10. Threats to Industrial Organization and Commerce
    a. Displaces existing businesses
    b. Requires new regulation
    c. Increases social costs
    d. Increases environmental costs
    e. Disrupts marketing practices
11. Threats to Agricultural Community
    a. Threatens plant and animal yields
    b. Displaces market for production
    c. Disrupts rural social organization
    d. Increases environmental costs
12. Threats to Monetary and Financial System
    a. Disrupts coinage and legal tender system
    b. Impedes recording of financial transactions
    c. Threatens savings, investment and credit systems
    d. Disrupts insurance systems
    e. Threatens international trade
13. Special Problems
    a. Challenges energy and power systems
    b. Causes waste disposal problems
    c. Impacts on national and regional goals
    d. Special impacts on urban life
    e. Conflicts over ethical values
    f. Changes criteria for standards in any area
    g. Affects conception of time or space
    h. Other effects

remedies for a problem often have exactly the opposite effect over the long range from that desired. Conversely, we tend to overlook the very influence points where remedies might be initiated.[28]

10. *Prepare Draft Final Report.* This will be a very difficult task—writing a book, really, that integrates all of the collected information and projected events into a meaningful and easy to use whole. Obtain numerous reviews of the draft by diverse outside parties.

11. *Complete the technology assessment.* Revise draft in light of reviewer's comments. Prepare an executive summary for busy decision makers.

12. *Report the results* to the sponsor, the potential users, and the public. For the technology assessment to have been worth while, its results must be made known to the widest possible audience, and must be used in the decision-making processes of our society. In truth, it must also be updated periodically as new information develops, new events occur, or society's goals change.

The Technology Assessment Team

Whatever the methodology that is selected to perform the TA, the success of the effort will probably depend greatly on obtaining many different views on the subject. An *interdisciplinary* project team, with experts from each of the areas that needs to be considered is usually needed, because most of us are severely limited in our outlooks by our academic training and occupational experience. The students of our engineering and business schools, for example, have not shown a notable propensity for recognizing the social consequences of their efforts. Nor have our doctors and lawyers, sociologists and economists, for understanding problems outside their narrow specialities. Nor have our journalists for explaining science to the public.

In order to apply the technology concept efficiently, a member of the project team would ideally be a bit of all of these and more—a multidisciplinary person. But persons who have mastered even a few subdisciplines are rare and the TA team can be well served by project leaders who can be classified as interface persons; that is, those who can work efficiently across disciplinary boundaries. Preferably, such a person has a strong technical background,* can grasp other technologies quickly, and also understands the so-

---

*A background in the specific technology would be clearly necessary for some TA's, but not at all for others where the project leader could absorb sufficient knowledge during the study.

cial sciences, the humanities, the business world, government, and the environment. Such individuals are not easy to find; our colleges and universities have rarely yielded such products. While many schools are becoming aware of the need for more broadly educated graduates, many other "technology assessors" will have to be developed on the job—in government, in industry, in labor unions, in medical schools, in banking circles, and other areas. Technology impinges today on almost every phase of our life—from birth to death, at work and at play, whether awake or asleep—on every industry, institution, and government agency, and on our courts and legislatures. The concept of technology assessment will have to be applied at every level in society where significant decisions are made, if we hope to obtain full benefit from it.

REFERENCES

1. "Technology Assessment Seminar," proceedings before the Subcommittee on Science, Research and Development of the Committee on Science and Astronautics, U.S. House of Representatives, September 1967.

2. "Technology Assessment," hearings before the Subcommittee on Science, Research and Development of the Committee on Science and Astronautics, U.S. House of Representatives, November and December 1969.

3. "Technology: Process of Assessment and Choice," report of the National Academy of Sciences to House Committee of Science and Astronautics, July 1969.

4. "A Study of Technology Assessment," report of the National Academy of Engineering, Committee on Public Engineering Policy to House Committee on Science and Astronautics, July 1969.

5. "Technical Information for Congress," report of the Legislative Reference Service, Library of Congress to the House Committee on Science and Astronautics, July 1969.

6. "Perspectives on Benefit-Risk Decision Making," report of a colloquium conducted by the Committee on Public Engineering Policy, National Academy of Engineering 26–27 April 1971. The National Academy of Engineering, Washington, D.C., 1972.

7. Brooks, Harvey, and Raymond Bowers, "The Assessment of Technology," Sci Am, 222(2), 13–21 (February 1970).

8. Lear, John, "Predicting the Consequences of Technology," Sat R, 53, 44 (28 March 1970).

9. Kiefer, D. M., "Technology Assessment," C&EN, 48, 42 (5 October 1970).

10. Cunningham, Donald E., and David M. Glancy, "Technology Assessment—Fond Hopes and Rational Expectations," paper presented to

the Operations Research Society of America, Anaheim, California, October 1971.

11. Coates, Joseph F., "Technology Assessment: The Benefits . . . The Costs . . . The Consequences," *The Futurist*, **5**, 225 (December 1971).

12. Kiefer, David M., "Assessing Technology Assessment," *The Futurist*, **5**, 234 (December 1971).

13. Bauer, Raymond A., Richard S. Rosenbloom, Laure Sharp, et al., *Second-Order Consequences: A Methodological Essay on the Impact of Technology*, MIT Press, Cambridge, Massachusetts, 1969.

14. *Technology Assessment: Understanding the Social Consequences of Technological Applications*, Raphael G. Kasper, Ed., Praeger Publishers, New York, N.Y., 291, 1972.

15. Cetron, Marvin J., Bodo Bartocha and Christine A. Ralph, *The Methodology of Technology Assessment*, Gordon and Breach, New York, N.Y., 1972.

16. "Science Policy, A Working Glossary," prepared by the Science Policy Research Division, Congressional Research Service, Library of Congress for the Subcommittee on Science, Research and Development Committee on Science and Astronautics. U.S. House of Representatives, April 1972.

17. "Technology and Public Policy: The Process of Technology Assessment in The Federal Government," by Vary T. Coates, The George Washington University. Final Report under NSF Grants GQ-4, GI-30422 and GI-30422#1, July 1972.

18. "A Survey of Technology Assessment Today," by Peat, Marwick, Mitchell, & Company, Final Report under Contract No. NSF-C631.

19. "A Technology Assessment Methodology," by The Mitre Corporation, Final Report, Vols. 1–6, on a program in cooperation with the Office of Science and Technology, June 1971.

20. "Technology Assessment of Winter Orographic Snowpack Augmentation in the Upper Colorado River Basin," by Stanford Research Institute, under NSF Contract No. C641000, Final Report to be published.

21. Kash, Don E., Irvin L. White, et al., *Energy Under the Oceans: A Technology Assessment of Outer Continental Shelf Oil and Gas Operations*, University of Oklahoma Press, Norman, Oklahoma, 1973.

22. Enzer, S., "A Case Study Using Forecasting as a Decision-Making Aid," *Futures*, **2** (4), 341 ff, December 1970.

23. Gordon, T. J., "Cross-Impact Matrices, An Illustration of Their Use for Policy Analysis," *Futures*, **1** (6), 527 December 1969.

24. "Cross Impact Assesses Corporate Ventures," *C&EN*, **51** (16), 8–9 (1973).

25. Helmer, O., "Analysis of the Future: The Delphi Method," in *Technological Forecasting for Industry and Government, Methods and Applications*, J. P. Bright, Ed., Prentice-Hall, Inc., Englewood Cliff, New Jersey, 1968.

26. Swartzman, G. L., and G. M. Van Dyne, "An Ecologically Based

Simulation-Optimization Approach to Natural Resource Planning," *Annual Review of Ecology and Systematics*, **3**, 347–398 (1972).

27. Jones, Martin V., "The Impact Assessment Scenario, A Planning Tool for Meeting the Nation's Energy Needs," The Mitre Corporation, McLean, Virginia, April 1972.

28. Forrester, J. W., "Counterintuitive Behavior of Social Systems," *Technology Review*, **73**(3), 53–68, January 1971.

Appendix D

List of Periodical Abbreviations*

Abbreviation	Periodical
Adv Age	Advertising Age
AMA Arch Ind Hyg Occ Med	AMA Archives of Industrial Hygiene and Occupational Medicine
America	America
Am City	American City
Amer Forest	American Forest
Am Home	American Home
Am J Dig Dis	American Journal of Digestive Diseases
Am J Psycho	American Journal of Psychotherapy
Am J Publ Health	American Journal of Public Health
Anal Chem	Analytical Chemistry
Ann New York Acad Sci	Annals of the New York Academy of Sciences
Arch Environ Health	Archives of Environmental Health
Atlan	Atlantic
Aviat W	Aviation Week and Space Technology
A/W PR	Air/Water Pollution Report
Barrons	Barron's
Bet Hom & Gard	Better Homes & Gardens
Biol Abs	Biological Abstracts
Biol Sci	Biological Science
Bioscience	Bioscience
Bull Atomic Sci	Bulletin of the Atomic Scientist
Bull Environ Contam Toxicol	Bulletin of Environmental Contamination and Toxicology
Bsns W	Business Week
Business News	Business News
Calif Fish Gam	California Fish and Game
C&EN	Chemical and Engineering News
Changing Times	Changing Times
Chem	Chemistry
Chem Eng	Chemical Engineering
Chem Ind	Chemical Industries
Chem Market Rep	Chemical Marketing Report
Chem Pharm Bull Japan	Chemical Pharmacology Bulletin of Japan
Chem W	Chemical Week
Christian Cent	Christian Century
Colliers	Colliers
Columbia Journ Rev	Columbia Journalism Review
Commonweal	Commonwealth
Consumer Bul	Consumer Bulletin
Consumer Rep	Consumer Reports
Consumers Res Bul	Consumers Research Bulletin
Coronet	Coronet
Current Therap Res	Current Therapeutic Research
Ed Digest	Education Digest
Elect N	Electronic News
Eng N	Engineering News-Record
Environ	Environment
Environ Health L	Environmental Health Letter
Environ Health Perspectives	Environmental Health Perspectives

*Abbreviations are primarily those used in *Reader's Guide to the Periodical Literature*, the *Business Periodicals Index*, or are according to *American Standard for Periodical Title Abbreviations*, American Standards Association, Inc., November 20, 1973.

Abbreviation	Periodical
Environ Pollut	Environmental Pollution
Environ Sci & Technol	Environmental Science and Technology
Esquire	Esquire
Facts on File	Facts on File
Family Circle	Family Circle
Farm J	Farm Journal
FDA Consumer	FDA Consumer
FDA Papers	FDA Papers
FDA Rep	FDA Report
Fed Reg	Federal Register
Field and Stream	Field and Stream
Food H	Food Health
Forbes	Forbes
Fortune	Fortune
Good H	Good Housekeeping
Harper	Harper's Magazine
Home and Gard	Home and Garden
Horticulture	Horticulture
House Beaut	House Beautiful
House & Gard	House and Garden
Ind Res	Industry Research
Ind W	Industry Week
J Agr Food Chem	Journal of Agricultural Food Chemists
J Am Med Assoc	Journal of American Medical Association
J Am Vet Med Assoc	Journal of American Veterinarian Medicine Association
J Am Waterworks Assoc	Journal of American Waterworks Association
JAOAC	Journal of the Association of Official Agricultural Chemists
J Biol Chem	Journal of Biological Chemists
J Econ Entomol	Journal of Economic Entomologists
J Environ Quality	Journal of Environmental Quality
J Fish Res Bd Canada	Journal of the Fish Research Board of Canada
J Natl Cancer Inst	Journal of the National Cancer Institute
J Wildlife Manag	Journal of Wildlife Management
Labor Hyg Occupat Dis	Labor Hygiene and Occupational Diseases
Ladies Home J	Ladies Home Journal
Life	Life
Literary Dig	Literary Digest
Look	Look
McCalls	McCalls
Mech Illus	Mechanics Illustrated
Med Wor N	Medical World News
Merchand W	Merchandizing Week
Missiles and Rockets	Missiles and Rockets
Motor T	Motor Trend
Nation	Nation
Nations Bsns	Nation's Business
Nat Parks & Con Mag	National Parks and Conservation Magazine
Nat Parks Mag	National Parks Magazine
Nature	Nature.
Nat Wildlife	Natural Wildlife
New England J Med	New England Journal of Medicine
New Repub	New Republic
Newsweek	Newsweek
NY Times Mag	New York Times Magazine
New Yorker	New Yorker
Northwest Med	Northwest Medicine
Oil Paint & Drug Rep	Oil, Paint and Drug Reporter
Org Gard & Farm	Organic Gardening and Farming
Parents Mag	Parent's Magazine and Better Homemaking
Pesticide Monitoring J	Pesticide Monitoring Journal
Pop Mech	Popular Mechanics
Pop Sci	Popular Science Monthly
Prevention	Prevention
Proc Okla Acad Sci	Proceedings of the Oklahoma Academy of Science
Public Health Service Bull	Public Health Service Bulletin
Public Relations News	Public Relations News
Quick Frozen Foods	Quick Frozen Foods
Ramp Mag	Ramparts Magazine
Read Digest	Reader's Digest
Redbook	Redbook
Sales Manag	Sales Management
Sat Eve Post	Saturday Evening Post
Sat R	Saturday Review
Schol	Scholastica
Sci Am	Scientific American
Sci Digest	Science Digest
Science	Science
Sci Illus	Science Illustrated
Sci Mo	The Scientific Monthly
Sci N	Science News
Sci NL	Science Newsletter

Abbreviation	Periodical
Sea Front	Sea Frontiers
Sky & Tel	Sky and Telescope
Space World	Space World
Sports Illus	Sports Illustrated
Sr Schol	Senior Scholastic
Suc Farm	Successful Farming
Tenn Valley Perspective	Tennessee Valley Perspective
Time	Time
Today's Health	Todays Health
Tox Appl Pharm	Toxicology and Applied Pharmacology
Trans Am Fish Soc	Transactions of American Fisheries Society
TV Guide	TV Guide
US News	U.S. News and World Report
Vend	Vending
Weeds	Weeds
Womans Hom Companion	Woman's Home Companion
World W	World Week
Yachting	Yachting

INDEX

N

616